PRENTICE HALL INTERNATIONAL SERIES IN INDUSTRIAL AND SYSTEMS ENGINEERING

W. J. Fabrycky and J. H. Mize, Editors

AMOS AND SARCHET *Management for Engineers*
AMRINE, RITCHEY, MOODIE AND KMEC *Manufacturing Organization and Management, 6/E*
ASFAHL *Industrial Safety and Health Management, 3/E*
BABCOCK *Managing Engineering and Technology, 2/E*
BADIRU *Comprehensive Project Management*
BADIRU *Expert Systems Applications in Engineering and Manufacturing*
BANKS, CARSON AND NELSON *Discrete Event System Simulation, 2/E*
BLANCHARD *Logistics Engineering and Management, 5/E*
BLANCHARD AND FABRYCKY *Systems Engineering and Analysis, 3/E*
BROWN *Technimanagement: The Human Side of the Technical Organization*
BURTON AND MORAN *The Future Focused Organization*
BUSSEY AND ESCHENBACH *The Economic Analysis of Industrial Projects, 2/E*
BUZACOTT AND SHANTHIKUMAR *Stochastic Models of Manufacturing Systems*
CANADA AND SULLIVAN *Economic and Multi-Attribute Evaluation of Advanced Manufacturing Systems*
CANADA, SULLIVAN AND WHITE *Capital Investment Analysis for Engineering and Management, 2/E*
CHANG AND WYSK *An Introduction to Automated Process Planning Systems*
CHANG, WYSK AND WANG *Computer-Aided Manufacturing, 2/E*
DELAVIGNE AND ROBERTSON *Deming's Profound Changes*
EAGER *The Information Payoff: The Manager's Concise Guide to Making PC Communications Work*
EBERTS *User Interface Design*
EBERTS AND EBERTS *Myths of Japanese Quality*
ELSAYED AND BOUCHER *Analysis and Control of Production Systems, 2/E*
FABRYCKY AND BLANCHARD *Life-Cycle Cost and Economic Analysis*
FABRYCKY, THUESEN AND VERMA *Economic Decision Analysis, 3/E*
FISHWICK *Simulation Model Design and Execution: Building Digital Worlds*
FRANCIS, MCGINNIS AND WHITE *Facility Layout and Location: An Analytical Approach, 2/E*
GIBSON *Modern Management of the High-Technology Enterprise*
GORDON *Systematic Training Program Design*
GRAEDEL AND ALLENBY *Industrial Ecology*
HALL *Queuing Methods: For Services and Manufacturing*
HANSEN *Automating Business Process Reengineering*
HAMMER *Occupational Safety Management and Engineering, 4/E*
HAZELRIGG *Systems Engineering*
HUTCHINSON *An Integrated Approach to Logistics Management*
IGNIZIO *Linear Programming in Single- and Multiple-Objective Systems*
IGNIZIO AND CAVALIER *Linear Programming*
KROEMER, KROEMER AND KROEMER-ELBERT *Ergonomics: How to Design for Ease and Efficiency*
KUSIAK *Intelligent Manufacturing Systems*
LAMB *Availability Engineering and Management for Manufacturing Plant Performance*
LANDERS, BROWN, FANT, MALSTROM AND SCHMITT *Electronic Manufacturing Processes*
LEEMIS *Reliability: Probabilistic Models and Statistical Methods*
MICHAELS *Technical Risk Management*
MOODY, CHAPMAN, VAN VOORHEES AND BAHILL *Metrics and Case Studies for Evaluating Engineering Designs*
MUNDEL AND DANNER *Motion and Time Study: Improving Productivity, 7/E*
OSTWALD *Engineering Cost Estimating, 3/E*
PINEDO *Scheduling: Theory, Algorithms, and Systems*
PRASAD *Concurrent Engineering Fundamentals, Vol. I: Integrated Product and Process Organization*
PRASAD *Concurrent Engineering Fundamentals, Vol. II: Integrated Product Development*
PULAT *Fundamentals of Industrial Ergonomics*
SHTUB, BARD AND GLOBERSON *Project Management: Engineering Technology and Implementation*
TAHA *Simulation Modeling and SIMNET*
THUESEN AND FABRYCKY *Engineering Economy, 8/E*
TURNER, MIZE, CASE AND NAZEMETZ *Introduction to Industrial and Systems Engineering, 3/E*
TURTLE *Implementing Concurrent Project Management*
VON BRAUN *The Innovation War*
WALESH *Engineering Your Future*
WOLFF *Stochastic Modeling and the Theory of Queues*

LOGISTICS ENGINEERIN
AND MANAGEMEN

FIFTH EDITION

LOGISTICS ENGINEERING AND MANAGEMENT

BENJAMIN S. BLANCHARD
Virginia Polytechnic Institute and State University

PRENTICE HALL, Upper Saddle River, New Jersey 07458

Library of Congress Cataloging-in-Publication Data

Blanchard, Benjamin S.
Logistics engineering and management - 5th ed.
p. cm.
Includes bibliographical references and index.
ISBN 0-13-905316-6
1998
CIP data available

Acquisitions editor: ***Alice Dworkin***
Production editor: ***Rhodora V. Penaranda***
Editor-in-chief: ***Marcia Horton***
Managing editor: ***Bayani Mendoza de Leon***
Director of production and manufacturing: ***David W. Riccardi***
Copy editor: ***Andrea Hammer***
Cover designer: ***Bruce Kenselaar***
Manufacturing buyer: ***Donna Sullivan***
Editorial assistant: ***Nancy A. Garcia***
Compositor: ***Preparé, Inc.***

Simon & Schuster / A Viacom Company
Upper Saddle River, New Jersey 07458

Printed in the United States of America

10 9 8 7 6 5 4 3 2 1

ISBN 0-13-905316-6

Prentice-Hall International (UK) Limited, *London*
Prentice-Hall of Australia Pty. Limited, *Sydney*
Prentice-Hall Canada Inc., *Toronto*
Prentice-Hall Hispanoamericana, S.A., *Mexico*
Prentice-Hall of India Private Limited, *New Delhi*
Prentice-Hall of Japan, Inc., *Tokyo*
Simon & Schuster Asia Pte. Ltd., *Singapore*
Editora Prentice-Hall do Brasil, Ltda., *Rio de Janeiro*

CONTENTS

PREFACE

Through the past few decades, the field of logistics has grown significantly with emphasis in several different but related areas. In the commercial sector, the "business-oriented" functions of procurement, material flow, transportation, warehousing, and distribution have been predominant, and there has been much attention given to supply chain management. These activities have been directed primarily to the acquisition and delivery of *consumable* items, and the functions of product design, maintenance, and support have not been included in most instances. Conversely, in the defense sector the realm of logistics has been dealing with *systems*, and (in addition to the procurement and distribution functions) activities have included product design and sustaining maintenance and support. A system must first be designed to be supportable, produced (or constructed), distributed to the user, and ultimately maintained effectively and efficiently throughout its planned life cycle. This approach to logistics, which is life-cycle oriented, has in the past been directed primarily toward large, complex, and highly sophisticated defense systems. Yet, the concepts and principles of such may be applied to almost any category of systems whether a communications system, a chemical processing plant, an electrical power distribution system, a health care system, an information processing system, a production system, a transportation system, or any comparable type of functional entity. This book is about logistics as it applies to *systems*.

In dealing with systems, experience has indicated that the complexities of such have increased in many instances with the introduction of new technologies, the effectiveness and quality aspects for many of those in use have decreased, and the costs associated with system operation and maintenance over the life cycle have increased significantly. When addressing cause-and-effect relationships, a large percentage of the high costs of system operation and support is attributed to the engineering and management decisions made during the early stages of conceptual and preliminary system design. Early decisions associated with the selection of technologies, system/equipment packaging schemes, two versus three levels of maintenance, and the use of automation versus accomplishing functions manually have a great impact on logistic support and total life-cycle cost. Thus, with the current economic dilemma of decreasing budgets and upward inflationary trends, it has become ever more essential that the downstream aspects of logistics (i.e., system maintenance and support) be addressed in the early stages of the design and development of new systems (or the reengineering of existing systems). Of specific interest herein is the *design for supportability*.

Logistics can best be addressed during the early stages of the life cycle through proper implementation of the systems engineering process, commencing with the identification of a customer need and extending through the early definition of system requirements, functional analysis and allocation, synthesis, analysis, design optimization, evaluation, and validation. The maintenance and support infrastructure must be considered as a major element of the system if the resultant product output is to be cost-effective and meet the needs of the customer. Thus, the consideration of "design for supportability" must be inherent from the beginning.

This book addresses logistics from a system's perspective. Chapter 1 sets the stage by providing introductory material with some key terms and definitions. Chapter 2 describes some of the measures (i.e., metrics) of logistics. Chapter 3 presents the system engineering process and the framework within which logistics is addressed in the design process. Chapter 4 covers the broad spectrum of *supportability analysis*, as an inherent part of the ongoing overall systems analysis effort. This constitutes an iterative process, to include the use of numerous analytical methods/techniques/tools, applied with the objective of ensuring that a system is designed such that it can be effectively and efficiently supported throughout its planned life cycle. Chapter 5 addresses the many activities that are associated with logistics in the overall system design and development process. Chapters 6, 7, and 8 cover the many important logistics functions accomplished during the production, operational use, and system retirement and disposal phases. Chapter 9 addresses the aspects of logistics planning, organization, management, and control. Although the *technical* aspects of logistics may appear to be the best ever, the successful implementation of such is highly dependent on the *management* structure and an organizational environment that will allow it to happen.

The presentation of material in this text is more comprehensive than what has been provided in the previous editions. Of particular note is the emphasis on logistics from a *systems* perspective and the coverage of all phases of the life cycle (to include system retirement and material disposal). A new and separate chapter on the *systems engineering process* provides a foundation on which to build the material for subsequent chapters. A new chapter on *supportability analysis* provides a more integrated approach in meeting the design objectives for either new or reengineered systems. The *design for supportability* is emhasized to a greater extent herein. Further, the importance of design *integration* is highlighted, particularly in view of the trends toward greater "outsourcing," the increased utilization of commercial off-the-shelf (COTS) items, and the dealing with many different and varied suppliers from throughout the world. Although we may not be designing many small components these days, the challenges associated with the design of *systems* are even more greater than in the past. Those familiar with the earlier editions of this text will note that some of material has been retained, although the organization of such is quite different.

This book is designed for use in the classroom (at either the undergraduate or graduate program levels) or for the practicing professional in industry, business, or government. The concepts and techniques presented are applicable to any type of system, and the activities described within may be "tailored" to meet the needs of both large- and small-scale programs. Many practical problems are introduced, there are over 235 illustrations and 250 questions, and numerous references have been included. In addition, the text material has been arranged in such a manner as to guide the practicing engineer on a day-to-day basis in the performance of his or her job, and to serve as an authoritative source for those in management who must direct and control logistic support activities.

I wish to thank all of you practitioners in the logistics field who have, through the years, provided me with the "feedback" necessary for the ongoing improvement of this text in progressing from one edition to the next. Your valuable comments as to what's going on in the real world of industry, business, and government have been greatly appreciated. Thank you.

BENJAMIN S. BLANCHARD

LOGISTICS ENGINEERING AND MANAGEMENT

CHAPTER 1

INTRODUCTION TO LOGISTICS

A *system* may be considered as constituting a nucleus of elements combined in such a manner as to accomplish a function in response to an identified need. A system may vary in form, fit, and structure. One may be dealing with a group of aircraft accomplishing a mission at a specified geographical location, a communication network distributing information on a worldwide basis, a power distribution capability involving waterways and electrical power-generating units, a manufacturing plant producing x products in a designated time frame, or a small vehicle providing the transportation of cargo from one location to another. A system must have a *functional* purpose, may include a mix of products and processes, and may be contained within some form of hierarchy (i.e., an airplane within an airline, which is within an overall transportation capability, and so on).[1]

The elements of a system include a combination of resources in the form of materials, equipment, software, facilities, data, services, and personnel integrated in such a manner as to meet a specified requirement; that is, a self-sufficient entity operating in a satisfactory manner, in a defined user environment, throughout its planned life cycle. Inherent within the context of systems are the functions of materials procurement and flow, the distribution of products and services, and the sustaining maintenance and support of a system throughout its intended period of utilization. These initial product distribution and sustaining maintenance and support functions are included within the concept of *logistics*.

Through the past few decades, the field of logistics has grown significantly with emphasis in several different but related areas. In the commercial sector, the "business-

[1] The types and categories of *systems*, and their characteristics, are discussed in detail in Blanchard, B. S., and W. J. Fabrycky, *Systems Engineering and Analysis*, 3rd Ed., Prentice Hall, Inc., Upper Saddle River, N.J., 1998.

oriented" functions of procurement, transportation, material flow, warehousing, and distribution have been predominant, and there has been much attention given to supply-chain management. These activities have been directed primarily to the acquisition and delivery of *consumable* items to the customer in a timely, effective, and efficient manner. Functions associated with product design, maintenance and support have not been included in most instances.

Conversely, when dealing with *systems*, the realm of logistics has been expanded to include those activities pertaining to product design and support, in addition to the procurement and distribution functions mentioned earlier. A system must first be designed to be supportable, produced (or constructed), distributed to the user, and ultimately maintained effectively and efficiently throughout its planned life cycle. The elements of logistics, in this instance, include supply support (spares, repair parts, and associated inventories), test and support equipment, maintenance personnel, training, transportation and material handling, facilities, technical data, and computer resources (hardware and software). This approach to logistics, which is life-cycle oriented, has been implemented by the Department of Defense through the years and, although the emphasis has been directed primarily toward complex and highly sophisticated defense systems, the concepts and principles may be applied to any type of a system.

In evaluating these basic areas of thrust, one can see that there is a strong interrelationship between the two. To fulfill the logistics objectives for systems requires a good distribution network with the implementation of strong business practices in the handling of the many relatively small products needed for the design, production, and support of those systems. Further, in the interest of economy, there is a need to incorporate, to the maximum extent practicable, the commercially-oriented business processes for the handling of consumables within the logistics structure for systems. Thus, one can not separate these areas as being independent entities.

The objective of this text is to address the subject of logistics in the context of systems, from a broad life-cycle perspective, and to include reference to those business-related activities associated with the transportation and distribution of consumables. The purpose of this chapter is to provide the reader with a basic understanding of the scope of logistics, the need for logistics, logistics in the system life cycle, and some of the terms most commonly used in the language of logistics.

1.1 SCOPE OF LOGISTICS

Historically, the concept of logistics stems from specific facets of activity within the defense and commercial sectors of management. *Webster* defines *logistics* as

> The aspect of military science dealing with the procurement, maintenance, and transportation of military material, facilities, and personnel.[2]

[2] *Webster's Ninth New Collegiate Dictionary*, Merriam-Webster, Inc., Springfield, Mass., 1988.

and in the *American Heritage Dictionary*, it is defined as

> The procurement, distribution, maintenance, and replacement of material and personnel.[3]

In the defense sector, the emphasis on logistics has evolved through the concept of *integrated logistic support* (ILS), which may be defined as a[4]

> Disciplined, unified, and iterative approach to the management and technical activities necessary to (1) integrate support considerations into system and equipment design; (2) develop support requirements that are related consistently to readiness objectives, to design, and to each other; (3) acquire the required support; and (4) provide the required support during the operational phase at minimum cost.

Inherent within the context of this definition is the current requirement dealing with the *design for supportability*. This relates to the degree to which a system can be supported, both in terms of the built-in design characteristics of the prime mission-related components of the system and the characteristics of the overall maintenance and support infrastructure and its elements (e.g., supply support, test equipment, maintenance facilities, etc.). It pertains to such characteristics as standardization, interchangeability, accessibility, diagnostics, functional packaging, and compatibility among the various elements of support and between the elements of support and the prime mission-related elements of the system. The emphasis is on system design, the first two items in the above definition.

In the commercial sector, where the emphasis is on the "business-oriented" activities associated with the distribution of consumables, *logistics* may be defined as[5]

> "The process of planning, implementing, and controlling the efficient, cost-effective flow and storage of raw materials, in-process inventory, finished goods, and related information from point of origin to point of consumption for the purpose of conforming to customer requirements."

Although the orientation here is somewhat different, the concepts and principles are not limited to the commercial sector alone. There are materials handling, physical supply and distribution, warehousing and inventory control, and transportation activities across the board and in the defense sector as well. Additionally, the activities associated with the design, maintenance, and support of defense systems can be effectively

[3] *American Heritage Dictionary of the English Language*, Dell Publishing Co., New York, N.Y., 1978.

[4] DSMC, *Integrated Logistics Support Guide*, Defense Systems Management College, Fort Belvoir, Va., 1994.

[5] This definition was developed through the Council of Logistics Management (CLM), Oak Brook, Illinois, and deals with the various aspects of procurement, product distribution, transportation, warehousing, information systems, and supply-chain management. Two additional references are (1) Glaskowsky, N. A., D. R. Hudson, and R. M. Ivie, *Business Logistics*, 3rd Ed., The Dryden Press, Harcourt Brace Jovanovich Co., Orlando, Fl., 1991; and (2) Bowersox, D., D. Closs, and K. Helferich, *Logistical Management*, Macmillan Publishing Co., New York, N.Y., 1986.

applied to nondefense systems such as in commercial transportation, power generation and distribution, healthcare, and manufacturing.[6]

The activities associated with "business" logistics in the commercial sector have in the past been directed primarily toward production operations and the physical distribution of goods and services by the producer. On the other hand, emphasis in the defense environment has been placed on the sustaining life-cycle maintenance and support of a system or product while being utilized by the customer (i.e., user). These two areas of activity have basically been addressed on a relatively independent basis through the years. Further, in both situations, logistics has been considered as a "downstream" effort, addressed "after-the-fact" with regard to design and production, and the overall requirements for logistics have not been very well defined or integrated.

In recent years, systems have been increasing in complexity with the on-going introduction of new technologies, the industrial base has been changing and the availability of resources has been dwindling, the costs of acquiring new systems and maintaining and supporting existing systems have been increasing, and competition has been increasing worldwide. At the same time, the requirements for logistics, in both sectors, have been increasing at an alarming rate. Given the current economic dilemma of decreasing budgets with upward inflationary trends, there will be even less resources available for doing business in the future.

In view of these trends, one of the greatest challenges facing businesses, industries, government agencies, and the general consumer of products and services today is the growing need for a more effective and efficient method for the managing of our valuable resources. The requirement to increase overall productivity in a highly competitive resource-constrained environment has placed a great deal of emphasis on *systems*, in general, and on *all* aspects of the system life cycle. Logistics needs to be recognized as a significant factor throughout the life cycle and assume a major role comparable to research, design, production, and system performance during operational use. The need to address system *life-cycle cost* (in lieu of the short-term initial cost of procurement or acquisition) is evident, and experience has shown that logistics can have a major impact on overall life-cycle cost. Further, it has been indicated that much of the projected life-cycle cost for a given system can be greatly impacted by decisions made during the early phases of advanced planning and conceptual design. Management and design decisions at this point can have a major impact on the activities and operations in all subsequent phases of the life cycle. Thus, it is critical that logistics be addressed at program inception and in the early stages of planning and design, and not be relegated to a downstream after-the-fact activity.

Considering the current environment, and where the scope of logistics is expanding overall, the spectrum of activity as described earlier for *each* of the major thrust areas (i.e., defense and commercial business sectors) appears to be less than desirable if one is to view logistics in terms of the total system life cyle. In other words, the field of logistics has become much broader than initially defined. Logistics requirements must be initially planned from the beginning, and subsequently integrated into the sys-

[6] In the commercial sector, product *design* is often a function of an "engineering" or a "marketing" organization and is not considered as an element of logistics. Further, the activities associated with *maintenance* and *support* are usually considered separate and aside from logistics.

tem design process. Systems must be developed and produced (or constructed) such that they can be operated and maintained in an effective and efficient manner, with complete customer (user) satisfaction as a major objective. Logistics activities are inherent in each phase of the life cycle, with emphasis being required at the early stages.

The subject of *logistics*, as presented in this text, deals with "systems." Although the approach described through the definition of defense systems is adopted herein (see footnote 4), the concepts and principles are applicable to any type of system whether in the defense or commercial sectors. At the *system level*, there are requirements pertaining to the design for supportability, there are requirements involving the procurement and flow of materials in production, there are requirements for the transportation and distribution of products to the customer (user), there are requirements associated with customer service and the maintenance and support of systems throughout their period of utilization, and there are requirements associated with system retirement and the recycling and/or disposal of obsolete materials. In the fulfillment of these requirements, maximum use of the best available commercial business-oriented practices should be made (see footnote 5). In other words, the two major facets of logistics described earlier should be integrated to the maximum extent practicable, taking advantage of the benefits of each approach, in order to respond to the logistics challenges of tomorrow.

1.2 ELEMENTS OF LOGISTICS

In addressing the elements of logistics, one should commence with a good understanding of the overall operational, maintenance and support infrastructure. Referring to Figure 1.1, there is an *outward* flow of activities where materials and services are provided, evolving from the source of supply to the point where the system is being utilized by the customer (user) and a *reverse* flow when items are returned for maintenance. Planning is initiated, system/product design and development are accomplished, items are produced with suppliers providing the needed materials, system elements are installed at the user's operational site(s), and the system is then utilized throughout its planned life cycle. Throughout this process, there are activities involving the initial provisioning and procurement of components, materials handling and processing, transportation and distribution, warehousing, and so on. Many of the business-related logistics activities within the commercial sector are included here.

As the system is being utilized, there is an on-going maintenance and support capability that needs to be installed and in-place to ensure that the system continues to be available when required. As failures occur, faulty items will be returned as necessary for intermediate-level or depot/producer-level maintenance. Additionally, there are scheduled (or preventive) maintenance requirements that may be necessary to keep the system in an operational state. In any event, there is a flow of activities, evolving from the user's operational site back to the depot, producer, or applicable supplier of the item in question. At the same time, there is a forward flow of personnel, test equipment, spares and repair parts, consumables, documentation, and the like, necessary to support those scheduled and unscheduled maintenance actions that are performed at the user's operational site, at the intermediate-level facility, or at the

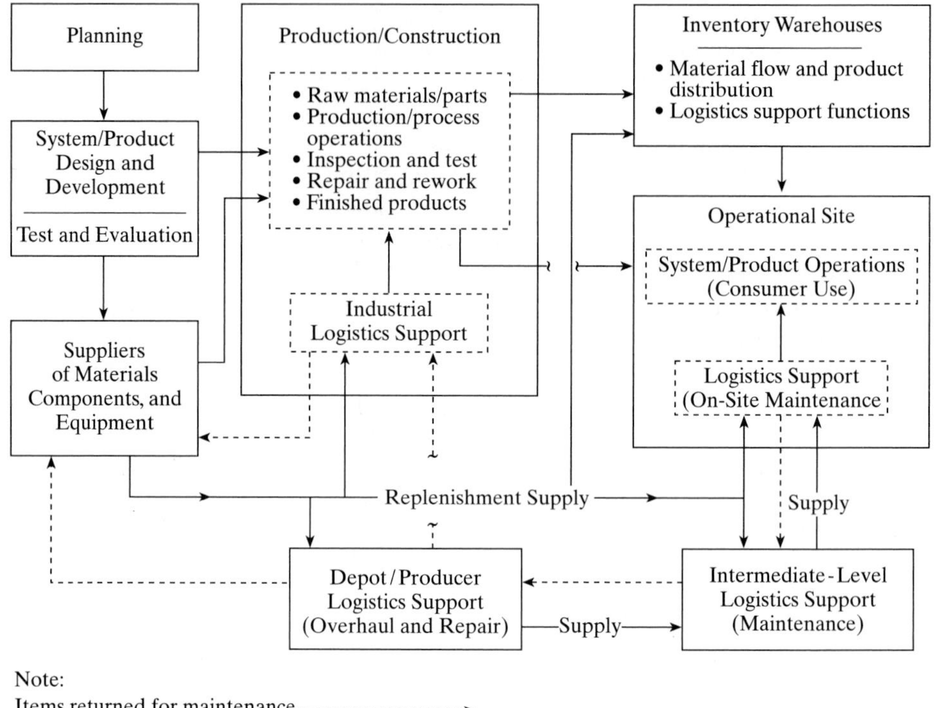

Figure 1.1 System operational and maintenance flow.

producer/supplier facility. Included within this spectrum are activities pertaining to system maintenance and support along with the provisioning, procurement, materials handling, transportation and distribution, and warehousing of those items that are necessary to ensure that an effective maintenance and support infrastructure is provided throughout the system utilization period. These activities are included within the overall spectrum of logistics when dealing with *systems*.[7]

Figure 1.2 reflects the maintenance and support infrastructure as part of the logistics concept for systems. It is this segment (and its associated activities) that often represents a major portion of the total life-cycle cost for a given system. Thus, much of the emphasis throughout this text is directed toward this area; i.e., the design of the prime mission-related elements of the system such that they can be effectively and efficiently supported, and the design of the maintenance and support infrastructure in response to this objective (the first two objectives in the ILS definition—see footnote 4).

[7] Although the specific nomenclature may vary from one application to the next, the concepts illustrated in Figures 1.1 to 1.3 and discussed herein may be applied to any type of system. For example, the same basic elements may be applicable in the implementation of total productive maintenance (TPM), total asset management (TAM) and integrated maintenance management (IMM), which are comparable with ILS but are being implemented in the commercial sector (e.g., manufacturing, railroads, and facilities).

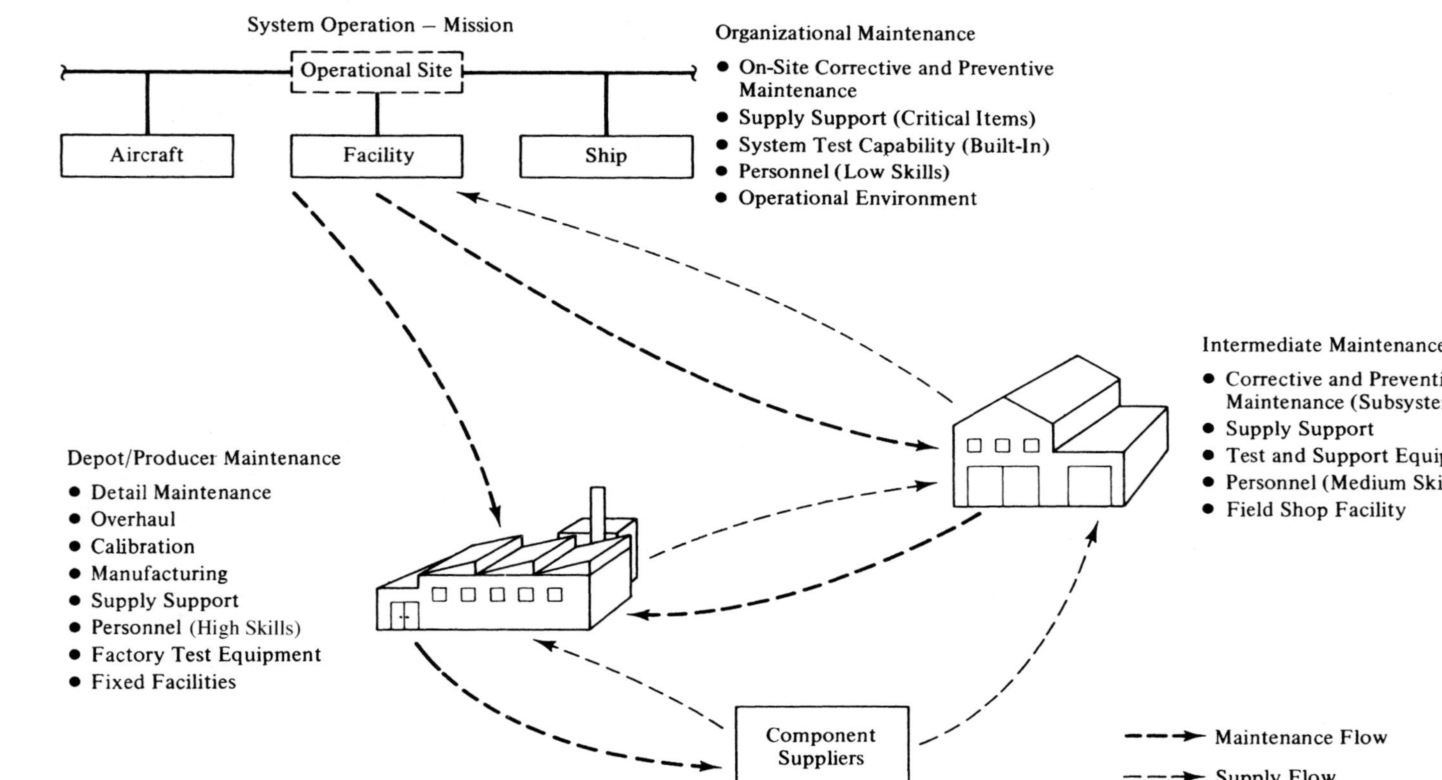

Figure 1.2 System support infrastructure.

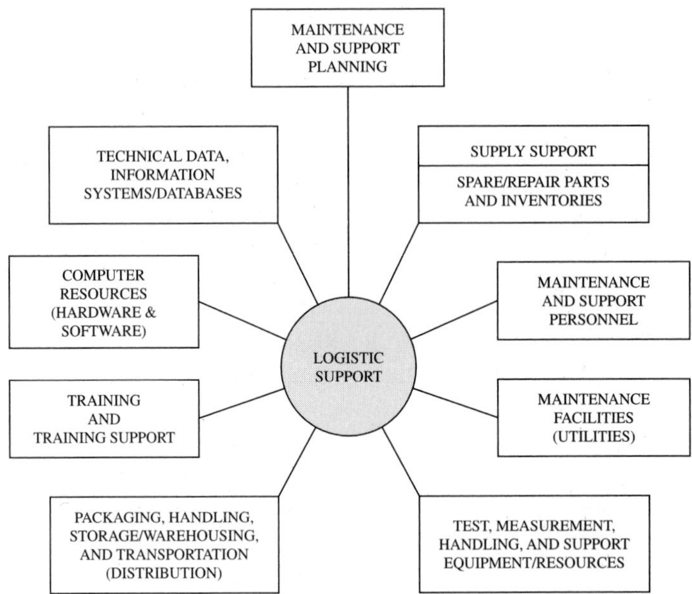

Figure 1.3 The basic elements of logistic support.

Referring to Figure 1.2, the elements of logistics required at each level of maintenance may be included under the categories noted in Figure 1.3. These elements are described below.[8]

1. *Maintenance and support planning*. This includes all planning and analysis associated with the establishment of requirements for the overall support of a system throughout its life cycle. Maintenance planning constitutes a sustaining level of activity commencing with the development of the maintenance concept and continuing through the accomplishment of supportability analyses during system design and development, the procurement and acquisition of support items, the system utilization phase when an ongoing maintenance and support capability is required to sustain operations, and during the retirement phase when materials are being recycled or phased-out for disposal. Maintenance planning should result in the integration of the various facets of support with each other, with the prime mission-related elements of the system, and should lead to the definition and development of the infrastructure illustrated in Figure 1.2.
2. *Supply support (spare/repair parts and associated inventories)*. This includes all spares (repairable units, assemblies, modules, etc.), repair parts (non-repairable components), consumables (liquids, lubricants, disposable items), special supplies, and related inventories needed to maintain the prime mission-related

[8] The categories presented here represent only an example of how one may wish to break down the resources required for system maintenance and support. Although the particular category descriptors may vary from one organization to the next, the critical issue is to ensure that *all* of the applicable resource requirements are included somewhere.

equipment, computers and software, test and support equipment, transportation and handling equipment, training equipment, and facilities. Also included are the provisioning and procurement activities and documentation associated with material acquisition, handling, distribution, recycling, and disposal.

3. *Maintenance and support personnel.* Personnel required for the installation, checkout, and sustaining maintenance and support of the system, its prime mission-related elements and the other elements of support (e.g., test equipment, transportation and handling equipment, and facilities), are included in this category. This includes personnel at all levels (refer to Figure 1.2), mobile teams, and operators at test facilities and calibration laboratories.
4. *Training and training support.* This includes all personnel, equipment, facilities, data/documentation, and associated resources necessary for the training of system operational and maintenance personnel, to include both *initial* and *replenishment* training. Training equipment (e.g., simulators, mockups, special devices), data, and software are developed and utilized as necessary to support both the informal day-to-day training and that of a more formal nature.
5. *Test, measurement, handling, and support equipment.* This category includes all tools, condition monitoring equipment, diagnostic and checkout equipment, special test equipment, metrology and calibration equipment, maintenance fixtures and stands, and special handling equipment required to support all scheduled and unscheduled maintenance actions associated with the system. Test and support equipment requirements at each level of maintenance must be addressed as well as the overall requirements for test *traceability* to a secondary standard and ultimately to a primary standard of some type.
6. *Packaging, handling, storage/warehousing, and transportation.* This element of logistics includes all materials, equipment, special provisions, containers (reusable and disposable), and supplies necessary support the packaging, preservation, storage, handling, and/or transportation of the prime mission-related elements of the system, personnel, spares and repair parts, test and support equipment, technical data, software, and mobile facilities. This category basically covers the initial and sustaining transportation requirements in support of the distribution of materials and personnel and the maintenance cycle illustrated in Figure 1.1 (i.e., the *outward* and *reverse* flows described earlier). The primary modes of transportation include *air, highway, railway, waterway*, and *pipeline*.[9]
7. *Maintenance facilities.* This category includes all facilities required to support scheduled and unscheduled maintenance actions at all levels (see Figure 1.2). Physical plant, portable buildings, mobile vans, housing, intermediate-level maintenance shops, calibration laboratories, and special repair shops (depot, overall, material suppliers) must be considered. Capital equipment and utilities (heat,

[9] The area of *transportation*, a major element of logistics in the commercial sector (refer to footnote 5), is covered extensively in a number of "business" logistics references in Appendix H. Two good sources are (1) Coyle, J. J., E. J. Bardi, and J. L. Cavinato, *Transportation*, 4th Ed., West Publishing Co., St. Paul, Minn., 1992; and (2) Glaskowsky, N. A., D. R. Hudson, and R. M. Ivie, *Business Logistics*, 3rd Ed., The Dryden Press, Harcourt Brace Jovanovich Co., Orlando, Fl., 1992.

power, energy requirements, environmental controls, communications, etc.) are generally included as part of facilities.

8. *Computer resources (hardware and software).* This covers all computers, associated software, interfaces, and the networks necessary to support scheduled and unscheduled activities at each level of maintenance. This may include condition monitoring programs, diagnostic tapes, and associated requirements for the implementation of a computer-aided integrated maintenance management (CIMM) capability.
9. *Technical data, information systems, and database structures.* Technical data may include system installation and checkout procedures, operating and maintenance instructions, inspection and calibration procedures, overhaul instructions, facilities data, modification instructions, engineering design data (specifications, drawings, materials and parts lists, digital data), supplier data, and logistics provisioning and procurement data that are necessary in the performance of system development, production, operation, maintenance, and retirement functions. Such data should not only cover the prime mission-oriented elements of the system but the other elements of the support infrastructure as well (i.e., test and support equipment, transportation and handling equipment, training equipment, and facilities). Included within this category are the information system capabilities, and associated databases, that allow for the implementation of effective *electronic data interchange* (EDI) processes and the requirements associated with *continuous acquisition and life-cycle support* (CALS).[10]

The objective is to provide the right *balance* of resources applied throughout the system support infrastructure illustrated in Figures 1.1 and 1.2. A cost-effective approach must be reached which will require the proper mix of best commercial practices, supplemented by new developmental items where necessary and when justified. As new technologies are introduced and current processes are improved, the specific resource requirements at each level may shift somewhat. For example, the need for large spares/repair parts inventories may not be required as transportation times and costs decline. The availability of over-night express, combined with good communication processes, can help to solve the high-cost large-inventory problem. The need for large data packages (e.g., technical manuals, drawings) at each maintenance location is reduced with the advent of new EDI data formats and processes. The development of computer-based information systems provides for faster and more accurate information, greater visibility relative to the type and location of various assets at a given point in time, and the decision-making and communications processes may be enhanced accordingly. Through the supportability analysis (refer to Chapter 4), numerous trade-off studies are conducted involving various mixes and combinations of the logistics elements identified herein which, hopefully, will lead to the design and development of an effective support infrastructure.

[10]The concept of CALS was introduced in the late 1980s, initially as *computer-aided logistics support* (CALS) and later as *computer-aided acquisition and logistics support* (CALS). Independent of what it is called, the concepts and principles are the same.

1.3 LOGISTICS IN THE SYSTEM LIFE CYCLE

Logistics in the context of a *system* involves planning, analysis and design, testing and evaluation, production and/or construction, distribution, and the sustaining maintenance and support of the system throughout its period of utilization and later during retirement as materials are phased out, recycled, or subject to disposal. For the purposes of illustration, the basic phases of the system life cycle are identified in Figure 1.4. Logistics activities are inherent within each phase.

In the development of systems, the basic phases in Figure 1.4 must be addressed concurrently as conveyed in Figure 1.5. Given that there is a need for a new system (i.e., the top life cycle), one must evolve through the design and development of the prime mission-related elements of the system, the production of multiple quantities of

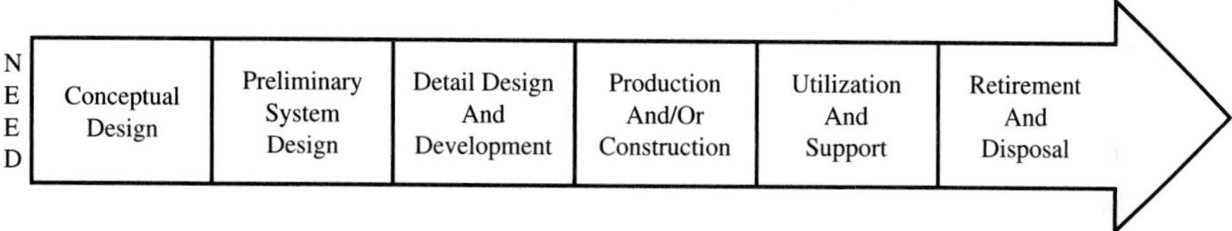

Figure 1.4 The system life-cycle phases.

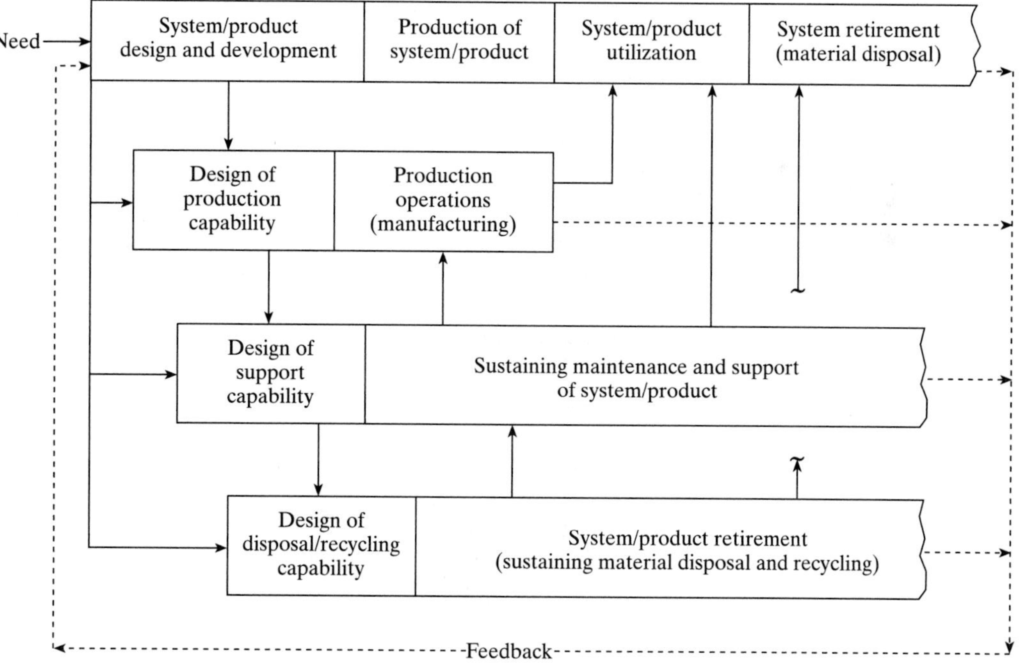

Figure 1.5 The interrelationships of the life-cycle phases in system development. (Source: Blanchard, B. S., and W. J. Fabrycky, *Systems Engineering and Analysis*, 3rd ed., Prentice Hall, Upper Saddle River, NJ, 1998).

these items (or the construction of a single entity), distribution and intallation of the system and its components at designated user sites, utilization and the sustaining maintenance and support of the system throughout the planned life cycle, and retirement and material phase-out. As the prime elements of the system are being developed, consideration must be given to the incorporation of producibility, supportability, and disposability characteristics into the design. This, in turn, leads to the design of the production capability (the second life cycle), design of the maintenance and support infrastructure (the third life cycle), and design of the material recycling and disposal capability (the fourth life cycle). Further, there are *feedback* effects, as the results of the design of the second, third, and fourth life cycles can have a detrimental (or positive) impact on the activities in the top life cycle.[11]

Referring to Figure 1.5, the emphasis throughout this text is on the first and third life cycles. The prime mission-related elements of the system must be designed for *supportability* and the maintenance and support infrastructure must be designed to be compatible with the overall functional objectives of the system in question. This actually becomes an integrated "back-and-forth" process of evolutionary development. In the interest of simplification, Figure 1.6 identifies the major steps within the overall system development process, along with the necessary feedback provisions for *continuous product/process improvement.* Figure 1.7 provides an example of the interface relationships that often exist between the basic mainstream design activities and the major logistics functions, all of which are inherent within the process illustrated in Figure 1.6. The basic steps in Figure 1.7 are highlighted subsequently.

1. Given a specific need, the system operational characteristics, mission profiles, deployment, utilization, effectiveness figures of merit, maintenance constraints, and environmental requirements are defined. Effectiveness figures of merit may include factors for cost effectiveness, availability, dependability, reliability, maintainability, supportability and so on. Using this information, the system maintenance concept is defined, and a system-level specification is developed. Operational requirements and the maintenance concept are the basic determinants of logistic support resources (Figure 1.7, blocks 1 and 2).
2. Major operational, test, production, and support functions are identified, and qualitative and quantitative requirements for the system are allocated as design criteria (or constraints) for significant indenture levels of the prime mission-related elements as well as applicable elements of support (i.e., test and support equipment, facilities, etc.). Those requirements that include logistics factors also form boundaries for the design (Figure 1.7, blocks 3 and 4).
3. Within the boundaries established by the design criteria, alternative prime element and support configurations are evaluated through trade-off studies, and a preferred approach is selected. For each alternative, a preliminary supportability analysis is accomplished to determine the anticipated required resources associated with that alternative. Through numerous trade study iterations, a cho-

[11] The activities within and the interrelationships between these life cycles are discussed extensively in Blanchard, B. S., and W. J. Fabrycky, *Systems Engineering and Analysis*, 3rd Ed., Prentice Hall, Inc., Upper Saddle River, N.J., 1998. Additionally, the principles of *concurrent engineering* must be applied in order to meet the objectives inherent in GOOD system design.

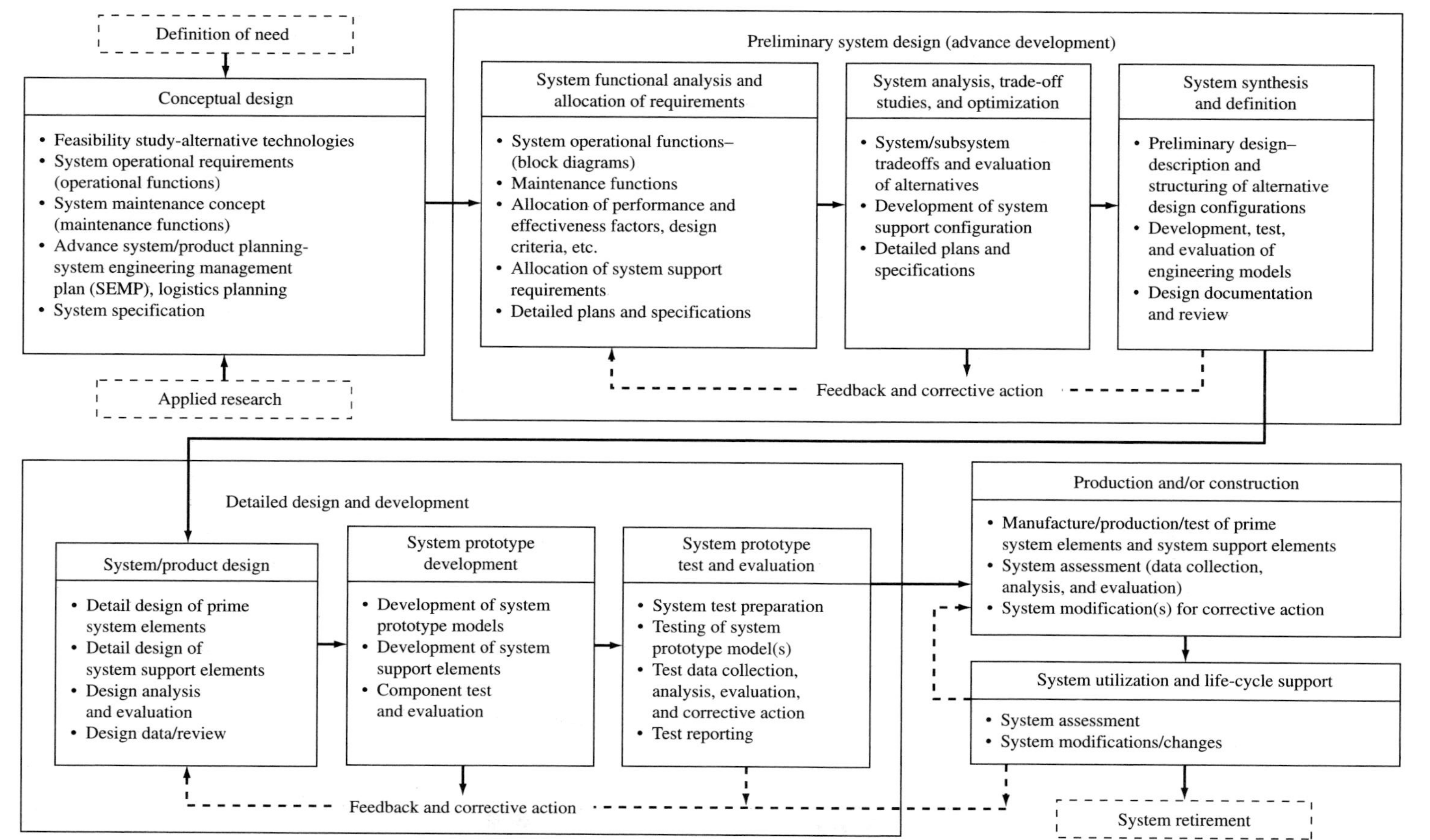

Figure 1.6 The major steps in system design and development.

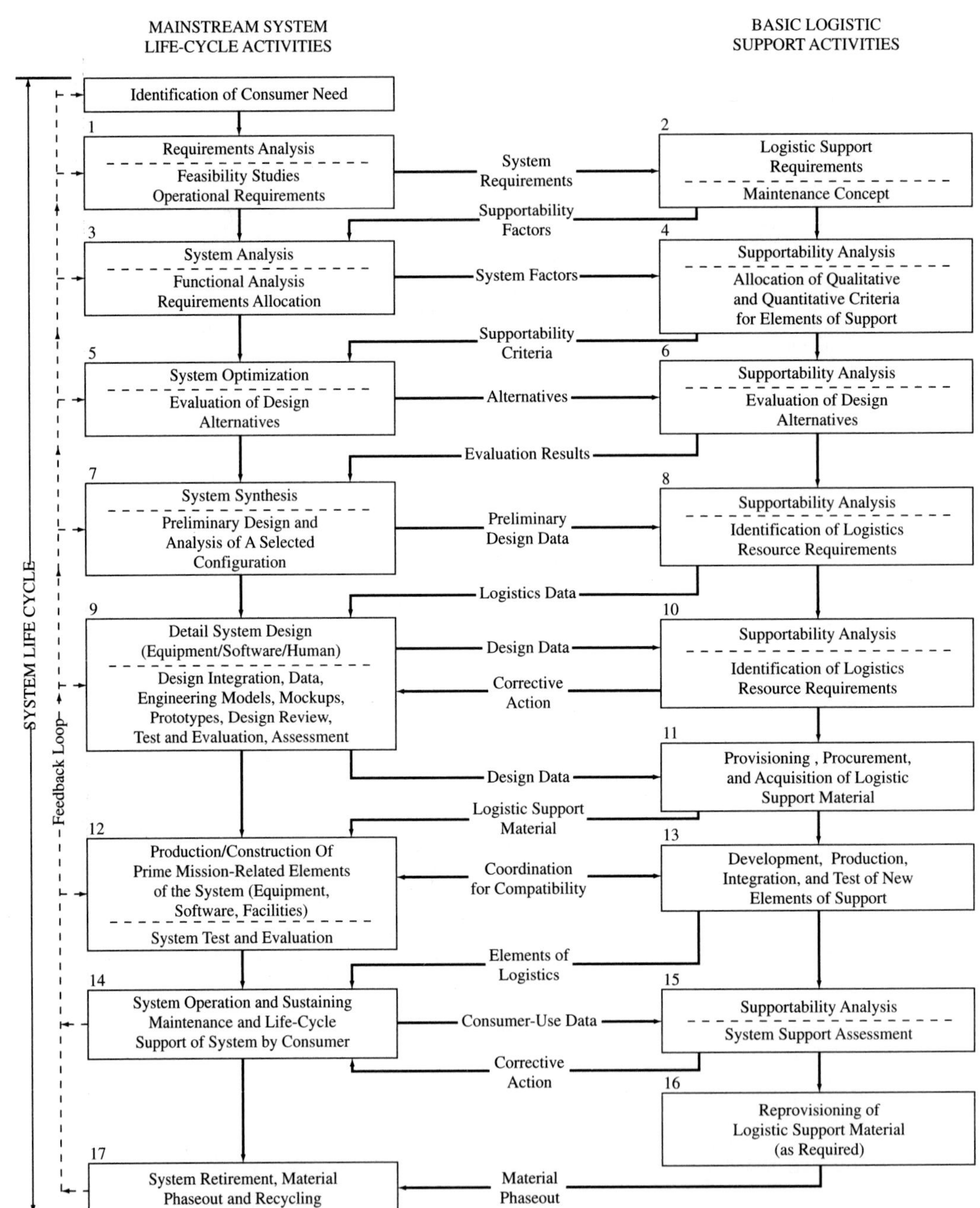

Figure 1.7 The interface relationships between basic design and logistics functions.

sen prime system architecture and support policy are identified (Figure 1.7, blocks 5 and 6).

4. The chosen prime system configuration is evaluated through a supportability analysis effort which leads to the gross identification of logistics resources. The system configuration (prime mission equipment and support elements) is reviewed in terms of its expected overall effectiveness and compliance with the initially specified qualitative and quantitative requirements (i.e., its capability to cost-effectively satisfy the statement of need). The ultimate output leads to the generation of subsystem specifications (and lower-level specifications) forming the basis for detail design (Figure 1.7, blocks 7 and 8).
5. During the design process, direct assistance is provided to design engineering personnel. These tasks include the interpretation of criteria; accomplishment of special studies; participation in the selection of equipment and suppliers; accomplishment of predictions (reliability and maintainability); participation in progressive formal and informal design reviews; and participation in the test and evaluation of engineering models and prototype equipment. An in-depth supportability analysis, based on released design data, results in the identification of specific support requirements in terms of tools, test and support equipment, spare/repair parts, personnel quantities and skills, training requirements, technical data, facilities, transportation, packaging, and handling requirements. The supportability analysis at this stage provides (a) an assessment of the system design for supportability and potential cost/system effectiveness, and (b) a basis for the provisioning and acquisition of specific support items (Figure 1.7, blocks 9 to 12).[12]
6. The prime mission-related elements of the system are produced and/or constructed, tested, and distributed or phased into full-scale operational use. Logistic support elements are acquired, tested, and phased into operation on an as-needed basis. Throughout the operational life cycle of the system, logistics data are collected to provide (a) an assessment of system/cost effectiveness and an early identification of operating or maintenance problems, and (b) a basis for the reprovisioning of support items at selected times during the life cycle (Figure 1.7, blocks 13 to 16).
7. As obsolescence occurs and elements of the system are retired and material items are phased out of the inventory, the appropriate processes must be inplace where such material items can either be recycled for other uses or disposed of in such a way as not to cause any degradation on the environment. The supportability analysis is updated to cover the support resources that are necessary for system retirement and material phaseout (Figure 1.7, block 17).

Consideration of the basic steps in Figure 1.7 is essential in the development of any system. However, the extent and level of activity within each block is "tailored" to the specific requirement (e.g., type of system, extent of development, associated risks,

[12] Provisioning and acquisition refer to the process of source coding, identification of potential suppliers, preparing the procurement package and establishing a contractual arrangement, receiving and inspection of materials, and the ultimate distribution of items to the desired locations.

mission and operational needs, etc.). The presentation in the figure represents a "thought process" in which logistic support considerations are presented in the context of system engineering requirements.

As one proceeds through the system life cycle, which is based on the need to perform a designated function over time, there may be additional logistics requirements associated with design changes and the modification of a system for one reason or another. Referring to Figure 1.8, the projected life cycle for selected *technologies* may be of limited duration and shorter than the life cycle of the system overall. Thus, the system will require modification with the insertion of replacement Technologies A to C at the points indicated. Depending on the system architecture (and whether an "open-architecture" approach has been incorporated into the design), the modification process may be relatively simple or highly complex! In any event, the supportability analysis must be updated to reflect any system modifications.[13]

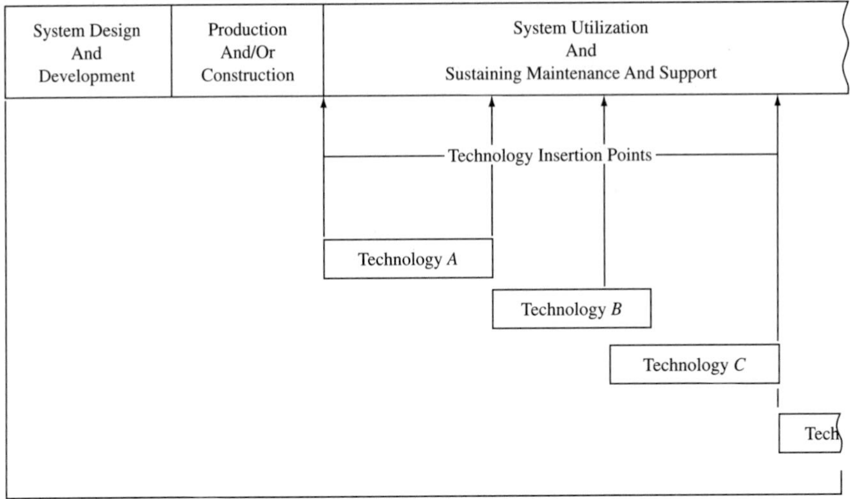

Figure 1.8 System life cycle with new technology insertions.

1.4 NEED FOR LOGISTICS ENGINEERING

Experience in recent years has indicated that the complexity and the costs of systems, in general, have been increasing. A combination of introducing new technologies in response to a constantly changing set of performance requirements, the increased external social and political pressures associated with environmental issues, the requirements to reduce the time that it takes to develop and deliver a new system to

[13] It is not uncommon in today's environment to extend the life of a system and to accomplish the necessary reengineering, upgrading, modification, etc., with the replacement of obsolete technologies as necessary. The system life cycle may be extended over a 15- to 30-year period; yet, many of the technologies initially selected in the design process may have life cycles of only 3 to 5 years (if that long). Thus, it is important to be able to plan for future growth. In any event, the requirements for logistics throughout the system life cycle are continuous and the posture is highly "dynamic."

the customer, and the requirement to extend the life cycle of systems already in operation constitutes a major challenge. Further, many of the systems currently in use today are not adequately responding to the needs of the user, nor are they cost-effective in terms of their operation and support. This is occurring at a time when available resources are dwindling and international competition is increasing worldwide.

In addressing the issue of cost-effectiveness, one often finds that there is a lack of *total cost visibility*, as illustrated by the "iceberg" in Figure 1.9. For many systems, the costs associated with design and development, construction, the initial procurement and installation of capital equipment, production, etc., are relatively well known. We deal with, and make decisions based on, these costs on a regular basis. However, the costs associated with utilization and the maintenance and support of the system throughout its planned life cycle are somewhat hidden. This has been particularly true through the past decade or so when systems have been modified to include the "latest and greatest technology," without first having considered the cost impact downstream. In essence, we have been relatively successful in addressing the short-term aspects of cost, but have not been very responsive to the long-term effects.

At the same time, it has been indicated that a large percentage of the total life-cycle cost for a given system is attributed to operating and maintenance activities (e.g., up to 75% for some systems). When addressing "cause-and-effect" relationships, one

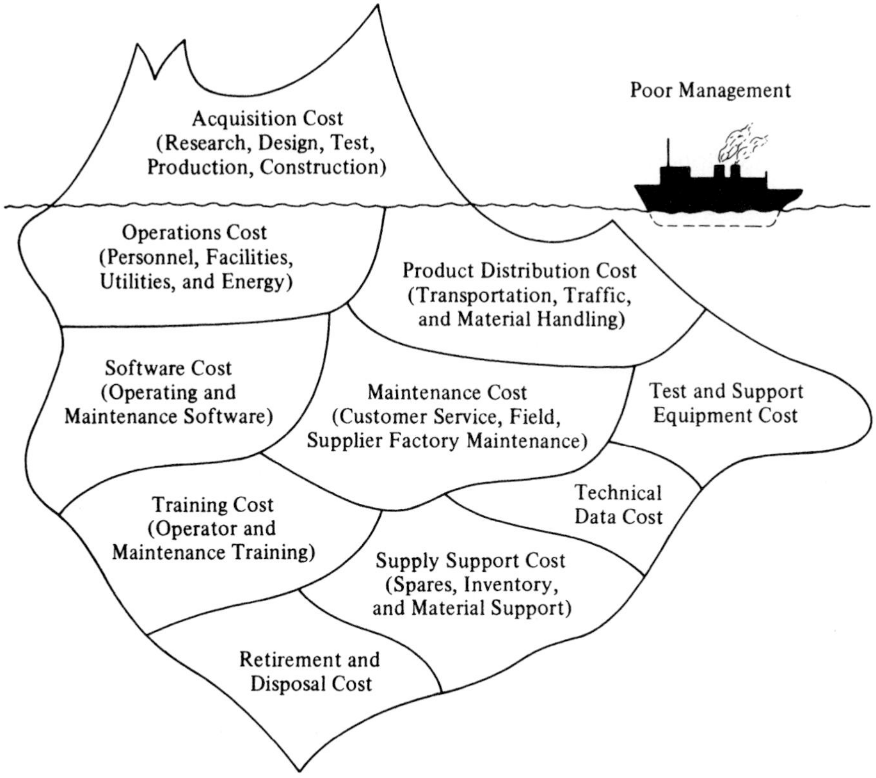

Figure 1.9 Total cost visibility.

often finds that a significant portion of this cost stems from the consequences of decisions made during the early phases of advance planning and conceptual design. Decisions pertaining to the selection of technologies, the selection of materials, the design of a manufacturing process, equipment packaging schemes and diagnostic routines, the accomplishment of functions manually versus the incorporation of automation, the design of maintenance and support equipment, etc., have a great impact on the "downstream" costs and, hence, life-cycle cost. Additionally, the ultimate maintenance and support infrastructure selected for a system throughout its period of utilization can significantly impact the overall cost-effectiveness of that system. Thus, including *life-cycle* considerations in the decision-making process from the beginning is critical. Referring to Figure 1.10, although improvements can be initiated for cost reduction purposes at any stage, it can be seen that the greatest impact on life-cycle cost (and hence, maintenance and support costs) can be realized during the early phases of system design and development. In other words, logistics and the *design for supportability* must be inherent within early system design and development process if the results are to be cost-effective.

Through the years, logistics has been considered "after-the-fact," and the activities associated with logistics have not been very popular, have been implemented downstream in the system life cycle, and have not received the appropriate level of management attention. Experience has indicated that these prevailing practices have been detrimental in many instances and the results have been costly, as conveyed in Figure 1.9. Although much has been accomplished in considering logistics in the design process, such coverage is still occurring rather late. Figure 1.11 provides a rough comparison showing the effects of early life-cycle planning as compared to programs that address supportability issues later on. Hence, it is imperative that for future system design and development (and/or reengineering) efforts emphasis should be placed on

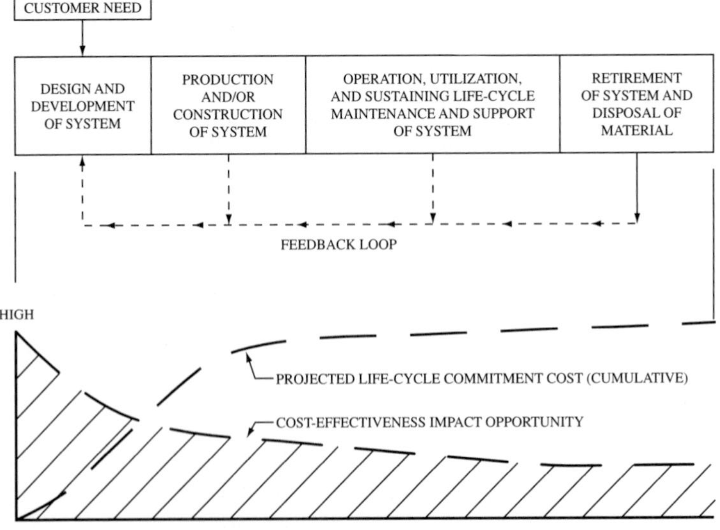

Figure 1.10 Opportunity for impacting logistics and system cost-effectiveness.

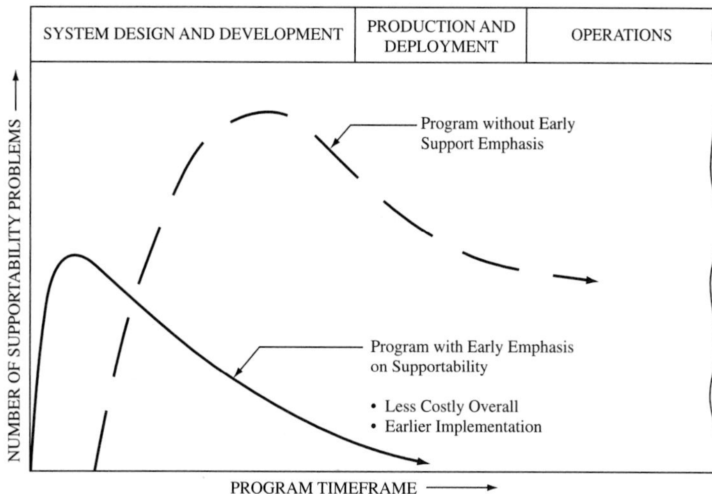

Figure 1.11 The consequences of not addressing supportability from the beginning.

(1) improving our methods for defining system requirements as related to *true* customer needs, early in the conceptual design phase, and addressing performance, effectiveness, and *all* essential characteristics of the system on an integrated basis (to include the specific requirements for logistics); (2) addressing the *total* system, its prime mission-oriented components and its elements of support from a life-cycle perspective; (3) organizing and integrating the appropriate and necessary logistics-related activities into the mainstream system design effort, concurrently and in a timely manner; and (4) establishing a *disciplined* approach, with the necessary review, evaluation, and feedback provisions to ensure that logistics (and the design for supportability) is adequately considered in the overall system acquisition process.

In summary, logistics (and the design for supportability) must be considered as an integral part of the engineering process. By identifying and defining the proper requirements in the beginning, a good foundation should be established. This should, in turn, lead to the effective implementation of all subsequent logistics activities as described in Section 1.3. The emphasis throughout this text is directed to the two shaded blocks in Figure 1.12; i.e., defining the requirements for supportability as part of the systems engineering process in the beginning, and designing the prime mission-oriented elements of the system and the support infrastructure in response to these requirements. This level of activity is consistent with the first two items in the ILS definition presented earlier (refer to footnote 4).

1.5 RELATED TERMS AND DEFINITIONS

With the objective of further clarifying the field of logistics and its many interfaces, it seems appropriate to direct some attention to the language. A few terms and definitions are discussed to provide the reader with the fundamentals necessary to better understand the material presented in this text.

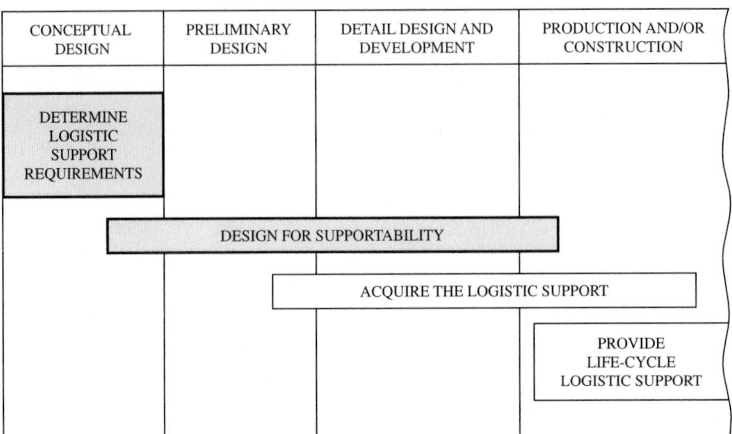

Figure 1.12 Logistics activities in the system life cycle.

System Engineering

Broadly defined, *system engineering* is "the effective application of scientific and engineering efforts to transform an operational need into a defined system configuration through the top-down iterative process of requirements analysis, functional analysis, allocation, synthesis, design optimization, test, and evaluation." The system engineering process, in its evolving of functional detail and design requirements, has as its goal the achievement of the proper balance between operational (i.e., performance), economic, and logistics factors.

A slightly different definition (and one that is preferred by the author) is "the application of scientific and engineering efforts to (a) transform an operational need into a description of system performance parameters and a system configuration through the use of an iterative process of definition, synthesis, analysis, design, test, and evaluation; (b) integrate related technical parameters and ensure compatibility of all physical, functional, and program interfaces in a manner that optimizes the total system definition and design; and (c) integrate reliability, maintainability, safety, human, supportability, producibility, disposability and other such factors into the total engineering effort to meet cost, schedule, and technical performance objectives."[14]

Basically, system engineering is *good* engineering with certain designated areas of emphasis, a few of which are noted:

1. A top-down approach is required, viewing the system as a *whole*. While engineering activities in the past have very adequately covered the design of various system components, the necessary overview and an understanding of how these components effectively fit together have not always been present.

[14] This is a modified version of the definition of systems engineering included in the original version of MIL-STD-499, "Systems Engineering," Department of Defense, Washington, D.C., July 1969. Also, refer to Blanchard, B. S., *System Engineering Management*, 2nd Ed., John Wiley & Sons, Inc., New York, N.Y. 1998.

2. A *life-cycle* orientation is required, addressing all phases to include system design and development, production and/or construction, distribution, operation, sustaining support, and retirement and phaseout. Emphasis in the past has been placed primarily on system design activities, with little (if any) consideration of their impact on production, operations, and logistic support.
3. A better and more complete effort is required relative to the initial *identification of system requirements*, relating these requirements to specific design goals, the development of appropriate design criteria, and the follow-on analysis effort to ensure the effectiveness of early decisions in the design process. In the past, the early front-end analysis effort, as applied to many new systems, has been minimal. This, in turn, has required greater individual design efforts downstream in the life cycle, many of which are not well integrated with other design activities and require later modification.
4. An *interdisciplinary* effort (or team approach) is required throughout the system design and development process to ensure that all design objectives are met in an effective manner. This necessitates a complete understanding of the many different design disciplines and their interrelationships, particularly for large projects.

When referring to the system life cycle, one should view not only the prime mission-oriented elements of the system (e.g., a radar set, a communications network), but the applicable production process, the support infrastructure, and the retirement and material disposal process as well. Basically, the four life cycles presented in Figure 1.5 must be addressed concurrently, as the interactions between the four categories of activity are numerous. As the prime equipment design materializes, the key questions are the following: Is the design configuration producible? Is it supportable? Is it disposable? Is the approach economically feasible?

In summary, system engineering per se is not to be considered as being an engineering discipline in the same context as electrical engineering, mechanical engineering, reliability engineering, or any other design specialty area. Actually, system engineering involves the integrated and coordinated efforts of many different design and related disciplines, applied as part of a top-down/bottom-up process, evolving from the point when a consumer need is first identified, through development, construction and/or production, distribution, utilization and support, and the ultimate system retirement. Figure 1.13 provides an illustration of the system development process, and the shaded area is where logistics engineering can play a major role.[15]

Concurrent Engineering

Recognizing some of the current trends in system acquisition, the Department of Defense introduced the concept of *concurrent engineering*, which is defined as "a systematic approach to the integrated, concurrent design of products and their related processes, including manufacture and support. This approach is intended to cause the

[15] The system engineering process is described further in Chapter 2. It is within this process where logistics engineering activities play a major role, with the *design for supportability* being a prime objective.

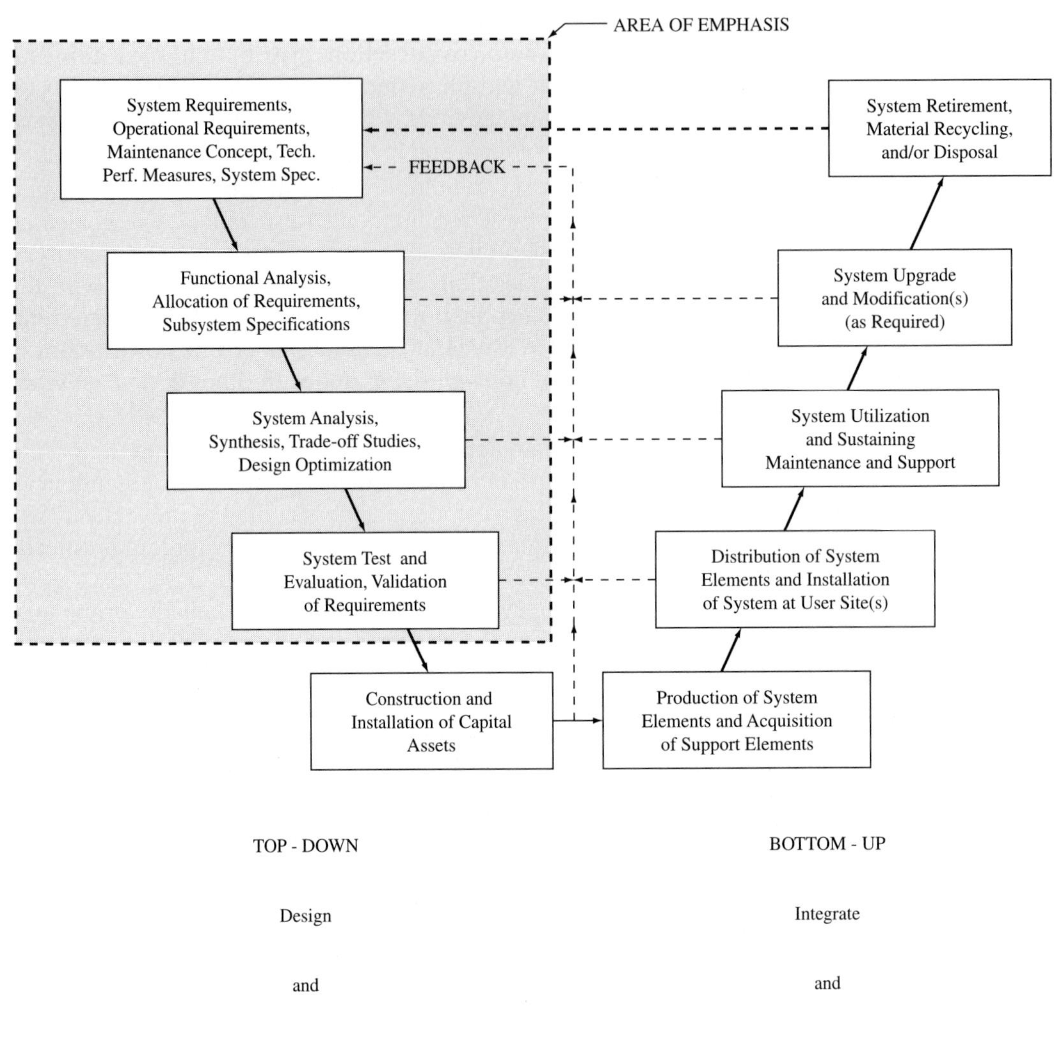

Figure 1.13 Top-down/bottom-up system development process.

developers, from the outset, to consider all elements of the product life cycle from conception through disposal, including quality, cost, schedule, and user requirements." The objectives of concurrent engineering include (1) improving the quality and effectiveness of systems/ products through a better integration of requirements, and (2) reducing the system/product development cycle time through a better integration of activities and processes. This, in turn, should result in a reduction in the total life-cycle cost for a given system.

From your author's perspective, the requirements associated with concurrent engineering are inherent within the concept of system engineering (and the system engineering process described in Chapter 2). The requirements in both instances are to address the four life cycles in Figure 1.5 concurrently and as an integrated entity.[16]

Integrated Logistic Support (ILS)

Integrated logistic support (ILS) is basically a management function that provides for the initial planning, funding, and controls which help to assure that the ultimate customer (or user) will receive a system that will not only meet performance requirements, but one that can be expeditiously and economically supported throughout its programmed life cycle. A major objective of ILS is to assure that the prime mission-related elements of the system are designed to be supportable, and that the elements of the support infrastructure (described in Section 1.2) are designed to be compatible with the prime system elements and with each other. A more precise definition of ILS is included in Section 1.1 (refer to footnote 4).[17]

Logistics Engineering

Logistics engineering includes those basic design-related functions, implemented as necessary to meet the objectives of ILS. This may include

1. The initial definition of system support requirements (as part of the requirements analysis task in systems engineering)
2. The development of criteria as an input to the design of not only those mission-related elements of the system but for the support infrastructure as well (input to design and procurement specifications)
3. The ongoing evaluation of alternative design configurations through the accomplishment of trade-off studies, design optimization, and formal design review (i.e., the day-to-day design integration tasks pertaining to system supportability)
4. The determination of the resource requirements for support based on a given design configuration (i.e., personnel quantities and skill levels, spares and repair parts, test and support equipment, facilities, transportation, data, and computer resources)
5. The ongoing assessment of the overall support infrastructure with the objective of continuous improvement through iterative process of measurement, evaluation, and recommendations for enhancement (i.e., the data collection, evaluation, and process improvement capability)

[16] IDA Report R-338, *The Role of Concurrent Engineering in Weapons System Acquisition*, Institute for Defense Analysis, Alexandria, Va., December 1988. Refer to Appendix H for additional references on concurrent engineering.

[17] The Department of Defense popularized the concept of ILS in the mid 1960s through the publication of 4100.35G, "ILS Planning Guide for Systems and Equipment." Subsequently, there have been numerous applications across the military services. Basically, the concept covers a life-cycle approach to system maintenance and support and includes such activities as logistics engineering, logistic support analysis, the provisioning and procurement of the various elements of support, and the sustaining support of the system throughout its planned period of use.

Although the overall spectrum of logistics includes many additional functions (e.g., procurement, distribution, transportation, maintenance, and so on), the emphasis here is on the *design for supportability*.

Acquisition Logistics

Acquisition logistics, a term currently being emphasized in the defense sector, can be defined as "a multifunctional technical management discipline associated with the design, development, test, production, fielding, sustainment, and improvement modifications of cost-effective systems that achieve the user's peacetime and wartime readiness requirements. The principal objectives of acquisition logistics are to ensure that support considerations are an integral part of the system's design requirements, that the system can be cost effectively supported throughout its life cycle, and that the infrastructure elements necessary to the initial fielding and operational support of the system are identified and developed and acquired." This definition is included in MIL-HDBK-502, "Department of Defense Handbook—Acquisition Logistics," May 1997, and constitutes those activities that the author has included under "Logistics Engineering".

Supportability Analysis (SA)

The *supportability analysis* (SA) is an iterative analytical process by which the logistic support necessary for a new (or modified) system is identified and evaluated. The SA constitutes the application of selected quantitative methods to (1) aid in the initial determination and establishment of supportability criteria as an input to design; (2) aid in the evaluation of various design alternatives; (3) aid in the identification, provisioning, and procurement of the various elements of maintenance and support; and (4) aid in the final assessment of the system support infrastructure throughout the utilization phase. The SA constitutes a design analysis process, which is part of the overall system engineering analysis effort, applied during the early phases in the life cycle and often includes the maintenance task analysis (MTA), level of repair analysis (LORA), FMECA/FTA, reliability-centered maintenance (RCM) analysis, transportation analysis, life-cycle cost analysis (LCCA), and logistics modeling. An output of the SA is the identification and justification of the logistics resource requirements described in Section 1.2. The SA is described further in Chapter 4.[18]

[18] Through the years, there have been many different terms used to describe to the same level of effort. These include the *logistics support analysis (LSA)*, a concept initiated in 1973 and still being applied on many programs today. Additionally, what is currently being defined within the scope of the SA has been covered in the past under a *maintenance engineering analysis (MEA), maintenance level analysis (MLA), maintenance task analysis (MTA), maintenance engineering analysis record (MEAR), maintenance analysis data system (MADS)*, and so on. Independent of what it is called, the concepts and principles remain the same.

Continuous Acquisition and Life-Cycle Support (CALS)

CALS pertains to the application of computerized technology to the entire spectrum of logistics. Of particular emphasis in recent years is the development and processing of data, primarily in a digital format, with the objectives of reducing preparation and processing times, eliminating redundancies, shortening the system acquisition process, and reducing overall program costs. Specific applications thus far have included the automation of technical publications, the preparation of digital data for spares/repair parts provisioning and procurement, and in the development of design data defining products in a digital format.[19]

Reliability (R)

Reliability can be defined simply as the probability that a system or product will perform in a satisfactory manner for a given period of time when used under specified operating conditions. This definition stresses the elements of *probability*, *satisfactory performance*, *time*, and *specified operating conditions*. These four elements are extremely important, since each plays a significant role in determining system/product reliability.

Probability, the first element in the reliability definition, is usually stated as a quantitative expression representing a fraction or a percent signifying the number of times that an event occurs (successes), divided by the total number of trials. For instance, a statement that the probability of survival (P_s) of an item for 80 hours is 0.75 (or 75%) indicates that we can expect that the item will function properly for at least 80 hours, 75 times out of 100 trials.

When there are a number of supposedly identical items operating under similar conditions, it can be expected that failures will occur at different points in time; thus, failures are described in probabilistic terms. In essence, the fundamental definition of reliability is heavily dependent on the concepts derived from probability theory.

Satisfactory performance, the second element in the reliability definition, indicates that specific criteria must be established that describe what is considered to be satisfactory system operation. A combination of qualitative and quantitative factors defining the functions that the system or product is to accomplish, usually presented in the context of a system specification, are required.

The third element, *time*, is one of the most important since it represents a measure against which the degree of system performance can be related. One must know the "time" parameter in order to assess the probability of completing a mission or a given function as scheduled. Of particular interest is being able to predict the probability of an item surviving (without failure) for a designated period of time (sometimes designated as R or P). Also, reliability is frequently defined in terms of mean time between failure (MTBF), mean time to failure (MTTF), or mean time between maintenance (MTBM); thus, the aspect of time is critical in reliability measurement.

[19] MIL-HDBK-59, *Computer-Aided Acquisition and Logistics Support (CALS) Program Implementation Guide*, Department of Defense, Washington, D.C.

The *specified operating conditions* under which we expect a system or product to function constitute the fourth significant element of the basic reliability definition. These conditions include environmental factors such as geographical location where the system is expected to operate, the operational profile, the transportation profile, temperature cycles, humidity, vibration, shock, and so on. Such factors must not only address the conditions for the period when the system or product is operating, but the conditions for the periods when the system (or a portion thereof) is in a storage mode or being transported from one location to another. Experience has indicated that the transportation, handling, and storage modes are sometimes more critical from a reliability standpoint than the conditions experienced during actual system operational use.

The four elements discussed above are critical in determining the reliability of a system or product. System reliability (or unreliability) is a key factor in the frequency of maintenance, and the maintenance frequency obviously has a significant impact on logistic support requirements. Reliability predictions and analyses are required as an input to the supportability analysis.

Reliability is an inherent characteristic of design. As such, it is essential that reliability be adequately considered at program inception, and that reliability be addressed throughout the system life cycle.

Maintainability ($\underline{M}$)

Maintainability, like reliability, is an inherent characteristic of system or product design. It pertains to the ease, accuracy, safety, and economy in the performance of maintenance actions. A system should be designed such that it can be maintained without large investments of time, cost, or other resources (e.g., personnel, materials, facilities, test equipment) and without adversely affecting the mission of that system. Maintainability is the *ability* of an item to be maintained, whereas maintenance constitutes a series of actions to be taken to restore or retain an item in an effective operational state. Maintainability is a design parameter. Maintenance is a result of design.

Maintainability can also be defined as a characteristic in design that can be expressed in terms of maintenance frequency factors, maintenance times (i.e., elapsed times and labor-hours), and maintenance cost. These terms may be presented as different figures of merit; therefore, maintainability may be defined on the basis of a combination of factors, such as:

1. A characteristic of design and installation which is expressed as the probability that an item will be retained in or restored to a specified condition within a given period of time, when maintenance is performed in accordance with prescribed procedures and resources.
2. A characteristic of design and installation which is expressed as the probability that maintenance will not be required more than x times in a given period, when the system is operated in accordance with prescribed procedures. This may be analogous to reliability when the latter deals with the overall frequency of maintenance.
3. A characteristic of design and installation which is expressed as the probability that the maintenance cost for a system will not exceed y dollars per designated

period of time, when the system is operated and maintained in accordance with prescribed procedures.

Maintainability requires the consideration of many different factors involving all aspects of the system, and the measures of maintainability often include a combination of the following:

1. MTBM: mean time between maintenance, which includes both preventive (scheduled) and corrective (unscheduled) maintenance requirements. It includes consideration of reliability MTBF and MTBR. MTBM may also be considered as a reliability parameter.
2. MTBR: mean time between replacement of an item due to a maintenance action (usually generates a spare part requirement).
3. $\overline{M}$: mean active maintenance time (a function of $\overline{M}ct$ and $\overline{M}pt$).
4. $\overline{M}ct$: mean corrective maintenance time. Equivalent to mean time to repair (MTTR).
5. $\overline{M}pt$: mean preventive maintenance time.
6. $\tilde{M}ct$: median active corrective maintenance time.
7. $\tilde{M}pt$: median active preventive maintenance time.
8. $MTTR_g$: geometric mean time to repair.
9. M_{max}: maximum active corrective maintenance time (usually specified at the 90% and 95% confidence levels).
10. MDT: maintenance downtime (total time during which a system is not in condition to perform its intended function). MDT includes active maintenance time ($\overline{M}$), logistics delay time (LDT), and administrative delay time (ADT).
11. MLH/OH: maintenance labor hours per system operating hour.
12. Cost/OH: maintenance cost per system operating hour.
13. Cost/MA: maintenance cost per maintenance action.
14. Turnaround time (TAT): that element of maintenance time needed to service, repair, and/or check out an item for recommitment. This constitutes the time that it takes an item to go through the complete cycle from operational installation through a maintenance shop and into the spares inventory ready for use.
15. Self-test thoroughness: the scope, depth, and accuracy of testing.
16. Fault isolation accuracy: accuracy of system diagnostic routines in percent.

Maintainability, as an inherent characteristic of design, must be properly considered in the early phases of system development, and maintainability activities are applicable throughout the life cycle.[20]

[20] Blanchard, B. S., D. Verma, and E. L. Peterson, *Maintainability: A Key to Effective Serviceability and Maintenance Management*, John Wiley & Sons, Inc., New York, N.Y., 1995.

Maintenance

Maintenance includes all actions necessary for retaining a system or product in, or restoring it to, a serviceable condition. Maintenance may be categorized as corrective maintenance or preventive maintenance.

1. *Corrective maintenance:* includes all unscheduled maintenance actions performed, as a result of system/product failure, to restore the system to a specified condition. The corrective maintenance cycle includes failure identification, localization and isolation, disassembly, item removal and replacement or repair in place, reassembly, checkout and condition verification. Also, unscheduled maintenance may occur as a result of a suspected failure, even if further investigation indicates that no actual failure occurred.
2. *Preventive maintenance:* includes all scheduled maintenance actions performed to retain a system or product in a specified condition. Scheduled maintenance includes the accomplishment of periodic inspections, condition monitoring, critical item replacements, and calibration. In addition, servicing requirements (e.g., lubrication, fueling, etc.) may be included under the general category of scheduled maintenance.

Maintenance Level

Corrective and preventive maintenance may be accomplished on the system itself (or an element thereof) at the site where the system is used by the customer, in an intermediate shop near the customer's operational site, and/or at a depot, supplier or manufacturer's plant facility. *Maintenance level* pertains to the division of functions and tasks for each area where maintenance is performed. Task complexity, personnel-skill-level requirements, special facility needs, economic criteria, and so on, dictate to a great extent the specific functions to be accomplished at each level. In support of further discussion, maintenance may be classified as *organizational, intermediate*, and *depot.*

Maintenance Concept

The *maintenance concept* (as defined in this book) constitutes a series of statements and/or illustrations defining criteria covering maintenance levels (i.e., two levels of maintenance, three levels of maintenance, etc.), major functions accomplished at each level of maintenance, basic support policies, effectiveness factors (e.g., MTBM, $\overline{\text{M}}$ct, MLH/OH, cost/MA, etc.), and primary logistic support requirements. The maintenance concept is defined at program inception and is a prerequisite to system/product design and development. The maintenance concept is also a required input to the supportability analysis.

Maintenance Plan

The *maintenance plan* (as compared to the maintenance concept) is a detailed plan specifying the methods and procedures to be followed for system support throughout the life cycle and during the utilization phase. The plan includes the identification and use of the required elements of logistics necessary for the sustaining support of the system. The maintenance plan is developed from the supportability analysis (SA) data, and is usually prepared during the detail design phase.

Total Productive Maintenance (TPM)

Total productive maintenance (TPM) is a Japanese concept involving an integrated, top-down, system life-cycle approach to maintenance, with the objective of maximizing productivity. TPM is directed primarily to the commercial manufacturing environment, utilizing many of the principles inherent within the ILS concept. More specifically, TPM

1. Aims to maximize overall equipment effectiveness (to improve overall efficiency).
2. Establishes a complete preventive maintenance program for the entire life cycle of equipment.
3. Is implemented on a team basis and involves various departments, such as engineering, production operations, and maintenance.
4. Involves every employee, from top management to the workers on the floor. Even equipment operators are responsible for maintenance of the equipment they operate.
5. Is based on the promotion of preventive maintenance through *motivational management* (autonomous small-group activities).

TPM, often defined as productive maintenance implemented by all employees, is based on the principle that equipment improvement must involve everyone in the organization, from line operators to top management. The objective is to eliminate equipment breakdowns, speed losses, minor stoppages, and so on. It promotes defect-free production, just-in-time (JIT) production, and automation. TPM includes continuous improvement in maintenance.[21]

[21] Refer to Nakajima, S. (Ed.), *TPM Development Program*, Productivity Press, Portland, Ore., translated into English in 1989. Subsequently, there have been numerous additional publications on the subject, also published by Productivity Press. The concept of TPM was first introduced in Japan in the early 1970s, and is being implemented throughout industry under the guidance of the Japanese Institute for Plant Maintenance (JIPM). The concept became popular in the United States, the American Institute of Total Productive Maintenance (AITPM) was established, and there are many companies currently applying the principles of TPM in one form or another.

Human Factors

Human factors pertain to the human element of the system and the interface(s) between the human being, the machine, facilities, and associated software. The objective is to assure complete compatibility between the system physical and functional design features and the human element in the operation, maintenance, and support of the system. Considerations in design must be given to anthropometric factors (e.g., the physical dimensions of the human being), human sensory factors (e.g., vision and hearing capabilities), physiological factors (e.g., impacts from environmental forces), psychological factors (e.g., human needs, expectations, attitude, motivation), and their interrelationships. Human factors (like reliability and maintainability) must be considered early in system development through the accomplishment of functional analysis, operator and maintenance task analysis, error analysis, safety analysis, and related design support activities. Operator and maintenance personnel requirements (i.e., personnel quantities and skill levels) and training program needs evolve from the task analysis effort. Maintenance personnel requirements are also identified in the supportability analysis.[22]

Software Engineering

With today's trends and the continuing development of computer technology, software is becoming (if not already) a significant element in the configuration of many systems. Current experience indicates that software considerations are inherent in more than 50% of the system design and development efforts in being. Software may be viewed in three areas.

1. Software that is included as a mission-related component of the system and is required for the operation of that system. From a logistics perspective, there is a requirement to maintain this software throughout its planned life cycle.
2. Software that is required to accomplish maintenance functions on the system (e.g., diagnostic routines, condition monitoring programs). A logistics engineering function includes the initial development and the subsequent maintenance of this software.
3. Software that is required in support of program-oriented activities (e.g., the software associated with various computer-based models used for design analyses, the software associated with the preparation and processing of various categories of design data such as required to meet the requirements for CALS).

The development of software must be properly integrated with the development of the hardware, human, and other elements of the system. Further, these activities, as

[22] This area of activity may also be included under such general terms as *human engineering, ergonomics, engineering psychology*, and *system psychology*. Refer to the bibliography in Appendix H.

they apply to system support, must be properly integrated with the activities accomplished in logistics engineering.[23]

Producibility

Producibility is a measure of the relative ease and economy of producing a system or a product. The characteristics of design must be such that an item can be produced easily and economically, using conventional and flexible manufacturing methods and processes without sacrificing function, performance, effectiveness, or quality. *Simplicity* and *flexibility* are the underlying objectives, and it is the goal to minimize the use of critical materials and critical processes, the use of proprietary items, the use of special production tooling and facilities, the application of unrealistic tolerances in fabrication and assembly, the use of special test systems, and the use of high personnel skills in manufacturing. Additionally, production and procurement lead times should be minimized to the extent possible. Producibility objectives should apply to the elements of logistic support, as well as to the main components of the system.[24]

Disposability

Disposability pertains to the degree to which an item can be recycled for some other use or disposed of without causing any degradation to the environment; i.e. the generation of solid waste, toxic substances (air pollution), water pollution, noise pollution, radiation, and so on. Should this area not be addressed in the design, the requirements for logistics may turn out to be rather extensive and costly in order to comply with the environmental requirements currently being imposed. For example, a large incineration facility may be required for material decomposition. This, in turn, may include large amounts of capital equipment which requires maintenance and could be very costly to support.

Total Quality Management (TQM)

Total quality management (TQM) can be described as a total integrated management approach that addresses system/product quality during all phases of the life cycle and at each level in the overall system hierarchy. It provides a before-the-fact orientation to quality, and it focuses on system design and development activities, as well as production, manufacturing, assembly, construction, logistic support, and related functions. TQM is a unification mechanism linking human capabilities to engineering, production, and support processes. Some specific characteristics of TQM are noted.

[23] Refer to the bibliography in Appendix H. Two good references are (1) Boehm, B. W., *Software Engineering Economics*, Prentice Hall, Inc., Upper Saddle River, N.J., 1981; and (2) Shere, K. D., *Software Engineering and Management*, Prentice Hall, Inc., Upper Saddle River, N.J., 1988.

[24] Design for *producibility* and *disposability* are addressed in Blanchard, B. S., and W. J. Fabrycky, *Systems Engineering and Analysis*, 3rd Ed., Prentice Hall, Inc., Upper Saddle River, N.J., 1998.

1. Total customer satisfaction is the primary objective, as compared to the practice of accomplishing as little as possible in conforming to the minimum requirements.
2. Emphasis is placed on the iterative practice of "continuous improvement" as applied to engineering, production, and support processes. The objective is to seek improvement on a day-to-day basis, as compared to the often-imposed last-minute single thrust initiated to force compliance with a standard (the Japanese version of this approach, known as *Kaizen*, is increasing in popularity).
3. In support of item 2, an individual understanding of processes, the effects of variation, the application of process control methods, and so on, is required. If individual employees are to be contributors relative to continuous improvement, they must be knowledgeable of the various processes and their inherent characteristics.
4. TQM emphasizes a total organizational approach, involving every group in the organization, not just the quality control group. Individual employees are motivated from within and are recognized as being key contributors to meeting TQM objectives.

As part of the initial system design and development effort, consideration must be given to (1) the design of the processes that will be used to produce the system and its components, and (2) the design of the support capability that will be used to provide the necessary ongoing maintenance and support for that system. As illustrated in Figure 1.5, these facets of program activity interact, and the results (in terms of ultimate customer satisfaction) will depend heavily on the level of quality attained.[25]

Configuration Management (CM)

CM is a management approach used to identify the functional and physical characteristics of an item in the early phases of its life cycle, control changes to those characteristics, and record and report change processing and implementation status. CM involves four functions to include (1) configuration identification, (2) configuration control, (3) configuration status accounting, and (4) configuration audits. CM is *baseline management* (i.e., the planning, design, and providing logistic support for a system with a given and known baseline configuration versus attempting to provide a support infrastructure for many different and constantly changing baselines which can be very expensive as conveyed in Figure 1.11). The big issue is the control and justification of *changes*. CM is a major factor in the implementation of system engineering requirements and *functional*, *allocated*, and *product* baselines are usually established as the system design and development effort evolves.

System Effectiveness

System effectiveness can be expressed as one or more figures of merit representing the extent to which the system is able to perform the intended function. The figures of

[25] Two references are (1) DOD 5100.51G, "Total Quality Management: A Guide for Implementation," Department of Defense, Washington, D.C., and (2) RAC SOAR-7, "A Guide for Implementing Total Quality Management," Rome Air Development Center, New York, N.Y. 1990.

merit used may vary considerably depending on the type of system and its mission requirements, and should consider the following:

1. *System performance parameters*, such as the capacity of a power plant, range or weight of an airplane, destructive capability of a weapon, quantity of letters processed through a postal system, amount of cargo delivered by a transportation system, and the accuracy of a radar capability.
2. *Availability*, or the measure of the degree a system is in the operable and committable state at the start of a mission when the mission is called for at an unknown random point in time. This is often called *operational readiness*. Availability is a function of operating time (reliability) and downtime (maintainability/supportability).
3. *Dependability*, or the measure of the system operating condition at one or more points during the mission, given the system condition at the start of the mission (i.e., availability). Dependability is a function of operating time (reliability) and downtime (maintainability/supportability).

A combination of these (and perhaps other) measures represents the technical characteristics of a system, as opposed to cost and the economic aspects. By inspection, one can see that logistics affects the various elements of system effectiveness to a significant degree, particularly with regard to availability and dependability. System operation is highly dependent on support equipment (handling equipment), operating personnel, data, and facilities. Maintenance and system downtime are based on the availability of test and support equipment, spare/repair parts, maintenance personnel, data, and facilities. The effect of the type and quantity of logistic support is measured through the parameters of system effectiveness.

Life-Cycle Cost (LCC)

LCC involves all costs associated with the system life cycle, to include the following:

1. *Research and development* (R&D) *cost*. The cost of feasibility studies; system analyses; detail design and development, fabrication, assembly, and test of engineering models; initial system test and evaluation; and associated documentation.
2. *Production and construction cost*. The cost of fabrication, assembly, and test of operational systems (production models); operation and maintenance of the production capability; and associated *initial* logistic support requirements (e.g., test and support equipment development, spare/repair parts provisioning, technical data development, training, entry of items into the inventory, facility construction, etc.).
3. *Operation and maintenance cost*. The cost of sustaining operation, personnel and maintenance support, spare/repair parts and related inventories, test and support equipment maintenance, transportation and handling, facilities, modifications and technical data changes, and so on.
4. *System retirement and phaseout cost*. The cost of phasing the system out of the inventory due to obsolescence or wearout, and subsequent equipment item recycling, disposal, and reclamation as appropriate.

Life-cycle costs may be categorized many different ways, depending on the type of system and the sensitivities desired in cost-effectiveness measurements.[26]

Cost-Effectiveness (CE)

The development of a system or product that is cost-effective, within the constraints specified by operational and maintenance requirements, is a prime objective. Cost-effectiveness relates to the measure of a system in terms of mission fulfillment (system effectiveness) and total life-cycle cost. Cost-effectiveness, which is similar to the standard cost-benefit analysis factor employed for decision-making purposes in many industrial and business applications, can be expressed in various terms (i.e., one or more figures of merit), depending on the specific mission or system parameters that one wishes to measure. The prime ingredients of cost-effectiveness are illustrated in Figure 1.14. Although there are different ways of presenting cost-effectiveness, this illustration is used for the purposes of showing the many influencing factors and their relationships. Further discussion of cost-effectiveness is presented in Chapters 2 and 4, and in Appendix E.

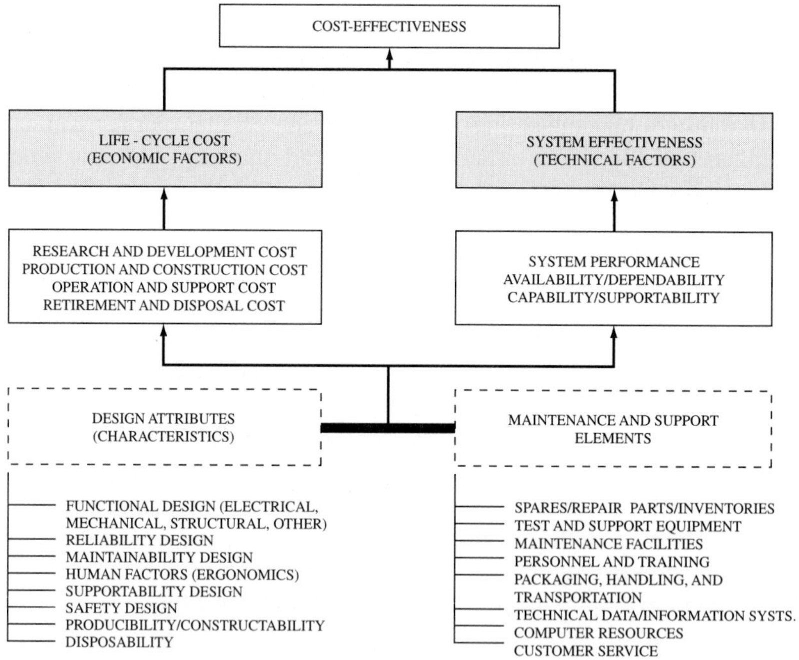

Figure 1.14 Basic ingredients of cost-effectiveness.

[26] Fabrycky W. J., and B. S. Blanchard, *Life-Cycle Cost and Economic Analysis*, Prentice Hall, Inc., Upper Saddle River, N.J., 1991. Refer to the cost breakdown structures in Chapter 4 and in Appendix E of this text. Also, there are additional references in Appendix H.

QUESTIONS AND PROBLEMS

1. What is meant by a *system*? What are its characteristics? Give some examples.
2. Would you consider the maintenance and support infrastructure as an element of a system? If so, why? If not, why not?
3. What is meant by system *hierarchy*?
4. Describe the system *life cycle* and its phases. How might this be different from the life cycle of a given technology (if at all)?
5. Define *system engineering*. How does it differ from *system science* and *system analysis*? Why is system engineering important?
6. How would you define *logistics*? What are the basic differences between logistics as it is practiced in the business-oriented commercial sector and logistics as it is practiced in the defense sector? Identify some of the functions that are common across the board.
7. Identify the elements of logistics (as defined in this text).
8. What is *logistics engineering* and how does it relate to system engineering?
9. What is meant by *design for supportability*? When should it be considered? Why is it important?
10. What is meant by the maintenance and support *infrastructure*? Select a system of your choice and provide an illustration.
11. Refer to Figure 1.5. Describe some of the interrelationships between the four life cycles.
12. What is the relationship (impact of one on another) between reliability and maintainability? Reliability and human factors? Maintainability and human factors? Reliability and logistic support? Maintainability and logistic support? Human factors and logistic support?
13. Describe the interrelationships and some of the trade-offs that may be required in determining spares/repair parts and test equipment? Test and support equipment and facilities? Personnel training and technical data? Spares/repair parts and transportation? Facilities and personnel? Information systems, spares/repair parts, and transportation?
14. What is the difference between *maintainability* and *maintenance*?
15. What is the difference between the *maintenance concept* and the *maintenance plan*? What type of information may be included in each?
16. Refer to Figure 1.13. Identify and describe the logistics activities associated with each of the blocks.
17. Define *integrated logistic support (ILS)*.
18. Define *supportability analysis (SA)*. What is included? When is it accomplished in the system life cycle?
19. Describe *CALS*. What is it, and when is it applied?
20. Describe *concurrent engineering*. How does it relate to system engineering and logistics?
21. Describe *total quality management* (TQM). How does it relate to logistics?
22. What is *configuration management* (CM)? Why is it important regarding logistics?
23. How does the *design for producibility* (or lack of) affect maintenance and logistic support? Provide some examples.
24. How does the *design for disposability* (or lack of) affect maintenance and logistic support? Provide some examples.
25. What is meant by *life-cycle cost* (LCC)? What is included? When in the system life cycle should LCC analysis be applied?
26. Describe *total productive maintenance* (TPM). What are some of the characteristics of TPM? What is *Kaizen*? Compare TPM with ILS.

CHAPTER 2

THE MEASURES OF LOGISTICS

Logistics may be viewed as the composite of all considerations necessary to assure the *effective* and *economical* support of a system throughout its life cycle. It is an integral part of all aspects of system planning, design and development, test and evaluation, production and/or construction, customer utilization, and system retirement. The elements of logistics, as depicted in Figure 1.3, must be developed on an integral basis with all other segments of the system.

To ensure that logistics is properly addressed throughout the life cycle, one must establish the appropriate logistic support *requirements* in the early stages of conceptual design (refer to Figure 1.6). Logistics requirements must be initially specified, both in quantitative and qualitative terms. As system development progresses, the configuration defined must be evaluated against the specified requirements, and modifications for improvement must be incorporated as necessary to ensure effective results. This evaluation task, which is an iterative process, is accomplished through a combination of predictions, analyses, and the use of physical models for conducting tests and demonstrations.

Intuitive within the process of requirements definition, specification, and system evaluation is the aspect of identifying the appropriate quantitative measures of logistics for a given system configuration. These measures may, of course, vary from system to system, as the customer need and mission requirements will vary from one application to the next. Further, there may be multiple factors for any given situation. Thus, it is impossible to cover all conditions and certainly not feasible within the confines of this text. Nevertheless, the qualitative measures of logistics must be addressed.

The intent of this chapter is to introduce some of the more commonly employed quantitative factors applicable in the development and evaluation of the maintenance and logistic support infrastructure for systems. Of particular significance are reliability and maintainability factors, supply support factors, test and support equipment factors, organizational factors, transportation and handling factors, facility factors, effectiveness and economic factors, and so on. Knowledge of the material presented in the various sections of this chapter is essential if one is to plan for, design, produce, and implement a logistic support capability in an effective and efficient manner. Because much of the material included herein is presented in terms of an overview, the review of additional text material as listed in Appendix H is recommended for more detailed coverage.[1]

2.1 RELIABILITY FACTORS

In determining system support requirements, the frequency of maintenance becomes a significant parameter. The frequency of maintenance for a given item is highly dependent on the reliability of that item. In general, as the reliability of a system increases, the frequency of maintenance will decrease and, conversely, the frequency of maintenance will increase as system reliability is degraded. Unreliable systems will usually require extensive maintenance. In any event, logistic support requirements are highly influenced by reliability factors. Thus, a basic understanding of reliability terms and concepts is required. Some of the key reliability quantitative factors used in the system design process and for the determination of logistic support requirements are briefly defined herein.

Reliability Function

As specified in Chapter 1, *reliability* can be defined simply as the probability that a system or product will perform in a satisfactory manner for a given period of time when used under specified operating conditions. The reliability function, $R(t)$, may be expressed as

$$R(t) = 1 - F(t) \tag{2.1}$$

where $F(t)$ is the probability that the system will fail by time t. $F(t)$ is basically the failure distribution function, or the *unreliability* function. If the random variable t has a density function of $f(t)$, then the expression for reliability is

$$R(t) = 1 - F(t) = \int_t^{\infty} f(t)\,dt \tag{2.2}$$

Assuming that the time to failure is described by an exponential density function, then

$$f(t) = \frac{1}{\theta} e^{-t/\theta} \tag{2.3}$$

where θ is the mean life, t is the time period of interest, and e is the natural logarithm base (2.7183). The reliability at time t is

[1] Although the purpose of this chapter is to introduce selected quantitative measures, specific applications in the context of system engineering and the life cycle are presented in Chapter 3.

$$R(t) = \int_t^\infty \frac{1}{\theta} e^{-t/\theta} dt = e^{-t/\theta} \tag{2.4}$$

Mean life (θ) is the arithmetic average of the lifetimes of all items considered. The mean life (θ) for the exponential function is equivalent to mean time between failure (MTBF). Thus,

$$R(t) = e^{-t/M} = e^{-\lambda t} \tag{2.5}$$

where λ is the instantaneous failure rate and M is the MTBF. If an item has a constant failure rate, the reliability of that item at its mean life is approximately 0.37. In other words, there is a 37% probability that a system will survive its mean life without failure. Mean life and failure rates are related in Equation (2.6).

$$\lambda = \frac{1}{\theta} \tag{2.6}$$

Figure 2.1 illustrates the exponential reliability function where time is given in units of t/M.

The illustration presented here focuses primarily on the reliability function in terms of the exponential distribution, which is commonly assumed in many applications. The presumption is that all like components are being utilized in exactly the same manner with the same stresses being imposed. Actually, the failure characteristics of different items may vary considerably depending on their usage. Also, the failure distribution for mechanical items may be different than that for electronic components. In many instances, the Weibull distribution may be more realistic. In any event, one should become familiar with different density functions to include the normal, binomial, exponential, Poisson, gamma, and Weibull distributions.[2]

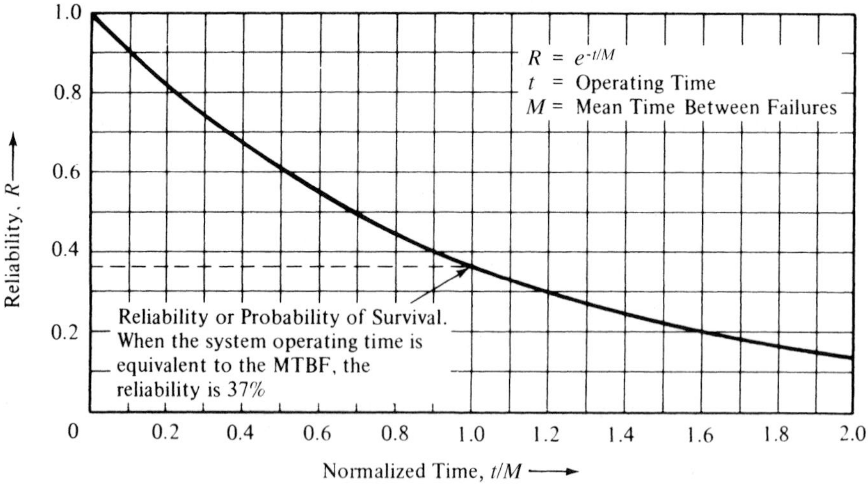

Figure 2.1 Exponential reliability function.

[2]Three excellent references in reliability are (1) Ireson, W. G., and C. F. Coombs, Eds., *Handbook of Reliability Engineering and Management*, McGraw-Hill Book Co., New York, N.Y., 1988; (2) Knezevic, J., *Reliability, Maintainability, and Supportability: A Probabilistic Approach*, McGraw-Hill Book Co., New York, 1993; and (3) Kececioglu, D., *Reliability Engineering Handbook*, Vols 1 and 2, Prentice Hall, Inc., Upper Saddle River, N.J., 1991.

Failure Rate

The rate at which failures occur in a specified time interval is called the *failure rate* during that interval. The failure rate (λ) is expressed as

$$\lambda = \frac{\text{number of failures}}{\text{total operating hours}} \tag{2.7}$$

The failure rate may be expressed in terms of failures per hour, percent failures per 1,000 hours, or failures per million hours. As an example, suppose that 10 components were tested under specified operating conditions. The components (which are not repairable) failed as follows:

- Component 1 failed after 75 hours.
- Component 2 failed after 125 hours.
- Component 3 failed after 130 hours.
- Component 4 failed after 325 hours.
- Component 5 failed after 525 hours.

There were five failures and the total operating time was 3805 hours. Using Equation (2.7), the calculated failure rate per hour is

$$\lambda = \frac{5}{3805} = 0.001314$$

As a second example, suppose that the operating cycle for a given system is 169 hours, as illustrated in Fig. 2.2. During that time six failures occur at the points indicated. A failure is defined as an instance when the system is not operating within a specified set of parameters. The failure rate, or corrective maintenance frequency, per hour is

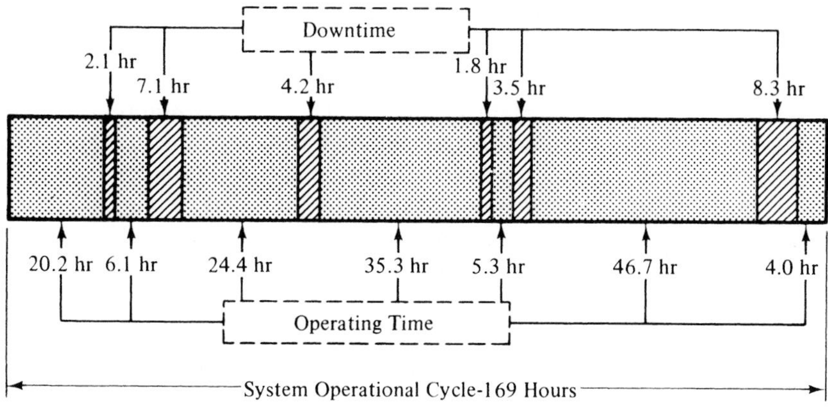

Figure 2.2 Operational cycle.

$$\lambda = \frac{\text{number of failures}}{\text{total mission time}} = \frac{6}{142} = 0.0422535$$

Assuming an exponential distribution, the system mean life or the mean time between failure (MTBF) is

$$\text{MTBF} = \frac{1}{\lambda} = \frac{1}{0.0422535} = 23.6667 \text{ hours}$$

Figure 2.3 presents a reliability nomograph (for the exponential failure distribution) which facilitates calculations of MTBF, λ, $R(t)$, and operating time. For example, if the MTBF is 200 hours ($\lambda = 0.005$), and the operating time is 2 hours, the nomograph gives a reliability value of 0.99.

When determining the overall failure rate, particularly with regard to estimating corrective maintenance actions (i.e., the frequency of corrective maintenance), one must address all system failures to include failures resulting from primary defects, failures due to manufacturing defects, failures due to operator and maintenance errors, and so on. The overall failure rate should cover all factors that will cause the system to be inoperative at a time when satisfactory system operation is required. A combined failure rate is presented in Table 2.1.

TABLE 2.1 Combined Failure Rate

Consideration	Assumed factor (instances/hour)
(a) Inherent reliability failure rate	0.000392
(b) Manufacturing defects	0.000002
(c) Wearout rate	0.000000
(d) Dependent failure rate	0.000072
(e) Operator-induced failure rate	0.000003
(f) Maintenance-induced failure rate	0.000012
(g) Equipment damage rate	0.000005
Total combined factor	0.000486

When assuming the negative exponential distribution, the failure rate is considered to be relatively constant during normal system operation if the system design is mature. That is, when equipment is produced and the system is initially distributed for operational use, there are usually a higher number of failures due to component variations and mismatches, manufacturing processes, and so on. The initial failure rate is higher than anticipated, but gradually decreases and levels off during the *debugging* or *burn-in* period, as illustrated in Figure 2.4. Similarly, when the system reaches a certain age, there is a *wear-out* period where the failure rate increases. The relatively level portion of the curve in Figure 2.4 is the constant failure rate region where the exponential failure law applies.

Figure 2.4 illustrates certain relative relationships. Actually, the curve may vary considerably depending on the type of system and its operational profile. Further, if the system is continually being modified for one reason or another, the failure rate may

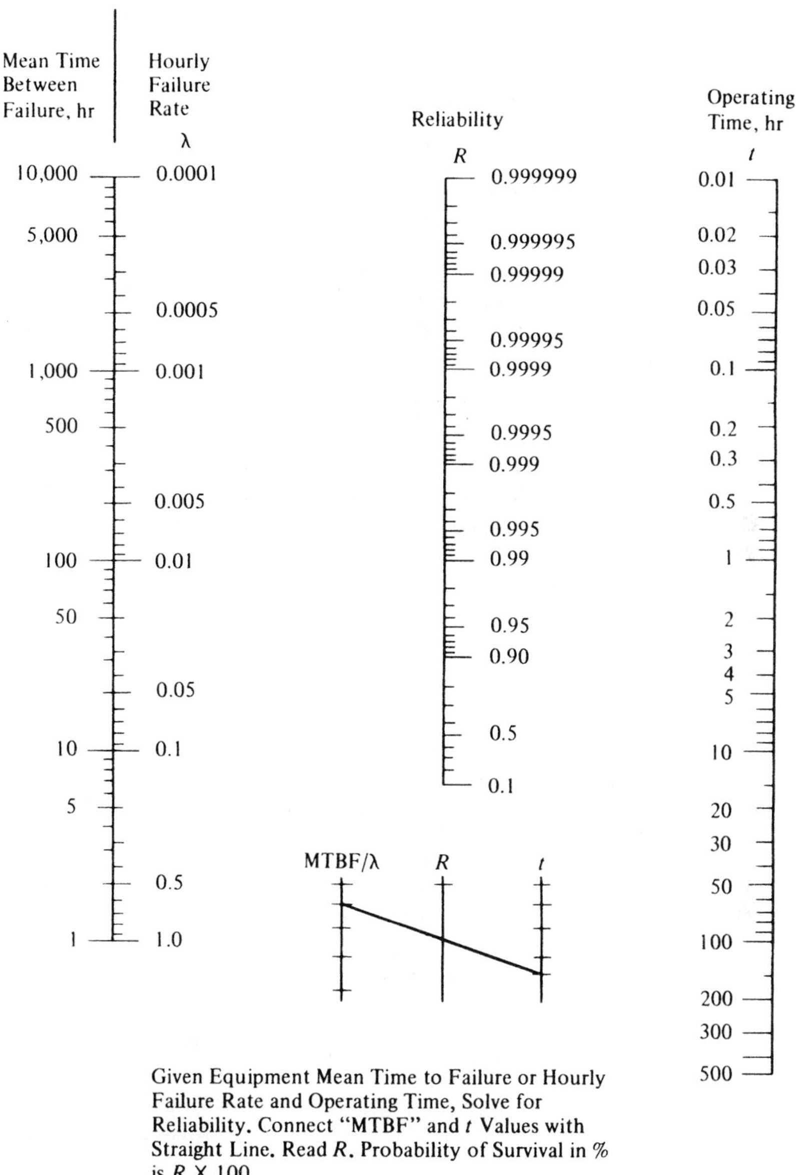

Figure 2.3 Reliability nomograph for the exponential failure distribution. (NAVAIR 01-1A-32, *Reliability Engineering Handbook*, Naval Air Systems Command, U.S. Navy, Washington, D.C., 1977.)

not be constant. In any event, the illustration does provide a good basis for considering failure-rate trends on a relative basis.

In the world of software, failures may be related to calender time, processor time, the number of transactions per period of time, the number of faults per module of

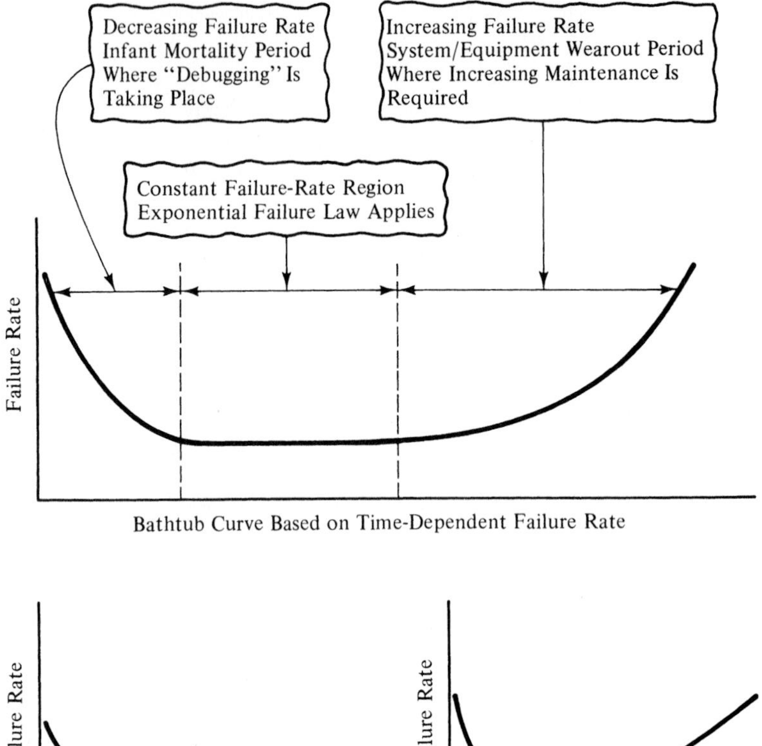

Failure Rate

Electronic Equipment

Failure Rate

Mechanical Equipment

Figure 2.4 Typical failure-rate curve relationships.

code, and so on. Expectations are usually based on an operational profile and criticality to the mission. Thus, an accurate description of the mission scenario(s) is required. As the system evolves from the design and development stage to the operational utilization phase, the ongoing maintenance of software often becomes a major issue. Whereas the failure rate of equipment generally assumes the profiles in Figure 2.4, the maintenance of software often has a negative effect on the overall system reliability. The performance of software maintenance on a continuing basis, along with the incorporation of system changes in general, usually impacts the overall failure rate, as shown in Figure 2.5. When a change or modification is incorporated, "bugs" are usually introduced and it takes a while for these to be "worked out" of the system.

Reliability Component Relationships

Given the basic reliability function and the measures associated with failure rate, it is appropriate to consider their application in series networks, parallel networks, and combinations thereof. These networks are used in reliability block diagrams and in

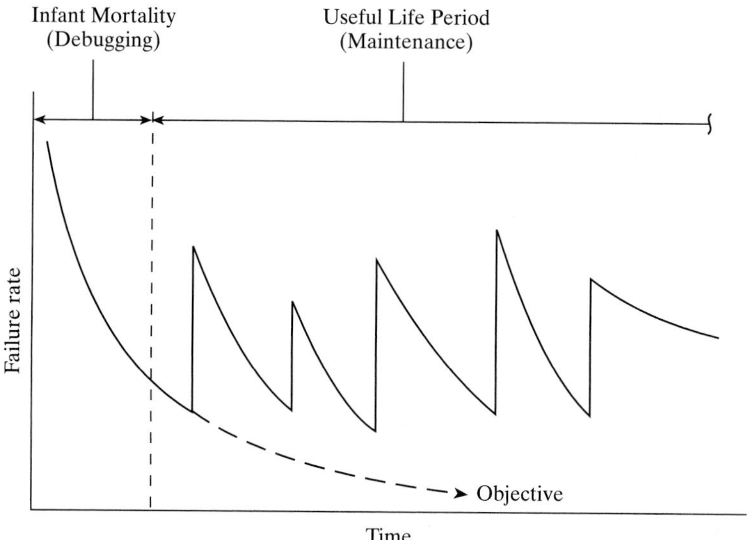

Figure 2.5 Failure-rate curve with maintenance (software application).

models employed for reliability prediction and analysis. Reliability prediction is a necessary input for supportability analyses.

1. Series networks. The series relationship, as illustrated in Figure 2.6, is probably the most commonly used and is the simplest to analyze. In a series network, all components must operate in a satisfactory manner if the system is to function properly. Assuming that a system includes Subsystem A, Subsystem B, and Subsystem C, the reliability of the system is the product of the reliabilities for the individual subsystems and may be expressed as

$$\text{Reliability } (R) = (R_A)(R_B)(R_C) \tag{2.8}$$

As an example, suppose that an electronic system includes a transmitter, a receiver, and a power supply. The transmitter reliability is 0.8521, the receiver reliability is 0.9712, and the power supply reliability is 0.9357. The overall reliability for the electronic system is

$$R_s = (0.8521)(0.9712)(0.9357) = 0.7743$$

If a series system configuration is expected to operate for a specified time period, its required overall reliability can be derived. Substituting Equation (2.5) into Equation (2.8) gives

$$R_s = e^{-(\lambda_1 + \lambda_2 + \lambda_3 + \cdots + \lambda_n)t} \tag{2.9}$$

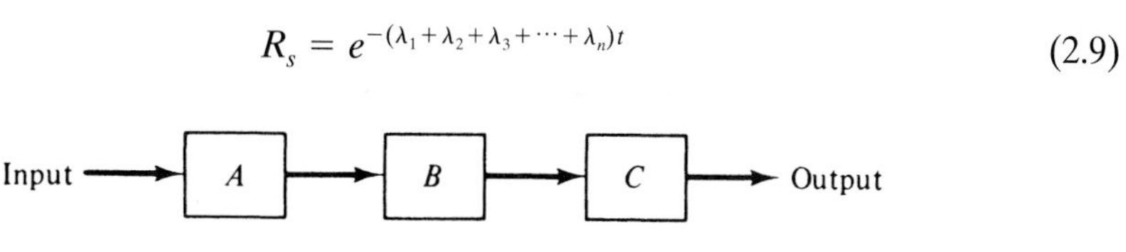

Figure 2.6 Series network.

Suppose that a series configuration consists of four subsystems and is expected to operate for 1000 hours. The four subsystems have the following MTBFs: Subsystem A, MTBF = 6000 hours; Subsystem B, MTBF = 4500 hours; Subsystem C, MTBF = 10,500 hours; Subsystem D, MTBF = 3200 hours. The objective is to determine the overall reliability of the series network where

$$\lambda_A = \frac{1}{6000} = 0.000167 \text{ failure per hour}$$

$$\lambda_B = \frac{1}{4500} = 0.000222 \text{ failure per hour}$$

$$\lambda_C = \frac{1}{10{,}500} = 0.000095 \text{ failure per hour}$$

$$\lambda_D = \frac{1}{3200} = 0.000313 \text{ failure per hour}$$

The overall reliability of the series network is found from Equation (2.9) as

$$R = e^{-(0.000797)(1000)} = 0.4507$$

This means that the probability of the system surviving (i.e., the reliability) for 1000 hours is 45.1%. If the requirement were reduced to 500 hours, the reliability would increase to about 67%.

2. Parallel networks. A pure parallel network is one where a number of the same components are in parallel and where all the components must fail in order to cause total system failure. A parallel network with two components is illustrated in Figure 2.7. The system will function if either A or B, or both, are working, and the reliability is expressed as

$$\text{Reliability } (R) = R_A + R_B - (R_A)(R_B) \tag{2.10}$$

Consider next a network with three components in parallel as shown in Figure 2.8. The network reliability is expressed as

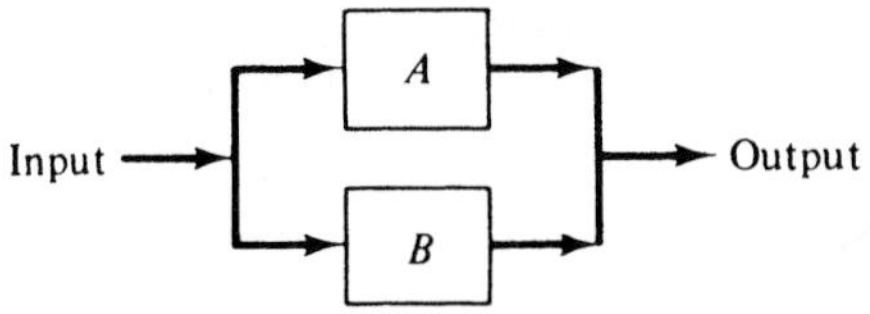

Figure 2.7 Parallel network.

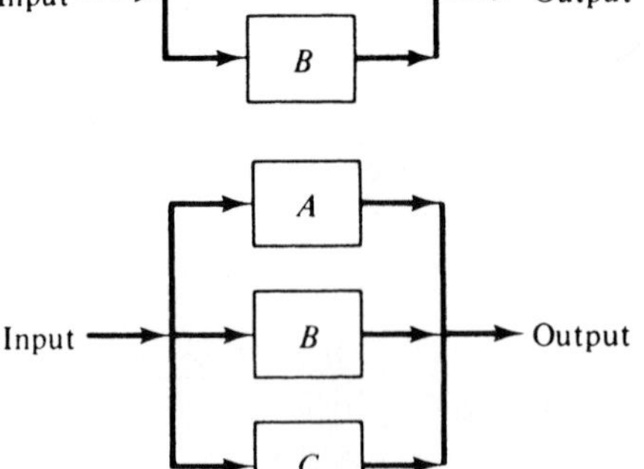

Figure 2.8 Parallel network with three components.

$$\text{Reliability } (R) = 1 - (1 - R_A)(1 - R_B)(1 - R_C) \tag{2.11}$$

If components A, B, and C are identical, then the reliability expression for a system with three parallel components can be simplified to

$$\text{Reliability } (R) = 1 - (1 - R)^3$$

For a system with n identical components, the reliability is

$$\text{Reliability } (R) = 1 - (1 - R)^n \tag{2.12}$$

Parallel redundant networks are used primarily to improve system reliability as Equations (2.11) and (2.12) indicate mathematically. For instance, assume that a system includes two identical subsystems in parallel and that the reliability of each subsystem is 0.95. The reliability of the system is found from Equation (2.10) as

$$\text{Reliability } (R) = 0.95 + 0.95 - (0.95)(0.95) = 0.9975$$

Suppose that the reliability of the system above needs improvement beyond 0.9975. By adding a third identical subsystem in parallel, the system reliability is found from Equation (2.12) to be

$$\text{Reliability } (R) = 1 - (1 - 0.95)^3 = 0.999875$$

Note that this is a reliability improvement of 0.002375 over the previous configuration, or that the *un*reliability of the system was improved from 0.0025 to 0.000125.

If the subsystems are not identical, Equation (2.10) can be used. For example, a parallel redundant network with two subsystems with $R_A = 0.75$ and $R_B = 0.82$ gives a system reliability of

$$\text{Reliability } (R) = 0.75 + 0.82 - (0.75)(0.82) = 0.955$$

3. Combined series–parallel networks. Various levels of reliability can be achieved through the application of a combination of series and parallel networks. Consider the three examples illustrated in Figure 2.9.

The reliability of the first network in Figure 2.9 is given by the equation

$$\text{Reliability } (R) = R_A(R_B + R_C - R_B R_C) \tag{2.13}$$

For the second network, the reliability is given by the equation

$$\text{Reliability } (R) = [1 - (1 - R_A)(1 - R_B)][1 - (1 - R_C)(1 - R_D)] \tag{2.14}$$

And for the third network the reliability is given by the equation

$$\text{Reliability } (R) = [1 - (1 - R_A)(1 - R_B)(1 - R_C)][R_D] \times [R_E + R_F - (R_E)(R_F)] \tag{2.15}$$

Combined series–parallel networks such as those in Figure 2.9 require that the analyst first evaluate the redundant (parallel) elements to obtain a unit reliability, and then combine the unit(s) with other elements of the system. Overall system reliability is then determined by finding the product of all *series* reliabilities.

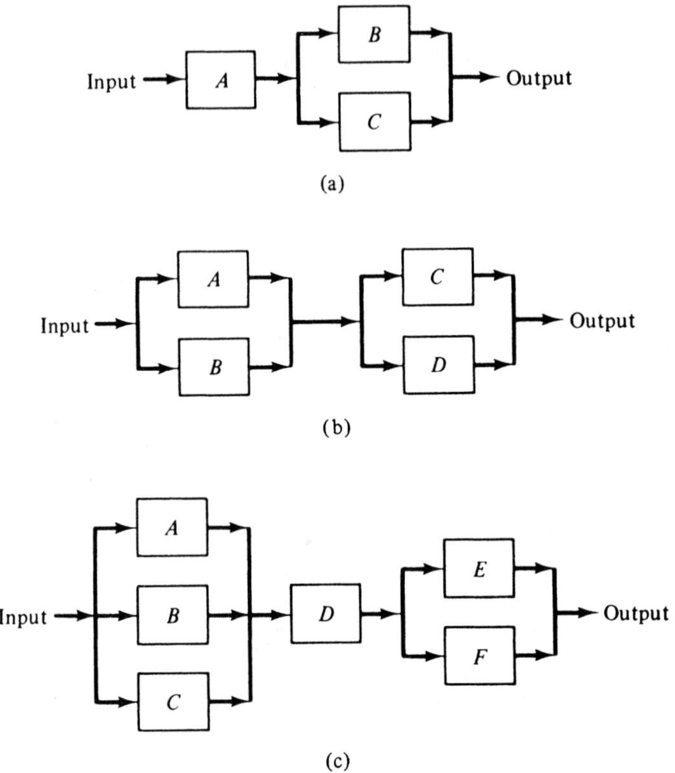

Figure 2.9 Some combined series–parallel networks.

Reliability block diagrams are generated, evolving from the functional block diagram for the system (refer to Chapter 3, Section 3.6) and leading through progressive expansion to individual system components. This process is illustrated in Figure 2.10. As a system design progresses, reliability block diagrams are used in reliability analyses and prediction functions.

2.2 MAINTAINABILITY FACTORS

Maintainability is an inherent design characteristic dealing with the ease, accuracy, safety, and economy in the performance of maintenance functions. Maintainability, defined in the broadest sense, can be measured in terms of a combination of elapsed times, personnel labor-hour rates, maintenance frequencies, maintenance cost, and related logistic support factors. The measures most commonly used are described in this section.

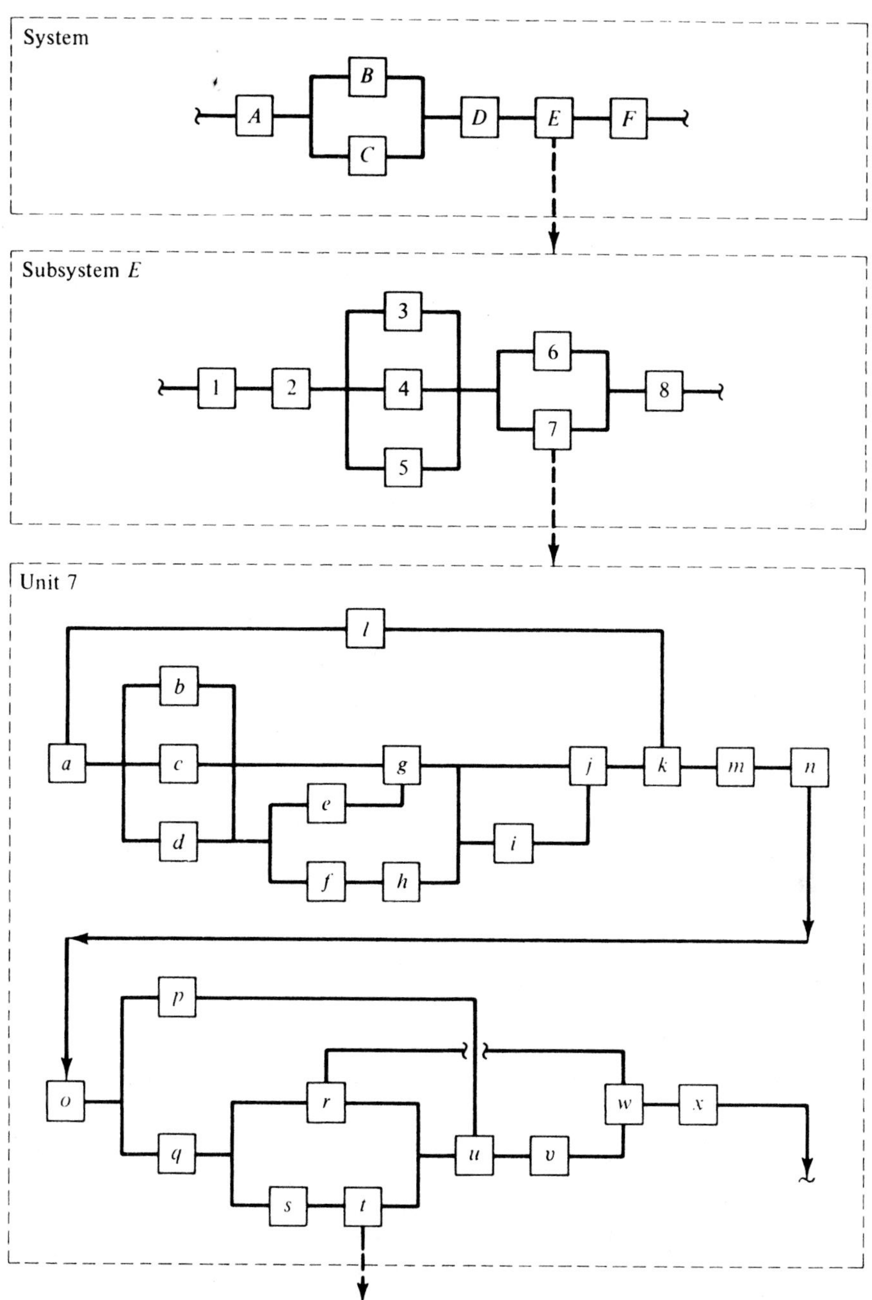

Figure 2.10 Progressive block diagram expansion.

Maintenance Elapsed-Time Factors

Maintenance can be classified in two categories:

1. *Corrective maintenance.* The unscheduled actions, initiated as a result of failure (or a perceived failure), that are necessary to *restore* a system to its required level of performance. Such activities may include troubleshooting, disassembly, repair, remove and replace, reassembly, alignment and adjustment, checkout, and so on. Additionally, this includes all software maintenance that is not initially planned; e.g., *adaptive* maintenance, *perfective* maintenance, and so on.
2. *Preventive maintenance.* The scheduled actions necessary to *retain* a system at a specified level of performance. This may include periodic inspections, servicing, calibration, condition monitoring, and/or the replacement of designated critical items.

Maintenance constitutes the act of diagnosing and repairing, or preventing, system failures. Maintenance time is made up of the individual task times associated with the required corrective and preventive maintenance actions for a given system or product. Maintainability is a measure of the ease and rapidity with which a system can be maintained, and is measured in terms of the time required to perform maintenance tasks. A few of the more commonly used maintainability time measures are defined subsequently.

1. Mean corrective maintenance time ($\overline{\text{M}}$ct). Each time that a system fails, a series of steps is required to repair or restore the system to its full operational status. These steps include failure detection, fault isolation, disassembly to gain access to the faulty item, repair, and so on, as illustrated in Figure 2.11. Completion of these steps for a given failure constitutes a corrective maintenance cycle.

Throughout the system use phase, there will be a number of individual maintenance actions involving the series of steps illustrated in Figure 2.11. The mean corrective maintenance time ($\overline{\text{M}}$ct), or the mean time to repair (MTTR) which is equivalent, is a composite value representing the arithmetic average of these individual maintenance cycle times.

For the purposes of illustration, Table 2.2 includes data covering a sample of 50 corrective maintenance repair actions accomplished on a typical equipment item. Each of the times indicated represents the completion of one corrective maintenance cycle illustrated in Figure 2.11. Based on the set of raw data presented, which constitutes a random sample, a frequency distribution table and frequency histogram may be prepared as illustrated in Table 2.3 and Figure 2.12, respectively.

Referring to Table 2.2, the range of observations is between 97 and 30 minutes, or a total of 67 minutes. This range can be divided into class intervals, with a class interval width of 10 assumed for convenience. A logical starting point is to select class intervals of 20–29, 30–39, and so on. In such instances, it is necessary to establish the dividing point between two adjacent intervals, such as 29.5, 39.5, and so on.

Given the frequency distribution of repair times, one can plot a histogram showing time values in minutes and the frequency of occurrence as in Figure 2.12. By determining the midpoint of each class interval, a frequency polygon can be developed as illustrated in Figure 2.13. This provides an indication of the form of the probability distribution applicable to repair times for this particular system.

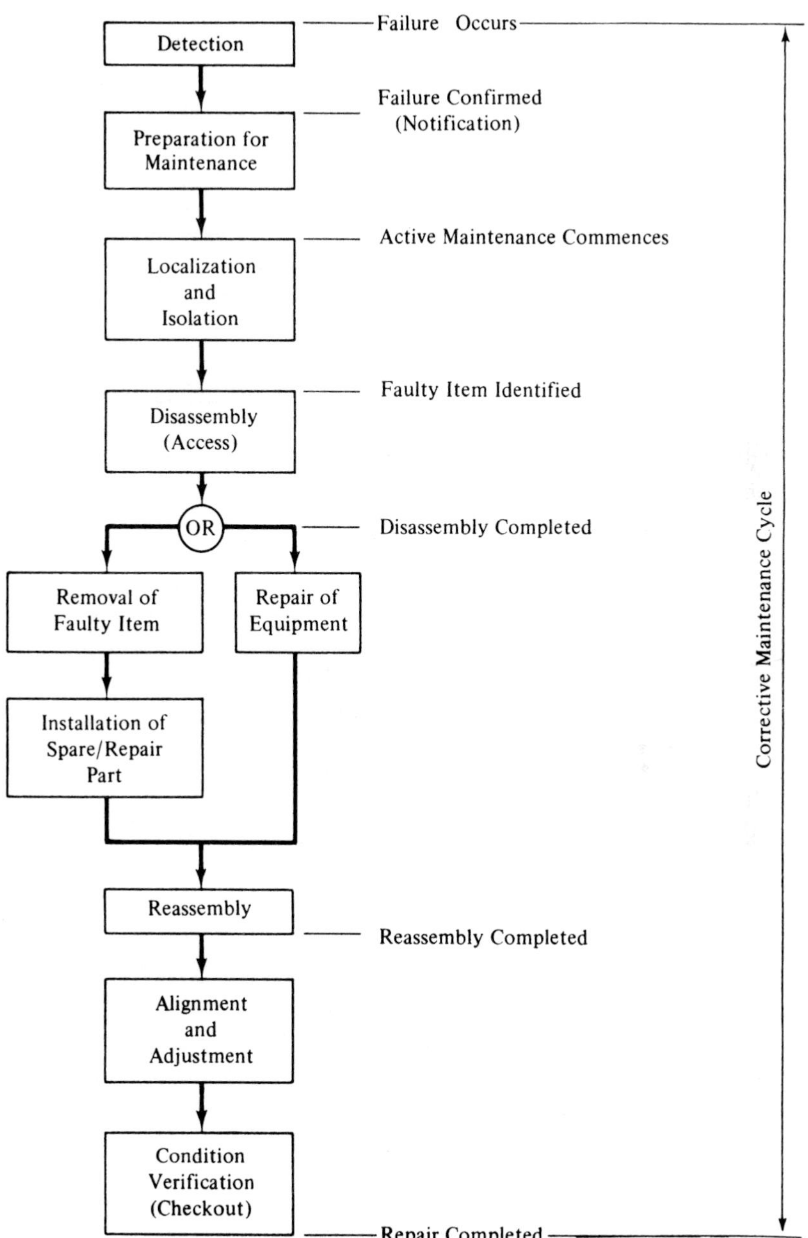

Figure 2.11 Corrective maintenance cycle.

TABLE 2.2 Corrective Maintenance Times (Minutes)

40	58	43	45	63	83	75	66	93	92
71	52	55	64	37	62	72	97	76	75
75	64	48	39	69	71	46	59	68	64
67	41	54	30	53	48	83	33	50	63
86	74	51	72	87	37	57	59	65	63

TABLE 2.3 Frequency Distribution

Class Interval	Frequency	Cumulative Frequency
29.5–39.5	5	5
39.5–49.5	7	12
49.5–59.5	10	22
59.5–69.5	12	34
69.5–79.5	9	43
79.5–89.5	4	47
89.5–99.5	3	50

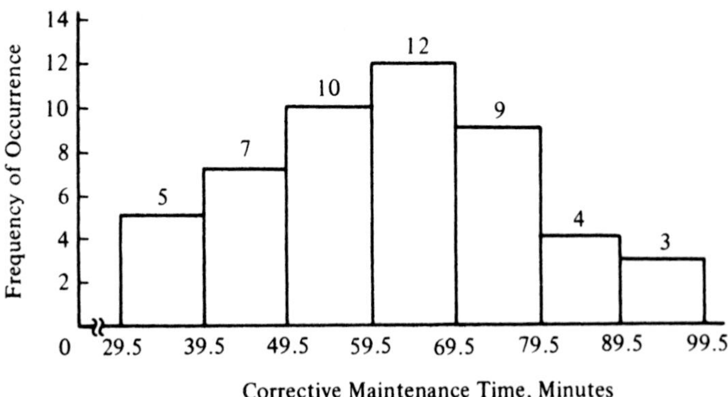

Figure 2.12 Histogram of maintenance actions.

The probability distribution function for repair times can usually be expected to take one of three common forms:

a. The *normal* distribution, which generally applies to relatively straightforward maintenance tasks and repair actions, such as simple remove and replace tasks which consistently require a fixed amount of time to accomplish with very little variation.

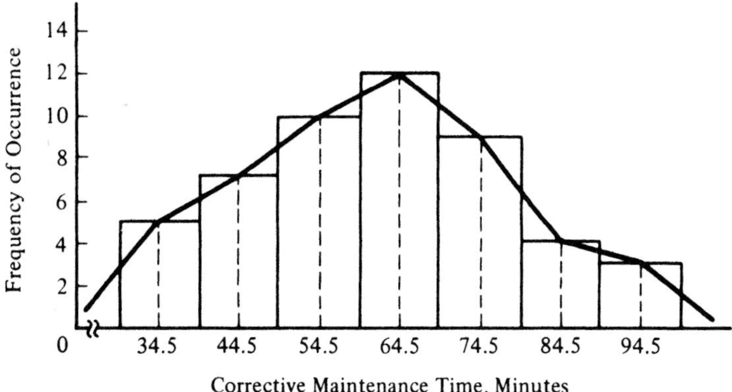

Figure 2.13 Frequency polygon.

b. The *exponential* distribution, which generally applies to equipment with an excellent built-in test capability and a remove and replace repair concept. The maintenance rate is constant.

c. The *log-normal* distribution, which applies to most maintenance tasks and repair actions where the task times and frequencies vary. Experience has indicated that in the majority of instances, the distribution of maintenance times for complex systems and equipment is log-normal.

Referring to Figure 2.13, as additional corrective maintenance actions occur and data points are plotted for the system in question, the curve may take the shape of the normal distribution. The curve is defined by the arithmetic mean ($\overline{X}$ or $\overline{\text{Mct}}$) and the standard deviation (σ). From the maintenance repair times presented in Table 2.2, the arithmetic average is determined as follows:

$$\overline{\text{Mct}} = \frac{\sum_{i=1}^{n} \text{Mct}_i}{n} = \frac{3095}{50} = 61.9 \quad \text{(assume 62)} \tag{2.16}$$

where Mct_i is the total active corrective maintenance cycle time for each maintenance action, and n is the sample size. Thus, the average value for the sample of 50 maintenance actions is 62 minutes.

The standard deviation (σ) measures the dispersion of maintenance time values. When a standard deviation is calculated, it is convenient to generate a table giving the deviation of each task time from the mean of 62. Table 2.4 illustrates this for only four individual task times, although all 50 tasks should be treated in a like manner. The total value of 13,013 does cover all 50 tasks.

The standard deviation (σ) of the sample normal distribution curve can now be determined as follows:

$$\sigma = \sqrt{\frac{\sum_{i=1}^{n} (\text{Mct}_i - \overline{\text{Mct}})^2}{n - 1}} = \sqrt{\frac{13{,}013}{49}} = 16.3 \text{ minutes} \quad \text{(assume 16)} \tag{2.17}$$

TABLE 2.4 Variance Data

Mct_i	$\text{Mct}_i - \overline{\text{Mct}}$	$(\text{Mct}_i - \overline{\text{Mct}})^2$
40	−22	484
71	+ 9	81
75	+13	169
67	+ 5	25
etc.	etc.	etc.
Total		13,013

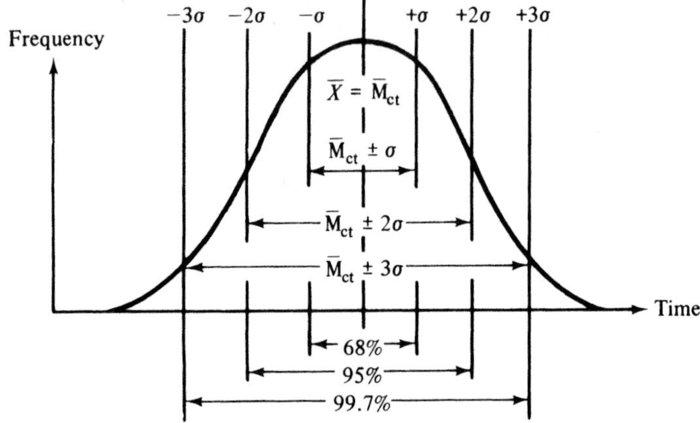

Figure 2.14 Normal distribution.

Assuming normal distribution, the characteristics displayed in Figure 2.14 will hold true. It can be stated that approximately 68% of the total population sample falls within the range 46 to 78 minutes. Also, it can be assumed that 99.7% of the sample population lies within the range of $\overline{\text{Mct}} \pm 3\sigma$, or 14 to 110 minutes.

As an example of a typical application, one may wish to determine the percent of total population of repair times that lies between 40 and 50 minutes. Graphically, this is represented in Figure 2.15. The problem is to find the percent represented by the shaded area. This can be calculated as follows:

a. Convert maintenance times of 40 and 50 minutes into standard values (Z), or the number of standard deviations above and below the mean of 62 minutes:

$$Z \text{ for 40 minutes} = \frac{X_1 - \overline{X}}{\sigma} = \frac{40 - 62}{16} = -1.37 \tag{2.18}$$

$$Z \text{ for 50 minutes} = \frac{X_2 - \overline{X}}{\sigma} = \frac{50 - 62}{16} = -0.75 \tag{2.19}$$

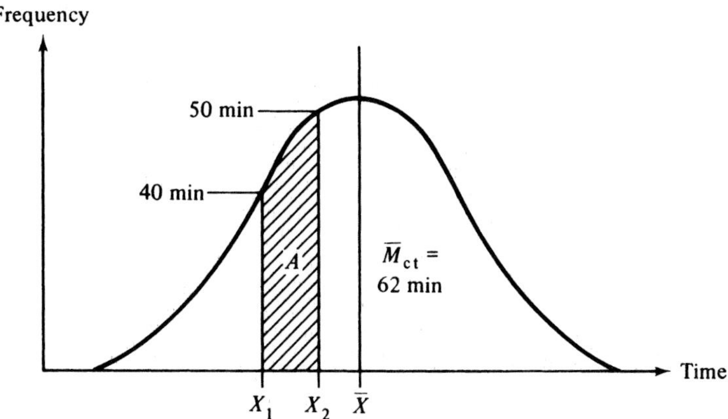

Figure 2.15 Normal distribution sample.

The maintenance times of 40 and 50 minutes represent 1.37 and 0.75 standard deviations below the mean since the values are negative.

b. Point $X_1(Z = -1.37)$ represents an area of 0.0853 and point $X_2(Z = -0.75)$ represents an area of 0.2266, as given in Appendix G. Table G-1.

c. The shaded area A in Figure 2.15 represents the difference in area, or area $X_2 - X_1 = 0.2266 - 0.0853 = 0.1413$. Thus, 14.13% of the population of maintenance times are estimated to lie between 40 and 50 minutes.

Next, confidence limits should be determined. Since the 50 maintenance tasks represent only a sample of all maintenance actions on the equipment being evaluated, it is possible that another sample of 50 maintenance actions on the same equipment could have a mean value either greater or less than 62 minutes. The original 50 tasks were selected at random, however, and statistically represent the entire population. Using the standard deviation, an upper and lower limit can be placed on the mean value ($\overline{\text{M}}$ct) of the population. For instance, if one is willing to accept a chance of being wrong 15% of the time (85% confidence limit), then

$$\text{upper limit} = \overline{\text{M}}\text{ct} + Z\left(\frac{\sigma}{\sqrt{N}}\right) \tag{2.20}$$

where $\sigma/\sqrt{N}$ represents the standard error factor.

The Z value is obtained from Appendix G. Table G-1, where 0.8508 is close to 85% and reflects a Z of 1.04. Thus,

$$\text{upper limit} = 62 + (1.04)\left(\frac{16}{\sqrt{50}}\right) = 64.35 \text{ minutes}$$

This means that the upper limit is 64.4 minutes at a confidence level of 85%, or that there is an 85% chance that $\overline{\text{M}}$ct will be less than 64.4. Variations in risk and upper limits are shown in Table 2.5. If a specified $\overline{\text{M}}$ct limit is established for the design of an equipment (based on mission and operational requirements) and it is known (or assumed) that maintenance times are normally distributed, then one would have to

TABLE 2.5 Risk/Upper Limit Variations

Risk	Confidence	Z	Upper Limit
5%	95%	1.65	65.72 minutes
10%	90%	1.28	64.89 minutes
15%	85%	1.04	64.35 minutes
20%	80%	0.84	63.89 minutes

compare the results of predictions and/or measurements (e.g., 64.35 minutes) accomplished during the development process with the specified value to determine the degree of compliance.

As indicated earlier, the maintenance task times for many systems and equipments do not fit within the normal curve. There may be a few representative maintenance actions where repair times are extensive, causing a skew to the right. This is particularly true for electronic equipment items, where the distribution of repair times often follows a log-normal curve, as shown in Figure 2.16. Derivation of the specific distribution curve for a set of maintenance task times is accomplished using the same procedure as given in the preceding paragraphs. A frequency table is generated and a histogram is plotted.

A sample of 24 corrective maintenance repair actions for a typical electronic equipment item is presented in Table 2.6. The arithmetic average provides a good early estimate of the mean. Using the data in the table, the arithmetic average is determined as follows:

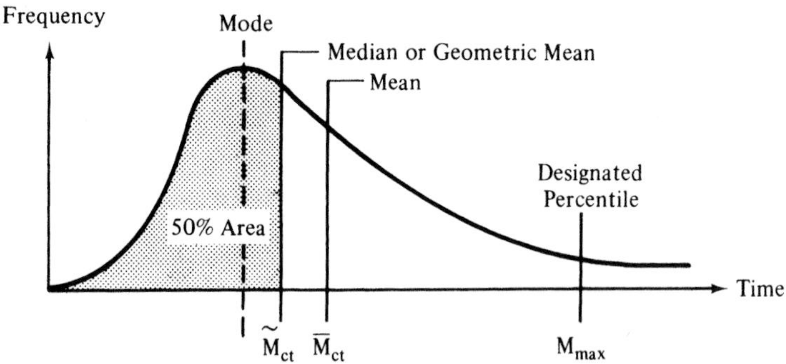

Figure 2.16 Log-normal distribution.

TABLE 2.6 Corrective Maintenance Repair Times (Minutes)

55	28	125	47	58	53	36	88
51	110	40	75	64	115	48	52
60	72	87	105	55	82	66	65

$$\overline{\text{Mct}} = \frac{\sum_{i=1}^{n} \text{Mct}_i}{n} = \frac{1637}{24} = 68.21 \text{ minutes}$$

When determining the mean corrective maintenance time ($\overline{\text{Mct}}$) for a specific sample population (empirically measured) of maintenance repair actions, the use of Equation (2.16) is appropriate. However, Equation (2.21) has a wider application since this equation has the individual task repair times weighted by the frequency with which each individual task is accomplished. Thus, this equation measures the "weighted" mean, which is the preferred method.

$$\overline{\text{Mct}} = \frac{\Sigma(\lambda_i)(\text{Mct}_i)}{\Sigma\lambda_i} \tag{2.21}$$

where λ_i is the failure rate of the individual (ith) element of the item being measured, usually expressed in failures per equipment operating hour. Equation (2.21) calculates $\overline{\text{Mct}}$ as a "weighted average" using reliability factors.

It should be noted that $\overline{\text{Mct}}$ considers only active maintenance time or that time which is spent working directly on the system. Logistics delay time and administrative delay time are not included. Although all elements of time are important, the $\overline{\text{Mct}}$ factor is oriented primarily to a measure of the supportability characteristics in equipment design.

2. Mean preventive maintenance time ($\overline{\text{Mpt}}$). Preventive maintenance consists of the actions required to retain a system at a specified level of performance and may include such functions as periodic inspection, servicing, scheduled replacement of critical items, calibration, overhaul, and so on. $\overline{\text{Mpt}}$ is the mean (or average) elapsed time to perform preventive or scheduled maintenance on an item, and is expressed as

$$\overline{\text{Mpt}} = \frac{\Sigma(\text{fpt}_i)(\text{Mpt}_i)}{\Sigma\text{fpt}_i} \tag{2.22}$$

where fpt_i is the frequency of the individual (ith) preventive maintenance action in actions per system operating hour, and Mpt_i is the elapsed time required for the ith preventive maintenance action.

Preventive maintenance may be accomplished while the system is in full operation, or could result in downtime. In this instance, the concern is for preventive maintenance actions which result in system downtime. Again, $\overline{\text{Mpt}}$ includes only active system maintenance time and not logistic delay and administrative delay times.

3. Median active corrective maintenance time ($\widetilde{\text{Mct}}$). The median maintenance time is that value which divides all of the downtime values so that 50% are equal to or less than the median and 50% are equal to or greater than the median. The median will usually give the best average location of the data sample. The median for a normal distribution is the same as the mean, while the median in a log-normal distribution is the same as the geometric mean (MTTRg) illustrated in Figure 2.16. $\widetilde{\text{Mct}}$ is calculated as

TABLE 2.7 Calculation for $\overline{\text{M}}\text{ct}$

Mct_i	Log Mct_i	$(\text{Log Mct}_i)^2$	Mct_i	Log Mct_i	$(\text{Log Mct}_i)^2$
55	1.740	3.028	64	1.806	3.262
28	1.447	2.094	115	2.061	4.248
125	2.097	4.397	48	1.681	2.826
47	1.672	2.796	52	1.716	2.945
58	1.763	3.108	60	1.778	3.161
53	1.724	2.972	72	1.857	3.448
36	1.556	2.241	87	1.939	3.760
88	1.945	3.783	105	2.021	4.084
51	1.708	2.917	55	1.740	3.028
110	2.041	4.166	82	1.914	3.663
40	1.602	2.566	66	1.819	3.309
75	1.875	3.516	65	1.813	3.287
Total				43.315	78.785

$$\widetilde{\text{M}}\text{ct} = \text{antilog}\,\frac{\sum_{i=1}^{n} \log \text{Mct}_i}{n} = \text{antilog}\,\frac{\Sigma(\lambda_i)(\log \text{Mct}_i)}{\Sigma \lambda_i} \tag{2.23}$$

For illustrative purposes, the maintenance time values in Table 2.6 are presented in the format illustrated in Table 2.7. The median is computed as follows:

$$\widetilde{\text{M}}\text{ct} = \text{antilog}\,\frac{\sum_{1}^{24} \log \text{Mct}_i}{24}$$

$$= \text{antilog}\,\frac{43.315}{24} = \text{antilog } 1.805 = 63.8 \text{ minutes}$$

4. Median active preventive maintenance time ($\widetilde{\text{M}}\text{pt}$). The median active preventive maintenance time is determined using the same approach as for calculating $\widetilde{\text{M}}\text{ct}$. $\widetilde{\text{M}}\text{pt}$ is expressed as

$$\widetilde{\text{M}}\text{pt} = \text{antilog}\,\frac{\Sigma(\text{fpt}_i)(\log \text{Mpt}_i)}{\Sigma \text{fpt}_i} \tag{2.24}$$

5. Mean active maintenance time ($\overline{M}$). $\overline{M}$ is the mean or average elapsed time required to perform scheduled (preventive) and unscheduled (corrective) maintenance. It excludes logistics delay time and administrative delay time, and is expressed as

$$\overline{M} = \frac{(\lambda)(\overline{Mct}) + (fpt)(\overline{Mpt})}{\lambda + fpt} \tag{2.25}$$

where λ is the corrective maintenance rate or failure rate, and *fpt* is the preventive maintenance rate.

6. Maximum active corrective maintenance time (M_{max}). M_{max} can be defined as that value of maintenance downtime below which a specified percent of all maintenance actions can be expected to be completed. M_{max} is related primarily to the log-normal distribution, and the 90th or 95th percentile point is generally taken as the specified value, as shown in Figure 2.16. It is expressed as

$$M_{max} = \text{antilog}\,[\overline{\log Mct} + Z\sigma_{\log Mct_i}] \tag{2.26}$$

where $\overline{\log Mct}$ is the mean of the logarithms of Mct_i, Z is the value corresponding to the specific percentage point at which M_{max} is defined (see Table 2.5, +1.65 for 95%) and

$$\sigma_{\log Mct_i} = \sqrt{\frac{\sum^{N} (\log Mct_i)^2 - \left(\sum_{i=1}^{N} \log Mct_i\right)^2 / N}{N - 1}} \tag{2.27}$$

or the standard deviation of the sample logarithms of average repair times, Mct_i.

For example, determining M_{max} at the 95th percentile for the data sample in Table 2.6 is accomplished as follows:

$$M_{max} = \text{antilog}\,[\log \widetilde{Mct} + (1.65)\,\sigma_{\log Mct_i}] \tag{2.28}$$

where, referring to Equation (2.27) and Table 2.7,

$$\sigma_{\log Mct_i} = \sqrt{\frac{78.785 - (43.315)^2/24}{23}} = 0.163$$

Substituting the standard deviation factor and the mean value into Equation (2.28), one can obtain

$$M_{max} = \text{antilog}\,[\log \widetilde{Mct} + (1.65)(0.163)]$$
$$= \text{antilog}\,(1.805 + 0.269) = 119 \text{ minutes}$$

If maintenance times are distributed log-normally, M_{max} cannot be derived directly by using the observed maintenance values. However, by taking the logarithm of each repair value, the resulting distribution becomes normal, facilitating usage of the data in a manner identical to the normal case.

7. Logistic delay time (LDT). Logistics delay time is the maintenance downtime that is expended as a result of waiting for a spare part to become available, waiting for the availability of an item of test equipment in order to perform maintenance, waiting for transportation, waiting to use a facility required for maintenance, and so on. LDT does not include active maintenance time, but does constitute a major element of total maintenance downtime (MDT).

8. Administrative delay time (ADT). Administrative delay time refers to that portion of downtime during which maintenance is delayed for reasons of an administrative nature: personnel assignment priority, labor strike, organizational constraint, and so on. ADT does not include active maintenance time, but often constitutes a significant element of total maintenance downtime (MDT).

9. Maintenance downtime (MDT). Maintenance downtime constitutes the total elapsed time required (when the system is not operational) to repair and restore a system to full operating status, and/or to retain a system in that condition. MDT includes mean active maintenance time ($\overline{\text{M}}$), logistics delay time (LDT), and administrative delay time (ADT). The mean or average value is calculated from the elapsed times for each function and the associated frequencies (similar to the approach used in determining $\overline{\text{M}}$).

Figure 2.17 presents an overview of the various elapsed time factors.

Maintenance Labor-Hour Factors

The maintainability factors covered in the previous paragraphs relate to elapsed times. Although elapsed times are extremely important in the performance of maintenance, one must also consider the maintenance labor-hours expended in the process. Elapsed times can be reduced (in many instances) by applying additional human resources in the accomplishment of specific tasks. However, this may turn out to be an expensive tradeoff, particularly when high skill levels are required to perform tasks that result in less overall clock time. In other words, maintainability is concerned with the *ease* and *economy* in the performance of maintenance. As such, an objective is to obtain the proper balance between elapsed time, labor time, and personnel skills at a minimum maintenance cost.

When considering measures of maintainability, it is not only important to address such factors as $\overline{\text{M}}\text{ct}$ and MDT, but it is also necessary to consider the labor-time element. Thus, some additional measures must be employed such as the following:

1. Maintenance labor hours per system operating hours (MLH/OH)
2. Maintenance labor hours per cycle of system operation (MLH/cycle)
3. Maintenance labor hours per month (MLH/month)
4. Maintenance labor hours per maintenance action (MLH/MA)

Any of these factors can be specified in terms of mean values. For example, $\overline{\text{MLH}_c}$ = mean corrective maintenance labor hours, expressed as

$$\overline{\text{MLH}_c} = \frac{\Sigma(\lambda_i)(\text{MLH}_i)}{\Sigma\lambda_i} \tag{2.29}$$

where λ_i is the failure rate of the ith item (failures/hour), and MLH_i is the average maintenance labor hours necessary to complete repair of the ith item.

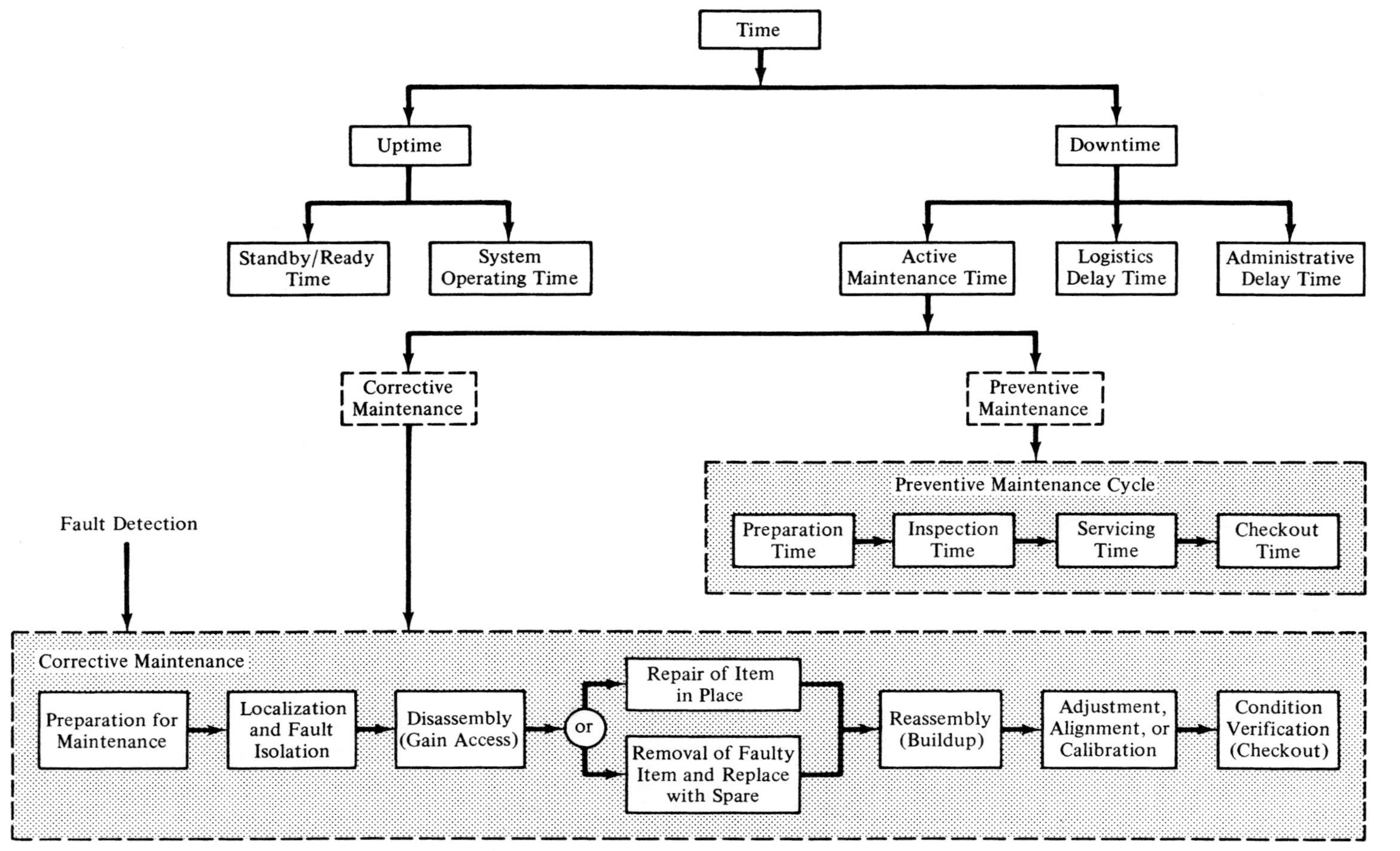

Figure 2.17 Composite view of uptime/downtime factors.

Additionally, the values for mean preventive maintenance labor hours and mean total maintenance labor hours (to include preventive and corrective maintenance) can be calculated on a similar basis. These values can be predicted for each level of maintenance, and are employed in determining specific support requirements and associated costs.

Maintenance Frequency Factors

Section 2.1 covers the measures of reliability, with MTBF and λ being key factors. Based on the discussion thus far, it is obvious that reliability and maintainability are very closely related. The reliability factors, MTBF and λ, are the basis for determining the frequency of corrective maintenance. Maintainability deals with the characteristics in system design pertaining to minimizing the corrective maintenance requirements for the system when it assumes operational status later on. Thus, in this area, reliability and maintainability requirements for a given system must be compatible and mutually supportive.

In addition to the corrective maintenance aspect of system support, maintainability also deals with the characteristics of design which minimize (if not eliminate) preventive maintenance requirements for that system. Sometimes, preventive maintenance requirements are added with the objective of improving system reliability (e.g., reducing failures by specifying selected component replacements at designated times). However, the introduction of preventive maintenance can turn out to be quite costly if not carefully controlled. Further, the accomplishment of too much preventive maintenance (particularly for complex systems/products) often has a degrading effect on system reliability, as failures are frequently induced in the process. Hence, an objective of maintainability is to provide the proper balance between corrective maintenance and preventive maintenance at least overall cost.

1. Mean time between maintenance (MTBM). MTBM is the mean or average time between *all* maintenance actions (corrective and preventive) and can be calculated as

$$\text{MTBM} = \frac{1}{1/\text{MTBM}_u + 1/\text{MTBM}_s} \tag{2.30}$$

where MTBM_u is the mean interval of unscheduled (corrective) maintenance and MTBM_s is the mean interval of scheduled (preventive) maintenance. The reciprocals of MTBM_u and MTBM_s constitute the maintenance rates in terms of maintenance actions per hour of system operation. MTBM_u should approximate MTBF, assuming that a combined failure rate is used which includes the consideration of primary inherent failures, dependent failures, manufacturing defects, operator and maintenance induced failures, and so on. The maintenance frequency factor, MTBM, is a major parameter in determining system availability and overall effectiveness.

2. Mean time between replacement (MTBR). MTBR, a factor of MTBM, refers to the mean time between item replacement and is a major parameter in determing spare part requirements. On many occasions, corrective and preventive maintenance actions are accomplished without generating the requirement for the replacement of a component part. In other instances, item replacements are required which, in turn, necessitates the availability of a spare part and an inventory requirement. Additionally, higher levels of maintenance support (i.e., intermediate shop and depot levels) may be required.

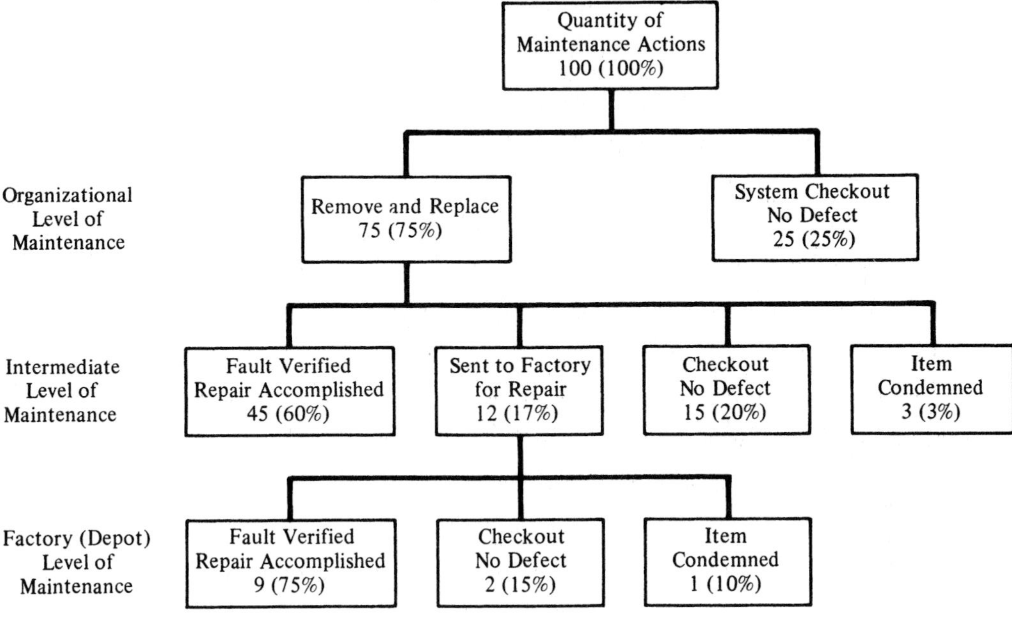

Figure 2.18 System *XYZ* unscheduled maintenance actions.

In essence, MTBR is a significant factor, applicable in both corrective and preventive maintenance activities involving item replacement, and is a key parameter in determining logistic support requirements. A maintainability objective in system design is to maximize MTBR where feasible.

When relating these factors to experiences in the field, one often finds the situation illustrated in Figure 2.18. In this example, there are 100 corrective maintenance actions at the system level, stimulated from reports of system failure (perceived or actual). Each maintenance action, which is counted in the MTBM measure, results in the consumption of personnel resources (MLH). Some of these maintenance actions, or 75 in this case, result in the removal and replacement of system components. These actions, which are included in the MTBR measure, generate the requirements for spare/repair parts. Some of the items that are removed and returned to the intermediate shop for higher-level maintenance are considered to be faulty, where a failure has actually been verified. These confirmed failures are included in the reliability MTBF figure of merit. In the example presented, only 58 of the 100 unscheduled maintenance actions resulted in confirmed failures. Where reliability is concerned primarily with the MTBF measure, maintainability is concerned with the MTBM and MTBR factors. These factors will be discussed further in later chapters.

Maintenance Cost Factors

For many systems/products, maintenance cost constitutes a major segment of total life-cycle cost. Further, experience has indicated that maintenance costs are significantly impacted by design decisions made throughout the early stages of system development. Thus, it is essential that total life-cycle cost be considered as a major design parameter, beginning with the definition of system requirements.

Of particular interest is the aspect of *economy* in the performance of maintenance actions. In other words, maintainability is directly concerned with the characteristics of system design that will ultimately result in the accomplishment of maintenance at minimum overall cost.

When considering maintenance cost, the following cost-related indices may be appropriate as criteria in system design:

1. Cost per maintenance action ($/MA)
2. Maintenance cost per system operating hour ($/OH)
3. Maintenance cost per month ($/month)
4. Maintenance cost per mission or mission segment ($/mission)
5. The ratio of maintenance cost to total life-cycle cost

A presentation of maintenance cost, in the context of life-cycle cost, is included in Appendix E.

2.3 SUPPLY SUPPORT FACTORS

Supply support includes the spare parts and the associated inventories necessary for the accomplishment of unscheduled and scheduled maintenance actions. At each maintenance level, one must determine the type of spare parts (by manufacturing part number) and the quantity of items to be purchased and stocked. Also, it is necessary to know how often various items should be ordered and the number of items that should be procured in a given purchasing transaction.

Spare part requirements are initially based on the system maintenance concept and are subsequently defined and justified through the supportability analysis (SA). Essentially, spare part quantities are a function of demand rates and include consideration of:

1. Spares and repair parts covering actual item replacements occurring as a result of corrective and preventive maintenance actions. Spares are major replacement items that are repairable, whereas repair parts are nonrepairable smaller components.
2. An additional stock level of spares to compensate for repairable items in the process of undergoing maintenance. If there is a backup (lengthy queue) of items in the intermediate maintenance shop or at the depot manufacturer awaiting repair, these items obviously will not be available as recycled spares for subsequent maintenance actions; thus, the inventory is further depleted (beyond expectation), or a stock-out condition results. In addressing this problem, it becomes readily apparent that the test equipment capability, personnel, and facilities directly impact the maintenance turnaround times and the quantity of additional spare items needed.

3. An additional stock level of spares and repair parts to compensate for the procurement lead times required for item acquisition. For instance, prediction data may indicate that 10 maintenance actions requiring the replacement of a certain item will occur within a 6-month period and it takes 9 months to acquire replacements from the supplier. One might ask: What additional repair parts will be necessary to cover the operational needs and yet compensate for the long supplier lead time? The added quantities will, of course, vary depending on whether the item is designated as repairable or will be discarded at failure.
4. An additional stock level of spares to compensate for the condemnation or scrapage of repairable items. Repairable items returned to the intermediate maintenance shop or depot are sometimes condemned (i.e., not repaired) because, through inspection, it is decided that the item was not economically feasible to repair. Condemnation will vary depending on equipment utilization, handling, environment, and organization capability. An increase in the condemnation rate will generally result in an increase in spare part requirements.

In reviewing the foregoing considerations, of particular significance is the determination of spares requirements as a result of item replacements in the performance of corrective maintenance. Major factors involved in this process are (1) the reliability of the item to be spared, (2) the quantity of items used, (3) the required probability that a spare will be available when needed, (4) the criticality of item application with regard to mission success, and (5) cost. Use of the reliability and probability factors are illustrated in the examples presented subsequently.

Probability of Success with Spares Availability Considerations

Assume that a single component with a reliability of 0.8 (for time t) is used in a unique system application and that one backup spare component is purchased. Determine the probability of system success having a spare available in time t (given that failures occur randomly and are exponentially distributed).

This situation is analogous to the case of an operating component and a parallel component in standby (i.e., standby redundancy) discussed in Section 2.1. The applicable expression is stated as

$$P = e^{-\lambda t} + (\lambda t)e^{-\lambda t} \tag{2.31}$$

With a component reliability of 0.8, the value of λt is 0.223. Substituting this value into Equation (2.31) gives a probability of success of

$$P = e^{-0.223} + (0.223)e^{-0.223}$$

$$= 0.8 + (0.223)(0.8) = 0.9784$$

Assuming next that the component is supported with two backup spares (where all three components are interchangeable), the probability of success during time t is determined from

$$P = e^{-\lambda t} + (\lambda t)e^{-\lambda t} + \frac{(\lambda t)^2 e^{-\lambda t}}{2!}$$

or

$$P = e^{-\lambda t}\left[1 + \lambda t + \frac{(\lambda t)^2}{2!}\right] \tag{2.32}$$

With a component reliability of 0.8 and a value of λt of 0.223, the probability of success is

$$\begin{aligned} P &= 0.8\left[1 + 0.223 + \frac{(0.223)^2}{(2)(1)}\right] \\ &= 0.8(1.2479) = 0.9983 \end{aligned}$$

Thus, adding another spare component results in one additional term in the Poisson expression.[3] If two spare components are added, two additional terms are added, and so on.

The probability of success for a configuration consisting of two operating components, backed by two spares, with all components being interchangeable can be found from the expression

$$P = e^{-2\lambda t}\left[1 + 2\lambda t + \frac{(2\lambda t)^2}{2!}\right] \tag{2.33}$$

With a component reliability of 0.8 and $\lambda t = 0.223$,

$$\begin{aligned} P &= e^{-0.446}\left[1 + 0.446 + \frac{(0.446)^2}{(2)(1)}\right] \\ &= 0.6402[1 + 0.446 + 0.0995] = 0.9894 \end{aligned}$$

These examples illustrate the computations used in determining system success with spare parts for three simple component configuration relationships. Various combinations of operating components and spares can be assumed, and the system success factors can be determined by using

$$1 = e^{-\lambda t} + (\lambda t)e^{-\lambda t} + \frac{(\lambda t)^2 e^{-\lambda t}}{2!} + \frac{(\lambda t)^3 e^{-\lambda t}}{3!} + \cdots + \frac{(\lambda t)^n e^{-\lambda t}}{n!} \tag{2.34}$$

Equation (2.34) can be simplified into a general Poisson expression

$$f(x) = \frac{(\lambda t)^x e^{-\lambda t}}{x!} \tag{2.35}$$

The objective is to determine the probability of x failures occurring if an item is placed in operating for t hours, and each failure is corrected (through item replacement) as it occurs. With n items in the system, the number of failures in t hours will be $n\lambda t$, and the general Poisson expression becomes

$$f(x) = \frac{(n\lambda t)^x e^{-n\lambda t}}{x!} \tag{2.36}$$

To facilitate calculations, a cumulative Poisson probability graph is presented in Figure 2.19 derived from Equation (2.36). The ordinate value can be viewed as a con-

[3] The Poisson and exponential distributions are equivalent except for the choice of the random variable. For the exponential, the random variable is the time to failure, whereas it is the number of failures per a given time period for the Poisson. The exponential variable is continuous and the Poisson variable is discrete.

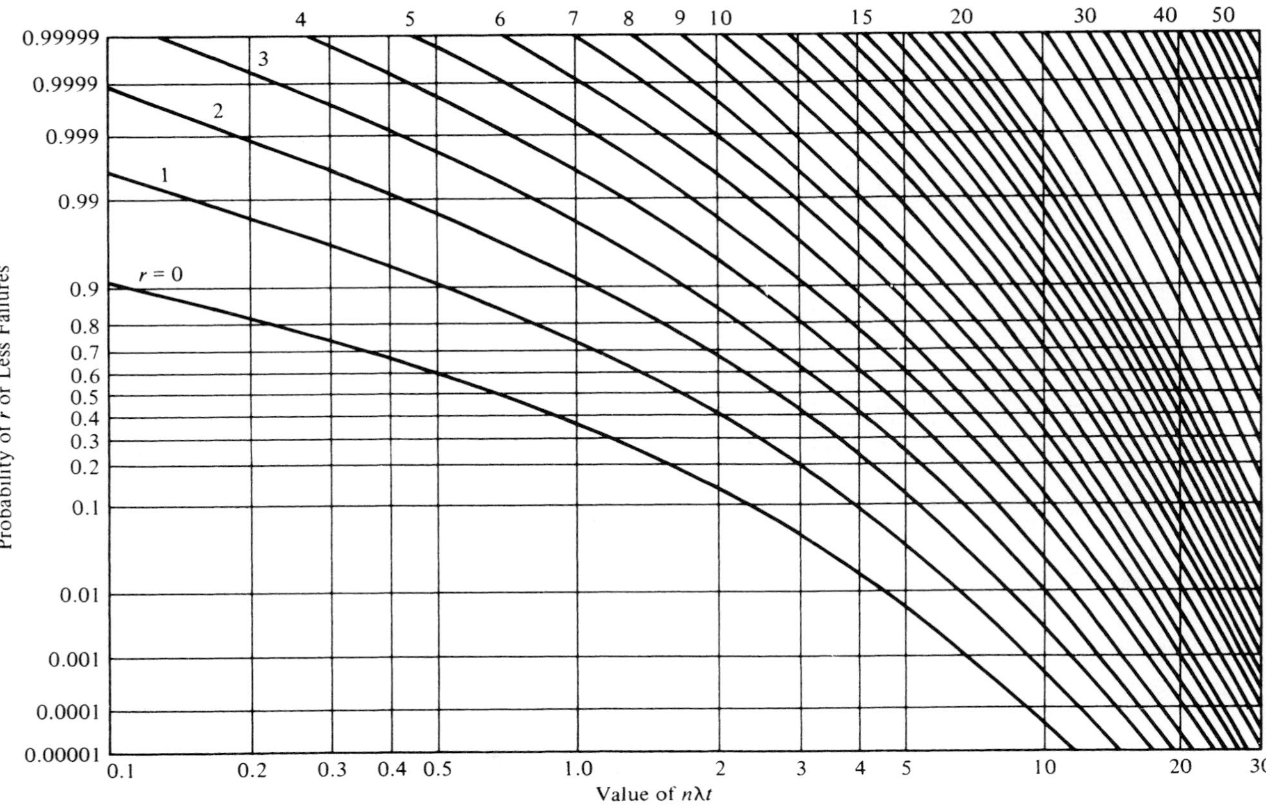

Figure 2.19 Poisson cumulative probabilities. (NAVAIR 01-1A-32, *Reliability Engineering Handbook*, Naval Air Systems Command, U.S. Navy, Washington, D.C., 1977.)

fidence factor. Several simple examples will be presented to illustrate the application of Figure 2.19.

Probability of Mission Completion

Suppose that one needs to determine the probability that a system will complete a 30-hour mission without a failure when the system has a known mean life of 100 hours. Let

$$\lambda = 1 \text{ failure per 100 hours or 0.01 failure per hour}$$

$$t = 30 \text{ hours}$$

$$n = 1 \text{ system}$$

$$n\lambda t = (1)(0.01)(30) = 0.3$$

Enter Figure 2.19 where $n\lambda t$ is 0.3. Proceed to the intersection where r equals zero and read the ordinate scale, indicating a value of approximately 0.73. Thus, the probability of the system completing a 30-hour mission is 0.73.

Assume that the system identified above is installed in an aircraft and that 10 aircraft are scheduled for a 15-hour mission. Determine the probability that at least 7 systems will operate for the duration of the mission without failure. Let

$$n\lambda t = (10)(0.01)(15) = 1.5$$

and

$$r = 3 \text{ failures or less (allowed)}$$

Enter Figure 2.19 when $n\lambda t$ equals 1.5. Proceed to the intersection where r equals 3, and read the ordinate scale indicating a value of approximately 0.92. Thus, there is a 92% confidence that at least 7 systems will operate successfully out of 10. If an 80% operational reliability is specified (i.e., 8 systems must operate without failure), the confidence factor decreases to about 82%.

Although the graph in Figure 2.19 provides a simplified solution, the use of Equation (2.34) is preferable for accurate results. Additionally, there are many textbooks that contain tables covering the Poisson expansion.

Spare-Part Quantity Determination

Spare-part quantity determination is a function of a probability of having a spare part available when required, the reliability of the item in question, the quantity of items used in the system, and so on. An expression, derived from the Poisson distribution, useful for spare part quantity determination is

$$P = \sum_{n=0}^{n=s} \left[\frac{R(-\ln R)^n}{n!} \right] \tag{2.37}$$

where

P = probability of having a spare of a particular item available when required

S = number of spare parts carried in stock
R = composite reliability (probability of survival); $R = e^{-K\lambda t}$
K = quantity of parts used of a particular type
$\ln R$ = natural logarithm of R

In determining spare part quantities, one should consider the level of protection desired (safety factor). The protection level is the P value in Equation (2.37). This is the probability of having a spare available when required. The higher the protection level, the greater the quantity of spares required. This results in a higher cost for item procurement and inventory maintenance. The protection level, or safety factor, is a hedge against the risk of stock-out.

When determining spare part quantities, one should consider system operational requirements (e.g., system effectiveness, availability) and establish the appropriate level at each location where corrective maintenance is accomplished. Different levels of corrective maintenance may be appropriate for different items. For instance, spares required to support prime equipment components which are critical to the success of a mission may be based on one factor; high-value or high-cost items may be handled differently than low-cost items; and so on. In any event, an optimum balance between stock level and cost is required.

Figures 2.20A and 2.20B present a nomograph which simplifies the determination of spare part quantities using Equation (2.37). The nomograph not only simplifies solutions for basic spare part availability questions, but provides information that can aid in the evaluation of alternative design approaches in terms of spares and in the determination of provisioning cycles. The following examples illustrate the use of the nomograph.[4]

Suppose that a piece of equipment contains 20 parts of a specific type with a failure rate (λ) of 0.1 failure per 1000 hours of operation. The equipment operates 24 hours a day, and spares are procured and stocked at 3-month intervals. How many spares should be carried in inventory to ensure a 95% chance of having a spare part when required?

Let

$$K = 20 \text{ part}$$
$$\lambda = 0.1 \text{ failure per 1000 hours}$$
$$T = 3 \text{ months}$$
$$K\lambda T = (20)(0.0001)(24)(30)(3) = 4.32$$
$$P = 95\%$$

Using the nomograph in Figures 2.20A and 2.20B as illustrated, approximately 8 spares are required.

As a second example, suppose that a particular part is used in three different equipments (A, B, C). Spares are procured every 180 days. The number of parts used, the part failure rate, and the equipment operating hours per day are given in Table 2.8.

[4] NAVSHIPS 94324, *Maintainability Design Criteria Handbook for Designers of Shipboard Electronic Equipment*, Naval Ship Systems Command, Department of the Navy, Washington, D.C., 1964.

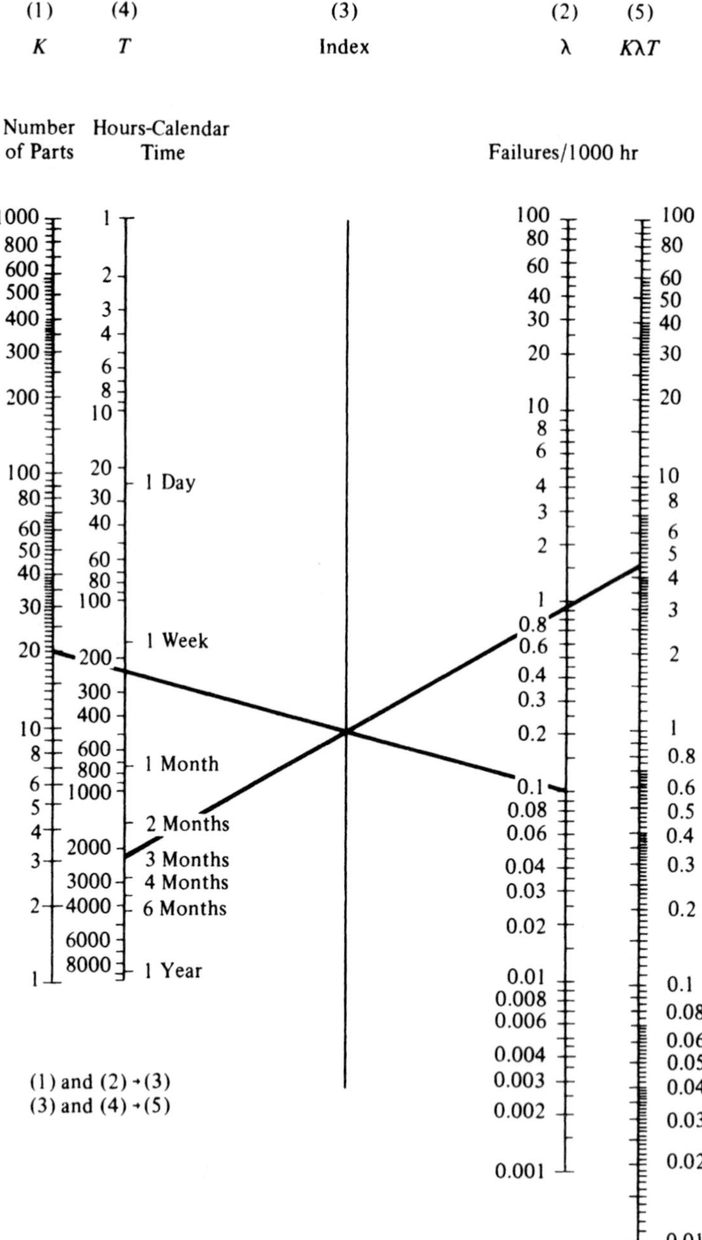

Figure 2.20(a) Spare-part requirement nomograph (sheet 1 of 1).

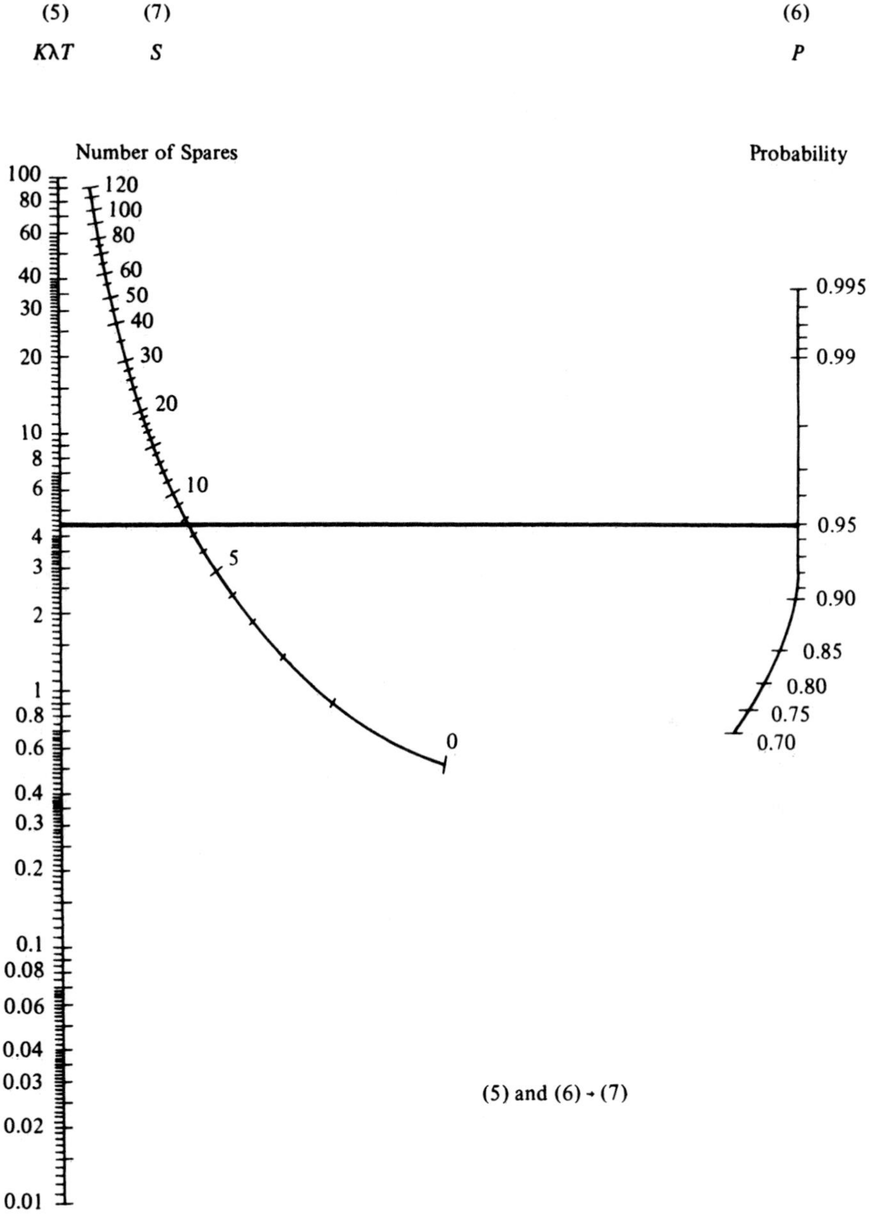

Number of Spares

Probability

100 80 60 40 30 20 10 8 6 4 3 2 1 0.8 0.6 0.4 0.3 0.2 0.1 0.08 0.06 0.04 0.03 0.02 0.01

120 100 80 60 50 40 30 20 10 5 0

0.995 0.99 0.95 0.90 0.85 0.80 0.75 0.70

(5) and (6) → (7)

Figure 2.20(b) Spare-part requirement nomograph (sheet 2 of 2). (NAVSHIPS 94324, *Maintainability Design Criteria Handbook for Designers of Shipboard Electronic Equipment*, U.S. Navy, Washington, D.C.)

TABLE 2.8 Data for Spares Inventory

Item	K	Failures per 1000 Hours (λ)	Operating Hours per Day (T)
Equipment A	25	0.10	12
Equipment B	28	0.07	15
Equipment C	35	0.15	20

The number of spares that should be carried in inventory to ensure a 90% chance of having a spare available when required is calculated as follows:

1. Determine the product of K, λ, and T as

$$A = (25)(0.0001)(180)(12) = 5.40$$
$$B = (28)(0.00007)(180)(15) = 5.29$$
$$C = (35)(0.00015)(180)(20) = 18.90$$

2. Determine the sum of the $K\lambda T$ values as

$$\Sigma K\lambda T = 5.40 + 5.29 + 18.90 = 29.59$$

3. Using sheet 2 of the nomograph (Figure 2.20B), construct a line from $K\lambda T$ value of 29.59 to the point where P is 0.90. The approximate number of spares required is 36.

Inventory Considerations

In progressing further, one needs to address not only the specific demand factors for spares, but to evaluate these factors in terms of the overall inventory requirements. Too much inventory may ideally respond to the demand for spares. However, this may be costly, with a great deal of capital tied up in the inventory. In addition, much waste could occur, particularly if system changes are implemented and certain components become obsolete. On the other hand, providing too little support results in the probability of causing the system to be inoperative due to stock-out, which also can be costly. In general, it is desirable to obtain an economic balance between the quantity of items in inventory at any given point in time, the frequency of purchase order transactions, and the quantity of items per purchase order.

Figure 2.21 presents a graphical portrayal of an inventory cycle. The illustration assumes a constant lead time and a constant item demand (e.g., failure rate). Stock depletions are represented by the sloping consumption line. When the stock is depleted to a designated level, additional items are ordered (represented at the order point) in enough time to allow for supply replenishment before a stock-out condition occurs. The terms identified in the figure are briefly defined as follows:[5]

[5] Inventory concepts may vary. A *fixed review time method* (when inventory is reviewed at fixed time intervals and depleted items are reordered in variable quantities) or a *fixed reorder point method* (standard orders are initiated when the inventory is depleted to a certain level) may be employed. However, the economic order principle should govern in most instances.

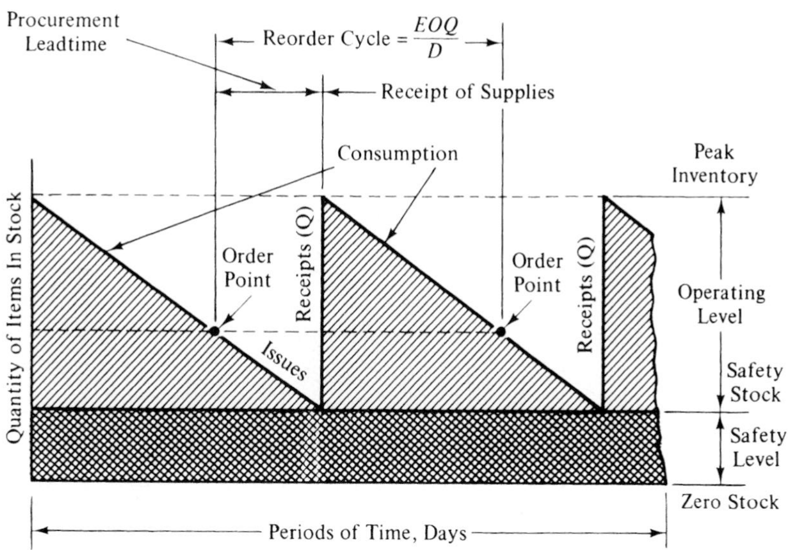

Figure 2.21 Theoretical inventory cycle.

1. *Operating level*—describes the quantity of material items required to support normal system operations in the interval between orders and the arrival of successive shipments of material.
2. *Safety stock*—additional stock required to compensate for unexpected demands, repair and recycle times, pipeline, procurement lead time, and unforeseen delays.
3. *Reorder cycle*—interval of time between successive orders.
4. *Procurement lead time*—the span of time from the date of order to receipt of the shipment in the inventory. This includes (a) administrative lead time from the date that a decision is made to initiate an order to the receipt of the order at the supplier; (b) production lead time or the time from receipt of the order by the supplier to completion of the manufacture of the item ordered; and (c) delivery lead time from completion of manufacture to receipt of the item in the inventory. Delivery lead time includes the pipeline.
5. *Pipeline*—the distance between the supplier and consumer, measured in days of supply. If a constant flow is assumed with a transit time of 30 days and the consumption rate is one item per day, then 30 items would be required in the pipeline at all times. An increase in the demand would require more in the pipeline.
6. *Order point (O.P.)*—the point in time when orders are initiated for additional quantities of spare/repair parts. This point is often tied to a given stock level (after the stock has been depleted to that level).

Figure 2.21 is a theoretical representation of an inventory cycle for a given item. Actually, demands are not always constant and quite often the reorder cycle changes with time. Figure 2.22 presents a situation that is more realistic.

Referring to Figure 2.21 (i.e., the theoretical inventory cycle), the ultimate goal is to have the correct amount and type of supplies available for the lowest total cost.

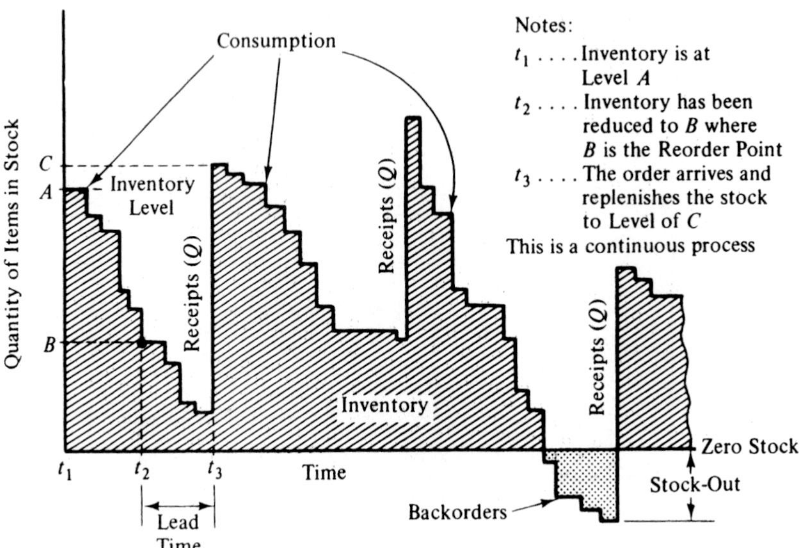

Figure 2.22 Representation of an actual inventory cycle.

Procurement costs vary with the quantity of orders placed. The economic inventory principle involves the optimization between the placing of many orders resulting in high material acquisition costs and the placing of orders less frequently while maintaining a higher level of inventory, causing increasing inventory maintenance and carrying costs. In other words, ordering creates procurement cost while inventory creates carrying cost. The economic order principle equates the *cost to order* to the *cost to hold* and the point at which the combined costs are at a minimum indicates the desired size of order. Figure 2.23 graphically illustrates this concept.

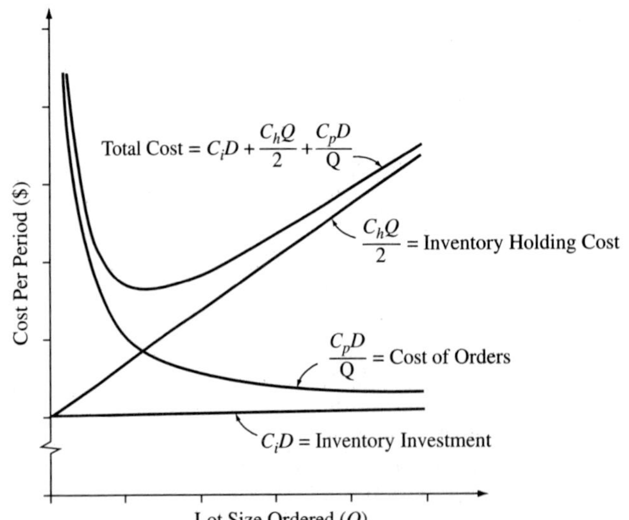

Figure 2.23 Economic inventory cost considerations. (Source: Blanchard, B. S., and W. J. Fabrycky, *Systems Engineering and Analysis*, 3rd ed., Prentice Hall, Upper Saddle River, NJ, 1998.)

Referring to the figure, the illustration forms the basis for the size of orders placed, which is known as the *economic order quantity* (EOQ).[6] The total cost curve in Figure 2.23 is expressed as

$$\text{Total cost (TC)} = C_i D + \frac{C_h Q}{2} + \frac{C_p D}{Q} \tag{2.38}$$

where

C_i = acquisition cost of inventory

C_p = average cost of ordering in dollars per order (includes cost of setup, processing orders, receiving, etc.)

C_h = cost of carrying an item in inventory (percent of the item price multiplied by the total acquisition price for the item)

D = annual item demand (this is assumed to be constant for the purposes of discussion)

Q = most economical quantity

Differentiating and solving for Q, which represents the quantity of units ordered, gives the following equation for EOQ:

$$Q^* = \text{EOQ} = \sqrt{\frac{2C_p D}{C_h}} \tag{2.39}$$

Where Equation (2.39) is the expression for determining the quantity of items per order, the number of purchase orders per year, N, can be determined from the equation

$$N = \frac{D}{\text{EOQ}} \tag{2.40}$$

The EOQ model is generally applicable in instances where there are relatively large quantities of common spares and repair parts. However, it may be feasible to employ other methods of acquisition for major high-value items and for those items considered to be particularly critical to mission success.

High-value items are those components with a high unit acquisition cost, and should be purchased on an individual basis. The dollar value of these components is usually significant and may even exceed the total value of the hundreds of other spares and repair parts in the inventory. In other words, a relatively small number of items may represent a large percentage of the total inventory value. Because of this, it may be preferable to maintain a given quantity of these items in the inventory to compensate for repair and recycle times, pipeline and procurement lead times, and so on, and order new items on a one-for-one basis as failures occur and a spare is withdrawn from the inventory. Thus, where it may appear to be economically feasible to purchase many

[6] The EOQ concept presented is a simplified representation of the major consideration in establishing an inventory. However, there are many variations and additional material should be reviewed before making a decision on a specific application.

of these items in a given purchase transaction, only a small quantity of items is actually procured because of the risks involved relative to tying up too much capital and the resultant high inventory maintenance cost.[7]

Another consideration in the spares acquisition process is that of *criticality*. Some items are considered more critical than others in terms of impact on mission success. For instance, the lack of a $100 item may cause the system to be inoperative, while the lack of a $10,000 item might not cause a major problem. The criticality of an item is generally based on its function in the system and not necessarily its acquisition cost.

A third factor influencing the procurement of spares is the availability of suppliers and the likelihood that the supplier(s) selected will be producing spares for the entire duration of the system life cycle. If only a single sole source of supply exists and it is anticipated that the supplier may not continue to provide the spares as needed, the purchase may be in terms of expected *life-cycle needs*. Conversely, if there are multiple sources available, periodic reprocurements can be expected.

As a final point, the spares acquisition process may vary somewhat between items of a comparable nature if the usage rates are significantly different. Fast-moving items may be procured locally near the point of usage, such as the intermediate maintenance shop, whereas slower-moving items stocked at the depot, or a central repair facility, may be acquired from a remotely located supplier as the pipeline and procurement lead times are not as critical.

2.4 TRANSPORTATION, PACKAGING, AND HANDLING FACTORS

Referring to Figures 1.1 and 1.2, it is evident that the area of *transportation* constitutes a major ingredient in the maintenance and logistic support infrastructure. Transportation requirements include the movement of humans and material resources, in support of both operational and maintenance activities, from one location to another. When evaluating the effectiveness of transportation, one must deal with such factors as

1. Transportation route, both national and international (distances, number of nationalities, customs requirements, political and social factors, and so on)
2. Transportation capability or capacity (volume of goods transported, number of loads, ton-miles per year, frequency of transportation, modes of transportation, legal forms, and so on)
3. Transportation time (short-haul versus long-haul time, mean delivery time, time per transportation leg, and so on)
4. Transportation cost (cost per shipment, cost of transportation per mile/kilometer, cost of packaging and handling, etc.)

The basic modes of transportation include *railroad*, *highway*, *waterway*, *air*, and *pipeline*. Additionally, there are various combinations of these modes that can be cat-

[7] The classification of high-value items will vary with the program, and may be established at a certain dollar value (i.e., all components whose initial unit cost exceeds x dollars are considered as high-value items).

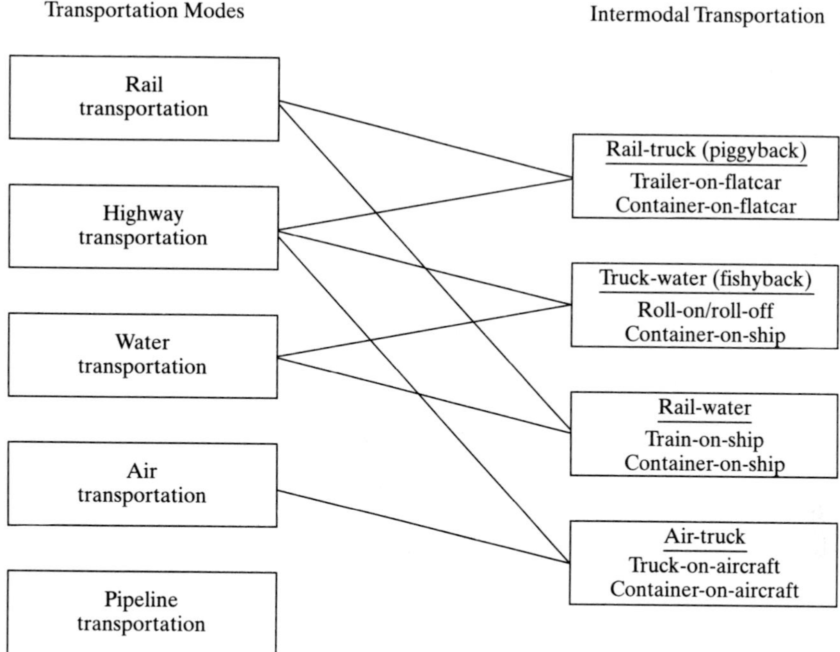

Figure 2.24 The various forms of transportation.

egorized as *intermodal*, as shown in figure 2.24. Depending on the geographical location of system elements and the applicable suppliers (i.e., the need for transportation), the degree of urgency in terms of delivery-time requirement, etc., alternative routing requirements can be identified. Through the process of conducting trade-offs considering time factors, available modes, and cost, a recommended transportation route is established. This may include transportation by a specific mode or through the use of an intermodal approach.[8]

Given the possible modes of transportation and the proposed routes that have been identified to support the *outward* and *reverse* flows pertaining to operational and maintenance activities (refer to Figure 1.1), along with the various environments in which an item may be subjected when being transported from one location to another, the key issue becomes that of *packaging* design or the design of an item for *transportability* or *mobility*. Products that are to be transported must be designed in such a

[8] This is an over-simplication of the transportation issue. One should consider many different factors in determining a specific transportation route to include the mode and intermodal possibilities, the legal forms of transportation (common, contract, private), the various rate structures, the many governing legislative acts and regulations, and the overall structure of the transportation industry in general. A good reference is Glaskowsky, N. A, D. R. Hudson, and R. M. Ivie, *Business Logistics*, 3rd Ed., The Dryden Press, Harcourt Brace Jonanovich Publishing Co., Orlando, Fl., 1992. Also, refer to the literature on business logistics listed in Appendix H.

way to eliminate damage, possible degradation, and so on. The following questions should be addressed:[9]

1. Does the package (in which the item being transported is contained) incorporate the desired strength and material characteristics?
2. Can it stand rough handling or long-term storage without degradation?
3. Does the package provide adequate protection against various environmental conditions such as rain, vibration and shock, temperature, sand/dust, salt spray?
4. Is the package compatible with existing transportation and handling methods? Does it possess good stacking qualities?
5. Has the package been designed for safety (i.e., to discourage pilferage or theft)?
6. Can the packaging materials be recycled for additional use. If not, will the materials meet the requirements for disposability?

2.5 TEST AND SUPPORT EQUIPMENT FACTORS

The general category of test and support equipment may include a wide spectrum of items, such as precision electronic test equipment, mechanical test equipment, ground handling equipment, special jigs and fixtures, maintenance stands, and the like. These items, in varying configurations and mixes, may be assigned to different maintenance locations and geographically dispersed throughout the country (or world). However, regardless of the nature and application, the objective is to provide the right item for the job intended, at the proper location, and in the quantity required.

Because of the likely diversification of the test and support equipment for any given system, it is difficult to specify quantitative measures that can be universally applied. Certain measures are appropriate for electronic test equipment, other measures are applicable to ground handling equipment, and so on. Further, the specific location and application of a given item of test equipment may also result in different measures. For instance, an item of electronic test equipment used in support of on-site organizational maintenance may have different requirements than a similar item of test equipment used for intermediate maintenance accomplished in a remote shop.

Although all of the test and support equipment requirements at each level of maintenance are considered to be important relative to successful system operation, the testers or test stations in the intermediate and depot manufacturer maintenance facilities are of particular concern, since these items are likely to support a number of system elements at different customer locations. That is, an intermediate maintenance facility may be assigned to provide the necessary corrective maintenance support for a large number of system elements dispersed throughout a wide geographical area. This means, of course, that a variety of items (all designated for intermediate-level maintenance) will arrive from different customer sites at different times.

[9] It is not uncommon to find that a product/item in transit will be subjected to a harsh or rigorous environment—one that is more rigorous than will be experienced throughout the accomplishment of its operational mission(s).

When determining the specific test equipment requirements for a shop, one must define (1) the type of items that will be returned to the shop for maintenance; (2) the test functions to be accomplished, including the performance parameters to be measured as well as the accuracies and tolerances required for each item; and (3) the anticipated frequency of test functions per unit of time. The type and frequency of item returns (i.e., shop arrivals) is based on the maintenance concept and system reliability data. The distribution of arrival times for a given item is often a negative exponential with the number of items arriving within a given time period following a Poisson distribution. As items arrive in the shop, they may be processed immediately or there may be a waiting line, or queue, depending on the availability of the test equipment and the personnel to perform the required maintenance functions. A servicing policy, with priorities, must be established.

When evaluating the test process itself, one should calculate the anticipated test equipment utilization requirements (i.e., the total amount of *on-station* time required per day, month, or year). This can be estimated by considering the repair time distributions for the various items arriving in the shop. However, the ultimate elapsed times may be influenced significantly depending on whether manual, semiautomatic, or automatic test methods are employed.

Given the test equipment utilization needs (from the standpoint of total test station time required for processing shop arrivals), it is necessary to determine the anticipated reliability and maintainability of the test equipment configuration being considered for the application. Thus, one must consider the MTBM and MDT values for the test equipment itself. Obviously, the test equipment configuration should be more reliable than the system component being tested. Also, in instances where the complexity of the test equipment is high, the logistic resources required for the support of the test equipment may be extensive (e.g., the frequent requirement to calibrate an item of test equipment against a secondary or primary standard in a "cleanroom" environment). In essence, there is a requirement to determine the time that the test equipment will be available to perform its intended function.

The final determination of the requirements for test equipment in a maintenance facility is accomplished through an analysis of various alternative combinations of arrival rates, queue length, test station process times, and/or quantity of test stations. Basically, one is dealing with a single-channel or multichannel queuing situation using queuing techniques. As the maintenance configuration becomes more complex, involving many variables (some of which are probabilistic in nature), then Monte Carlo analysis may be appropriate. In any event, there may be a number of feasible servicing alternatives, and a preferred approach is sought.

2.6 ORGANIZATIONAL FACTORS

The measures associated with a maintenance organization are basically the same as those factors which are typical for any organization. Of particular interest relative to logistic support are:

1. The direct maintenance labor time for each personnel category, or skill level, expended in the performance of system maintenance activities. Labor time may be broken down to cover both unscheduled and scheduled maintenance individually, and may be expressed in
   (a) Maintenance labor hours per system operating hour (MLH/OH).
   (b) Maintenance labor hours per mission cycle (or segment of a mission).
   (c) Maintenance labor hours per month (MLH/month).
   (d) Maintenance labor hours per maintenance action (MLH/MA).
2. The indirect labor time required to support system maintenance activities (i.e., overhead factor).
3. The personnel attrition rate or turnover rate (in percent).
4. The personnel training rate or the worker-days of formal training per year of system operation and support.
5. The number of maintenance work orders processed per unit of time (e.g., week, month, or year), and the average time required for work order processing.
6. The average administrative delay time, or the average time from when an item is initially received for maintenance to the point when active maintenance on that item actually begins.

When addressing the total spectrum of logistics (and the design for supportability), the organizational element is critical to the effective and successful life-cycle support of a system. The right personnel quantities and skills must be available when required, and the individuals assigned to the job must be properly trained and motivated. As in any organization, it is important to establish measures dealing with organizational effectiveness and productivity.

2.7 FACILITY FACTORS

Facilities are required to support activities pertaining to the accomplishment of active maintenance tasks, providing warehousing functions for spares and repair parts, and providing housing for related administrative functions. Although the specific quantitative measures associated with facilities may vary significantly from one system to the next, the following factors are considered to be relevant in most instances:

1. Item process time or turnaround time (TAT) (i.e., the elapsed time necessary to process an item for maintenance, returning it to full operational status).
2. Facility utilization (the ratio of the time utilized to the time available for use, percent utilization in terms of space occupancy, and so on).
3. Energy utilization in the performance of maintenance (i.e., unit consumption of energy per maintenance action, cost of energy consumption per increment of time or per maintenance action, and so on).
4. Total facility cost for system operation and support (i.e., total cost per month, cost per maintenance action, and so on).

2.8 SOFTWARE FACTORS

For many systems, software has become a major element of support. This is particularly true where automation, computer applications, digital databases, and the like are used in the accomplishment of maintenance and logistics functions. Software may be evaluated in terms of language levels or complexity, number of programs, program length on the basis of the number of source code lines, cost per maintenance subroutine, or something of a comparable nature.

As with equipment, reliability, maintainability, and quality are significant considerations in the development of software. Although software does not degrade in the same way as equipment, the reliability of software is still important and must be measured. There has been a great deal of discussion on how software reliability is to be measured, and there still is not complete agreement as to the specific measures or levels of acceptable performance. However, one definition of *software reliability* is "the probability of failure-free operation of a software component or system in a specified environment for a specified time." A *failure* is defined as "an unacceptable departure of program operation from program requirements," and a *fault* is "the software defect that causes a failure."[10]

In any event, errors occur in the initial development of software, and software reliability is a function of the number of inherent errors contained within the software that have not been eliminated. Such errors may be classified as faulty or omitted logic, addressability errors, missing commentary, regression or integration problems, counting or calculation problems, or general documentation problems. Usually, the overall measure is in terms of the number of errors per 1000 source code lines. Higher-level languages will probably contain less error since fewer lines of code are required (as compared with assembly languages). Conversely, language complexity may be introduced in the process.

In the specification and development of system software, applicable measurement factors must be identified and the system must be evaluated to include not only coverage of equipment and personnel factors, but software factors as well! The logistic support resource requirements for the overall system includes consideration of equipment, personnel, facilities, data, consumables, and software.

2.9 TECHNICAL DATA AND INFORMATION SYSTEM FACTORS

In recent years, the maintenance and logistic support infrastructure has been going through a significant evolution with the advent of computers and associated software, the availability of information networks (LANs and WANs), the implementation of CALS, the introduction of electronic data interchange (EDI) and electronic commerce (EC) methods, and related technologies. The objectives have included (1) simplifying

[10] Vick, C. R., and C. V. Ramamoorthy, *Handbook of Software Engineering*, Van Nostrand Reinhold Company, Inc., New York, N.Y., 1984. Refer to Appendix H for additional references covering the subject of software.

the task of generating and processing technical data through better packaging, by eliminating redundancies, reducing processing times, and making the information more accessible to all organizations in need; (2) providing an expeditious means for the introduction of design changes and for better implementation of configuration management requirements; (3) providing a mechanism for greater asset visibility relative to the traceability of components in transit and the location of items in inventories; and (4) enabling faster, timely, accurate, and more reliable communications between multiple locations on a current basis.

In essence, the *information age* has had a major impact on logistics. With this in mind, some of the measures that may be applicable include the following:

1. Logistics reponse time; i.e., the time that is consumed from the point when a system support requirement is first identified until that requirement has been satisfied (this may include the time required for the provisioning and procurement of a new item, the time to ship an item from inventory to the location of need, the time required to acquire the necessary personnel or test equipment for maintenance, and so on).
2. Data access time; i.e., the time to locate and gain access to the data/information element needed.
3. Item location time; i.e., the time required to locate a given asset whether in use, transit, or in some inventory (asset visibility).
4. Information processing time; i.e., the time required for the processing of messages.
5. Change implementation time; i.e., the time to process and implement a design change.
6. Cost; i.e., the cost of transmission per bit of data, the cost per data access incident, and so on.

2.10 AVAILABILITY FACTORS

The term *availability* is often used as a measure of system readiness (i.e., the degree, percent, or probability that a system will be ready or available when required for use—refer to system effectiveness factors in Chapter 1). Availability may be expressed differently, depending on the system and its mission. Three commonly used figures of merit (FOMs) are described subsequently.

Inherent Availability (A_i)

Inherent availability is the probability that a system or equipment, when used under stated conditions in an *ideal* support environment (i.e., readily available tools, spares, maintenance personnel, etc.), will operate satisfactorily at any point in time as required. It excludes preventive or scheduled maintenance actions, logistics delay time, and administrative delay time, and is expressed as

$$A_i = \frac{\text{MTBF}}{\text{MTBF} + \overline{\text{M}}\text{ct}} \tag{2.41}$$

where MTBF is the mean time between failure and $\overline{\text{Mct}}$ is the mean corrective maintenance time, as described in Sections 2.1 and 2.2.

Achieved Availability (A_a)

Achieved availability is the probability that a system or equipment, when used under stated conditions in an *ideal* support environment (i.e., readily available tools, spares, personnel, etc.), will operate satisfactorily at any point in time. This definition is similar to the definition for A_i except that preventive (i.e., scheduled) maintenance is included. It excludes logistics delay time and administrative delay time and is expressed by

$$A_a = \frac{\text{MTBM}}{\text{MTBM} + \overline{\text{M}}} \tag{2.42}$$

where MTBM is the mean time between maintenance and $\overline{\text{M}}$ is the mean active maintenance time. MTBM and $\overline{\text{M}}$ are a function of corrective (unscheduled) and preventive (scheduled) maintenance actions and times.

Operational Availability (A_o)

Operational availability is the probability that a system or equipment, when used under stated conditions in an *actual* operational environment, will operate satisfactorily when called upon. It is expressed as

$$A_o = \frac{\text{MTBM}}{\text{MTBM} + \text{MDT}} \tag{2.43}$$

where MDT is the mean maintenance downtime.

If one were to impose an availability figure of merit as a design requirement for a given equipment supplier, and the supplier has no control over the operational environment in which that equipment is to function, then A_a or A_i might be appropriate figures of merit against which the supplier's equipment can be properly assessed. Conversely, if one is to assess a system in a realistic operational environment, then A_o is a preferred figure of merit to employ for assessment purposes. Further, the term *availability* may be applied at any time in the overall mission profile representing a point estimate, or may be more appropriately related to a specific segment of the mission where the requirements are different from other segments. Thus, one must define precisely what is meant by *availability* and how it is to be applied.

2.11 ECONOMIC FACTORS

The recent combination of economic trends, rising inflation, cost growth experienced for many systems and products, the continuing reduction in buying power, budget limitations, increased competition, and so on, has created an awareness and interest in total system/product cost. Not only are the acquisition costs associated with new systems rising, but the costs of operating and maintaining systems already in use are increasing at alarming rates. The net result is that less money is available to meet new requirements, as well as maintaining systems that are already in being. In essence, many of the systems/products in existence today are not truly cost-effective.

In dealing with the aspect of cost, one must address total *life-cycle cost.* In the past, total system cost has not often been too visible, particularly those costs associated with system operation and support. The cost visibility problem can be related to the "iceberg effect" illustrated in Figure 1.9. One must consider not only system acquisition cost, but other costs as well. Further, when addressing total cost, experience has shown that a major portion of the projected life-cycle cost for a given system or product stems from the consequences of decisions made during the early phases of program planning and system conceptual design. Referring to Figure 2.25, while the greatest proportion of costs may result from activities occurring downstream in the system life cycle (e.g., system operation and support), the greatest opportunity for influencing these costs is realized during the early phases of a program, as reflected by the cross-hatched area. Decisions relating to the evaluation of alternative operational use profiles, maintenance and support policies, human-machine allocations, equipment packaging schemes, level of repair concepts, and so on, have a great impact on total cost. Thus, in dealing with economic factors, a life-cycle approach is required.[11]

In developing cost data, one of the initial steps is to define the system life cycle, and to project the various activities applicable for each phase. As defined in Chapter 1, life-cycle cost includes the costs associated with all system activities pertaining to research and development, design, test and evaluation, production, construction, prod-

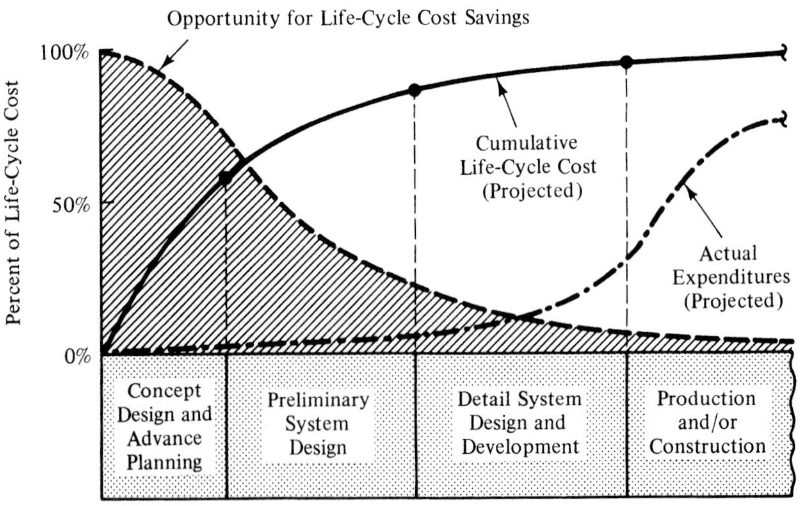

Figure 2.25 Commitment of life-cycle cost.

[11] Fabrycky, W. J., and B. S. Blanchard, *Life-Cycle Cost and Economic Analysis*, Prentice Hall, Inc., Upper Saddle River, N.J., 1991. Additional references on life-cycle costing are included in the bibliography in Appendix H.

uct distribution, system operation, sustaining maintenance and logistic support, and system retirement and material disposal. Although this information may change as one progresses through a program, a baseline needs to be established. With this information at hand, activity-generating costs may be derived.

In support of developing life-cycle cost factors, one of the first steps is to construct a cost breakdown structure (CBS), as illustrated in Figure 2.26. The CBS links objectives and activities with resources, and constitutes a logical subdivision of cost by functional activity area, major element of a system, and/or one or more discrete classes of common or like items.

The next step is to estimate costs, by category in the CBS, for each year in the system life cycle. Cost estimates must consider the effects of inflation, learning curves, and any other factors that are likely to cause changes in cost, upward or downward. Cost estimates are derived from a combination of accounting records (historical data), project cost projections, supplier proposals, and predictions. When cost data are not available, cost estimates are often based on analogous and/or parametric estimating methods.

Individual cost factors, estimated for each year in the life cycle in terms of the actual anticipated cost for that year (i.e., inflated cost), are totaled and projected in the context of a cost profile illustrated in Figure 2.27. Figure 2.28 provides a summary of costs by each applicable category in the CBS. The figures presented reflect future life-cycle cost budgetary requirements for the system.

The life-cycle cost analysis (LCCA) process is conveyed in Figure 2.29. Cost targets may be initially established as *design-to cost (DTC)* factors, predictions and analyses are accomplished as the system development progresses, and assessments are made later on. From these assessments, high-cost contributors can be identified (refer to the "percent of total" values in Figure 2.28), cause-and-effect relationships may be established, and recommendations for improvement may be initiated as appropriate; i.e., the implementation of a *continuous product/process improvement* approach with cost reduction as a major objective.

The entire life-cycle cost analysis process, including the detailed steps involved, is covered further in Section 4.2.1 and in Appendix E. The information presented in this section is very basic in nature, with the objective of briefly covering the economic side of the balance shown in Figure 1.14.

2.12 EFFECTIVENESS FACTORS

The aspect of effectiveness introduced in Chapter 1 can be quantified in terms of one or more figures of merit (FOMs), depending on the specific mission or system characteristics that one wishes to specify and measure. Effectiveness must consider

1. *System performance and physical parameters:* capacity, delivery rate, power output, range, accuracy, volume, speed, weight, and so on
2. *System operational and support factors:* availability, dependability, capability, operational readiness, reliability, maintainability, usability, supportability, transportability, producibility, disposability, and so on
3. *Total life-cycle cost:* research and development cost, production/construction cost, operation and maintenance cost, retirement and disposal cost, and so on

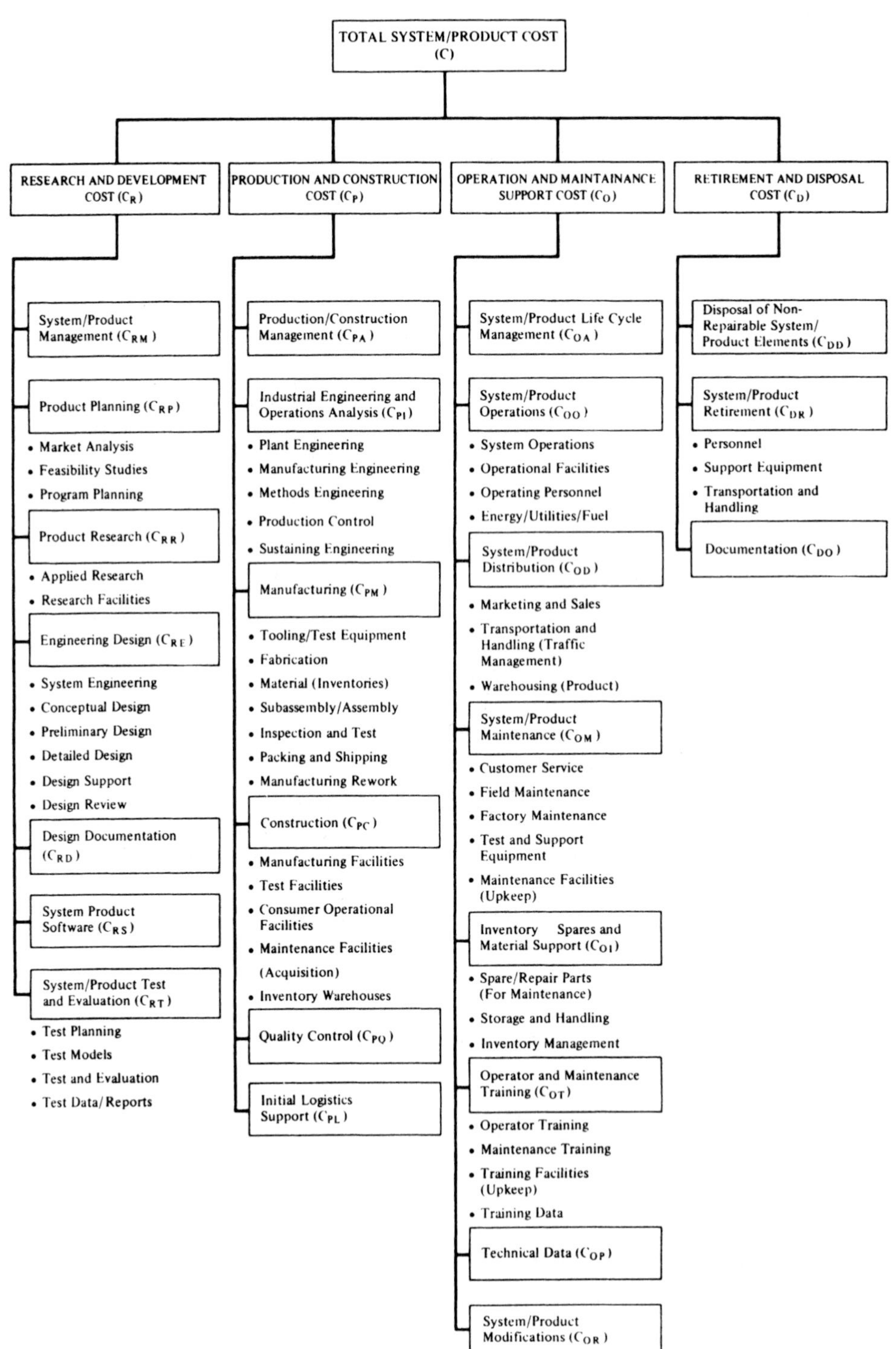

Figure 2.26 Cost breakdown structure (example).

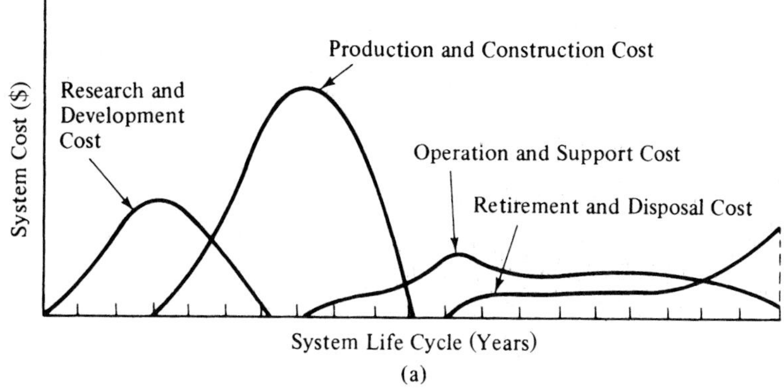

(a)

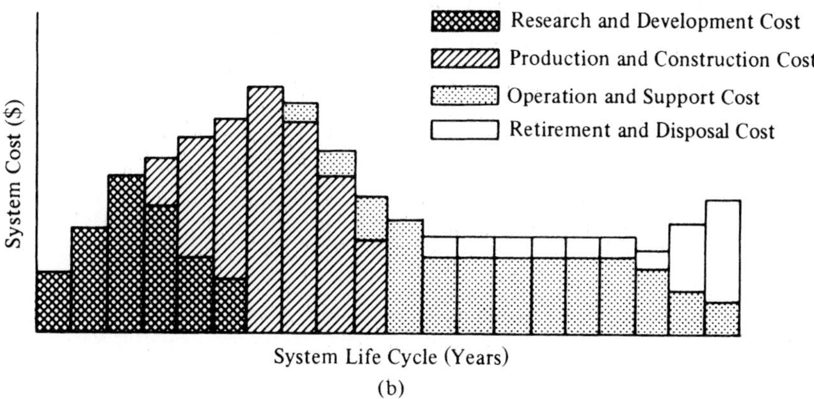

(b)

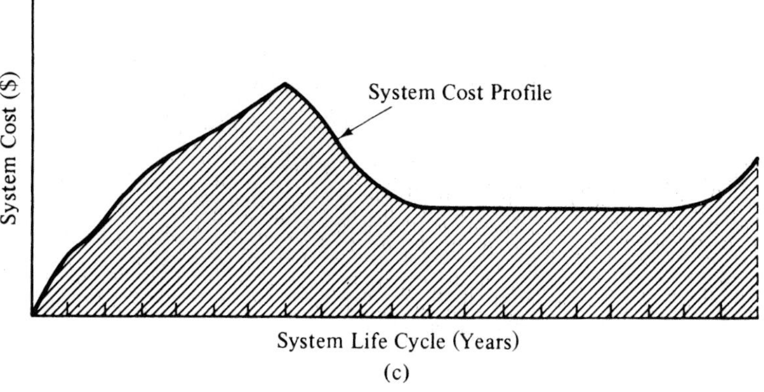

(c)

Figure 2.27 Development of a cost profile.

COST CATEGORY	Cost ($)	% of Total
1. Research and development cost (C_R)		
(a) System/product management (C_{RM})	247,071	9.8
(b) Product planning (C_{RP})	37,090	1.5
(c) Engineering design (C_{RE})	237,016	9.4
(d) Design data (C_{RD})	48,425	1.9
(e) System test and evaluation (C_{RT})	67,648	2.7
Subtotal	637,250	25.3
2. Production and construction cost (C_P)		
(a) System/product management (C_{RM})	112,497	4.5
(b) Industrial engineering and operations analysis (C_{PT})	68,200	2.8
(c) Manufacturing–recurring (C_{PMR})	635,100	25.3
(d) Manufacturing–nonrecurring (C_{PMN})	48,900	1.9
(e) Quality control (C_{PQ})	78,502	3.1
(f) Initial logistics support (C_{PL})		
(1) Supply support–initial (C_{PLS})	48,000	1.9
(2) Test and support equipment (C_{PLS})	70,000	2.8
(3) Technical data (C_{PLD})	5,100	0.2
(4) Personnel training (C_{PLP})	46,400	1.8
Subtotal	1,112,699	44.3
3. Operation and support cost (C_O)		
(a) Operating personnel (C_{OOP})	24,978	0.9
(b) Distribution–transportation (C_{OOR})	268,000	10.7
(c) Unscheduled maintenance (C_{OLA})	123,130	4.9
(d) Maintenance facilities (C_{OLM})	4,626	0.2
(e) Supply support (C_{OLS})	295,920	11.8
(f) Maintenance personnel training (C_{OLT})	9,100	0.4
(g) Test and support equipment (C_{OLE})	32,500	1.3
(h) Transportation and handling (C_{OLM})	5,250	0.2
Subtotal	763,504	30.4
Grand total	2,513,453	100

Figure 2.28 Life-cycle cost summary.

Establishing a relationship between a performance or an operational parameter and cost may constitute a desirable cost-effectiveness FOM. Other relationships may be equally as important. Some example FOMs are

$$\text{Effectiveness FOM} = \frac{\text{availability}}{\text{life-cycle cost}} \tag{2.48}$$

or

$$\text{Effectiveness FOM} = \frac{\text{reliability}}{\text{life-cycle cost}} \tag{2.49}$$

or

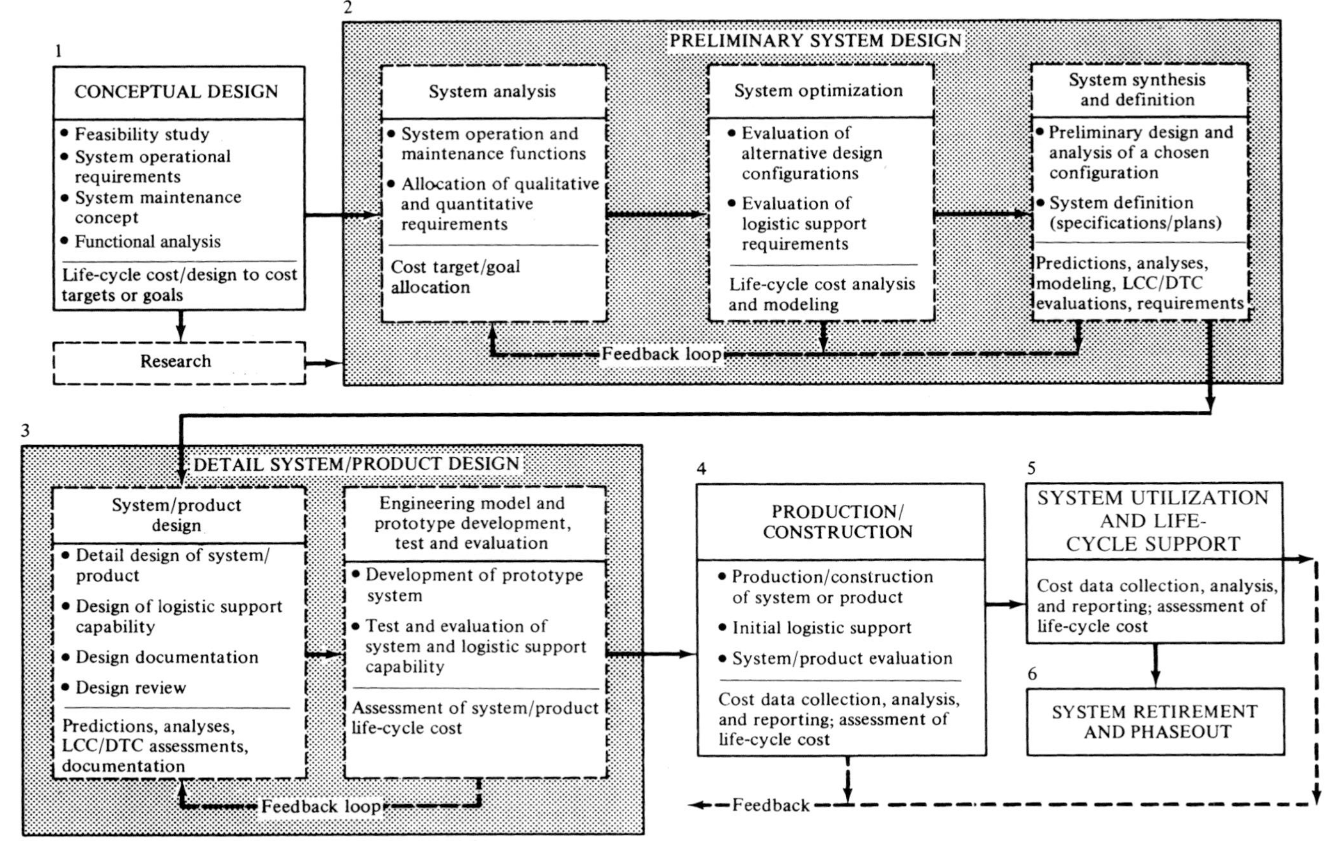

Figure 2.29 System life-cycle process with economic considerations (reference: Figure 1.6). (Source: Blanchard, B. S., and W. J. Fabrycky, *Systems Engineering and Analysis*, 3rd ed., Prentice Hall, Upper Saddle River, NJ, 1998.)

$$\text{Effectiveness FOM} = \frac{\text{system capacity}}{\text{range}} \tag{2.50}$$

or

$$\text{Effectiveness FOM} = \frac{\text{supportability}}{\text{life-cycle cost}} \tag{2.51}$$

or

$$\text{Effectiveness FOM} = \frac{\text{life-cycle cost}}{\text{facility space}} \tag{2.52}$$

Figure 2.30 illustrates a relationship between reliability (MTBF) and total life-cycle cost, where the objective is to design a system to meet a specified set of values (MTBF and a budget limitation) and yet be cost-effective. Through allocations, predictions, assessments, and so on, system design characteristics are evaluated in terms of reliability and cost. Design changes (as required) are recommended to the extent that the system configuration is represented at or near the minimum-cost point on the curve in Figure 2.30.

The use of effectiveness FOMs is particularly appropriate in the evaluation of two or more alternatives when decisions involving design and/or logistic support are necessary. Each alternative is evaluated in a consistent manner employing the same criteria for evaluation. In situations where the risks are high and available data for prediction are inadequate, one may wish to employ the three-level estimate approach using a pessimistic value, optimistic value, and expected value for performance, operational, and/or cost factors as appropriate. Using this approach, the cost-effectiveness relationship for the evaluation of two alternatives assumes the situation illustrated in Figure 2.31.

Referring to the figure, two alternatives are being evaluated on a comparable basis, and it appears that alternative *B* is the most cost-effective in the end. Prior to arriving at a final decision, however, one must address both the aspect of cost effectiveness (in terms of some quantitative FOM) and the point in time where alternative *B* becomes more cost-effective than alternative *A*. Thus, a break-even analysis is required to determine this point in time. Figure 2.32 illustrates an approach to a break-

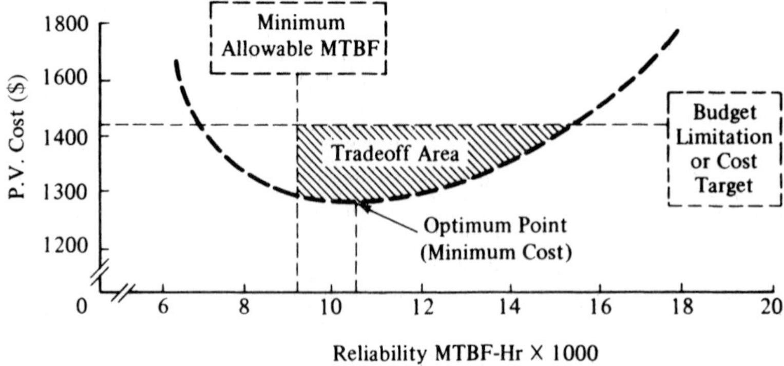

Figure 2.30 Reliability versus cost.

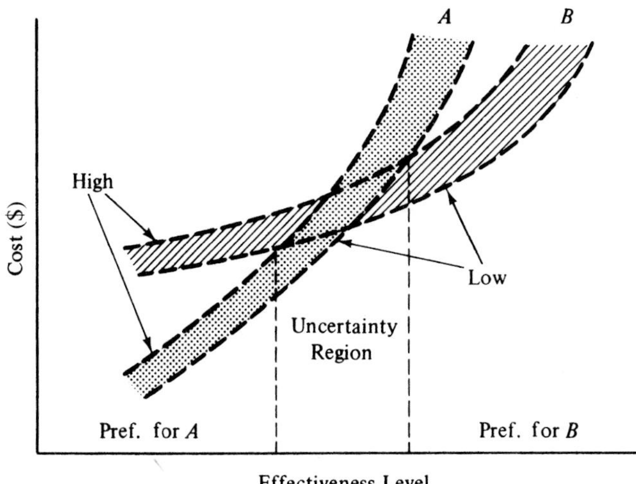

Figure 2.31 Range of estimates for two alternatives.

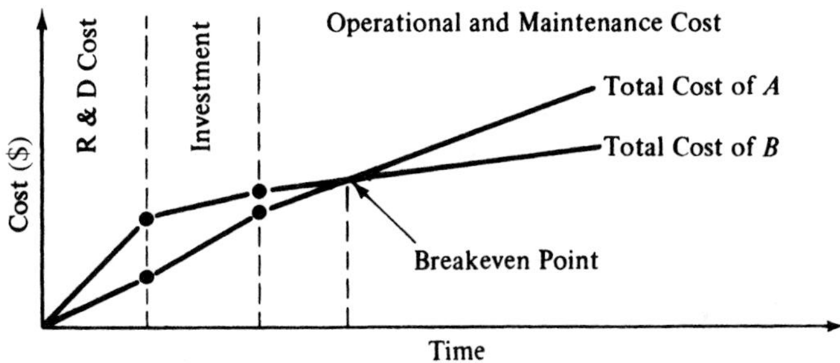

Figure 2.32 Break-even analysis.

even analysis where the cumulative costs for two programs (alternatives *A* and *B*) are estimated and projected.

Is the break-even point realistic in terms of expected system life or in view of possible obsolescence? The answer to this question will, of course, vary depending on the system and its intended mission. In the illustration, alternative *B* may be more cost-effective in the long run, but the advantages may be unrealistic in terms of time. This is an area that must be addressed in any analysis where two or more alternatives are being evaluated.

2.13 SUMMARY

Figure 2.33 (which is an extension of Figure 1.1) presents a basic activity flow (the blocks shown may or may not represent a separate physical location and will vary depending on the type and mission of the system). Development is accomplished and

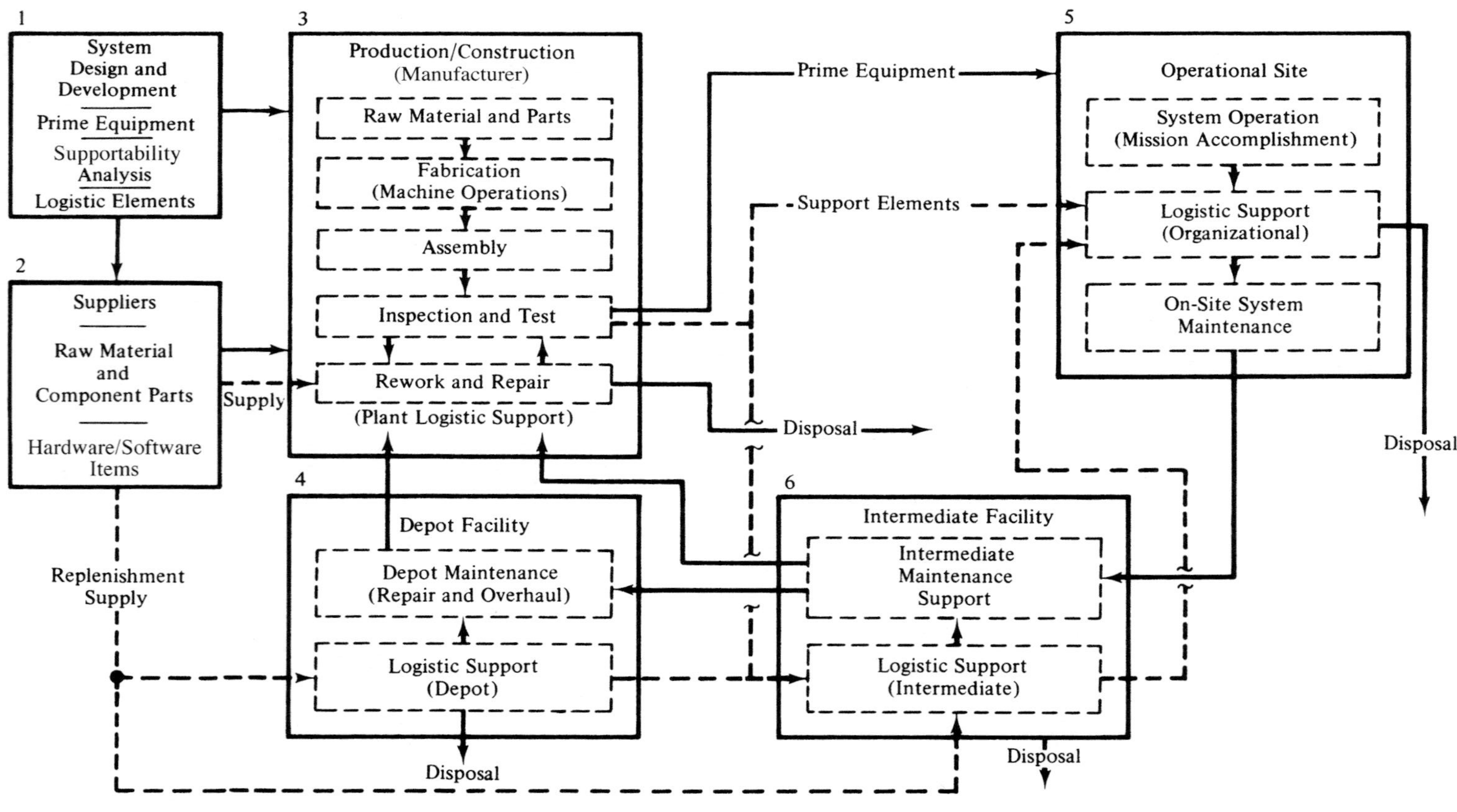

Figure 2.33 Operational/maintenance flow.

the results are translated into production/construction and the distribution or delivery of an operational system. Logistic support in the design process (block 1) is primarily reflected through the incorporation of reliability, maintainability, human factors, supportability, and economic factors in the design.

In production and/or construction, raw material and component parts (block 2) are transported to the plant (block 3), inspected, assembled, tested, and a finished product is delivered for operational use (block 5). Test and support equipment, spares/repair parts and associated inventories, technical data, facilities, and personnel are required to support production/construction and distribution functions.

When operational (block 5), the system in performing its mission requires operators, trained personnel, transportation and handling equipment, data, computer resources, and facilities. These elements of logistic support are necessary for successful mission fulfillment.

As the system progresses through its life cycle, corrective (unscheduled) and preventive (scheduled) maintenance actions are necessary to restore or retain the equipment in full operational status. Thus, Figure 2.33 presents the maintenance flow, including on-site or organization maintenance (block 5), intermediate maintenance (block 6), and depot or manufacturer support (blocks 3 and 4). Test and support equipment, spare/repair parts, maintenance personnel, data, computer resources, and facilities are required to accomplish the maintenance functions at each level.

When dealing with total logistic support, the entire flow process must be treated as an entity. Each block represented in the figure impacts on the others. The treatment of any single function must include consideration of the effects on other functions. For instance, a system is required to meet a particular operational effectiveness goal (e.g., availability of 90%). Considering that the system will fail at some point in time, a spares inventory is necessary at the organizational level (block 5) to ensure that the right spare item is available when needed. Assuming that the faulty item is *repairable*, one has to determine the spares required at the intermediate level (block 6) or depot level (block 4) to support the necessary repair actions. Also, it is necessary to determine the test and support equipment required to accomplish the fault isolation and checkout to the system indenture level desired. The goal is to develop an overall optimum logistic support capability by evaluating alternative configurations, including various mixes of the logistic support elements at each level.

Accomplishment of this goal requires an understanding of the various logistic support measures presented in this chapter. These measures are closely interrelated, and each area must be addressed in the context of the system as an entity (i.e., the activities represented in Figure 2.33).

QUESTIONS AND PROBLEMS

1. Define *reliability*. What are the major characteristics?
2. Refer to Figure 2.1. What is the probability of success for a system if the system MTBF is 600 hours and the mission operating time is 420 hours? (Assume exponential distribution.)
3. A system consists of four subassemblies connected in series. The individual subassembly reliabilities are as follows:

Subassembly $A = 0.98$
Subassembly $B = 0.85$
Subassembly $C = 0.90$
Subassembly $D = 0.88$

Determine the overall system reliability.

4. A system consists of three subsystems in parallel (assume operating redundancy). The individual subsystem reliabilities are as follows:

   Subsystem $A = 0.98$
   Subsystem $B = 0.85$
   Subsystem $C = 0.88$

   Determine the overall system reliability.

5. Refer to Figure 2.9(c). Determine the overall network reliability if the individual reliabilities of the subsystems are as follows:

   Subsystem $A = 0.95$
   Subsystem $B = 0.97$
   Subsystem $C = 0.92$
   Subsystem $D = 0.94$
   Subsystem $E = 0.90$
   Subsystem $F = 0.88$

6. A system consists of five subsystems with the following MTBFs:

   Subsystem A: MTBF = 10,540 hours
   Subsystem B: MTBF = 16,220 hours
   Subsystem C: MTBF = 9,500 hours
   Subsystem D: MTBF = 12,100 hours
   Subsystem E: MTBF = 3,600 hours

   The five subsystems are connected in series. Determine the probability of survival for an operating period of 1000 hours.

7. A lot contains 10,000 mechanical components. A sample of 1000 components was selected from that lot. During an inspection of the sample, it was learned that the sample contains 90 defective units. Determine the upper and lower limits of sampling error.

8. Four basic system configurations are illustrated in Figure 2.34. Determine the reliability of each and rank them in order (i.e., the configuration with the highest reliability first). The individual reliability values for each of the configuration components are:

   Subsystem $A = 0.95$
   Subsystem $B = 0.90$
   Subsystem $C = 0.85$
   Subsystem $D = 0.95$
   Subsystem $E = 0.80$

9. What statistical distribution is applicable in the prediction of reliability?

10. Define *maintainability*. What are some of the characteristics?

11. Assume that maintenance repair times are distributed normally and that the MTTR is 50 minutes. What percent of the total population lies between 30 and 40 minutes? (Assume that $\sigma = 15$).

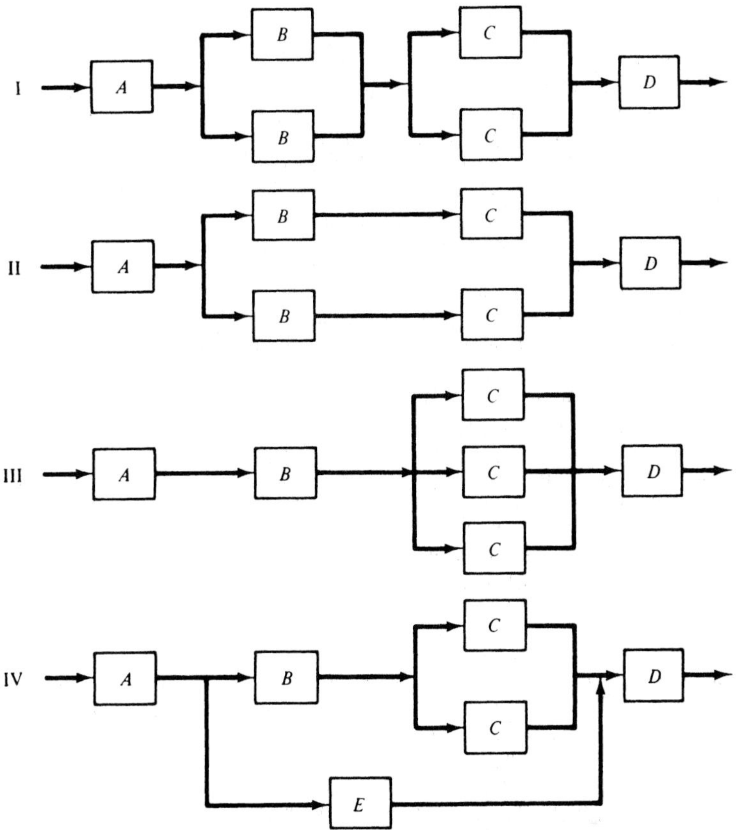

Figure 2.34 Reliability block diagrams.

12. The following corrective maintenance task times were observed:

Task Time (Min)	Frequency	Task Time (Min)	Frequency
41	2	37	4
39	3	25	10
47	2	35	5
36	5	31	7
23	13	13	3
27	10	11	2
33	6	15	8
17	12	29	8
19	12	21	14

(a) What is the range of observations?
(b) Using a class interval width of 4, determine the number of class intervals. Plot the data and construct a curve. What type of distribution is indicated by the curve?
(c) What is the $\overline{\text{M}}\text{ct}$?
(d) What is the geometric mean of the repair times?
(e) What is the standard deviation of the sample data?
(f) What is the M_{max} value? (Assume 90%.)

13. The following corrective-maintenance task times were observed:

Task Time (Min)	Frequency	Task Time (Min)	Frequency
35	2	25	12
17	6	19	10
12	2	21	12
15	4	23	13
37	1	29	8
27	10	13	3
33	3	9	1
31	6	—	—

(a) What is the range of observations?
(b) Assuming 7 classes with a class interval width of 4, plot the data and construct a curve. What type of distribution is indicated by the curve?
(c) What is the mean repair time?
(d) What is the standard deviation of the sample data?
(e) The system is required to meet a mean repair time of 25 minutes at a stated confidence level of 95%. Do the data reveal that the specification requirements will be met? Why?

14. A system is operated for 20,000 hours, and the demonstrated MTBF was 500 hours. For each maintenance action two technicians were assigned to accomplish corrective maintenance, and the $\overline{\text{M}}\text{ct}$ was 2 hours. Determine the MLH/OH expended.

15. Assuming that a single component with a reliability of 0.85 is used in a unique application in the system and that one backup spare component is purchased, determine the probability of success by having a spare available in time, t, when required.

16. Assuming that the component in Problem 15 is supported with two backup spares, determine the probability of success by having two spares available when needed. Determine the probability of success for a configuration consisting of two operating components backed by two spares. (Assume that the component reliability is 0.875.)

17. There are 10 systems located at a site scheduled to perform a 20-hour mission. The system has an expected MTBF of 100 hours. What is the probability that at least 8 of these systems will operate for the duration of the mission without failure?

18. An equipment contains 30 parts of the same type. The part has a predicted mean failure frequency of 10,000 hours. The equipment operates 24 hours a day and spares are provisioned at 90-day intervals. How many spares should be carried in the inventory to ensure a 95% probability of having a spare available when required?

19. Determine the economic order quantity of an item for spares inventory replenishment, where:
(a) The cost per unit is $100.
(b) The cost of preparing for a shipment and sending a truck to the warehouse is $25.
(c) The estimated cost of holding the inventory, including capital tied up, is 25% of the initial inventory value.
(d) The annual demand is 200 units. Assume that the cost per order and the inventory carrying charge is fixed.

20. Refer to Figure 2.21. What happens to the EOQ when the demand increases? What happens when there are outstanding backorders? What factors are included in procurement lead time?

21. Does the EOQ principle apply in the procurement of all spares? If not, describe some exceptions.

22. Calculate as many of the following parameters as you can from the information given:

Determine:		**Given:**
A_i	MTBM	$\lambda = 0.004$
A_a	MTBF	Total operation time = 10,000 hours
A_o	$\overline{M}$	Mean downtime = 50 hours
$\overline{\text{M}}\text{ct}$	MTTR_g	Total number of maintenance actions = 50
M_{max}		Mean preventive maintenance time = 6 hours
		Mean logistics plus administrative time = 30 hours

23. Select a system of your choice and construct a flowchart identifying basic operational and maintenance functions similar to that presented in Figure 2.33. Identify areas where you believe that logistic support has an impact (be specific). What type of impact?

24. Given alternatives *a* through *g* shown in Figure 2.35, which one would you select? Why?

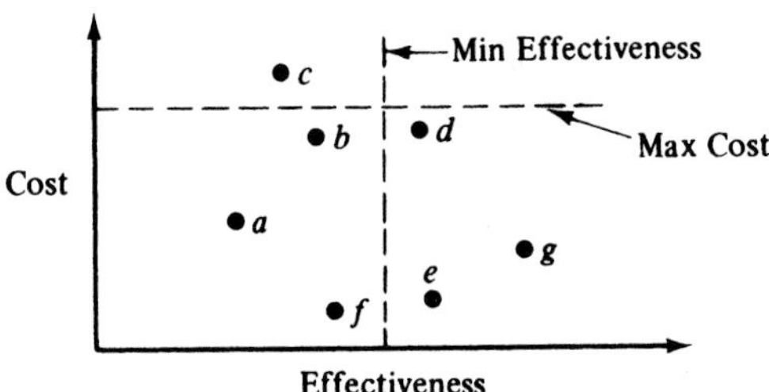

Figure 2.35 Effectiveness versus cost.

25. Define *self-test thoroughness*. How would you measure it?

26. Identify some of the more significant quantitative measures that are directly related with test and support equipment. Illustrate an application for each factor identified.

27. How should the reliability of the test and support equipment relate to the reliability of the prime-mission-oriented equipment being supported?

28. How does the length of the queue for items awaiting test influence spare/repair part requirements for the system? How does the on-station test time influence spare/repair part requirements?

29. Identify some quantitative measures that are directly associated with (1) individual personnel requirements for the projected maintenance and support of a system, and (2) the overall organization (as an entity) that is responsible for the ongoing support of that system. Briefly describe how these individual personnel factors and organizational factors interrelate.
30. Identify some quantitative measures that are directly associated with maintenance facilities. Provide an illustration of application for each factor identified. Briefly describe how facility factors can affect organizational factors.
31. Identify some quantitative measures that are directly related to transportation as it applies to system maintenance and support. Provide an illustration of application for each factor identified. Briefly describe how transportation factors affect supply support factors.
32. Identify and discuss some of the measures, associated with technical data and information systems.
33. Identify some of the new technologies that have recently been introduced, and describe how they can be applied to enhance logistics (select a few examples).
34. Describe how the applications of EDI/EC can impact logistics (provide five examples).

CHAPTER 3

SYSTEM ENGINEERING PROCESS

The *system engineering process* is inherent within the overall system life cycle, as illustrated in Figure 3.1. The initial emphasis is on a top-down, integrated, life-cycle approach to system design and development, conveyed through the activities depicted in blocks 0.1 to 4.6. This includes problem definition and the identification of customer need, the conductance of feasibility analysis, the development of operational requirements and the maintenance and support concept, functional analysis, requirements allocation, and so on. Subsequently, there is the iterative process of assessment and system validation, and the incorporation of changes for product/process improvement as illustrated in Figure 1.13. Although the process is more directed to the early stages of system design and development, maintaining cognizance of the activities in the latter phases of construction/production, operational use, and system maintenance and support is essential for understanding the consequences of earlier decisions and the establishment of benchmarks for the future. In other words, the *feedback loop* is critical and an integral part of the system engineering process.[1]

Referring to the figure, the phases and milestones as shown are not intended to convey a highly complex program with specified periods or levels of funding. The figure is intended to reflect an overall *process* that is applicable in system acquisition. Regardless of the type, size, and complexity of the system, there is a conceptual design requirement (to include requirements analysis), a preliminary design requirement, and so on. Every time that there is a new system requirement, one should evolve through a top-down process, properly "tailored" to the particular system being addressed.

[1] Refer to Blanchard, B. S., and W. J. Fabrycky, *Systems Engineering and Analysis*, 3rd Ed., Prentice Hall, Inc., Upper Saddle River, N.J., 1998; or Blanchard, B. S., *System Engineering Management*, 2nd Ed., John Wiley & Sons, Inc., New York, N.Y., 1998.

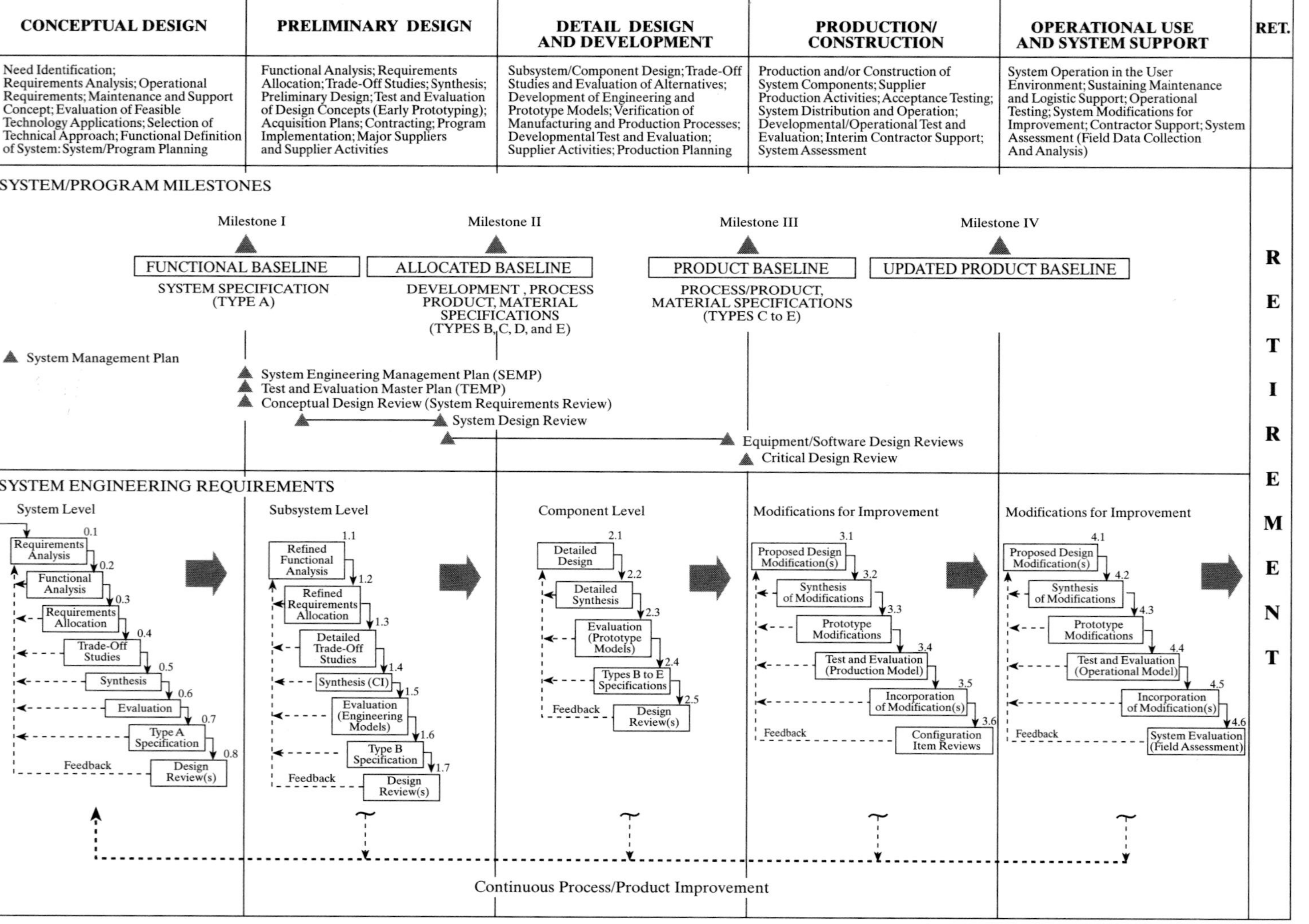

Figure 3.1 System engineering requirements within the life cycle. (Source: Blanchard, B. S., "The System Engineering Process: An Application for the Identification of Resource Requirements," *Systems Engineering*, Journal of the International Council on Systems Engineering, Volume 1, Number 1, July 1994.)

Inherent within the system engineering process (defined by blocks 0.1 through 0.8 for the system, blocks 1.1 through 1.7 for the subsystem, etc.) are the requirements dealing with logistics and the design for supportability as illustrated through the relationships identified in Figure 1.7. One must commence with problem definition and a needs analysis, the conductance of feasibility analysis and the selection of technologies based on life-cycle factors, the development of system operational requirements and the maintenance and support concept, functional analysis and the identification of maintenance and support functions, requirements allocation and the development of supportability design criteria, design optimization and the evaluation of alternatives considering the factors described in Chapter 2, and so on.

The objective of this chapter is to describe the system engineering process and those activities within this process that relate to and have an impact on logistics and the design for supportability. Figure 3.2 identifies the basic steps in the system life cycle, with the blocks shown in bold representing the key areas of emphasis.

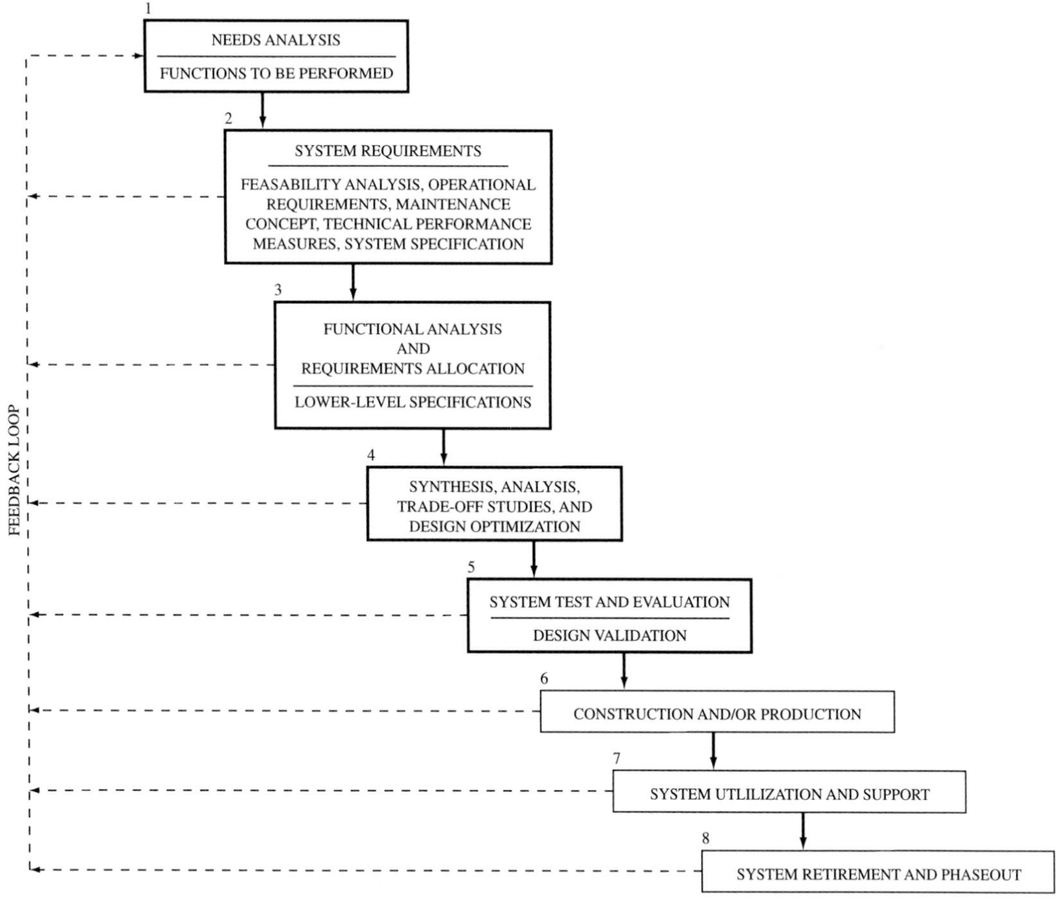

Figure 3.2 Steps in the system life cycle.

3.1 DEFINITION OF PROBLEM AND NEEDS ANALYSIS

The system engineering process generally commences with the identification of a "want" or "desire" for something, and is based on a real (or perceived) deficiency. For instance, the current capability is not adequate in terms of meeting certain required performance goals, is not available when needed, cannot be properly supported, is too costly in terms of operation, and so on. As a result, a new system requirement is defined along with the priority for introduction, the date when the new system capability is required for customer use, and an estimate of the resources necessary for acquiring the new system capability. To ensure a good start, a "statement of the problem" should be presented in specific qualitative and quantitative terms, in enough detail to justify progressing to the next step.

The requirement for identifying the need (as a starting point) may seem to be rather "basic," or self-evident. However, one often finds that a design effort is initiated as a result of a personal interest or a political whim, without first having adequately defined the requirements for such! In the software area, there is the tendency to accomplish a lot of coding before identifying the need. Additionally, there are instances where the engineer sincerely believes to know what the customer needs, without having involved the customer in the process. The "design-it-now-fix-it-later" philosophy prevails, which, in turn, can be rather costly.

Defining the problem is sometimes the most difficult part of the process, particularly if one is in a rush to "get going!" Yet, the number of "false starts" and the ultimate risks can be significant unless a good foundation is laid from the beginning. A complete description of the need, expressed in quantitatively-stated performance parameters where possible, is essential. It is important that the results reflect a true customer requirement, particularly in today's environment where resources are limited.

Given the problem definition, a *needs analysis* must be accomplished with the objective of translating a broadly defined "want" into a more specific system-level requirement. The questions are: What is required of the system in *functional* terms? What functions must the system perform? What are the *primary* functions? What are the *secondary* functions? What must be accomplished to alleviate the stated deficiency? When must this be accomplished? Where is this to be accomplished? How many times must this be accomplished? There are many basic questions of this nature, and it is important to describe the customer requirements in a *functional* manner in order to avoid a premature commitment to a specific design concept or configuration and, thus, the unnecessary expenditure of valuable resources. The ultimate objective is to define the "whats" and *not* the "hows."

Accomplishing the needs analysis in a satisfactory manner can best be realized through a "team" approach involving the customer, the ultimate consumer or user (if different from the customer), the contractor or producer, and major suppliers as appropriate. The objective is to ensure that the proper communications exist between the parties involved. The "voice of the customer" must be heard, and the system developer must respond accordingly.

3.2 SYSTEM FEASIBILITY ANALYSIS

Through the needs analysis, the functions that the system must perform are identified. There may be a single function such as "transport product *XYZ* from point *A* to point *B*," or "communicate among points *D*, *E*, and *F*," or "produce *X* quantity of *Y* products by time *Z*." Conversely, there may be a number of different functions to be performed, some primary and some secondary. To ensure a good design, all possible functions must be identified, the most rigorous functions being selected as the basis for defining system-level design requirements. It is important that *all* possibilities be addressed to ensure that the proper technologies and components are selected for design consideration.

The *feasibility analysis* is accomplished with the objective of evaluating the different technological approaches that may be considered in responding to the specified functional requirements. In considering different design approaches, alternative technology applications are investigated. For instance, in the design of a communications system, should one use fiber-optics technology, cellular, or the conventional hardwired approach? In designing an aircraft, to what extent should one incorporate composite materials? When designing an automobile, should one apply very high-speed integrated electronic circuitry in certain control applications, or should one select a more conventional electromechanical approach?

It is necessary to (1) identify the various possible design approaches that can be pursued to meet the requirements; (2) evaluate the most likely candidates in terms of performance, effectiveness, logistics requirements, and life-cycle economic criteria; and (3) recommend a preferred approach. The objective is to select an overall *technical* approach, and not to select specific components. There may be many different alternatives; however, the number of possibilities must be narrowed down to a few feasible options, consistent with the availability of resources (i.e., manpower, materials, and money).

It is at this early stage in the life cycle (i.e., the conceptual design phase) where major decisions are made relative to adopting a specific design approach. When there is not enough information available, a research activity may be initiated with the objective of developing new methods/techniques for specific applications. On some programs, the completion of applied research tasks and preliminary design activity is accomplished sequentially, whereas in other situations, there may be a number of different miniprojects underway at the same time.

The results of the feasibility analysis will have a significant impact not only on the operational characteristics of a system, but on the production and maintenance support requirements as well. The selection (and application) of a given technology has reliability and maintainability implications, may significantly impact the requirements for spare parts and test equipment, may impact manufacturing methods, and will certainly impact life-cycle cost.

3.3 SYSTEM OPERATIONAL REQUIREMENTS

Once a system need and technical approach have been identified, it is necessary to project that need in terms of anticipated operational requirements. At this point, the fol-

al requirements is best covered through
s. The first illustration is an aircraft sys-
d illustration is a communication system
e third deals with commercial airline
olitan area. These illustrations are dis-
e relationship with the various facets of

buted in multiple quantities throughout
ions, estimated quantities per location,
cted in Figure 3.3.

al Use Per Year

ber					Total Systems
6	7	8	9	10	
50	60	60	35	25	310
24	24	24	24	24	180
24	24	24	24	24	156
24	24	24	24	24	180
24	24	24	12	12	132
2	12	12	12	12	96
8	168	168	131	121	1054

Day, 365 Days per Year

al requirements.

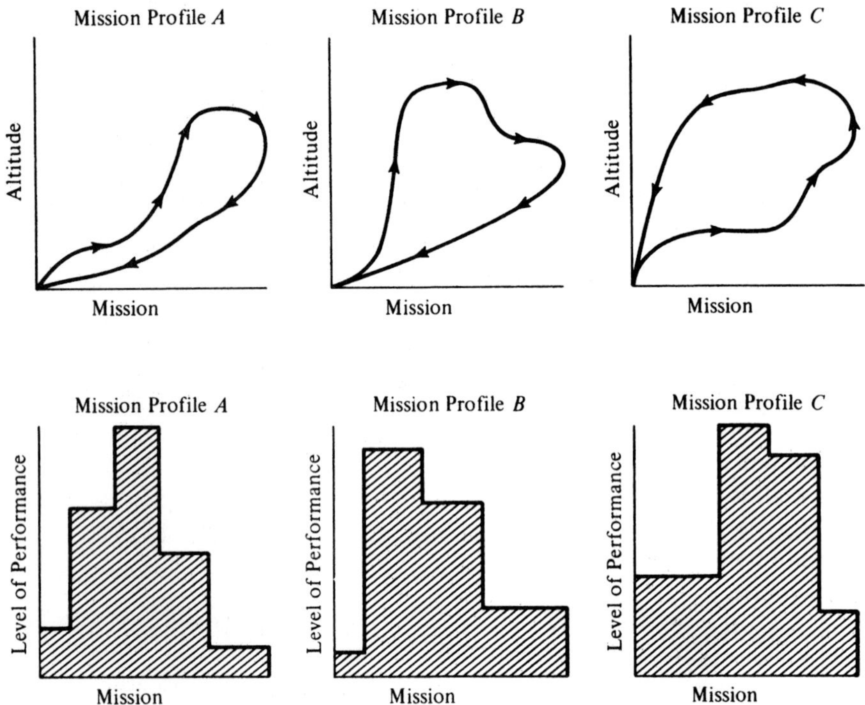

Figure 3.4 Typical system operational profiles.

In terms of mission, each aircraft will be required to fly three different mission profiles, as illustrated in Figure 3.4 and described in "Specification 12345." Basically, an aircraft will be prepared for flight, take off and complete a specific mission scenario, return to its base, undergo maintenance as required, and be returned to a ready status. For planning purposes, all aircraft will be required to fly at least one each of the three different mission profiles per week.

The aircraft shall meet system performance requirements in accordance with Specification 12345; the operational availability (A_o) shall be at least 90%; the maintenance downtime (MDT) shall not exceed 3 hours; the $\overline{\text{M}}$ct shall be 30 minutes or less; and Mmax at the 90th percentile shall be 2 hours; the MLH/OH for the aircraft shall be 10 or less; and the cost per maintenance action at the organizational level shall not exceed $1000. The aircraft will incorporate a built-in test capability that will allow for fault isolation to the unit level with an 85% self-test thoroughness. No special external support equipment will be allowed. Relative to the support infrastructure, there will be an intermediate-level maintenance capability located at each operational base. Additionally, there will be two depot-level maintenance facilities. The overall support concept is illustrated in Figure 3.5.

Although the information given above is rather cursory in nature considering the total spectrum of system operational requirements, it is necessary from the standpoint of logistics to define (1) the mission scenario(s) in order to identify operational sequences, stresses on the system, environmental requirements, and system reliability

t are the anticipated quantities of equipment, here are they to be located? How is the system is the system to be supported, and by whom? each operational site (i.e., user location)? The ons leads to the definition of system operating pport concept for the system, and the identifi- criteria. The operational concept, as defined ure 1.7, includes the following information:

f the prime mission of the system and alternate e system to accomplish? How will the system ssion may be defined through one or a set of It is important that the *dynamics* of the system fied.

ers. Definition of the operating characteristics e, weight, range, accuracy, bits, capacity, trans- ritical system performance parameters? How nario(s)?

ution. Identification of the quantity of equip- es, and so on, and the expected geographical and mobility requirements. How much equip- stributed, and where is it to be located? When ational?

nticipated time that the system will be in oper- tory profile throughout the system life cycle? and for what period of time?

ed usage of the system and its elements (e.g., ntage of total capacity, operational cycles per is the system to be used by the customer or

rements specified as figures of merit such as nal availability, readiness rate, dependability, time between maintenance (MTBM), failure MDT), facility utilization (in percent), opera- ment requirements, personnel efficiency, and erform, how effective or efficient is it? How sion scenario(s)?

ironment in which the system is expected to ty, arctic or tropics, mountainous or flat ter- tc.). This should include a range of values as portation, handling, and storage modes. How t? What will the system be subjected to dur- g? A complete environmental profile should

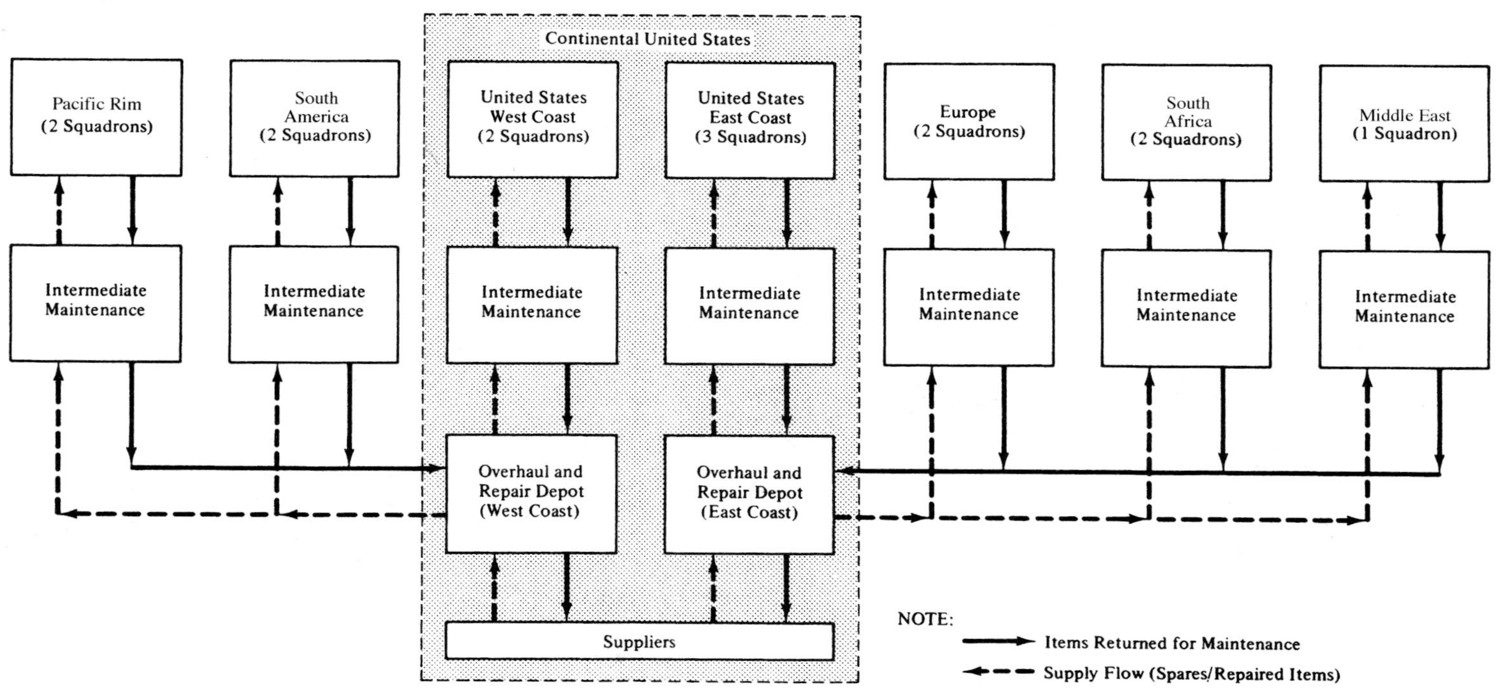

Figure 3.5 Top-level system support capability.

or the frequency of anticipated maintenance; (2) the quantity of items and the geographical distribution in order to determine the magnitude of support and the location where logistic support will be required; (3) the operational life cycle in order to determine the duration of support and the basic requirements for replenishment; and (4) the effectiveness factors in order to determine the frequency, level of support, and the anticipated logistic support resources that will be required. In essence, the overall concept conveyed in Figures 3.3, 3.4, and 3.5 (supported by specific qualitative and quantitative factors) must be defined in order to establish a future baseline for system support.

Illustration 2: Communication System

A new radio communication system with an increased range capability and improved reliability is required to replace several existing systems that are currently deployed in multiple quantities throughout the world. The system must accomplish three basic missions.

> *Mission Scenario 1.* The system is to be installed in low-flying light aircraft (10,000 feet altitude or less) in quantities of one per aircraft. The system shall enable communication with ground vehicles dispersed throughout mountainous and flat terrain, and with a centralized area communication facility. It is anticipated that each aircraft will fly 15 missions per month with an average mission duration of 2 hours. A typical mission profile is illustrated in Figure 3.6. The communication system utilization requirement is 110% (1.1 hours of system operation for every hour of aircraft operation, which includes air time plus some ground time). The system must be operationally available 99.5% of the time and have a reliability MTBF of not less than 2000 hours.

> *Mission Scenario 2.* The system is to be installed in ground vehicular equipment (e.g., car, light truck, or equivalent) in quantities of one per vehicle. The system shall enable communication with other vehicles at a range of 200 miles in relatively flat terrain, overhead aircraft at an altitude of 10,000 feet or less, and a centralized area communication facility. Sixty-five percent of the vehicles will be in operational use at any given point in time and the system shall be utilized 100%

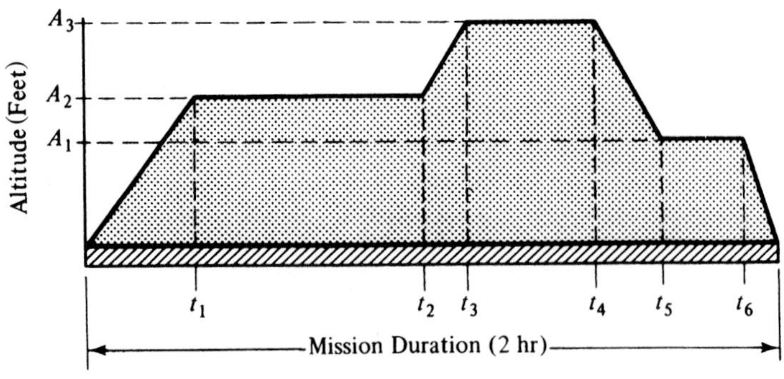

Figure 3.6 Mission profile.

of the time for those vehicles which are operational. The system must have a reliability MTBF of at least 1800 hours and a $\overline{M}ct$ of 1 hours or less.

Mission Scenario 3. The system is to be installed in 20 area communication facilities located throughout the world with 5 operational systems assigned to each facility. The system shall enable communication with aircraft flying at an altitude of 10,000 feet or less and within a radius of 500 miles from the facility, and with ground vehicles at a range of 300 miles in relatively flat terrain. Four of the systems are utilized 24 hours a day while the remaining system is a backup and used an average of 6 hours per day. Each operational system shall have a reliability MTBF of at least 2500 hours and a $\overline{M}ct$ of 30 minutes or less. Each communication facility shall be located at an airport.

In the interest of minimizing the total cost of support (e.g., test and support equipment, spares, personnel, etc.), the transmitter-receiver, which is a major element of the system, shall be a common design for the vehicular, airborne, and ground applications. The antenna configuration may be peculiar in each instance.

Operational prime equipment/software shall be introduced into the inventory commencing 4 years from this date, and a maximum complement is acquired by 8 years. The maximum complement must be sustained for 10 years, after which a gradual phaseout will occur through attrition. The last equipment is expected to phase out of the inventory in 25 years. The program schedule is illustrated in Figure 3.7.

As stated previously, the new system will replace several different systems currently located throughout the world. Specifically, the requirements dictate the need for

- 20 centralized communication facilities
- 11 aircraft assigned to each communication facility
- 55 vehicles assigned to each communication facility

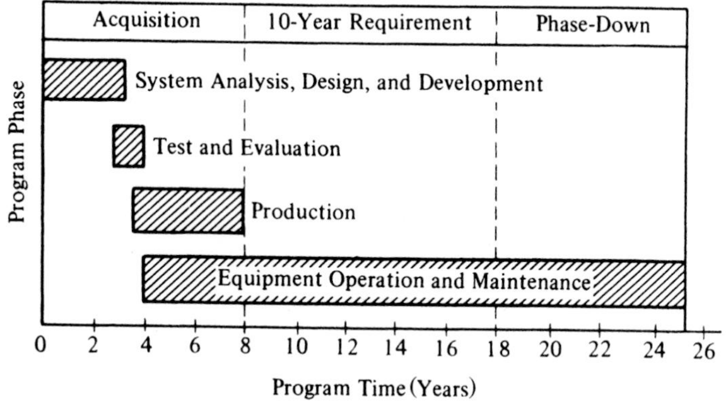

Figure 3.7 Program schedule.

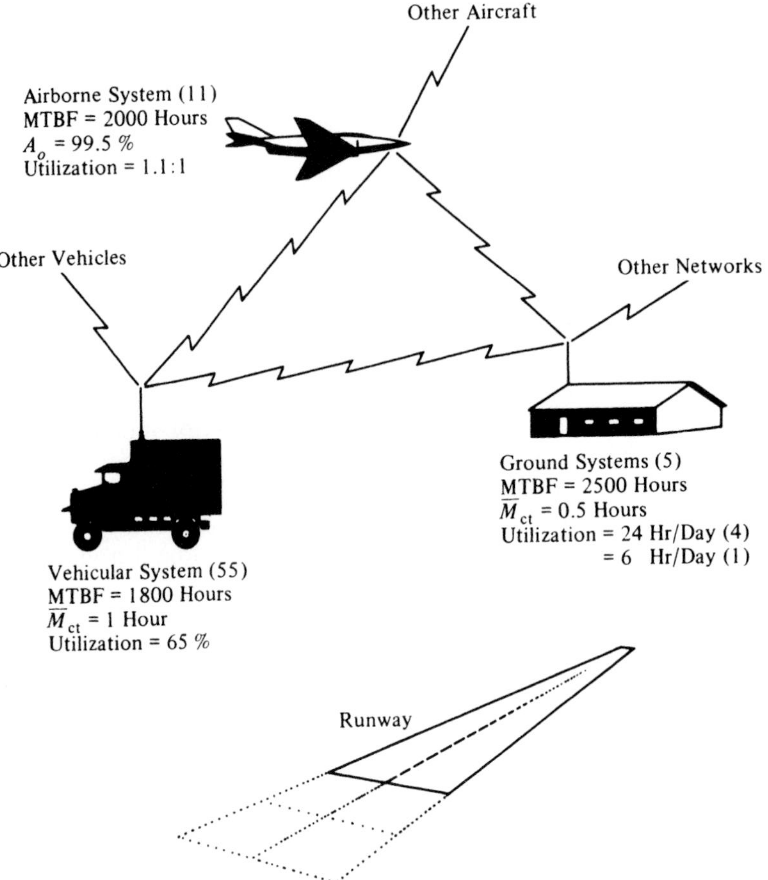

Figure 3.8 Typical communication network.

Based on the three mission scenarios just defined, there is a total requirement for 1420 prime equipments deployed in a series of communication networks (the total system) as illustrated in Figure 3.8.

A good definition of the planned system operational deployment (even if assumptions must be made) is necessary to determine what elements of logistic support are required and the location where the requirement will exist. For example, we know that 1420 prime equipments are required and will be deployed to certain locations (designated by the need) as they are produced. We need to know the quantities deployed by designated location, time of deployment, distances, and operational environment to determine:

1. Test and support equipment (to include transportation and handling requirements) needed to transport, install, and check out the system elements as they arrive at the operational site(s)
2. Personnel and training requirements for operating and maintaining the system

3. Spares and repair part types and quantities, inventory requirements, and so on, for maintaining the system
4. Technical data, including system installation and operating instructions
5. Facilities for equipment processing, installation, and operation

The logistic support requirements are identified with the outflow of equipment indicated by blocks 1, 2, 3, and 5 of Figure 2.33. The depth and timing of logistic support is dependent on system production and the delivery rate specified by the program schedule (Figure 3.7).

In support of the program schedule and the basic need, it is necessary to develop an equipment inventory profile as shown in Figure 3.9. This provides an indication of the total quantity of prime equipments in the user's inventory during any given year in the life cycle. The front end of the profile represents the production rate, which of course may vary considerably, depending on the type and complexity of equipment/software, the capacity of the production facility, and the cost of production. The total quantity of prime equipments produced is 1491, which assumes (1) that 5% of the equipments will be condemned during the 10-year full-complement period due to loss or damage beyond economical repair; and (2) that production is accomplished on a one-time basis (to avoid production startup and shutdown costs). In other words, assuming that production is continuous, 1491 equipments must be produced to cover attrition and yet maintain the operational requirements of 1420 systems through the 10-year period. After the 10-year period, the quantity of systems is reduced by attrition and/or phaseout due to obsolescence until the inventory is completely depleted.

When predicting logistics requirements for system maintenance (e.g., the inflow of equipment items illustrated in Figure 2.33), one must initially determine the demand for maintenance. Given the demand, it is necessary to develop the system maintenance concept as described in Section 3.4. The demand for system maintenance is derived through an analysis of the inventory profile, the location of equipment, system utilization, equipment reliability, and so on.

The profile in Figure 3.9 indicates the total number of equipments in the inventory at any given time (i.e., 1420 per year for the 10-year requirement). Operational

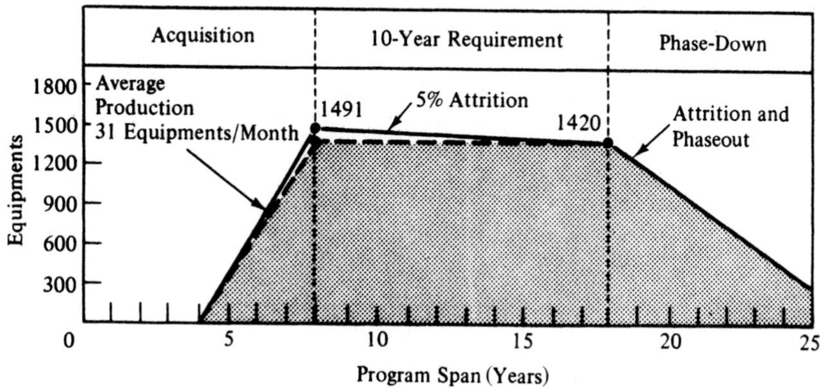

Figure 3.9 Equipment inventory profile.

deployment data will indicate the specific location of the equipment, and utilization factors will provide information on the hours of system operation. Although utilization will actually vary from one operational site to the next, the mission requirements described previously are used for planning purposes.

The intent is to determine the total hours of system operation for each year in the inventory profile (Figure 3.9). Considering each year of the 10-year full-complement period, one can calculate the total hours as

$$\text{total hours/year} = (\text{quantity of equipments}) \times (\text{yearly hours of usage per equipment}) \qquad (3.1)$$

or

$$(20 \text{ communication facilities}) \times (4 \text{ equipments}) \times (24 \text{ hours per day}) \times (360) = 691{,}200$$

and

$$(20 \text{ communication facilities}) \times (1 \text{ equipment}) \times (6 \text{ hours per day}) \times (360) = 43{,}200$$

and

$$(220 \text{ airborne equipments}) \times (30 \text{ hours per month}) \times (1.1 \text{ utilization}) \times (12) = 87{,}120$$

and

$$(1100 \text{ vehicular equipments}) \times (65\%) \times (24 \text{ hours per day}) \times (360) = \underline{6{,}177{,}600}$$

and

$$\text{total hours/year} = 6{,}999{,}120$$

Thus, the total usage for all prime equipments each year of the 10-year period is 6,999,120 hours. Usage during the introduction and phase-out periods is determined by the same method except that the quantity of systems is reduced. Using the required reliability MTBF factor specified for each mission (i.e., airborne, vehicular, and ground), one can predict the average number of corrective maintenance actions expected as a result of system failure. For instance,

$$\text{maintenance actions for vehicular systems} = \frac{\text{total vehicular hours}}{\text{MTBF}} \qquad (3.2)$$

or

$$\text{maintenance actions} = \frac{6{,}177{,}600}{1800} = 3432/\text{year}$$

For each centralized communication facility location where support is concentrated (55 vehicles per area communication facility), the expected quantity of maintenance actions due to failure is

$$\text{maintenance actions} = \frac{308{,}880}{1800} = 171/\text{year/location}$$

By employing Poisson factors discussed earlier, one can predict the quantity of spare/repair parts. Through an analysis of each maintenance action, one can determine MDT, $\overline{M}ct$ MLH/OH, and associated logistic support resources.

Definition of the operational requirements for the radio communication system (e.g., distribution, utilization, effectiveness factors, etc.) provides the basis for determining the maintenance concept (refer to Section 3.4) and the identification of specific reliability, maintainability, and logistics quantitative factors. These data, in turn, are employed as input factors for system design and supporting analyses. As system development progresses, the communication system operational requirements are further refined (on an iterative basis). The presentation of an operational concept at the inception of a program is mandatory for the establishment of a baseline for all subsequent program actions.

Illustration 3: Commercial Airline Requirement

Three commercial airline companies are proposing to serve a large metropolitan area 8 years hence. As future growth is expected, additional airline companies may become involved at a later time. For planning purposes, the combined anticipated passenger handling requirement follows the projection in Figure 3.10. The combined airline requirements are

1. Anticipated flight arrivals/departures:

	Anticipated Flights per Day			
Time Period	Point *A*	Point *B*	Point *C*	Point *D*
6:00 a.m. to 11:00 a.m.	33	65	95	105
11:00 a.m. to 4:00 p.m.	17	43	50	55
4:00 p.m. to 9:00 p.m.	33	60	90	100
9:00 p.m. to 6:00 a.m.	3	10	15	18
Total	86	178	250	278

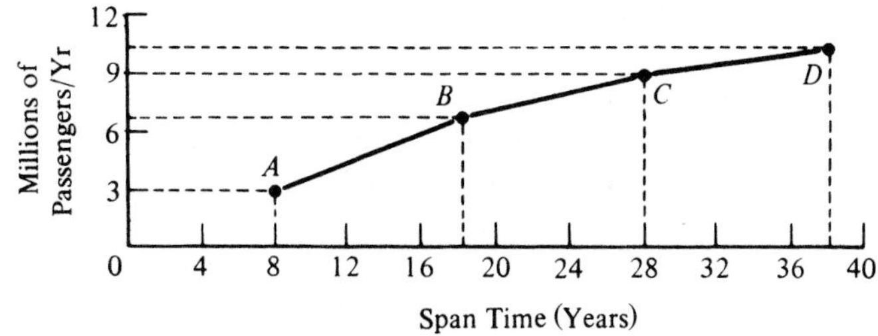

Figure 3.10 Projected passenger-handling requirement.

The flight arrivals are evenly spaced in the time periods indicated. It is assumed that 100 passengers constitute an average flight load.

2. The aircraft operational availability is 95%. In other words, 95% of all flights must be fully operational when scheduled (discounting aborts due to weather). Allowable factors for scheduled maintenance and passenger loading are:

Function	Frequency	Downtime
Through service	Each through flight	30 minutes
Turnaround service	Each turnaround	1 hour
Termination check	Each terminal flight	6 hours
Service check	15 Days	9 hours

Periodic and main base checkouts will be accomplished elsewhere.

3. Allowable unscheduled maintenance in the area shall be limited to the organizational level and will include the removal and replacement of line replaceable items, tire changes, and engine replacements as required. The specific maintenance downtime (MDT) limits are

Engine change	6 hours
Tire change	1 hour
Other items	1 hour

4. The metropolitan area must provide the necessary ground facilities to support the following types of aircraft (fully loaded): B-727, B-737, B-747, B-757, B-767, B-777, and L-1011. In addition, provisions must be made for cargo handling and storage.

The airline companies have identified a need to provide air transportation service for a metropolitan area. From the airline standpoint, the metropolitan area must provide the necessary logistics resources (facilities, test and support equipment, ground handling equipment, people movers, operating and maintenance personnel, etc.) to support this service. This involves selecting a site for an air transportation facility; accomplishing the design and construction of the facility; providing local transportation to and from the airline terminal; acquiring the test and support equipment, spare/repair parts, personnel, and data to support airline operations; and maintaining the total capability on a sustaining basis throughout the planning period.

Because of the size of the program and the growth characteristics projected in Figure 3.10, a three-phased construction effort is planned. The program schedule is presented in Figure 3.11.

The initial step is to select a location for the air transportation facility. The selection process considers available land, terrain, geology, wind effects, distance from the metropolitan area, access via highway and/or public transportation, noise and ecology requirements, and cost. Once a site has been established, the facility must include runways, holding area, flight control equipment, airline terminal, control tower, operations building, hangars, maintenance docks, fuel docks, cargo handling and storage capabil-

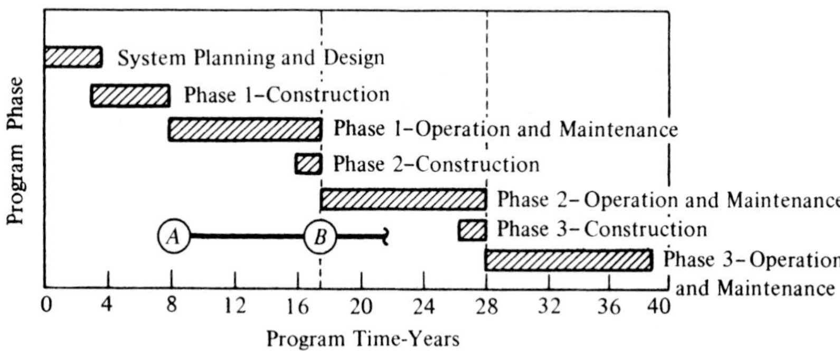

Figure 3.11 Program schedule.

ity, utilities, and all the required support directly associated with passenger needs and comfort. Construction is accomplished in three phases, consistent with passenger handling growth and the schedule in Figure 3.11.

The ultimate design configuration of the air transportation facility is based directly on the operational requirements—anticipated airline flight arrivals/departures, passenger loading, aircraft turnaround times, and maintenance and servicing requirements. For instance, at a point 18 years hence (Figure 3.10, point *B*), the anticipated average number of flights is 65 per day between the hours of 6:00 A.M. and 11:00 A.M. The number of through flights is 30 and the gate time for each is 30 minutes (to accomplish servicing and passenger loading), and the number of turnarounds is 35 with a gate time of 1 hour each. The total gate time required (considering no delays and assuming one gate for each aircraft arrival/departure) is 50 hours during the 6:00 A.M. to 11:00 A.M. time period; thus, at least 10 gates are required in the passenger terminal to satisfy the load. This, in turn, influences the size of the passenger waiting lobby and the number of airline personnel agents required. Further, the servicing and ground handling of the aircraft requires certain consumables (fuel, oil, lubricants), spare/repair parts, test and support equipment (towing vehicle, fuel truck, etc.), and technical data (operating and maintenance instructions). The possibility of unscheduled maintenance dictates the need for trained maintenance personnel, spare/repair parts, data, and a backup maintenance dock or hangar.

One can go on indefinitely identifying requirements to support the basic need of the metropolitan area and the commercial airline companies. It readily becomes obvious that logistic support plays a major role. There are logistic requirements associated with the aircraft, air transportation facility, and the transportation media between the air transportation facility and the metropolitan area. The commercial airline illustration presented herein merely touches on a small segment of the problem. The problem should be addressed from a total *systems approach* considering the functions associated with all facets of the operation. This would better emphasize the magnitude of the logistics involved.

Additional Applications

The three illustrations just described are representative of typical needs where system operational requirements must be defined at the inception of a program and must serve as the basis for all subsequent program actions. The methodology employed is basically the same for any system, whether the subject is a relatively small item installed in an aircraft or on a ship, is a factory, or a large one-of-a-kind project involving construction. In any event, the system must be defined in terms of its mission, performance, operational deployment, life cycle, utilization, effectiveness factors, and environment.

3.4 THE MAINTENANCE AND SUPPORT CONCEPT

In addressing system requirements, the normal tendency is to deal primarily with those elements of the system that relate directly to the "performance of the mission;" that is, prime equipment, operator personnel, operational software, and associated data. At the same time, there is very little attention given to system maintenance and support. In general, the emphasis in the past has been directed toward only *part* of the system, and not the entire system. This, of course, has led to some of the problems discussed in Section 1.4.

To meet the overall objectives of systems engineering, it is essential that *all* aspects of the system be considered on an integrated basis. This includes not only the prime mission-oriented segments, but the support capability as well. System support must be considered from the beginning (e.g., during the feasibility analysis when new technologies are being evaluated for possible application), and a before-the-fact *maintenance concept* must be developed on how the proposed system is to be designed and supported on a life-cycle basis.[2]

The maintenance concept, developed during conceptual design, evolves from the definition of system operational requirements, as illustrated by the flow chart in Figure 1.7. Initially, one must deal with the flow of activities and materials from design, through production, and to the customer's operational site(s) where the system is being utilized. In addition, there is a flow involving the system support capability. A maintenance flow exists when items are returned from the operational site to the intermediate and depot levels of maintenance. A second flow involves the distribution of spare parts, personnel, test equipment, and data from the various suppliers to the intermediate and depot levels of maintenance, and to the operational sites as required. It is the flow chart in Figure 3.12 (which is an extension of Figures 1.1 and 1.2) that reflects the activities that are related to the overall system support capability.

[2] The author defines the "maintenance concept" as being a before-the-fact series of illustrations and statements on how the system is to be designed for supportability, and the "maintenance plan" defines the requirements for system support based on a known configuration and on the results of the supportability analysis (or equivalent). The maintenance concept is an *input* to design, and the maintenance plan is the *result* of design.

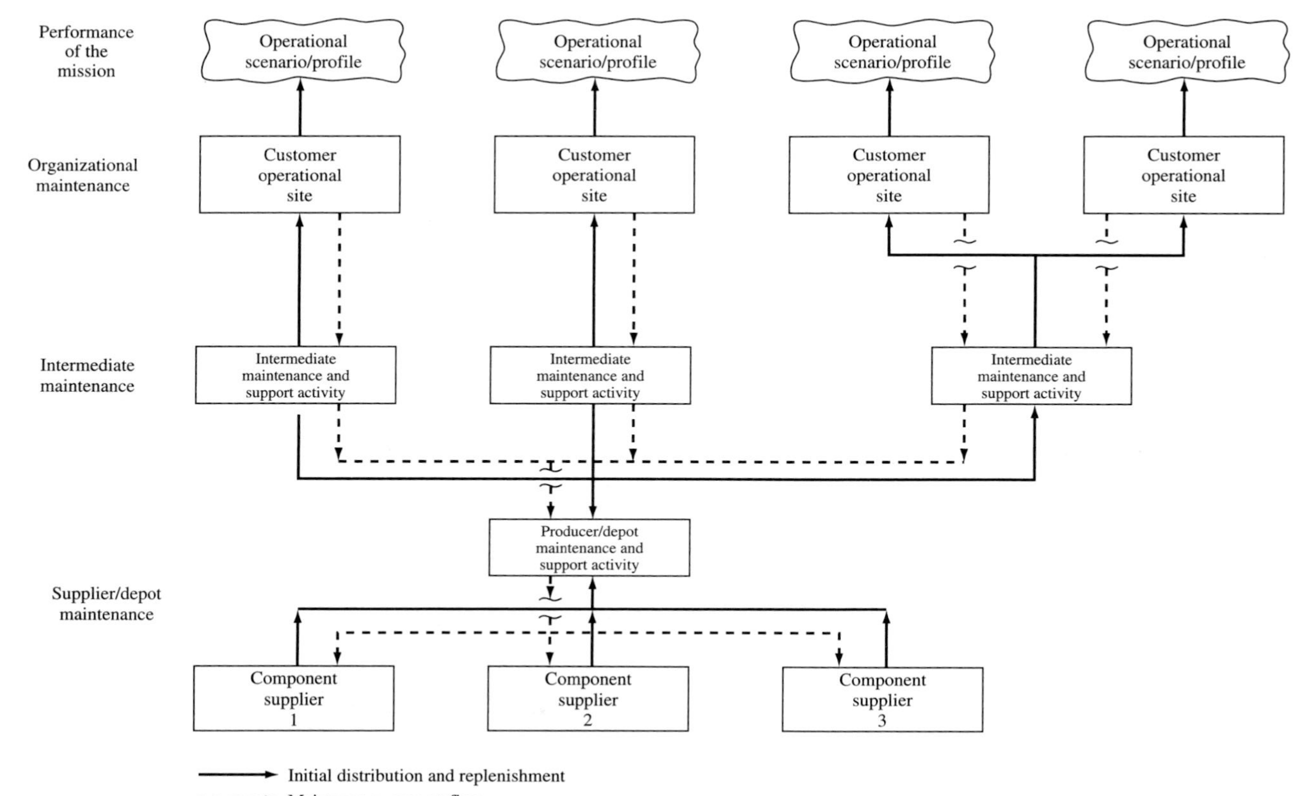

Figure 3.12 System maintenance flow (reference: Figure 1.2).

Although there are some variations as a function of the nature and type of system, the maintenance concept generally includes the following information:

1. *Levels of maintenance.* Corrective and preventive maintenance may be accomplished on the system itself (or an element thereof) at the site where the system is used by the customer, in an intermediate shop near the customer's operational site, and/or at a depot or manufacturer's facility. Maintenance level pertains to the division of functions and tasks for each area where maintenance is performed. Anticipated frequency of maintenance, task complexity, personnel skill-level requirements, special facility needs, and so on, dictate to a great extent the specific functions to be accomplished at each level. Depending on the nature and mission of the system, there may be two levels, three levels, or four levels of maintenance. However, for the purposes of further discussion, maintenance may be classified as "organizational," "intermediate," and "supplier/depot."
   (a) *Organizational maintenance.* Organizational maintenance is performed at the operational site (e.g., airplane, vehicle, manufacturing production line, or communication facility). Generally, it includes tasks performed by the using organization on its own equipment/software. Organizational-level personnel are usually involved with the operation and use of equipment/software, and have minimum time available for detail system maintenance. Maintenance at this level normally is limited to periodic checks of equipment performance, visual inspections, cleaning of system elements, verification of software, some servicing, external adjustments, and the removal and replacement of some components. Personnel assigned to this level generally do not repair the removed components, but forward them to the intermediate level. From the maintenance standpoint, the least skilled personnel are assigned to this function. The design of equipment/software must take this fact into consideration (e.g., design for simplicity).
   (b) *Intermediate maintenance.* Intermediate maintenance tasks are performed by mobile, semimobile, and/or fixed specialized organizations and installations. At this level, end items may be repaired by the removal and replacement of major modules, assemblies, or piece parts. Scheduled maintenance requiring equipment disassembly may also be accomplished. Available maintenance personnel are usually more skilled and better equipped than those at the organizational level and are responsible for performing more detail maintenance.

   Mobile or semimobile units are often assigned to provide close support to deployed operational systems. These units may constitute vans, trucks, or portable shelters containing some test and support equipment and spares. The mission is to provide on-site maintenance (beyond that accomplished by organizational-level personnel) to facilitate the return of the system to its full operational status on an expedited basis. A mobile unit may be used to support more than one operational site. A good example is the maintenance vehicle that is deployed from the airport hangar to an airplane parked at a commercial airline terminal gate and needing extended maintenance.

   Fixed installations (permanent shops) are generally established to support both the organizational-level tasks and the mobile or semimobile units.

Maintenance tasks that cannot be performed by the lower levels, due to limited personnel skills and test equipment, are performed here. High personnel skills, additional test and support equipment, more spares, and better facilities often enable equipment repair to the module and piece part level. Fixed shops are usually located within specified geographical areas. Rapid maintenance turnaround times are not as imperative here as at the lower levels of maintenance.

(c) *Depot, supplier, or manufactures maintenance.* The depot level constitutes the highest type of maintenance, and supports the accomplishment of tasks above and beyond the capabilities available at the intermediate level. Physically, the depot may be a specialized repair facility supporting a number of systems/equipments in the inventory or may be the equipment manufacturer's plant. Depot facilities are fixed and mobility is not a problem. Complex and bulky equipment, large quantities of spares, environmental control provisions, and so on, can be provided if required. The high-volume potential in depot facilities fosters the use of assembly-line techniques which, in turn, permits the use of relatively unskilled labor for a large portion of the workload with a concentration of highly skilled specialists in such certain key areas as fault diagnosis, calibration, and quality control.

The depot level of maintenance includes the complete overhauling, rebuilding, and calibration of equipment as well as the performance of highly complex maintenance actions. In addition, the depot provides an inventory supply capability. The depot facilities are generally remotely located to support specific geographical area needs or designated product lines.

The three levels of maintenance just discussed are covered in Figure 3.13.

2. *Repair policies.* Within the constraints illustrated in Figures 3.12 and 3.13, there may be a number of possible policies specifying the extent to which repair of a system component will be accomplished (if at all). A repair policy may dictate that an item should be designed to be nonrepairable, partially repairable, or fully repairable. Repair policies are initially established, criteria are developed, and system design progresses within the bounds of the repair policy that is selected. An example of a repair policy for System XYZ, developed as part of the maintenance concept during conceptual design, is illustrated in Figure 3.14.[3]

3. *Organizational responsibilities.* The accomplishment of maintenance may be the responsibility of the customer, the producer (or supplier), a third party, or a combination thereof. Additionally, the responsibilities may vary, not only with different components of the system, but as one progresses in time through the system operational use and sustaining support phase. Decisions pertaining to organizational responsibilities may impact system design from a diagnostic and packaging stand-point, as well as dictating repair policies, product warranty provisions, and the like. Although conditions may change, some initial assumptions are required at this point in time.

[3] Repair policies are ultimately verified through a level-of-repair analysis, the result of which leads into the maintenance plan. The level-of-repair analysis is usually accomplished as part of a maintainability analysis, a supportability analysis, or both. Refer to Chapter 4 for additional coverage.

Criteria	*Organizational Maintenance*	*Intermediate Maintenance*		*Depot, Supplier, or Manufacturer Maintenance*
Done where?	At the operational site or wherever the prime equipment is located	Mobile or semimobile units	Fixed units	Depot/Supplies
		Truck, van, portable shelter, or equivalent	Fixed field shop	Specialized repair activity, or manufacturer's plant
Done by whom?	System/equipment operating personnel (low maint. skills)	Personnel assigned to mobile, semimobile, or fixed units (intermediate maintenance skills)		Depot facility personnel or manufacturer's production personnel (mix of intermediate fabrication skills and high maintenance skills)
On whose equipment?	Using organization's equipment	Equipment owned by using organization		
Type of work accomplished?	Visual inspection Operational checkout Minor servicing External adjustments Removal and replacement of some components	Detailed inspection and system checkout Major servicing Major equipment repair and modifications Complicated adjustments Limited calibration Overload from organizational level of maintenance		Complicated factory adjustments Complex equipment repairs and modifications Overhaul and rebuild Detailed calibration Supply support Overload from intermediate level of maintenance

Figure 3.13 Major levels of maintenance.

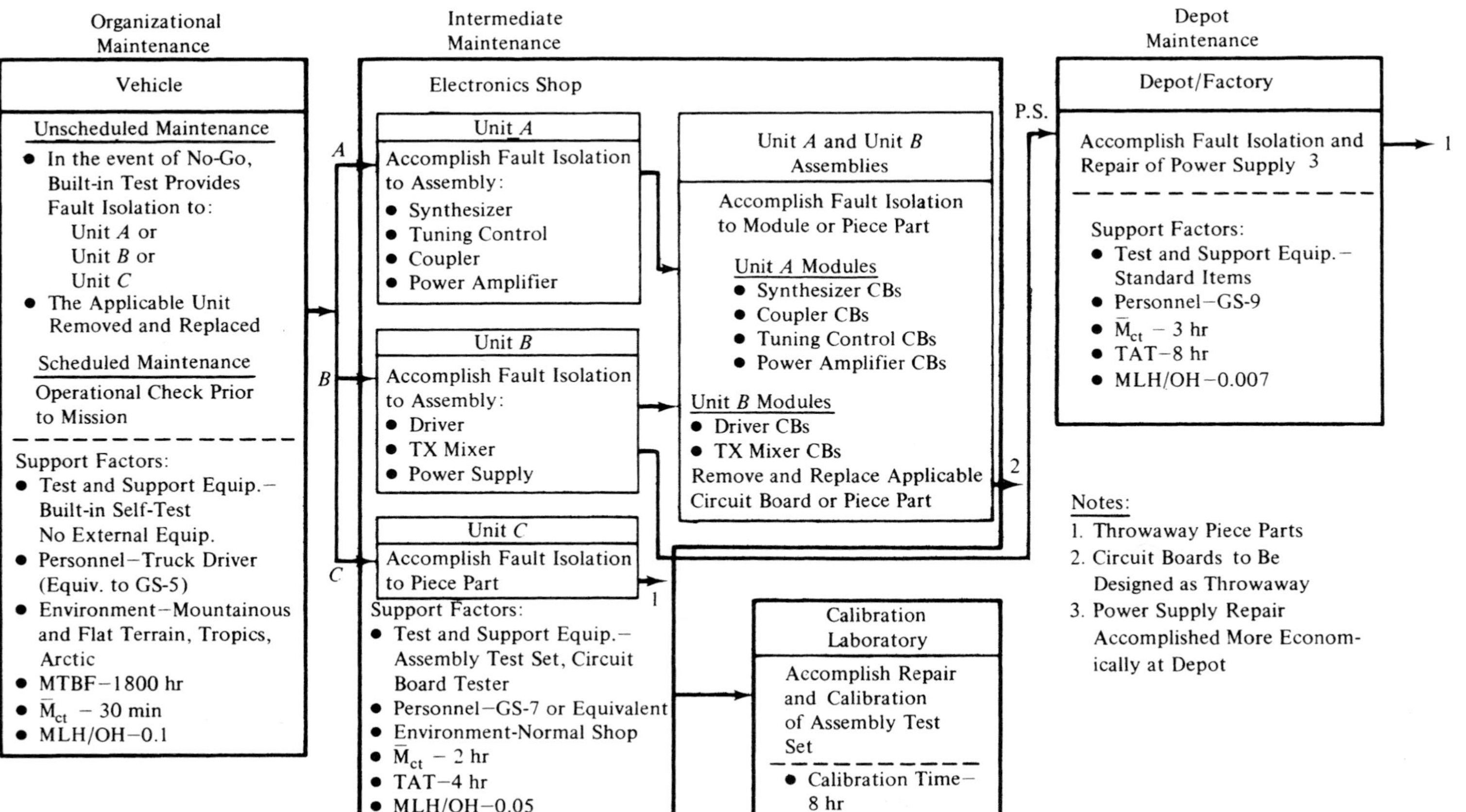

Figure 3.14 System maintenance concept flow (repair policy).

4. *Maintenance support elements.* As part of the initial maintenance concept, criteria must be established relating to the various elements of maintenance support. These elements include supply support (spares and repair parts, associated inventories, provisioning data), test and support equipment, personnel and training, transportation and handling equipment, facilities, data, and computer resources (refer to Figure 1.3, Section 1.2). Such criteria, as an input to design, may cover self-test provisions, built-in versus external test requirements, packaging and standardization factors, personnel quantities and skill levels, transportation and handling factors and constraints, and so on. The maintenance concept provides some initial system design criteria pertaining to the activities illustrated in Figure 3.12, and the final determination of specific logistic and maintenance support requirements will occur through the completion of a supportability analysis as design progresses.
5. *Effectiveness requirements.* This constitutes the effectiveness factors associated with the support capability. In the supply support area, this may include a spare part demand rate, the probability of a spare part being available when required, the probability of mission success given a designated quantity of spares in the inventory, and the economic order quantity as related to inventory procurement. For test equipment, the length of the queue while waiting for test, the test station process time, and the test equipment reliability are key factors. In transportation, transportation rates, transportation times, the reliability of transportation, and transportation costs are of significance. For personnel and training, one should be interested in personnel quantities and skill levels, human error rates, training rates, training times, and training equipment reliability. In software, the number of errors per mission segment, per module of software, or per line of code may be important measures. These factors, as related to a specific system-level requirement, must be addressed. It is meaningless to specify a tight quantitative requirement applicable to the repair of a prime element of the system when it takes 6 months to acquire a needed spare part. The effectiveness requirements applicable to the support capability must complement the requirements for the system overall.
6. *Environment.* Definition of the environment as it pertains to maintenance and support. This includes temperature, shock and vibration, humidity, noise, arctic versus tropical environment, mountainous versus flat terrain, shipboard versus ground conditions, and so on, as applicable to maintenance activities and related transportation, handling, and storage functions.

In summary, the maintenance concept provides the basis for the development of the support infrastructure and the specific design requirements for the various elements of support (refer to Figure 1.3). It evolves from the definition of system operational requirements as discussed in Section 3.3. For example, the flow presented in Figure 3.15 is an extension of the communication system flow shown in Figure 3.8. The figure illustrates the quantity of elements deployed, the levels of maintenance, and key effectiveness factors (e.g., MTBF, $\overline{M}ct$, TAT, and logistics flow times). These factors may be specified as design criteria for the prime elements of the system and the support structure as well.

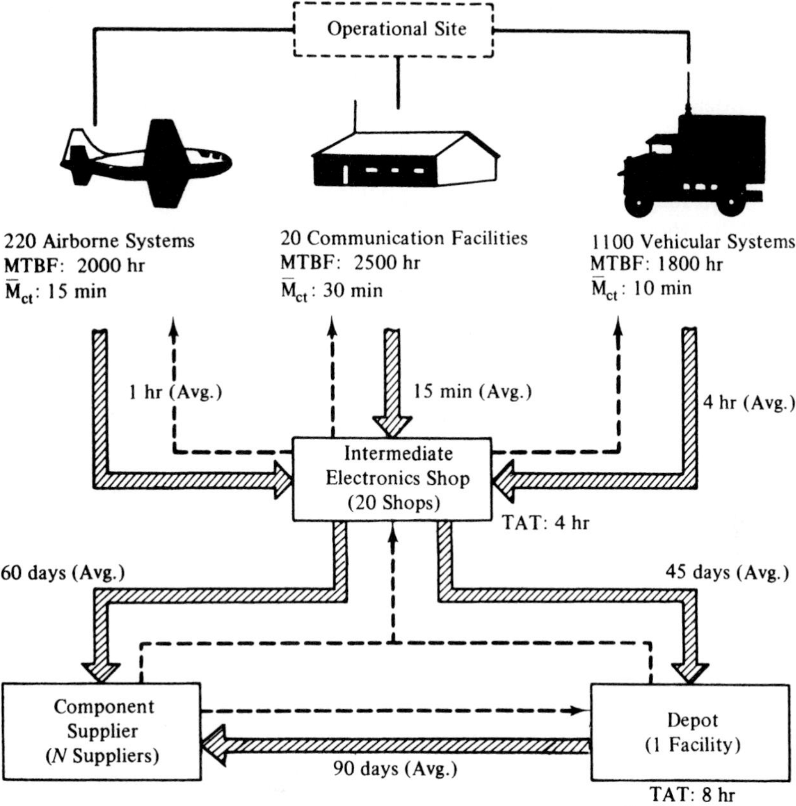

Figure 3.15 Operational/maintenance concept.

3.5 IDENTIFICATION AND PRIORITIZATION OF TECHNICAL PERFORMANCE MEASURES (TPMs)[4]

Through the definition of operational requirements and the maintenance concept for the system, specific *performance-related* factors are identified and applied with the objective of ensuring that the system will be designed and developed such that it will

[4] TPMs refer to those measures of a system, program activity/process, or product that are considered as being critical for the successful accomplishment of system engineering objectives. Specific *performance-oriented* quantitative parameters, reflecting the characteristics that should be inherent within the design, should initially be specified and later verified through test and evaluation. Refer to Blanchard, B. S., and W. J. Fabrycky, *Systems Engineering and Analysis*, 3rd Ed., Prentice Hall, Inc., Upper Saddle River, N.J., 1998.

satisfactorily accomplish its intended mission(s). These factors, identified as *technical performance measures (TPMs)*, may be applied as "design-to" criteria for the prime mission-related elements of the system and for those elements that are necessary for support. With regard to the maintenance and support infrastructure (and the design for supportability), those measures of logistics that are described in Chapter 2 and applicable through the definition of operational requirements and the maintenance concept may be identified as critical TPMs.

In the identification of TPMs, there may be a number of different metrics that are applicable, and priorities need to be established in order to determine the relative degrees of importance in the event that design trade-offs (i.e., compromises) are necessary. For example, in the design of a vehicle is *speed* more important than *size*? For a manufacturing plant, is *production quantity* more important than *product quality*? In a communication system, is *range* more important than *reliability* or *clarity of message*? For a computer capability, is *capacity* more important than *speed*? Is *reliability* more important than *maintainability*? Are *human factors* more important than *cost*? There may be a number of different design objectives and the designer needs to understand which are more important than others and the relationships that exist between such. Further, the designer needs to ensure that the ultimate system configuration does incorporate the necessary attributes or characteristics that are responsive to the TPM requirements and are consistent with the established priorities.

Figure 3.16 conveys the results from a TPM identification and prioritization effort involving a team of individuals representing the customer (user), the appropriate designer(s), the area of logistics, etc. Note that the critical TPMs are identified along with their relative degrees of importance. The performance factors of *velocity*, *availability*, and *size* are the most critical and where emphasis in design must be directed. The specific features or attributes to be incorporated into the design (e.g., the degree of standardization, diagnostic provisions, the use of reliable components, the use of lightweight materials, etc.) must be responsive to these requirements, and there must be a *traceability* of requirements from the system-level down to its various elements.

An excellent tool that can be applied to aid in the establishment and prioritization of TPMs is the *quality function deployment (QFD)* model. QFD constitutes a "team" approach to help ensure that the "voice of the customer" is reflected in the ultimate design. The purpose is to establish the necessary *requirements* and to translate those requirements into technical solutions. Customer requirements and preferences are defined and categorized as *attributes*, which are then weighted based on the degree of importance. The QFD method provides the design team an understanding of customer desires, forces the customer to prioritize those desires, and enables a comparison of one design approach against another. Each customer attribute is then satisfied by a technical solution.[5]

[5] The good references pertaining to the QFD process are (1) Yoji Akao (Ed.), *Quality Function Deployment: Integrating Customer Requirements Into Product Design*, Productivity Press, Portland, OR, translated into English, 1990; and (2) Lou Cohen, *Quality Function Deployment: How to Make QFD Work for You*, Addison-Wesley, Reading, MA, 1995. The QFD method was develop at the Kobe Shipyard of Mitsubishi Heavy Industries, Ltd., Japan, in the late 1960s and has evolved considerably since.

Technical performance measure	Quantitative requirement ("metric")	Current "benchmark" (competing systems)	Relative importance (customer desires) (%)
Process time (days)	30 days (maximum)	45 days (system M)	10
Velocity (MPH)	100 mph (minimum)	115 mph (system B)	32
Availability (operational)	98.5% (minimum)	98.9% (system H)	21
Size (feet)	10 feet long 6 feet wide 4 feet high (maximum)	9 feet long 8 feet wide 4 feet high (system M)	17
Human factors	Less than 1% error rate per year	2% per year (system B)	5
Weight (pounds)	600 pounds (maximum)	650 pounds (system H)	6
Maintainability (MTBM)	300 miles (minimum)	275 miles (system H)	9
			100

Figure 3.16 Prioritization of technical performance measures.

The QFD process involves constructing one or more matrices, the first of which is often referred to as the "House of Quality (HOQ)."[6] A modified version of the HOQ is presented in Figure 3.17. Starting on the left side of the structure is the identification of customer needs and the ranking of those needs in terms of priority, the levels of importance being specified quantitatively. This reflects the "WHATs" that must be addressed. A team, with representation from both customer and design organizations, determines the priorities through an iterative process of review, evaluation, revision, reevaluation, and so on. The top part of the HOQ identifies the designer's *technical* response relative to the attributes that must be incorporated into the design in order to respond to the needs (i.e., the "voice of the customer"). This constitutes the "HOWs," and there should be at least one technical solution for each identified customer need. The interrelationships among attributes (or technical correlations) are identified, as well as possible areas of conflict. The center part of the HOQ conveys

[6] Hauser, J. R., and D. Clausing, "The House of Quality." *Harvard Business Review,* May–June 1988, pp. 63–73.

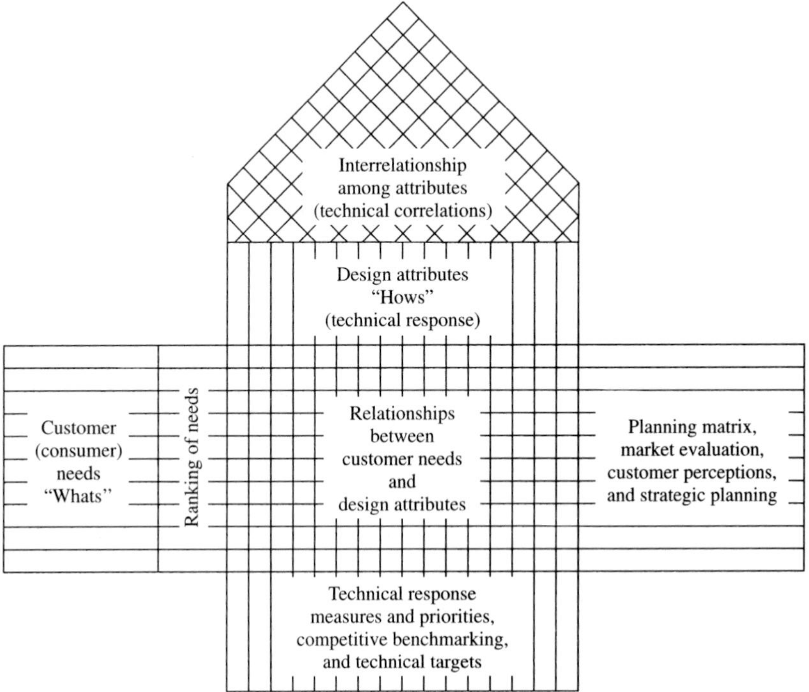

Figure 3.17 Modified House of Quality. (Source: Blanchard, B. S. *System Engineering Management*), 2nd ed., Wiley, New York, 1998.)

the strength or impact of the proposed technical response on the identified requirement. The bottom part allows for a comparison between possible alternatives, and the right side of the HOQ is used for planning purposes.

The QFD method is used to facilitate the translation of a prioritized set of subjective customer requirements into a set of *system-level* requirements during conceptual design. A similar approach may be used to subsequently translate system-level requirements into a more detailed set of requirements at each stage in the design and development process. In Figure 3.18, the "HOWs" from one house become the "WHATs" for a succeeding house. Requirements may be developed for the system, subsystem, component, the manufacturing process, the support infrastructure, and so on. The objective is to ensure the required justification and traceability of requirements from the top down. Further, requirements should be stated in *functional* terms.

Although the QFD method may not be the only approach used in helping to define the requirements for system design, it does constitute an excellent tool for creating the necessary visibility from the beginning. One of the largest contributors to "risk" is the lack of a good set of requirements and an adequate system specification. Inherent within the system specification should be the identification and prioritization of technical performance measures (TPMs). The TPM, its associated measure (i.e., "metric"), its relative importance, and "benchmark" objective in terms of what is currently available will provide designers with the necessary guidance for accomplishing

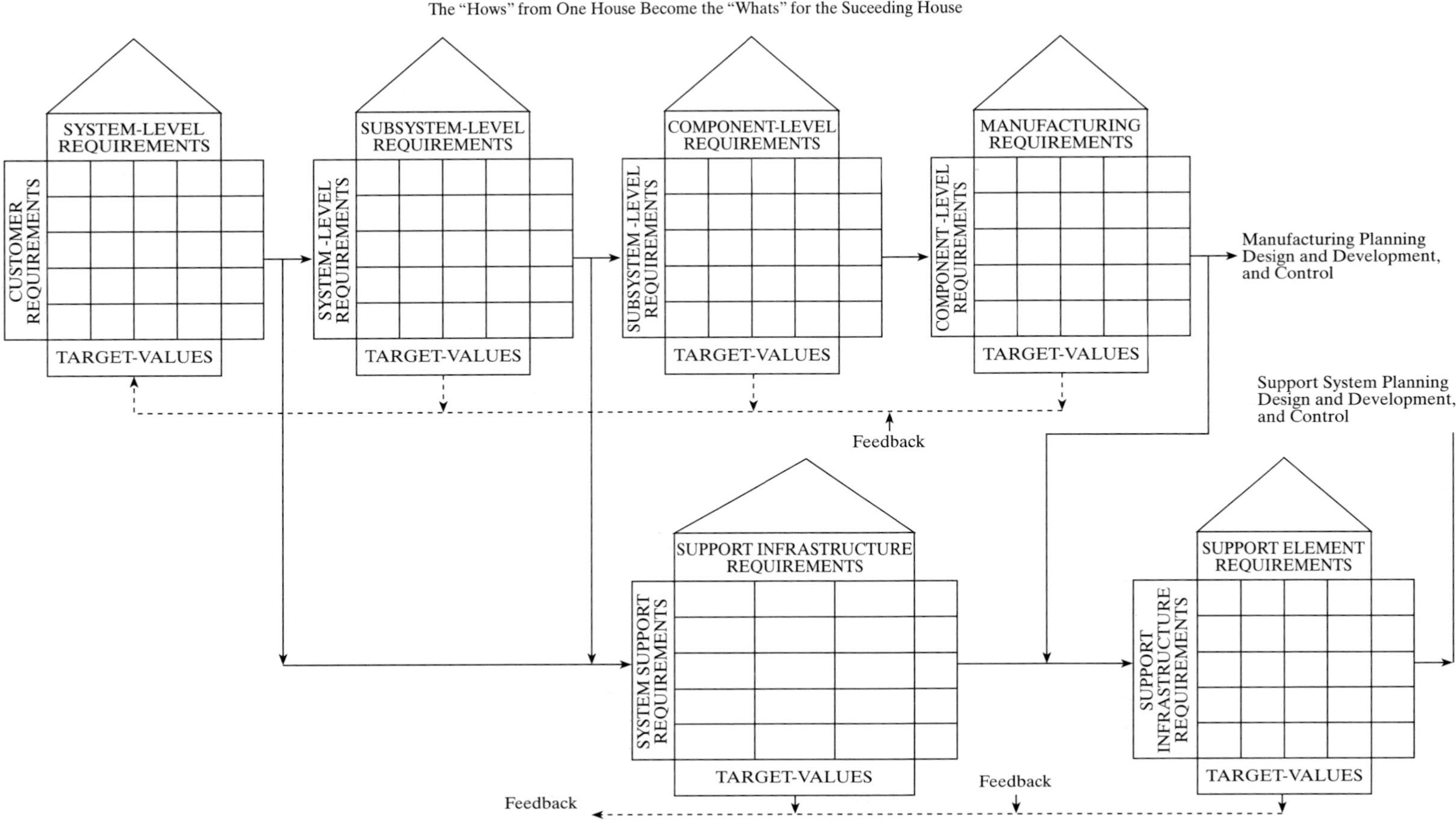

Figure 3.18 Family of houses—traceability of requirements. (Source: Blanchard, B. S. *System Engineering Management*, 2nd ed., Wiley, New York, 1998.)

their task. This is essential for establishing the appropriate levels of design emphasis, for defining the criteria as an input to the design, and for identifying the levels of possible risk should the requirements not be met.[7]

3.6 FUNCTIONAL ANALYSIS

An essential element of early conceptual and preliminary design is the development of a *functional* description of the system to serve as a basis for the identification of the resources necessary for the system to accomplish its objective(s). A "function" refers to a specific or discrete action (or series of actions) that is necessary to achieve a given objective; that is, an operation that the system must perform to accomplish its mission, or a maintenance action that is necessary to restore the system to operational use. Such actions may ultimately be accomplished through the use of equipment, people, software, facilities, data, or combinations thereof. However, at this point, the objective is to specify the "whats" and *not* the "hows;" that is, *what* needs to be accomplished versus *how* it is to be done![8] The functional analysis is an iterative process of breaking requirements down from the system level, to the subsystem, and as far down the hierarchical structure as necessary to identify input design criteria and/or constraints for the various elements of the system.

In Figure 3.1, the functional analysis may be initiated in the early stages of conceptual design as part of the problem definition and needs analysis task, and functions that the system must perform in order to fulfill the needs of the customer are identified. These *operating* functions are then expanded and formalized through the development of system operational requirements. Primary *maintenance and support* functions for the system, which evolve from the operational requirements, are identified as part of the maintenance concept development process. Subsequently, these functions must be expanded to include *all* of the activity from the initial identification of need to the retirement of the system.

The accomplishment of a functional analysis can be facilitated through the use of functional flow block diagrams, as illustrated in Figure 3.19. Block diagrams are developed primarily for the purpose of structuring system requirements into "functional terms." They are developed to illustrate basic system organization, and to identify functional interfaces. The functional analysis (and the generation of functional flow diagrams) is intended to enable the completion of the design, development, and the system definition process in a comprehensive and logical manner. Top level requirements are identified, partitioned to a second level, and on down to the depth required

[7] To gain a complete perspective of the QFD process and the advantages relative to its implementation, the reader is advised to review some of the literature in the field. The coverage presented herein is very cursory in nature and intended to provide only an "overview" of the process.

[8] In applying the principles of systems engineering, not one piece of equipment, or software, or data item, or element of support should be identified and purchased, without first having justified the need for such through the functional analysis. On many projects, items are often purchased based on what is initially perceived as being a "requirement" but that later turns out not to be needed in the end. This practice can turn out to be quite costly.

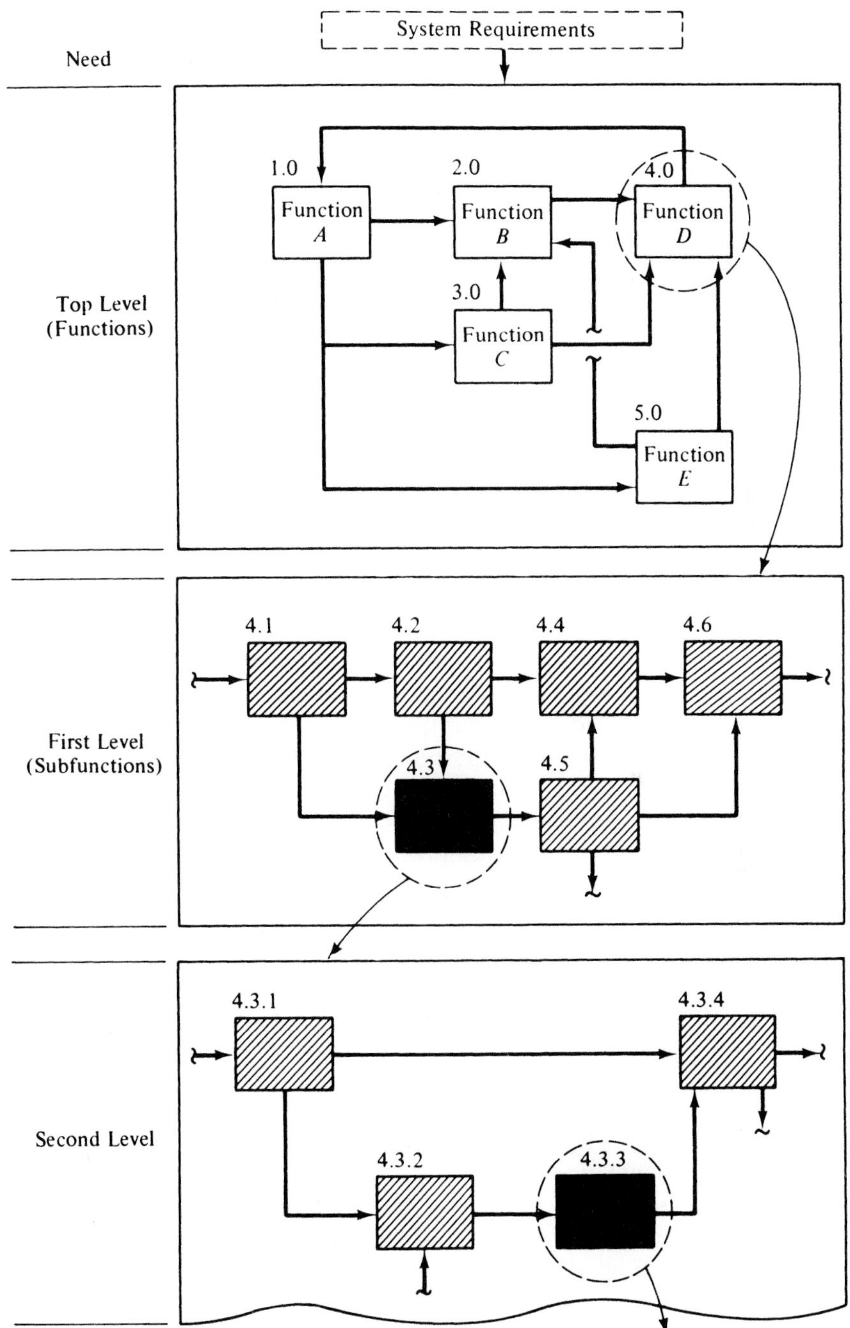

Figure 3.19 System functional indenture levels.

for the purposes of "definition." More specifically, the functional approach helps to ensure the following:

1. That all facets of system design and development, operation, support, and retirement are covered; that is, all significant activities within the system life cycle.
2. That all elements of the system are fully recognized and defined; that is, prime equipment, spare/repair parts, test and support equipment, facilities, personnel, data, and software.
3. That a means is provided for relating system packaging concepts and support requirements to specific system functions; that is, satisfying the requirements of good "functional" design.
4. That the proper sequences of activity and design relationships are established, along with critical design interfaces.

One of the objectives of functional analysis is to ensure traceability from the top system-level requirements down to the requirements for detail design. In Figure 3.20, it is assumed that there is a need for transportation between Cities A and B. Through the conductance of a feasibility analysis, trade-off studies are accomplished, and the results indicate that transportation by air is the preferred mode. Subsequently, through the definition of operational requirements, it was concluded that there is a requirement for a new aircraft system, demonstrating good performance and effectiveness characteristics with quantitative goals specified for size, weight, thrust, range, fuel capacity, reliability, maintainability, supportability, cost, and so on. An aircraft must be designed and produced that will accomplish its mission in a satisfactory manner, flying through a number of operational profiles such as the one illustrated in Figure 3.20. Further, the maintenance concept indicates that the aircraft will be designed for support at three levels of maintenance by the user, will incorporate built-in test provisions, and will be in operational use for a life cycle of 10 years.

With this basic information, following the general steps in Figure 3.2, one can commence with the structuring of the system in functional terms. A top-level functional flow diagram can be developed to cover the primary activities identified within the specified life cycle. Each of these designated activities can be expanded through a second-level functional flow diagram, a second-level activity into a third-level functional flow, and so on.

Through this progressive expansion of functional activities, directed to defining the "whats" (versus the "hows"), one can evolve from the mission profile in Figure 3.20 down to a specific aircraft capability such as "communications." A communications subsystem is identified, trade-offs are accomplished, and a detail design approach is selected. Specific resources that are necessary to respond to the stated functional requirement can be identified. In other words, one can drive downward from the system level to identify the resources needed to perform certain functions (e.g., equipment, people, facilities, and data). Also, given a specific equipment requirement, one can progress "upward" for *justification* of that requirement. The functional analysis provides the mechanism for "down-up" traceability.

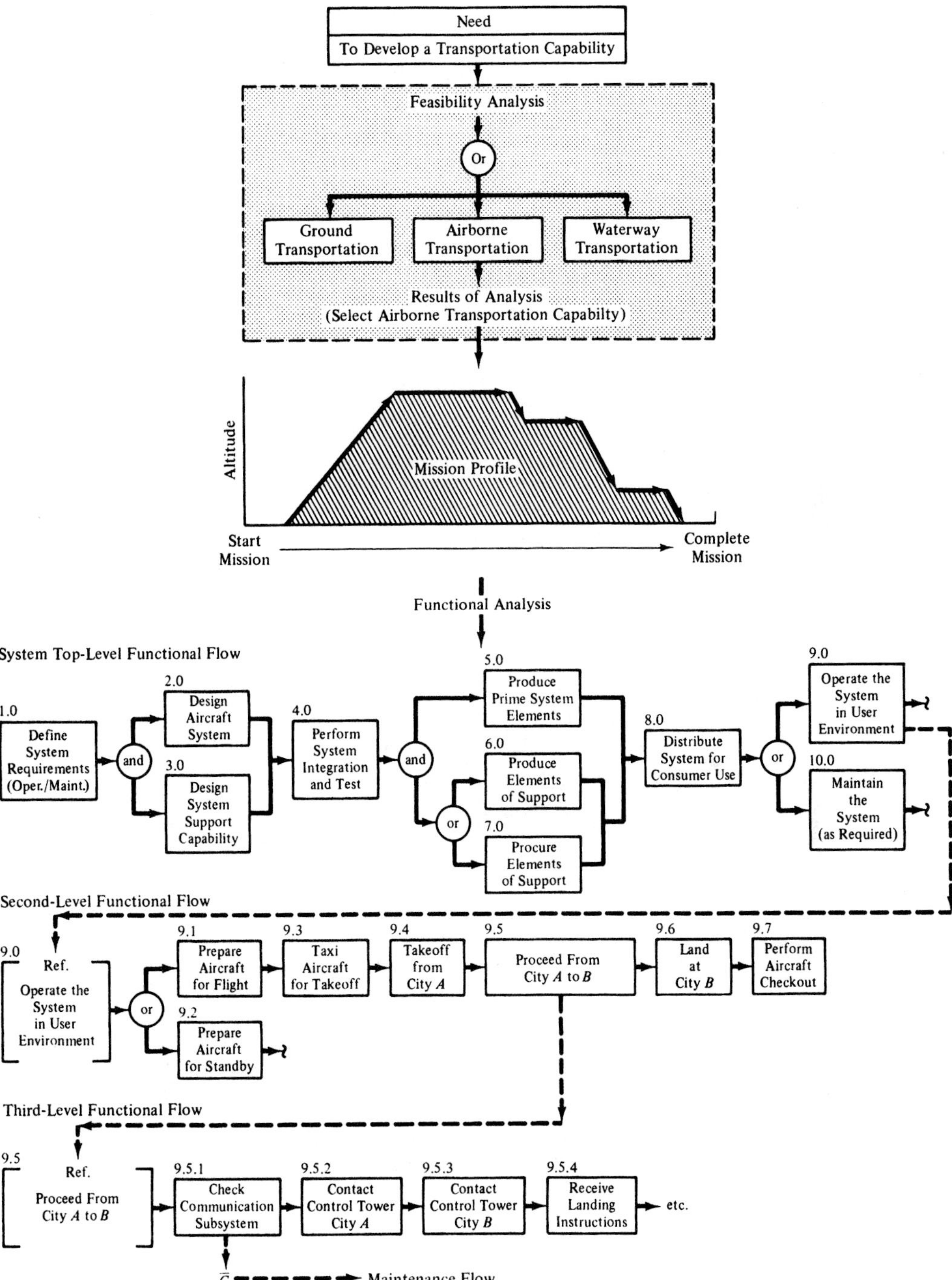

Figure 3.20 Evolutionary development of functional requirements.

Functional Flow Diagrams

In the development of functional flow diagrams, some degree of standardization is necessary, for the purpose of "communication," in defining the system. Thus, certain basic practices and symbols should be used, whenever possible, in the physical layout of functional diagrams. The following paragraphs provide some guidance in this direction:

1. *Function block.* Each separate function in a functional diagram should be presented in a single box enclosed by a solid line. Blocks used for reference to other flows should be indicated as partially enclosed boxes labeled "REF." Each function may be as gross or detailed as required by the level of functional diagram on which it appears, but it should stand for a definite, finite, discrete action to be accomplished by equipment, personnel, facilities, software, or any combination thereof. Questionable or tentative functions should be enclosed in dotted blocks.
2. *Function numbering.* Functions identified on the functional flow diagrams at each level should be numbered in a manner that preserves the continuity of functions and provides information with respect to function origin throughout the system. Functions in the top-level functional diagram should be numbered 1.0, 2.0, 3.0, and so on. Functions that further indenture these top functions should contain the same parent identifier and should be coded at the next decimal level for each indenture. For example, the first indenture of function 3.0 would be 3.1, the second 3.1.1, the third 3.1.1.1, and so on. For expansion of a higher-level function within a particular level of indenture, a numerical sequence should be used to preserve the continuity of function. For example, if more than one function is required to amplify function 3.0 at the first level of indenture, the sequence should be 3.1, 3.2, 3.3, ..., 3.*n*. For expansion of function 3.3 at the second level, the numbering shall be 3.3.1, 3.3.2, ..., 3.3.*n*. Where several levels of indentures appear on a single functional diagram, the same pattern should be maintained. Whereas the basic ground rule should be to maintain a minimum level of indentures on any one particular flow, it may become necessary to include several levels to preserve the continuity of functions and to minimize the number of flows required to functionally depict the system.
3. *Functional reference.* Each functional diagram should contain a reference to its next higher functional diagram through the use of a reference block. For example, function 4.3 should be shown as a reference block in the case where the functions 4.3.1, 4.3.2, ..., 4.3.*n*, and so on, are being used to expand function 4.3. Reference blocks shall also be used to indicate interfacing functions as appropriate.
4. *Flow connection.* Lines connecting functions should indicate only the functional flow and should not represent either a lapse in time or any intermediate activity. Vertical and horizontal lines between blocks should indicate that all functions so interrelated must be performed in either a parallel or series sequence. Diagonal lines may be used to indicate alternative sequences (cases where alternative paths lead to the next function in the sequence).
5. *Flow directions.* Functional diagrams should be laid out so that the functional flow is generally from left to right and the reverse flow, in the case of a feedback

functional loop, from right to left. Primary input lines should enter the function block from the left side; the primary output, or *GO* line, should exit from the right; and the *NO-GO* line should exit from the bottom of the box.

6. *Summing Gates.* A circle should be used to depict a summing gate. As in the case of functional blocks, lines should enter and/or exit the summing gate as appropriate. The summing gate is used to indicate the convergence, or divergence, or parallel or alternative functional paths and is annotated with the term AND or OR. The term AND is used to indicate that parallel functions leading into the gate must be accomplished before proceeding to the next function, or that paths emerging from the AND gate must be accomplished after the preceding functions. The term OR is used to indicate that any of several alternative paths (alternative functions) converge to, or diverge from, the OR gate. The OR gate thus indicates that alternative paths may lead or follow a particular function.
7. *Go and no-go paths.* The symbols G and $\overline{G}$ are used to indicate go and no-go paths, respectively. The symbols are entered adjacent to the lines leaving a particular function to indicate alternative functional paths.
8. *Numbering procedure for changes to functional diagrams.* Additions of functions to existing data should be accomplished by locating the new function in its correct position without regard to sequence of numbering. The new function should be numbered using the first unused number at the level of indenture appropriate for the new function.

The functions identified should not be limited strictly to those necessary for the operation of the system, but must consider the possible effects of maintenance on system design. In most instances, maintenance functional flows will evolve directly from operational flows.

Operational Functions

Operational functions, in this instance, constitute those that describe the activities that must be accomplished in order to fulfill the mission requirements. These may include both (1) those activities that involve the design, development, production, and distribution of a system for use; and (2) those activities that are related directly to the completion of a customer mission scenario. In the second category, these may include a description of the various modes of system operation and utilization. For instance, typical gross operating functions may entail (1) "prepare aircraft for flight," (2) "transport material from the factory to the warehouse," (3) "initiate communications between the producer and the user," (4) "produce '*x*' quantity of units in a seven-day timeframe," and (5) "process '*a*' data to eight company distribution outlets, in '*b*' time, with '*c*' accuracy, and in '*d*' format." System functions necessary to successfully complete the identified modes of operation are then described.

Figure 3.21 illustrates a simplified operational flow diagram. Note that the words in each block are "action-oriented" and the block numbering allows for the downward-upward traceability of resource requirements. The functions are broken down to the

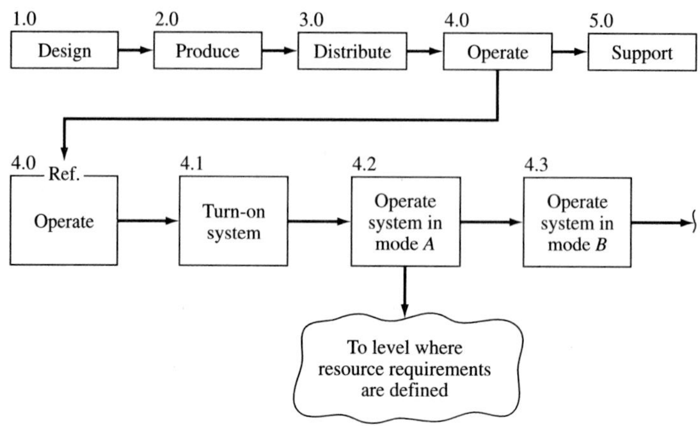

Figure 3.21 Functional block diagram (partial).

depth necessary to describe the resources that will be required to accomplish the function; that is, equipment, software, people, facilities, and so on.

Maintenance and Support Functions

Once operational functions are described, the system development process leads to the identification of *maintenance and support* functions. For instance, there are specific performance expectations or measures associated with each block in an operational functional-flow diagram. A check of the applicable functional requirement will indicate either a "go" or a "no-go" decision. A "go" decision leads to a check of the next operational function. A "no-go" indication (constituting a symptom of failure) provides a starting point for the development of a detailed maintenance functional flow diagram. The transition from an operational function to a maintenance function is illustrated in Figure 3.22. Figure 3.23 presents a more in-depth functional flow diagram.

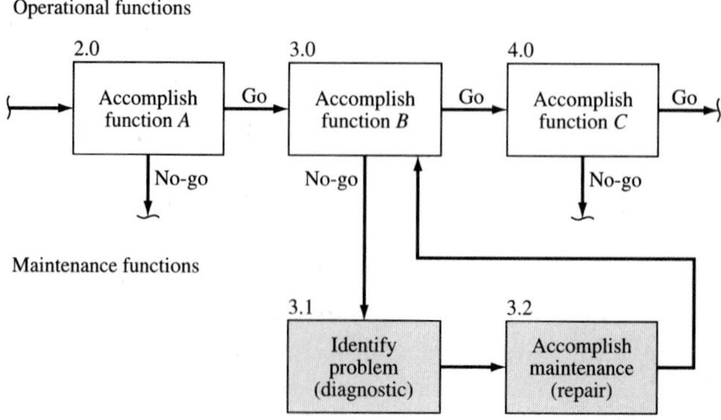

Figure 3.22 Transition from operational functions to maintenance functions.

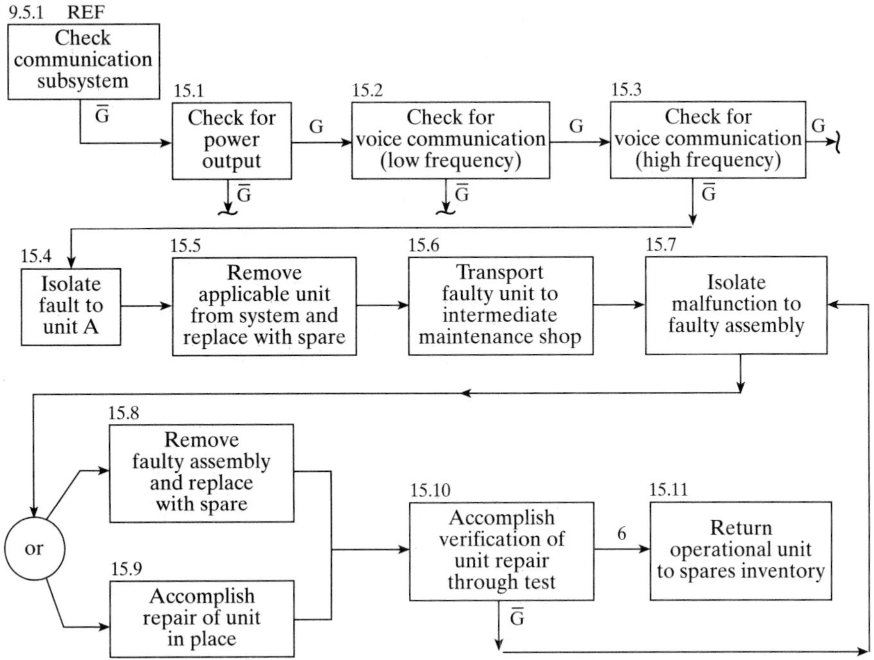

Figure 3.23 Maintenance functional flow diagram (reference: Figure 3.20).

Application of Functional Flow Diagrams

The functional analysis provides an initial description of the system and, as such, its applications are extensive. Figure 3.24 illustrates a top-level operational functional flow diagram for a manufacturing system, commencing with the identification of need (block 1.0) and extending through system retirement (block 7.0). In areas where a greater degree of definition is desirable, the applicable block(s) may be broken down to a second level, third level, and so on, in order to gain the appropriate level of visibility necessary for the determination of resource requirements. In this instance, the ultimate manufacturing "operating" functions have been identified in the breakout of block 5.1 (refer to page 135).

For each of the blocks in Figure 3.24, the analyst should be able to specify *input* requirements, expected *outputs*, external *controls* or *constraints*, and the *mechanisms* (or resources) necessary to accomplish the specific function in question. In the process of identifying the appropriate resource requirements, there may be a number of different alternative approaches that should be considered. Trade-off studies are conducted, alternatives are evaluated against criteria developed from the established technical performance measures (i.e., the TPMs derived in Section 3.5), and a preferred approach is recommended. It is at this point when one begins to identify the requirements for hardware, software, people, facilities, data, or combinations thereof. Figure 3.25 reflects the process that should be applied to each of the blocks in Figure 3.24.

In the evaluation of each functional requirement, the alternatives may include the selection of "commercial-off-the shelf (COTS)" items readily available from a number

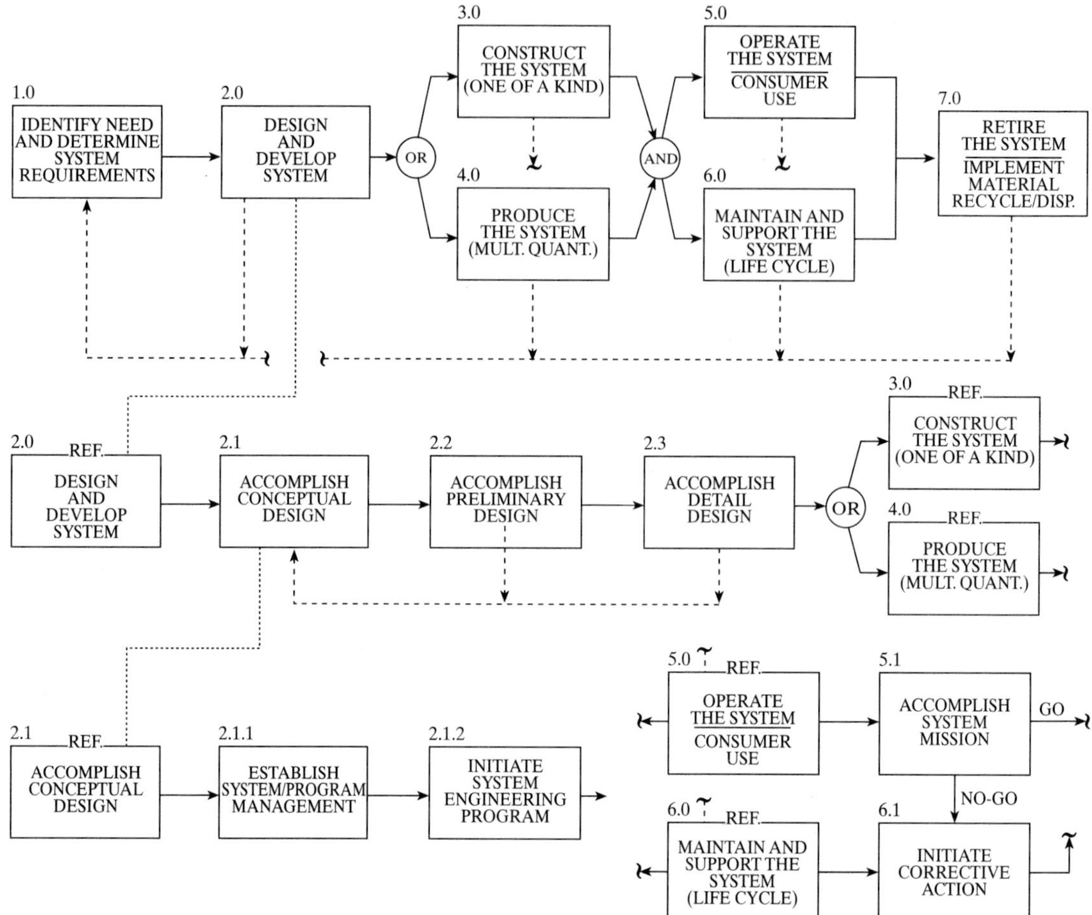

Figure 3.24 Functional flow diagram for a manufacturing system.

of different sources of supply, COTS items that may require some degree of modification, or "developmental" items that are unique to a particular application and where some new design is required. Past experience has indicated that extensive time and cost savings can be realized through the selection of readily available COTS equipment, reusable software, the utilization of existing facilities, and so on. Figure 3.26 illustrates the various options in this area.[9]

[9] In recent years, the Department of Defense (DOD) has placed considerable emphasis on the preferred use of COTS items versus the pursuit of new design and development efforts. The objectives are to reduce the time involved in the development and acquisition of new systems, improve system supportability/serviceability through the utilization of standard components that can be easily backed up with readily available spares and repair parts, and to reduce costs from a life-cycle perspective. A good reference is the technical report, *American Defense Preparedness Association Commercial-off-the-Shelf (COTS) Supportability Study*, ADPA, Arlington, Va., 1994.

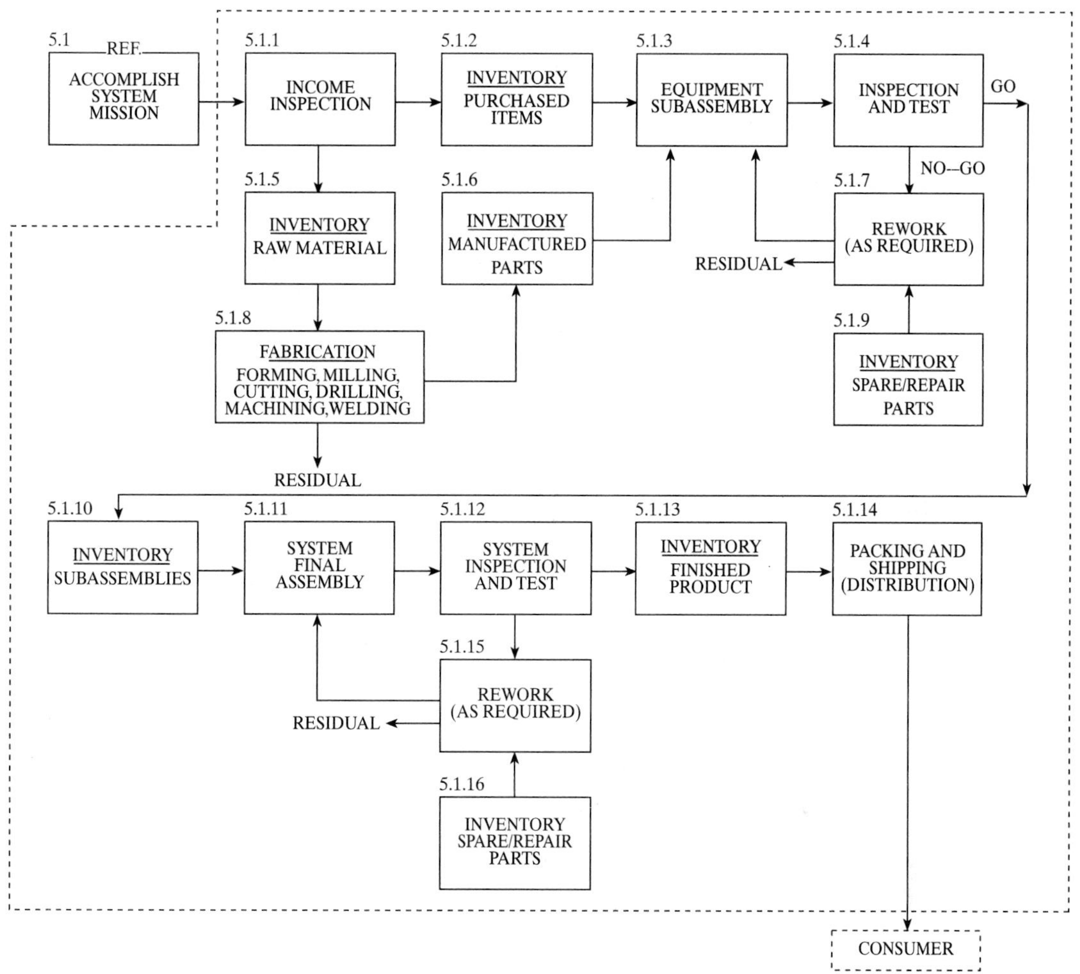

Figure 3.24 *(continued)* Functional flow diagram for a manufacturing system.

Figure 3.26 shows that it is essential that a *good* definition of the inputs and outputs (and the applicable metrics) be established if one is to fully understand not only the *interfaces* between the different functions identified in Figure 3.24, but the precise requirements in the process of resource identification. If these input-output requirements are not well defined, the decision-making process as to a preferred approach becomes difficult, thus, leading to the possibility of initiating a new costly design and development effort when, in actuality, an existing off-the-shelf item could fulfill the need.

The functional analysis can facilitate an "open-architecture" approach to system design. A good comprehensive functional description of the system, with the interfaces well defined (both qualitatively and quantitatively), can lead to a structure that will not only allow for the rapid identification of resource requirements, but for the possible

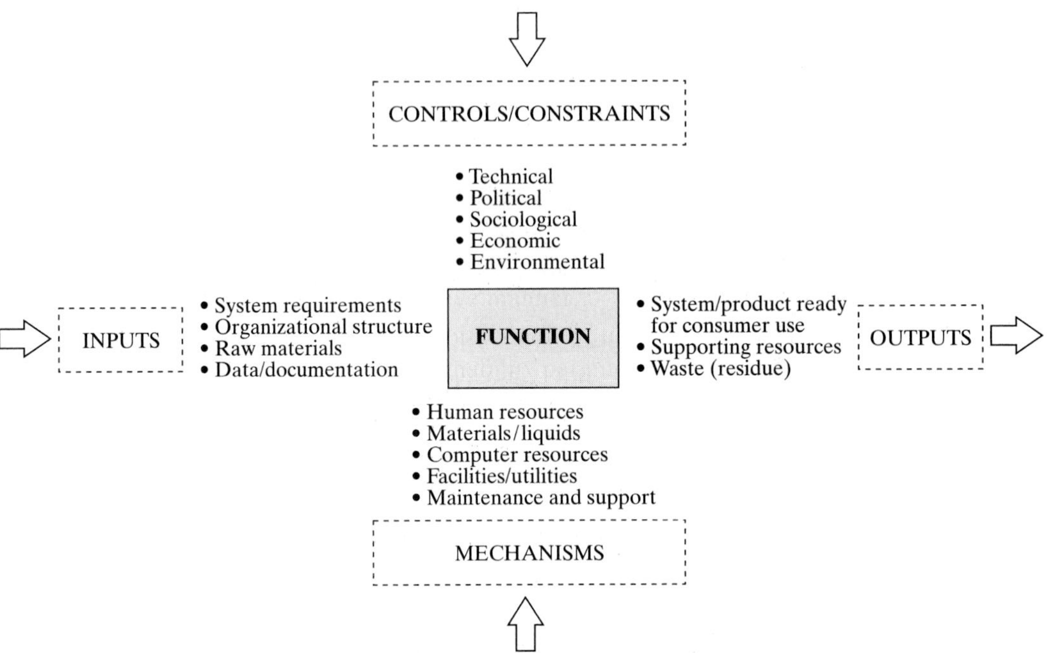

Figure 3.25 Identification of resource requirements (i.e. "mechanisms").

incorporation of new technologies later on. The objective is to design and develop a system that can be easily modified, through the insertion of new technologies, without causing a "costly" redesign of all of the elements of the system in the process (refer to Figure 1.8).

In many current situations, the requirements in design are changing from a detailed "design to the component level" to the design of systems using a "black-box-integration" approach. Given the need to reduce acquisition times, while responding to an ever-changing set of requirements on a continuing basis and with many more suppliers involved, the system *architecture* must allow for the "ease of upgrade and/or modification." In other words, the system *structure* must be such as to facilitate design on an *evolutionary* basis, and with minimum cost. This can be enhanced through a good and comprehensive functional definition of the system in the early conceptual design phase of the life cycle.

Figure 3.27 illustrates a manufacturing system where there are many suppliers (from various locations throughout the world) who produce components for a consumer product that must be effectively integrated and tested. There are fabrication functions, subassembly functions, assembly functions, and test functions. Where, in many instances in the past, the manufacturing activity involved a bottom-up "build" approach, the challenges today relate to the *integration* of the various components into the end product. Without a good early definition and specification of the functional interfaces, the final integration and test activity may result in a costly "trial-by-error" process. In the figure, the example reflects a factory where the sub-processes were

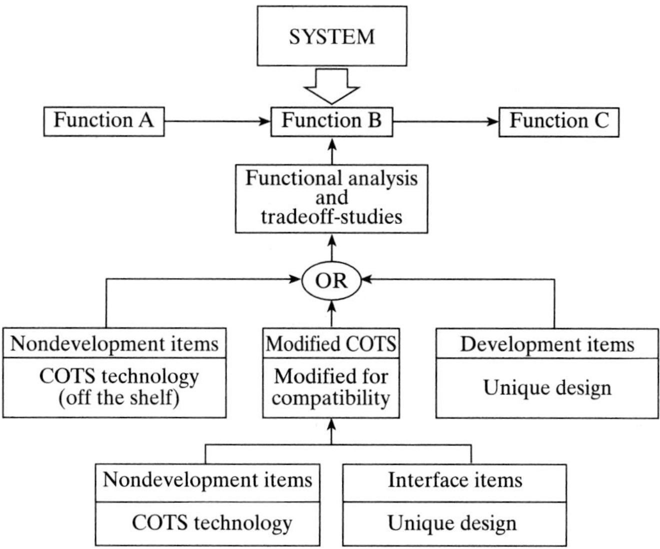

Figure 3.26 Identification of commercial off-the-shelf (COTS) items from functional analysis.

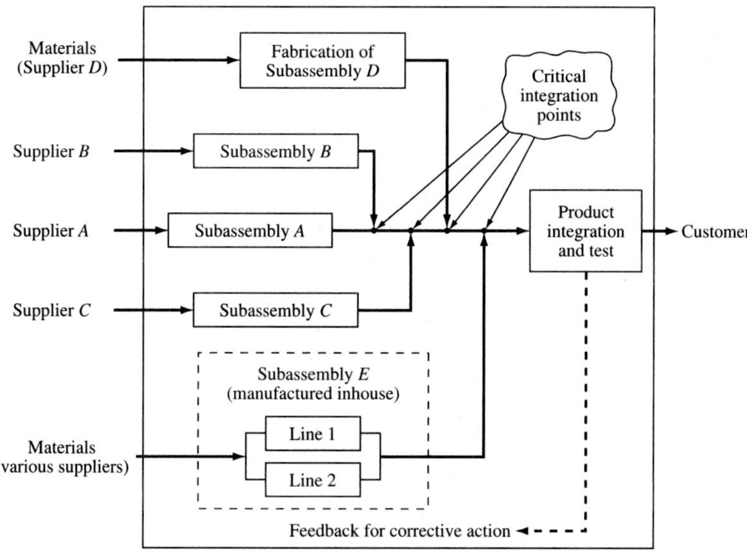

Figure 3.27 Manufacturing system (critical integration points).

being accomplished effectively and efficiently; however, there were considerable problems associated with the "integration" activities; that is, the four critical integration points. The functional interfaces were not well defined from the beginning, causing a great deal of modification and rework downstream.

In completing a functional analysis, care should be taken to ensure that the required resources are properly identified for each function. A timeline analysis may be accomplished to determine whether the functions are to be accomplished in series or in parallel. It may be possible to share resources in some instances; that is, the same resources may be utilized to accomplish more than one function. The identified resources may be combined and integrated to the extent possible. Every effort should be made to avoid the specification of resources that are not necessary.

In summary, the functional analysis constitutes a critical step in the early system design and development effort, and it forms a baseline for many activities that are conducted subsequently. For instance, it serves as a basis in the development of the following:

1. Electrical and mechanical design for functional packaging, condition monitoring, and diagnostic provisions
2. Reliability models and block diagrams
3. Failure mode, effect, and criticality analysis (FMECA)
4. Fault-tree analysis (FTA)
5. Reliability-centered maintenance (RCM) analysis
6. System safety/hazard analysis
7. Maintainability analysis
8. Level-of-repair analysis
9. Maintenance task analysis (MTA)
10. Operator task analysis (OTA)
11. Operational sequence diagrams (OSDs)
12. Supportability analysis
13. Operating and maintenance procedures
14. Producibility and disposability analyses

In the past, the functional analysis has not always been completed in a timely manner, if completed at all. As a result, the various design disciplines assigned to a given program have had to generate their own analyses in order to comply with program requirements. In many instances, these efforts were accomplished independently, and many design decisions were made without the benefit of a common baseline to follow. This, of course, resulted in design discrepancies and costly modifications occurring later in the system life cycle.

The functional analysis provides an excellent and very necessary baseline, and all applicable design activities must "track" the same data source to meet the objectives for system engineering. For this reason, the functional analysis is considered as being a key activity in the system engineering process.

3.7 ALLOCATION OF REQUIREMENTS

Given a top-level definition of the system through the functional analysis, the next step is to break the system down into components by *partitioning*.[10] This involves a breakdown of the system into subsystems and lower-level elements. The challenge is to identify and group closely related functions into packages, employing a common resource (e.g., equipment, software) to accomplish multiple functions to the extent possible. Although it may be relatively easy to identify individual functional requirements and associated resources on an independent basis, this may turn out to be rather costly when it comes to system packaging, weight, size, and so on. The questions are: What hardware or software can be selected that will perform multiple functions? How can new functions be added without adding any new physical elements to the system structure?

The partitioning of the system into elements is evolutionary in nature. Common functions may be grouped or combined in such a way as to provide a system packaging scheme with the following objectives in mind:

1. System elements may be grouped by geographical location, a common environment, or by similar types of equipment.
2. Individual system "packages" should be as independent as possible with a minimum of "interaction effects" with other packages. A design objective is to be able to remove and replace a given package without having to remove and replace other packages in the process, or requiring an extensive amount of alignment and adjustment in the process.
3. In breaking down a system into subsystems, select a configuration where the communications between the subsystems is minimized. In other words, whereas the subsystem *internal* complexity may be high, the *external* complexity should be low. Breaking down the system into packages where there are high rates of information exchange between these packages should be avoided.

An overall design objective is to break down the system into elements such that only a very few critical events can influence or change the inner workings of the various packages that make up the system architecture.

As a result of partitioning, a system may be broken down into components such as shown in Figure 3.28. The process utilized in the identification and packaging of elements is illustrated in Figure 3.29. System functions are identified, broken down into subfunctions, and grouped into three equipment units; that is, Unit A, Unit B, and Unit C. The design should be such that any one of the three units can be removed and replaced without impacting the other units. In other words, there should be a minimum of interaction effects between the three units.

Given the identification of system elements, the next step is to *allocate* or *apportion* the requirements specified for the system down to the level desired to provide a meaningful *input* to design. This involves a top-down distribution of the quantitative

[10] The concepts of system *architecture* and *partitioning* are presented in Rechtin, E., *Systems Architecting: Creating and Building Complex Systems*, Prentice Hall, Inc., Upper Saddle River, N.J., 1991.

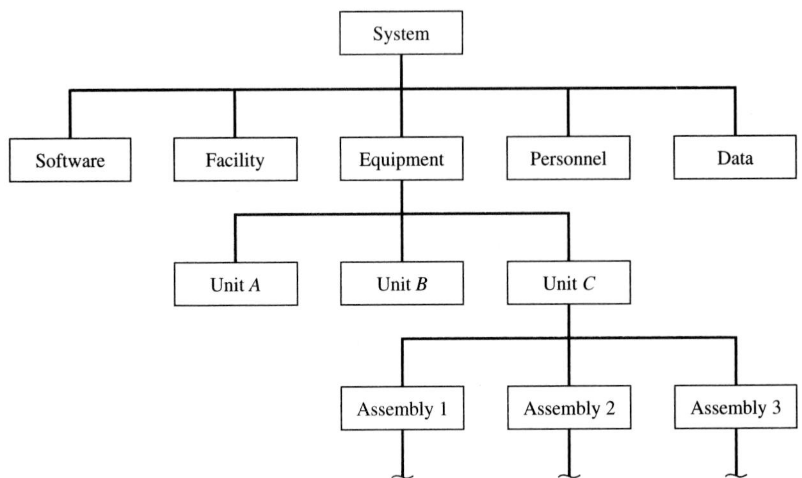

Figure 3.28 Hierarchy of systems components.

and qualitative criteria developed in Section 3.5. From the prioritized technical performance measures (TPMs) in Figure 3.16, what should be specified for the unit level in Figure 3.29 in order to meet the system-level requirements?

Reliability Allocation

After an acceptable reliability factor (e.g., probability of survival) or failure rate has been established for the system, it must be allocated among the various subsystems, units, assemblies, and so on. The allocation commences with the generation of a reliability block diagram. The block diagram is a further extension of the functional flow diagrams presented in Section 3.6. The intent is to develop a reasonable approximation of those elements or items that must function for successful operation of the system. To the extent practicable, the diagram should be structured so that each block represents a functional entity that is relatively independent of neighboring blocks.

In the development of a block diagram, items that are predominantly electronic in function are noted as electronic elements and items that are basically mechanical are identified accordingly. Item redundancy contemplated at this stage of system planning should be illustrated along with any planned provisions for alternative operating mode capability.

Figure 3.30 is a simplified reliability block diagram and the progressive expansion of such from the system level down as design detail becomes known (refer to Figure 2.10). Generally, levels I and II are available through conceptual design activity, while levels III and on are defined in preliminary system design.

Referring to the figure, the reliability requirement for the system (e.g., λ, R, MTBF) is specified for the entire network identified in Level I, and an individual requirement is specified for each individual block in the network. For instance, the reliability of block 4, Function *X*, may be expressed as a probability of survival of 0.95 for a 4-hour period of operation at Level I. Similar requirements are specified for blocks

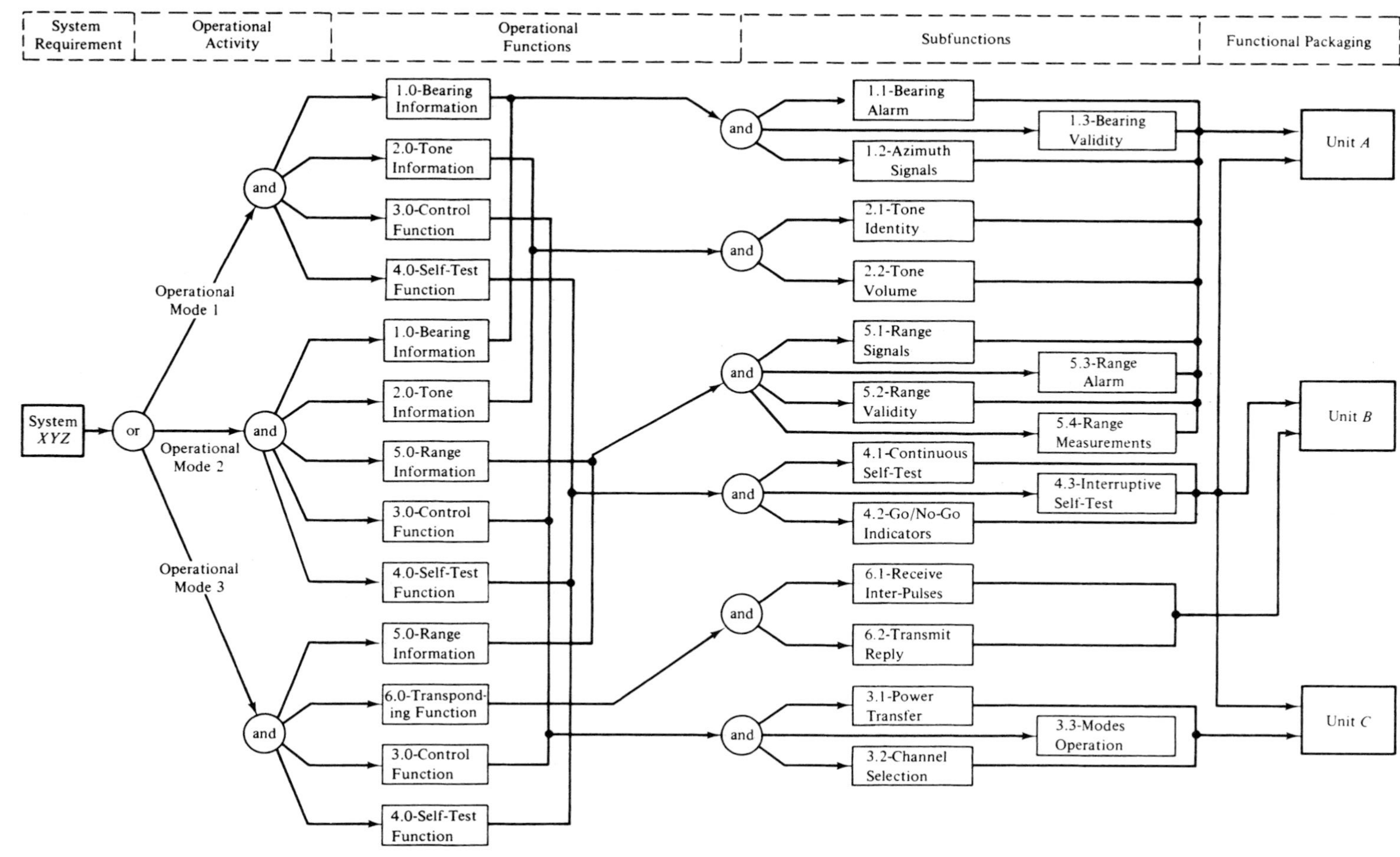

Figure 3.29 System *XYZ* operational functional-flow diagram.

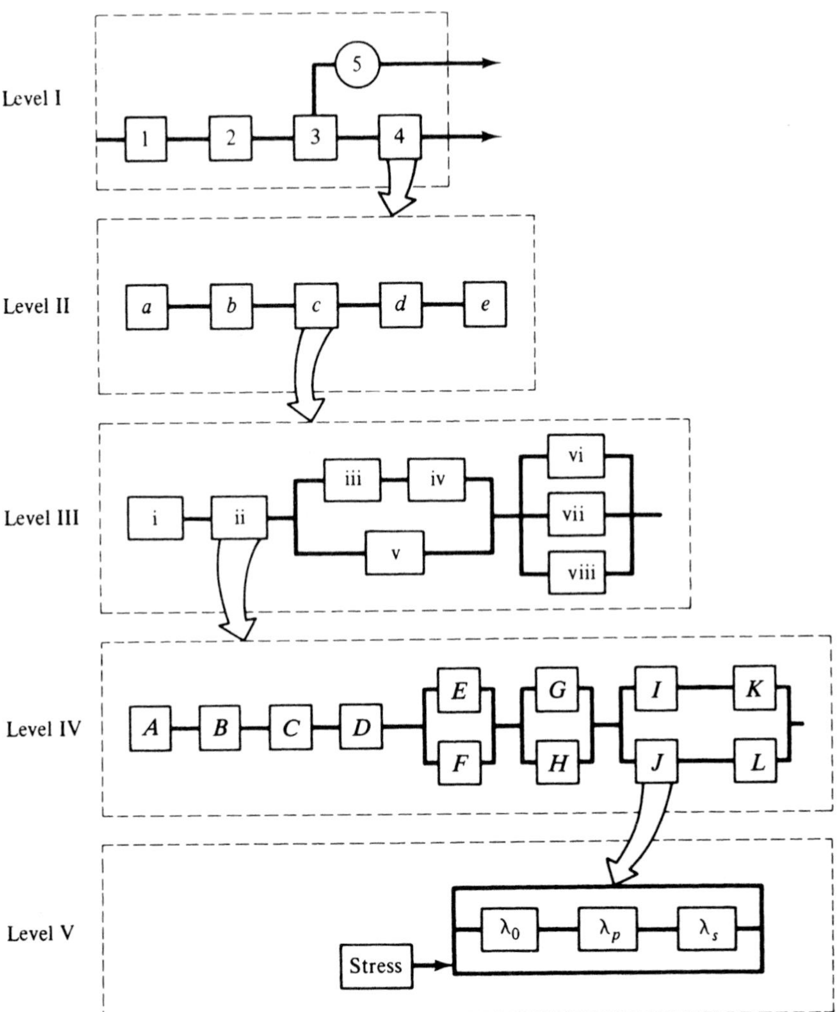

Figure 3.30 Reliability block diagram approach (*N* levels). (NAVAIR 00-65-502/NAVORD OD 41146, *Reliability Engineering Handbook*, U.S. Navy, Washington, D.C., revised March 1968.)

1, 2, 3, and 5. These, when combined, will indicate the system reliability, which in turn is evaluated in terms of the overall requirement.

Block diagrams are generated to cover each of the major functions identified in Figure 3.29. Success criteria (go/no-go parameters) are established and failure rates (λ) are estimated for each block, the combining of which provides an overall factor for a series of blocks constituting a function or subfunction. Depending on the function, one or more of these diagrams can be related to a physical entity such as Unit *A* in Figure 3.29, or an assembly of Unit *A*. The failure-rate information provided at the unit/assembly level represents a reliability design goal. This, in turn, represents the anticipated frequency of corrective maintenance that is employed in the determination of logistic resource requirements.

The approach used in determining failure rates may vary depending on the maturity of system definition. Failure rates may be derived from direct field or test experience covering like items, reliability prediction reports covering items that are similar in nature, or engineering estimates based on judgment. In some instances, weighting factors are used to compensate for system complexity and environmental stresses.

When accomplishing reliability allocation, the following steps are considered appropriate:

1. Evaluate the system functional flow diagram(s) and identify areas where design is known and failure-rate information is available or can be readily assessed. Assign the appropriate factors and determine their contribution to the top-level system reliability requirement. The difference constitutes the portion of the reliability requirement which can be allocated to the other areas.
2. Identify the areas which are new and where design information is not available. Assign complexity weighting factors to each functional block. Complexity factors may be based on an estimate of the number and relationship of parts, the equipment duty cycle, the mode of operation and criticality of the path, whether an item will be subjected to temperature extremes, and so on. That portion of the system reliability requirement which is not already allocated to the areas of known design is allocated using the assigned weighting factors.

The end result should constitute a series of lower-level values which can be combined to represent the system reliability requirement initially specified (i.e., MTBF of 450 hours for System XYZ). The combining of these values is facilitated through the application of a reliability mathematical model.

A reliability mathematical model is developed to relate the individual "block" reliability to the reliabilities of its constituent blocks or elements. The procedure simply consists of determining a mathematical expression that represents the probability of survival for a small portion of the proposed configuration. Multiple applications of this process will eventually reduce the original complex system to an equivalent serial configuration. It is then possible to represent the system with a single probability statement. Some of the mathematical relationships used in this instance are described in Section 2.1.

When allocating a system level requirement (i.e., MTBF of 450 hours), one should construct a simplified functional breakdown as illustrated in Figure 3.31. The diagram must reflect series-parallel relationships.

Initially, failure rates are identified for items of known design and are deducted from the overall system requirement. A complexity factor may be established for each of the remaining items.[11] The complexity factors are used to apportion failure rates to the next lowest level and on down. As a check, failure rates at the assembly level are totaled to obtain the unit failure rate, and unit failure rates support the system failure

[11] In Figure 3.31, complexity factors are assumed for all items.

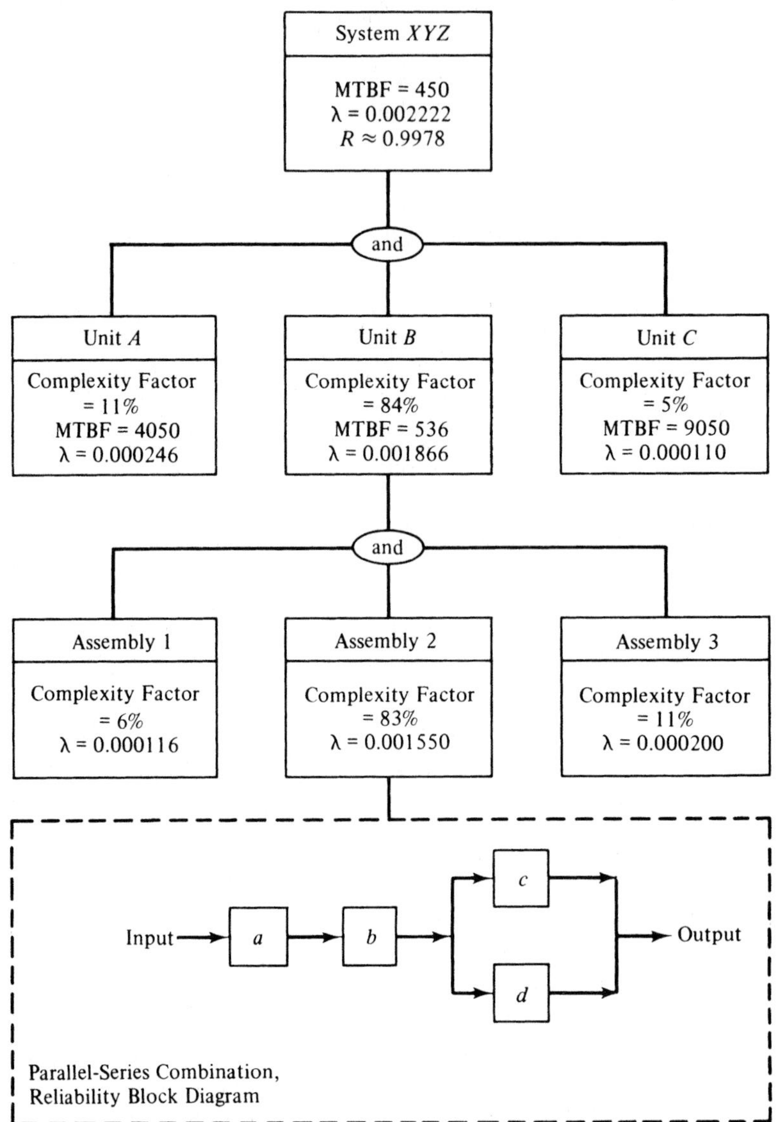

Figure 3.31 System *XYZ* functional breakdown.

rate (note that Units *A* to *C* represent a series operation). The MTBF is usually assumed to be the reciprocal of the failure rate, and the reliability (R) of the system or unit may be determined from Equation (2.5) or the nomograph in Figure 2.3.

Referring to Figure 3.31, a reliability block diagram showing the functional relationship of four elements (a, b, c, and d) illustrates the "makeup" of assembly 2. The mathematical expression for the four elements is

$$R_{\text{assy 2}} = (R_a)(R_b)[R_c + R_d - (R_c)(R_d)] \tag{3.3}$$

Using the general relationships in Equation (2.5), reliability factors and failure rates can be determined. In this instance, the total failure rate for Assembly 2 should not exceed 0.00155.

Review of the equipment breakdown configuration illustrated in Figure 3.31 indicates a top-level system requirement supported by factors established at the unit level and on down. Unless otherwise specified, the requirements at the unit level may be altered or traded off as long as the combined unit-level requirements support the system objective. In other words, the failure rate of Unit *B* may be higher and the failure rate of Unit *A* may be lower than indicated without affecting the requirement of 0.002222, and so on! The techniques of trading off different parameters to meet an overall requirement are discussed further in Chapter 4.

The reliability factors established for the various items identified in Figure 3.31 serve as design criteria. For instance, the engineer responsible for Unit *B* shall design Unit *B* such that the failure rate (λ) shall not exceed 0.001866. As design progresses, reliability predictions are accomplished and the predicted value is compared against the requirement of 0.001866. If the predicted value does not meet the requirement (i.e., higher failure rate or lower MTBF), then the design configuration must be reviewed for reliability improvement and design changes are implemented as appropriate.

The allocated factors not only provide the designer with a reliability criterion, but serve as an indicator of the frequency of corrective maintenance due to anticipated equipment failure. Assume that System *XYZ* is projected into an operational posture similar to the one described for the vehicular communication system in Section 3.3, and that the total system operating time per year is 60,000 hours for a 10-year period. The expected quantity of system maintenance actions due to failure is

$$\text{Expected maintenance actions} = \frac{\text{total operating hours per year}}{\text{MTBF}} \tag{3.4}$$

or

$$\text{Expected maintenance actions} = \frac{60{,}000}{450} = 133/\text{year}$$

Assuming that each of the units is operating or energized on a full-time basis when the system is operational, then the quantity of expected maintenance actions for each unit can be determined from Equation (3.4). The results are 15 maintenance actions per year for Unit *A*, 112 maintenance actions per year for Unit *B*, and 6 maintenance actions per year for Unit *C*. The frequency of maintenance is a necessary input in the determination of logistic resource requirements (material and cost) for a system or equipment.

Maintainability Allocation

The process of translating system maintainability requirements (e.g., MTBM, $\overline{\text{M}}\text{ct}$, $\overline{\text{M}}\text{pt}$, MLH/OH) into lower-level design criteria is accomplished through maintainability allocation. The allocation requires the development of a simplified functional breakdown as illustrated in Figure 3.31. The functional breakdown is based on the maintenance concept, functional analysis data, and a description of the basic repair policy—whether a system is to be repaired through the replacement of a unit, an assembly, or a part.

For the purpose of illustration, it is assumed that system XYZ must be designed to meet an inherent availability requirement of 0.9989, a MTBF of 450, and a MLH/OH (for corrective maintenance) of 0.2 and a need exists to allocate $\overline{\text{M}}$ct and MLH/OH to the assembly level.[12] The $\overline{\text{M}}$ct equation is

$$\overline{\text{M}}\text{ct} = \frac{\text{MTBF}(1 - A_i)}{A_i} \tag{3.5}$$

or

$$\overline{\text{M}}\text{ct} = \frac{450(1 - 0.9989)}{0.9989} = 0.5$$

Thus, the system's $\overline{\text{M}}$ct requirement is 0.5 hour, and this requirement must be allocated to Units A to C, and the assemblies within each unit. The allocation process is facilitated through the use of a format similar to that illustrated in Table 3.1.

TABLE 3.1 System XYZ Allocation

1 Item	2 Quantity of Items per System (Q)	3 Failure Rate (λ) × 1000 hr	4 Contribution of Total Failures $C_f = (Q)(\lambda)$	5 Percent Contribution $C_p = C_f/\Sigma C_f \times 100$	6 Estimated Corrective Maint. Time $\overline{\text{M}}\text{ct}_i$ (hr)	7 Contribution of Total Corrective Maint. Time $C_t = (C_f)(\overline{\text{M}}\text{ct})$
1. Unit A	1	0.246	0.246	11%	0.9	0.221
2. Unit B	1	1.866	1.866	84%	0.4	0.746
3. Unit C	1	0.110	0.110	5%	1.0	0.110
Total			$\Sigma C_f = 2.222$	100%		$\Sigma C_t = 1.077$

$$\overline{\text{M}}\text{ct for System } XYZ = \frac{\Sigma C_t}{\Sigma C_f} = \frac{1.077}{2.222} = 0.485 \text{ hour (requirement: 0.5 hour)}$$

Referring to Table 3.1, each item type and the quantity (Q) of items per system are indicated. Allocated reliability factors are specified in column 3, and the degree to which the failure rate of each unit contributes to the overall failure rate (represented by C_f) is entered in column 4. The average corrective maintenance time for each unit is estimated and entered in column 6.[13] These times are ultimately based on the inherent characteristics of equipment design, which are not known at this point in the system life cycle. Thus, corrective maintenance times are initially derived using a complexity factor which is indicated by the failure rate. As a goal, the item that con-

[12] MTBM and $\overline{\text{M}}$pt may be allocated on a comparable basis as MTBF and $\overline{\text{M}}$ct, respectively.

[13] It should be noted that the Mct_i values are not allocated using the weighting factors as was the case for reliability, cost, and MLH/OH. Rather, the Mct_i values are *estimated* and then a calculation is made to see if the system $\overline{\text{M}}$ct meets the specification requirement.

tributes the highest percentage to the anticipated total failures (Unit *B* in this instance) should require a low $\overline{\text{Mct}}$, and those with low contributions may require a higher $\overline{\text{Mct}}$. On certain occasions, however, the design costs associated with obtaining a low $\overline{\text{Mct}}$ for a complex item may lead to a modified approach which is feasible as long as the end result ($\overline{\text{Mct}}$ at the system level) falls within the quantitative requirement.[14]

The estimated value of C_t for each unit is entered in column 7, and the sum of the contributions for all units can be used to determine the overall system's $\overline{\text{Mct}}$ as

$$\overline{\text{Mct}} = \frac{\Sigma C_t}{\Sigma C_f} = \frac{1.077}{2.222} = 0.485 \tag{3.6}$$

In Table 3.1, the calculated $\overline{\text{Mct}}$ for the system is within the requirement of 0.5 hour. The $\overline{\text{Mct}}$ values for the units provide corrective maintenance downtime criteria for design, and the values are included in equipment design specifications.

Once allocation is accomplished at the unit level, the resultant $\overline{\text{Mct}}$ values can be allocated to the next lower equipment indenture item. For instance, the 0.4-hour $\overline{\text{Mct}}$ value for Unit *B* can be allocated to Assemblies 1, 2, and 3, and the procedure for allocation is the same as employed in Equation (3.6). An example of allocated values for the assemblies of Unit *B* is included in Table 3.2.

TABLE 3.2 Unit *B* Allocation

1	2	3	4	5	6	7
Assembly 1	1	0.116	0.116	6%	0.5	0.058
Assembly 2	1	1.550	1.550	83%	0.4	0.620
Assembly 3	1	0.200	0.200	11%	0.3	0.060
Total			1.866	100%		0.738

$$\overline{\text{Mct}} \text{ for Unit B} = \frac{\Sigma C_t}{\Sigma C_f} = \frac{0.738}{1.866} = 0.395 \text{ hour (Requirement: 0.4 hour)}$$

The $\overline{\text{Mct}}$ value covers the aspect of *elapsed* or *clock* time for restoration actions. Sometimes this factor, when combined with a reliability requirement, is sufficient to establish the necessary maintainability characteristics in design. On other occasions, specifying $\overline{\text{Mct}}$ by itself is not adequate since there may be a number of design approaches which will meet the $\overline{\text{Mct}}$ requirement but not necessarily in a cost-effective manner. Meeting a $\overline{\text{Mct}}$ requirement may result in an increase in the skill levels of personnel accomplishing maintenance actions, increasing the quantity of personnel for given maintenance functions, and/or incorporating automation for manual operations. In each instance, there are costs involved; thus, one may wish to specify additional constraints such as the skill level of personnel at each maintenance level and the maintenance labor hours per operating hour (MLH/OH) for significant equipment items. In other words, a requirement may dictate that an item be designed such that it can be

[14] Note that, in any event, the maintainability parameters are dependent upon the reliability parameters. Also it will frequently occur that reliability allocations are incompatible with maintainability allocations (or vice versa). Hence, a close feedback relationship between these activities is mandatory.

repaired within a specified elapsed time with a given quantity of personnel possessing skills of a certain level. This will influence design in terms of accessibility, packaging schemes, handling requirements, diagnostic provisions, and so on, and is perhaps more meaningful in terms of designing for logistic support.

The factor MLH/OH is a function of task complexity and the frequency of maintenance. The system-level requirement is allocated on the basis of system operating hours, the anticipated quantity of maintenance actions, and an estimate of the number of manhours per maintenance action. Experience data are used where possible.

Following the completion of quantitative allocations for each indenture level of equipment, all values are included in the functional breakdown illustrated in Figure 3.32. The illustration provides an overview of major system design requirements.

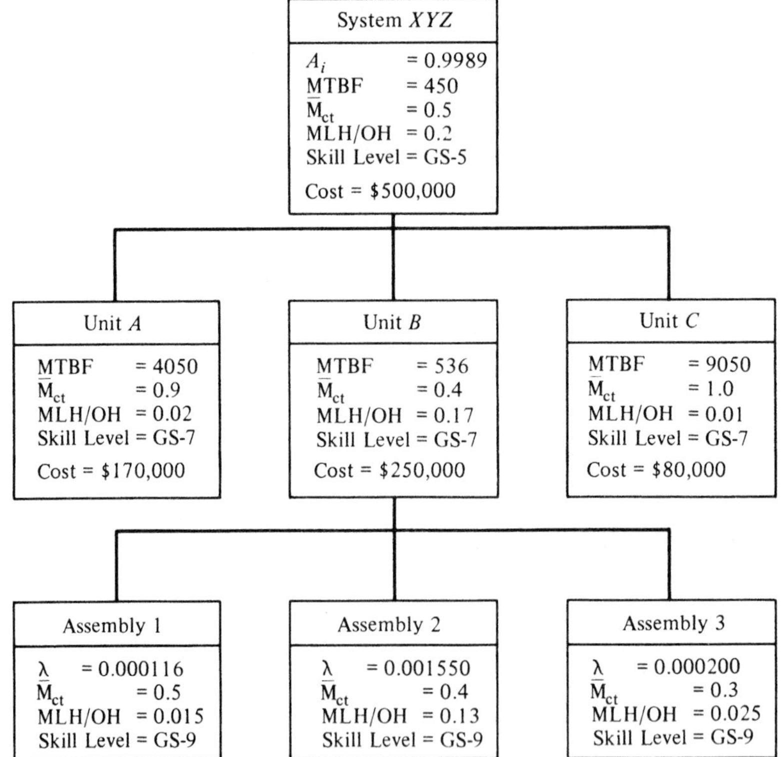

Figure 3.32 System *XYZ* functional requirements.

Allocation of Logistics Factors

In addition to reliability and maintainability parameters and their impact on design, one must also consider other factors that are critical to successful system operation. These factors, some of which were introduced in Chapter 2, deal with supply support, test and support equipment, personnel and maintenance organization, facilities, and transportation.

As mentioned earlier, *all* elements of the system must be addressed to include the various activities depicted in Figure 2.33. Thus, it may be necessary to establish some additional design criteria covering the various elements of logistic support. A few examples are noted below.

1. Test equipment utilization in the intermediate maintenance shop shall be at least 80%, and test equipment reliability shall be at least 90%.
2. Self-test thoroughness for the system (using the built-in test capability) shall be 95% or better.
3. Personnel skill levels at the organizational level of maintenance shall be equivalent to grade *x* or below.
4. The maintenance facility at the intermediate level shall be designed for a minimum of 75% utilization.
5. The transportation time between the location where organizational maintenance is accomplished and the intermediate maintenance shop shall not exceed 4 hours.
6. The turnaround time in the intermediate maintenance shop shall be 2 days (or less), and 10 days (or less) in the depot maintenance facility.
7. The probability of spares availability at the organizational level of maintenance shall be at least 95%.

In essence, in defining system operational requirements and the maintenance concept (described in Sections 3.3 and 3.4), system supportability factors must be determined along with performance parameters. These factors, established at the system level, may be allocated to the extent necessary to influence design activities.

Allocation of Economic Factors

Using an approach similar to that described in previous sections, cost factors may be allocated as appropriate to system needs. If the ultimate product is to be cost-effective, it may be desirable to assign cost targets for various system elements. For instance, an objective might be to design System *XYZ* such that the unit system cost is no more than $500,000, based on a production quantity of 300 and an operational life of 10 years. Unit cost constitutes total life-cycle cost (to include research and development, production, operation and maintenance, and retirement cost) divided by the quantity of systems. This cost factor, specified at the top level, can be apportioned to lower equipment indenture levels as cost targets for design. Cost targets combined with reliability (or equivalent) requirements may create a boundary situation for design as illustrated in Figure 2.30. In other words, one can *design to a cost.*

3.8 SYNTHESIS, ANALYSIS, AND DESIGN OPTIMIZATION

Synthesis refers to the combining and structuring of components in such a way as to represent a feasible system configuration. The requirements for a system have been established, some preliminary trade-off studies have been completed, and a baseline configuration needs to be developed to demonstrate the concepts discussed earlier. Synthesis is *design*. Initially, synthesis is employed to develop preliminary concepts and to establish basic relationships among the various components of the system. Later, when sufficient functional definition and decomposition have occurred, synthesis is used to further define the "hows," in response to the "what" requirements. Synthesis involves the selection of a configuration that could be representative of the form that the system will ultimately take, although a final configuration is certainly not to be assumed at this point.[15]

The synthesis process usually leads to the definition of several possible alternative design approaches, which will be the subject of further analysis, evaluation, refinement, and optimization. As these alternatives are initially structured, it is essential that the appropriate technical performance parameters be properly aligned to applicable components of the system. For instance, technical performance parameters may include factors such as weight, size, speed, capacity, accuracy, volume, range, processing time, along with reliability and maintainability and other factors as presented in Figure 3.33. These parameters, or measures, must be prioritized and aligned to the appropriate elements of the system (e.g., an equipment, unit or assembly, item of software).

When defining the initial requirements for the system, technical performance measures (TPMs) are established based on their relationship and criticality to the accomplishment of the mission; that is, the impact that a given factor has on cost-effectiveness, or system effectiveness, or performance. These applicable TPMs are prioritized, and their relative relationships are presented in the form of design considerations presented in a hierarchical tree, as illustrated in Figure 3.33. The ranking of TPMs (and supporting design considerations), which will be built into the program management and review structure, will likely vary from one system to the next. A top-level measure for one system may be "reliability," whereas "availability" may be of greater importance in another example. In any event, the appropriate measures need to be established, prioritized, and included in the specifications. As the design process evolves, these measures will be used for the purposes of analysis and evaluation.

Given a number of alternatives, the evaluation procedure progresses through the general steps illustrated in Figure 3.34 and described as follows:

1. *Definition of analysis goals.* An initial step requires the clarification of objectives, the identification of possible alternative solutions to the problem at hand, and a description of the analysis approach to be employed. Relative to alternatives, all possible candidates must be initially considered; however, the more alternatives considered, the more complex the analysis process becomes. Thus, it is desirable to first list *all* possible candidates to ensure against inadvertent omissions, and then eliminate those candidates that are clearly unattractive, leaving only a few for evaluation. Those few candidates are then evaluated with the intent of selecting a preferred approach.
2. *Selection and weighting of evaluation parameters.* The criteria used in the evaluation process may vary considerably depending on the stated problem, the sys-

[15] Synthesis is covered further in Lacy, J., *Systems Engineering Management*, McGraw-Hill, New York, N.Y., 1992.

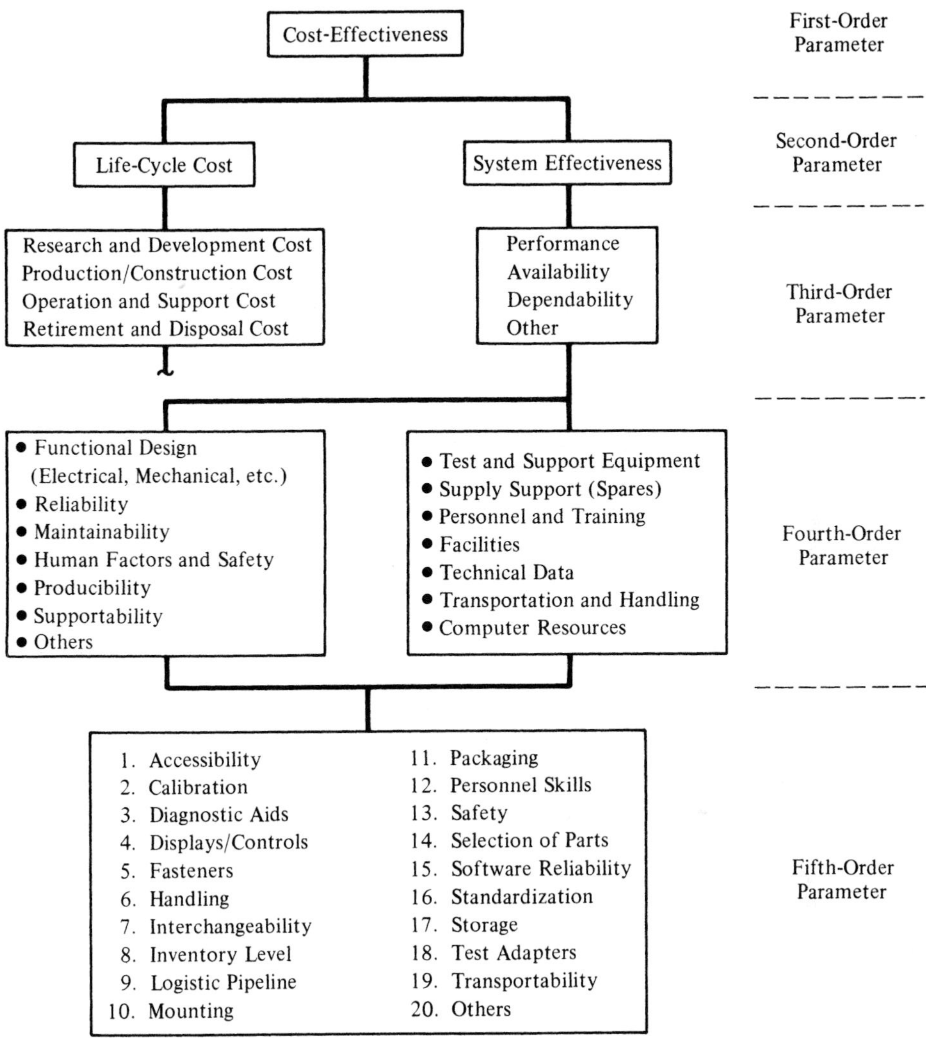

Figure 3.33 Order of evaluation parameters (reference Figure 1.14).

tem being evaluated, and the depth and complexity of the analysis. From Figure 3.33, parameters of primary significance include cost, effectiveness, performance, availability, dependability and so on. At the detail level, the order of parameters will be different. In any event, parameters are selected, weighted in terms of priority of importance, and are tailored to the system in a meaningful manner.

3. *Identification of data needs.* When evaluating a particular system configuration, it is necessary to consider operational requirements, the maintenance concept, major design features, production and/or construction plans, and anticipated system utilization and product support requirements. Fulfilling this need requires a variety of data, the scope of which depends on the type of evaluation being performed and the program phase during which the evaluation is accomplished. In

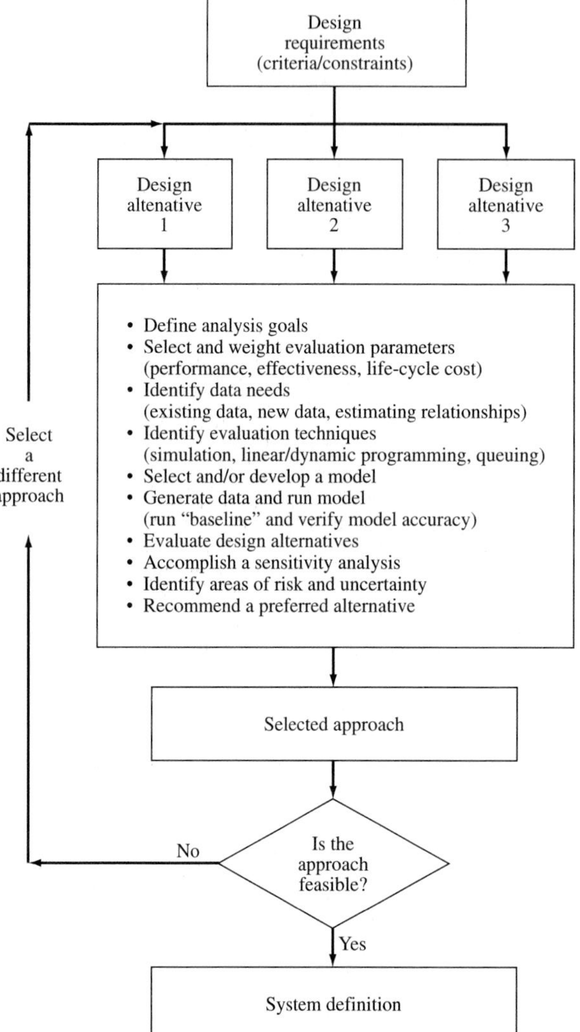

Figure 3.34 Evaluation of alternatives.

the early stages of system development, available data are limited; thus, the analyst must depend on the use of various estimating relationships, projections based on past experience covering similar system configurations, and intuition. As the system development progresses, improved data are available (through analyses and predictions) and are used as an input to the evaluation effort. At this point, it is important to initially determine the specific needs for data (i.e., type, quantity, and the time of need), and to identify possible data sources. The nature and validity of the data input for a given analysis could have a significant impact on the risks associated with the decisions made based on the analysis results. Thus, one needs to accurately assess the situation as early as practicable.

4. *Identification of evaluation techniques.* Given a specific problem, it is necessary to determine the analytical approach to be used and the techniques that can be

applied to facilitate the problem-solving process. Techniques may include the use of Monte Carlo simulation in the prediction of random events downstream in the life cycle, the use of linear programming in determining transportation resource requirements, the use of queuing theory in determining production and maintenance shop requirements, the use of networking in establishing distribution needs, the use of accounting methods for life-cycle costing purposes, and so on. Assessing the problem itself and identifying the available tools that can possibly be used in attacking the problem are necessary prerequisites to the selection of a model.

5. *Selection or the development of a model.* The next step requires the combining of various analytical techniques into the form of a model, or a series of models, as illustrated in Figure 3.35.[16] A model, as a tool used in problem solving, aids in the development of a simplified representation of the real world as it applies to the problem being solved. The model should (a) represent the dynamics of the system configuration being evaluated; (b) highlight those factors that are most relevant to the problem at hand; (c) be comprehensive by including *all* relevant factors and be reliable in terms of repeatability of results; (d) be simple enough in structure so as to enable its timely implementation in problem solving; (e) be designed such that the analyst can evaluate the applicable system configuration as an entity, analyze different components of the system on an individual basis, and then integrate the results into the whole; and (f) be designed to incorporate provisions for easy modification and/or expansion to permit the evaluation of additional factors as required. An important objective is to select or develop a tool that will help to evaluate the *overall* system configuration, as well as the *interrelations* of its various components. Models (and their applications) are discussed further in Chapter 4.
6. *Generation data and model application.* With the identification of analytical techniques and the model selection task accomplished, the next step is to "verify" or "test" the model to ensure that it is responsive to the analysis requirement. Does the model meet the stated objectives? Is it sensitive to the major parameters of the system configuration(s) being evaluated? Evaluation of the model can be accomplished through the selection of a *known* system entity, and the subsequent comparison of analysis results with historical experience. Input parameters may be varied to ensure that the model design characteristics are sensitive to these variations and will ultimately reflect an accurate output as a result.
7. *Evaluation of design alternatives.* Each of the alternatives being considered is then evaluated using the techniques and the model selected. The required data are collected from various sources such as existing data banks, predictions based on current design data, and/or gross projections using analogous and parametric estimating relationships. The required data, which may be taken from a wide

[16] There are many types of models including physical models, symbolic models, abstract models, mathematical models, and so on. Model, as defined here, refers primarily to a mathematical (or analytical) model. The development and application of various analytical methods are covered futher in most texts on operations research. Two excellent references are (1) Hillier, F. S., and G. J. Lieberman, *Introduction to Operations Research*, 6th Ed., McGraw-Hill, New York, N.Y. 1995; and (2) Taha, H. A., *Operations Research: An Introduction*, 5th Ed., Macmillan, New York, N.Y. 1992.

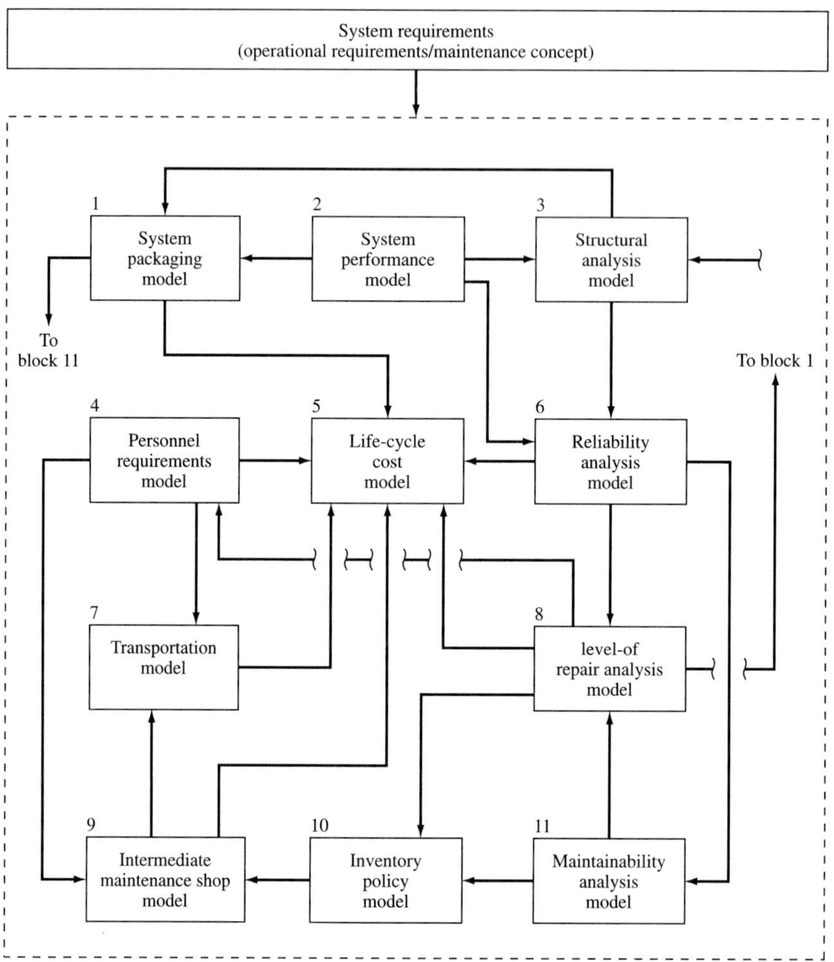

Figure 3.35 Example application of models.

variety of sources, must be applied in a consistent manner. The results are then evaluated in terms of the initially specified requirements for the system. Feasible alternatives are considered further. Figure 3.36 illustrates some considerations where possible feasible solutions fall within the desired shaded areas.

8. *Accomplishment of a sensitivity analysis.* In the performance of an analysis, there may be a few key system parameters about which the analyst is uncertain because of inadequate data input, poor prediction procedures, "pushing" the state of the art, and so on. There are several questions that need to be addressed: How sensitive are the results of the analysis to possible variations of these uncertain input parameters? To what extent can certain input parameters be varied before the choice of alternatives shifts away from the initially selected approach? From experience, there are certain key input parameters in a life-cycle cost analysis, such as the reliability MTBF and the maintainability $\overline{\text{M}}\text{ct}$, that are considered to

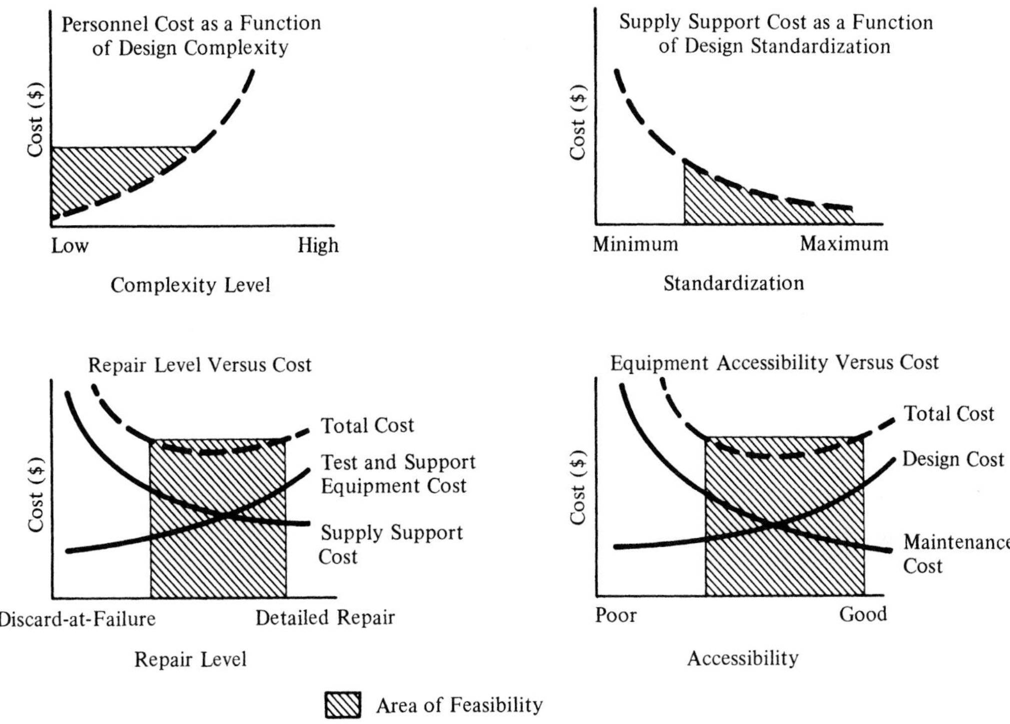

Figure 3.36 System design versus cost considerations (sample trends).

be critical in determining system maintenance and support costs. With good historical field data being very limited, there is a great deal of dependence placed on current prediction and estimating methods. Thus, with the objective of minimizing the risks associated with making an incorrect decision, the analyst may wish to vary the input MTBF and $\overline{M}ct$ factors over a designated range of values (or a distribution) to see what impact this variation has on the output results. Does a relatively *small* variation of an input factor have a *large* impact on the results of the analysis? If so, then these parameters might be classified as being critical TPMs in the overall design review and evaluation process, monitored closely as design progresses, and an additional effort might be generated to modify the design for improvement and to improve the reliability and maintainability prediction methods. In essence, a sensitivity analysis is directed toward determining the relationships between design decisions and output results.

9. *Identification of risk and uncertainty.* The process of design evaluation leads to decisions having a significant impact on the future. The selection of evaluation criteria, the weighting of factors, the selection of the life cycle period, the use of certain data sources and prediction methods, and the assumptions made in interpreting analysis results will obviously influence these decisions. Inherent within this process are the aspects of "risk" and "uncertainty" because the future is, of course, unknown. Although these terms are often used jointly, *risk* actually

implies the availability of discrete data in the form of a probability distribution around a certain parameter. *Uncertainty* implies a situation that may be probabilistic in nature, but one that is not supported by discrete data. Certain factors may be measurable in terms of risk, or may be stated under conditions of uncertainty. The aspects of risk and uncertainty, as they apply to the system design and development process, must be integrated into a program risk management plan.

10. *Recommendation of preferred approach.* The final step in the evaluation process is the recommendation of a preferred alternative. The results of the analysis should be fully documented and made available to all applicable project design personnel. A statement of assumptions, a description of the evaluation procedure that was followed, a description of the various alternatives that were considered, and an identification of potential areas of risk and uncertainty should be included in this analysis report.

In Figure 3.1, requirements for the system are established in conceptual design, functional analysis and allocation are accomplished either late in conceptual design or at the start of preliminary system design, and detail design is accomplished on a progressive basis from thereon. Throughout this overall series of steps, there is an ongoing effort involving synthesis, analysis, and design optimization. In the early stages of design, trade-off studies may entail the evaluation of alternative operational profiles, alternative technology applications, distribution schemes, or maintenance concepts. During early preliminary design, alternative methods for accomplishing a given function or alternative equipment packaging schemes may be the focus of analysis. In detail design, the problems will be at a lower level in the overall hierarchical structure of the system.

In any event, the process illustrated in Figure 3.34 is applicable throughout the system design and development effort. The only difference lies in the depth of analysis, the type of data required, and the model used in accomplishing the analysis. For instance, one can perform a life-cycle cost analysis early in conceptual design or late in detail design. The process is the same in either case; however, the depth of analysis and the data requirements are different. The synthesis, analysis, and optimization process must be tailored to the problem at hand. Too little effort will result in greater risks associated with decision making in design, and too much analysis effort will be expensive.

3.9 SUMMARY

The *system engineering* process, discussed throughout this chapter, is presented in the form of an "overview," with the objective of establishing a frame of reference for the material in subsequent chapters. Chapter 1 introduces the concepts and principles of logistics, the design for supportability, and includes some related terms and definitions. A "systems approach" is assumed. Chapter 2 provides coverage of some of the measures (i.e., "metrics") most closely associated with system maintenance and logistic support, supplementing the concepts introduced earlier. This chapter discusses the framework for logistics engineering and the *design for supportability*, providing illustrations on how logistics should be considered in system design and development. Although logistics activities are applicable in all phases of the system life cycle, the emphasis herein is on the early stages when the decision-making processes have the greatest impact on logistics overall and on life-cycle cost. The system engineering process, to be implemented successfully, requires that system maintenance and logistic support be considered from the beginning and included as an inherent factor within the various activities discussed in Sections 3.1 to 3.8. In other words, *logistics engineering* is a major ingredient within the systems engineering process.

As one proceeds through the the subsequent sections of this text, the concepts introduced here are extended and amplified to a much greater degree. The systems engineering process, with logistics as a major ingredient, continues through the activities of supportability analysis, design integration and review, system test and evaluation, system utilization and support, assessment, and the incorporation of modifications as required for continuous product/process improvement. The information presented in this chapter provides the framework for much of what is discussed from here on.

QUESTIONS AND PROBLEMS

1. Identify the basic steps in the system engineering process, and describe some of the *inputs* and *outputs* associated with each step.
2. Why is the initial *requirements definition process* so critical in system design and development?
3. What is the purpose of *feasibility analysis?* What information is desired from such an analysis?
4. Why is the definition of system *operational requirements* important? What is included?
5. Why is the definition of the system *maintenance concept* important? What is included? How does the maintenance concept relate to the *maintenance plan*?
6. Identify a specific *problem* that you wish to solve through the design and development of a new system. For your system:
   (a) Describe the current deficiency and identify the *need* for the new system.
   (b) Accomplish an abbreviated feasibility analysis and discuss the various alternative technical approaches that you may wish to consider in designing the new system.
   (c) Define the basic operational requirements for the new system.
   (d) Define the maintenance concept for the new system.
   (e) Identify the critical technical performance measures (TPMs) based on the defined operational requirements and maintenance concept. Describe the process leading from the identification of TPMs to the identification of specific design characteristics.

   How does logistics (and the design for supportability) tie in with each of items (a) through (e)?
7. What is meant by *quality function deployment (QFD)*? What are some of the benefits that can be derived from its application?
8. What is meant by *functional analysis?* When should it be accomplished (if at all)? Why is it important in system engineering? What purpose(s) does it serve?
9. For the system selected in Problem 6, accomplish a functional analysis. Construct a functional block diagram showing three levels of *operational* functions. From one of the blocks

in the operational functional flow diagram, show two levels of *maintenance* functions. Show how the operational functions and the maintenance functions relate.

10. Select one block from the operational functional diagram and one block from the maintenance functional diagram in Problem 9, and show inputs-outputs and how specific resource requirements are identified (e.g., hardware, software, people, facilities, data, etc.).
11. Why is the identification and description of system-level *functional interfaces* important? What can happen if these interfaces are not well defined?
12. Identify some applications of functional analysis.
13. Describe what is meant by *allocation* or *partitioning?* What is its purpose? To what depth should it be applied? How can the process of allocation influence system design?
14. For the system configuration described in Problem 6, show a breakdown of the system into its subsystems and lower-level elements. Accomplish an allocation of requirements specified through the TPMs as at the system level to the next level below. Show how supportability factors are allocated.
15. What are the basic steps involved in *system analysis?* Construct a basic flow diagram illustrating the process, showing the steps, and including feedback provisions.
16. Describe what is meant by *synthesis*. How do the functions of *synthesis*, *analysis*, and *evaluation* relate to each other?
17. What is a *model?* Identify some of the basic characteristics of a model. List some of the benefits associated with the usage of mathematical models in system analysis. What are some of the problems/concerns?
18. What is meant by *sensitivity analysis?* What are some of the objectives of performing a sensitivity analysis? Benefits?
19. In your opinion, what are some of the major problems in implementing the process described in Figure 3.34? Identify at least three areas of concern.
20. How does the maintenance concept affect system/equipment design?
21. When evaluating alternative repair policies, what measures would you use? Why?
22. Reliability and maintainability quantitative factors defined as part of the maintenance concept serve what purpose?
23. What factors would be considered in determining which maintenance functions should be accomplished at the organization level, at the intermediate level, and at the depot level?
24. Referring to Figure 3.15, what is the impact of the logistics pipeline times and facility turnaround times on total logistic support? What impact does the $\overline{M}ct$ value of 15 minutes for the airborne system have on the maintenance concept?
25. Referring to Figure 3.15, what factors must be considered in determining spare/repair part requirements at the intermediate electronics shop?
26. Referring to Figure 3.15, if the shop TAT is increased to 10 hours, how does this increase affect logistic support?
27. When developing the maintenance concept, all applicable levels of maintenance must be considered on an integrated basis. Why?
28. Select a system of your choice and assign top-level requirements. Accomplish a reliability allocation to the second indenture level. Accomplish a maintainability allocation to the same level. Allocate supportability factors as appropriate.

29. What is meant by "design criteria"? How are criteria developed? How are criteria applied to the design process?
30. From the allocations in Problem 28, develop design criteria for the system.
31. Referring to Figure 3.32, System *XYZ* has the following requirements: MTBF = 650, $\overline{\text{M}}\text{ct}$ = 0.6, MMH/OH = 0.7, and unit cost = \$100,000. Allocate these requirements to Units *A*, *B*, *C*, and to Assemblies 1, 2, and 3 of Unit *B*.
32. In Figure 3.37, allocate the quantitative factors to the unit level as indicated.

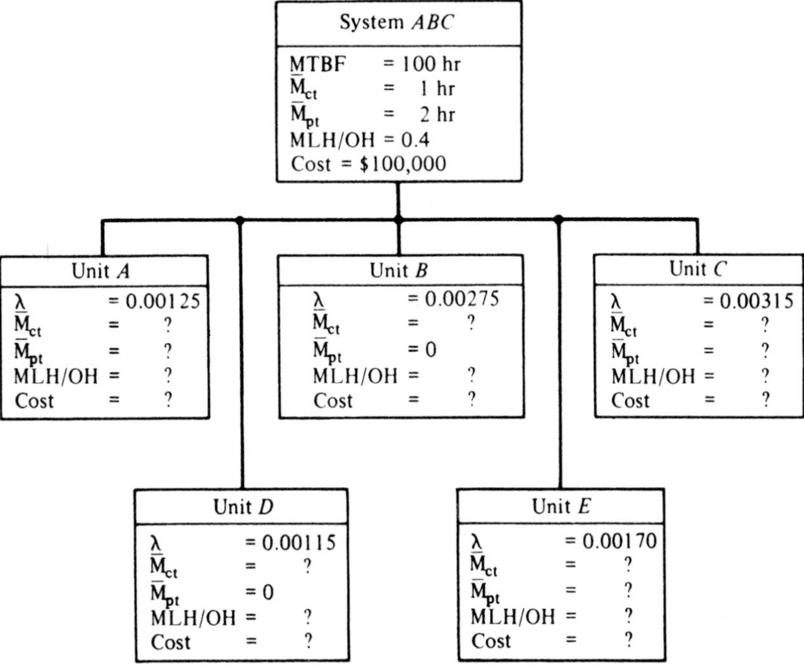

Figure 3.37 Allocation of requirements.

CHAPTER 4

SUPPORTABILITY ANALYSIS

The *supportability analysis (SA)* constitutes an iterative and continuous analytical process, which is inherent within the systems engineering process described in Chapter 3, that includes the integration and application of various techniques and methods to ensure that supportability requirements are considered in both the development of new systems and in the reengineering of existing systems currently in operational use. It involves the utilization of different analytical methods to solve a wide variety of problems of varying magnitudes, and its depth of application must be appropriately "tailored" to the specific program need. The supportability analysis is not a unique separate activity addressed as an independent entity, but represents a vehicle directed toward the design, development, and assessement of both the prime mission-related elements of the system and the maintenance and support infrastructure (illustrated in Figures 1.2 and 1.3). The emphasis is on maximizing system effectiveness (to include supportability) while minimizing life-cycle cost.[1]

More specifically, the supportability analysis addresses itself to a variety of objectives. Basically, the SA is accomplished to

1. Aid in the initial establishment of supportability requirements during conceptual design. In the selection of alternative technological approaches, accomplished as

[1] For defense systems, refer to MIL-HDBK-502, "DOD Handbook-Acquisition Logistics," Department of Defense, Washington, DC, 30 May 1997. The principles and concepts associated with *supportability analysis*, presented in this chapter, are not new and have been previously covered under such titles as the *logistics support analysis (LSA), maintenance engineering analysis (MEA), maintenance level analysis (MLA), maintenance task analysis (MTA), maintenance engineering analysis record (MEAR), maintenance analysis data system (MADS)*, and so on. While the nomenclature may change from time to time, the objectives and the processes utilized in each instance are basically the same.

part of a feasibility analysis (refer to Section 3.2), supportability-related factors must be addressed in the decision-making process. Alternatives must be evaluated from a total *life-cycle* perspective.

2. Aid in the early establishment of supportability "design-to" criteria through the definition of system operational requirements and the maintenance and support concept, the identification and prioritization of technical performance measures (TPMs), the accomplishment of functional analysis, and in the allocation of requirements (refer to Sections 3.3 through 3.7). System-level design-to factors are established that lead to the development of the maintenance concept, the identification of maintenance functions, and the subsequent allocation of requirements to the appropriate lower-level elements of the system. The applicable measures of logistics, as identified in Chapter 2, must be defined as design goals for the system.
3. Aid in the synthesis, analysis, and design optimization effort through the conductance of trade-off studies and the evaluation of various design alternatives (refer to Section 3.8). Specific applications may entail the evaluation of
   a. Alternative repair policies. Within the constraints dictated by the maintenance concept and the allocated criteria (defined in Sections 3.4 and 3.7, respectively), the SA supports design decisions pertaining to repair versus discard-at-failure and, when repaired, the maintenance level at which repair is to be accomplished.
   b. Specific reliability and maintainability characteristics in system/equipment design. This may cover alternative packaging schemes, test approaches and the extent to which diagnostic provisions are incorporated, built-in test versus external test levels, the incorporation of manual versus automatic maintenance provisions, transportation and handling characteristics, accessibility features, and so on.
   c. Two or more commercial off-the-shelf (COTS) items being considered for a single system application. Assuming that new design is not appropriate (due to cost, production lead time, etc.), which item, when installed as part of the system, will reflect the least overall logistics burden? The objective is to influence system design for supportability through the appropriate selection of system components and potential suppliers. The results may affect future procurement decisions as well as the type of contract structure imposed.
4. Aid in the evaluation of a *given* design configuration relative to the determination of specific maintenance and logistic support resource requirements (i.e., a new design configuration described through a data package or an existing system configuration already being utilized). Once that system requirements are known and design data are available, it is possible to determine the requirements for spares and repair parts, test and support equipment, personnel quantities and skill levels, training requirements, facilities, transportation and handling requirements, computer resources, and so on. Through the supportability analysis process, alternative candidates for system support may be evaluated, leading to the development of an optimum maintenance and support infrastructure. The results are generally integrated into a supportability analysis (SA) data package.

5. Aid in the measurement and evaluation, or *assessment*, of an operating system to determine its effectiveness and the degree of supportability being exihibited in the user's environment. Given a fully operational capability, can the system be effectively and efficiently supported through its planned life cycle? Field data are collected, evaluated, and the results are analyzed and compared against the initially specified requirements for the system. The objective is not only to verify that the system is successfully accomplishing its intended mission, but to identify the high-cost/high-risk areas and to incorporate modifications as part of a *continuous product/process improvement (CPPI)* effort.

When relating the supportability analysis (SA) effort to the overall system development process in Figure 1.7, each of the blocks to the right in the flowchart is affected. However, the major areas of emphasis are represented by blocks 4, 6, 8, 10, and 15. Referring to Figure 4.1, in the early phases of system development the emphasis is on *design influence,* whereas the objective later on is that of *resource identification*. Further, as the system is utilized, the emphasis is on *supportability assessment.*

4.1 THE ANALYSIS PROCESS

The supportability analysis (SA) process commences with problem identification and the needs analysis, and the subsequent definition of analysis goals, groundrules, constraints, and so on. As the name infers, a significant proportion of the effort constitutes that of *analysis*. A wide variety of problems needs to be addressed and requires the

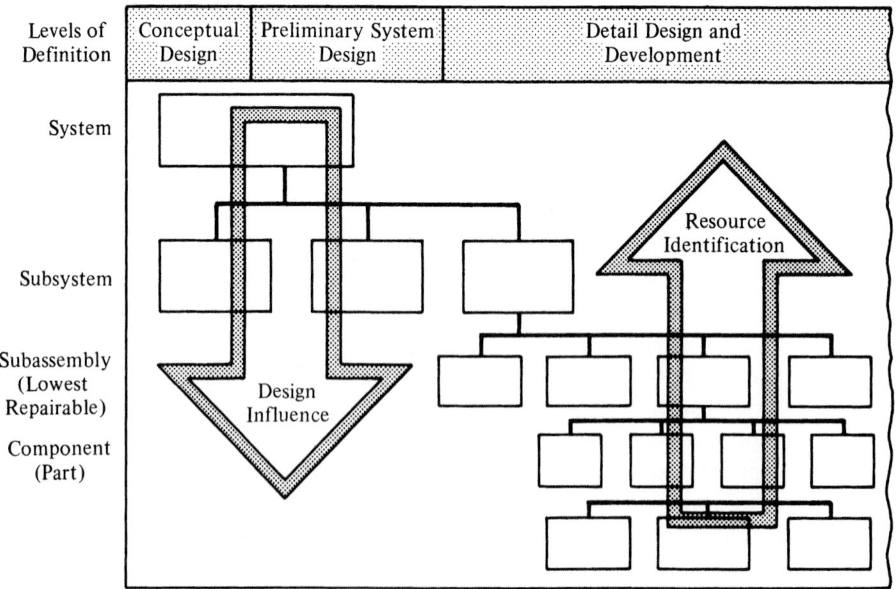

Figure 4.1 Supportability analysis thrusts.

application of analytical methods. For instance, the analyst may wish to use computerized simulation methods in the development of a maintenance concept and the identification of major support functions. A Monte Carlo analysis may be employed in determining maintenance frequency factors, or a queuing analysis may be used in developing test and support equipment requirements. Reliability and maintainability analyses and predictions may be accomplished through the use of statistical processes. Poisson distribution factors and inventory control theory may be applied in determining spares requirements. Linear and dynamic programming may be used in the determination of transportation and material handling requirements. Network theory may be applied in the development of a computerized information system capability. Accounting methods may be utilized in accomplishing life-cycle cost analyses. Many of these factors are discussed in Chapter 2.

In essence, the SA constitutes a continuous effort involving the use of different analytical methods in solving a wide variety of problems (i.e., those dealing with the initial definition of requirements, the evaluation of various system design alternatives, the determination of logistics resource requirements for a given configuration, and the ultimate assessment of a system while in use). During the early phases of conceptual design, the level of analysis is limited in depth (using rough estimates for input data) because the various elements of the system have not been adequately defined. As the design progresses, the analysis assumes a more refined approach because firm design data or engineering models are available. Detailed predictions, maintenance task analyses, and test results are often available and provide an input to the SA effort. In any event, the critical issue is to properly "tailor" the level of analysis to the problem(s) at hand. Quite open, there is a tendency to either do too little or too much. If the level of analysis is inadequate, then the risks associated with follow-on decisions that are based on the analysis results may be high. Conversely, going down to a depth beyond that which is required may result in a lot of meaningless data, provided too late, and that is costly. The key issue is to first gain a good understanding of the problem and then be able to define the analysis goals in a complete and specific manner.

The basic steps in a typical analysis are illustrated in Figure 4.2, although the extent of effort and depth of coverage will vary depending on the specific problem being addressed. Although these steps generally follow the systems analysis approach described in Section 3.8 (Figure 3.34), a few additional comments are warranted.

Selection of Evaluation Criteria

Referring to Figure 4.2, block 3, a critical step in the SA process is the establishment of the proper measures, or metrics, for the purposes of evaluation. These measures will, of course, vary in terms of magnitude and degrees of importance depending on the system and its intended mission. What may be an excellent high-level metric for the system in one case may not be as critical in another situation. The determining factors are the establishment and prioritization of the technical performance measures (TPMs), described in Section 3.5. Through application of such methods as QFD (or equivalent), specific customer expectations relative to the quantitative output requirements for the system are noted, and the results will dictate the critical measures, the characteristics

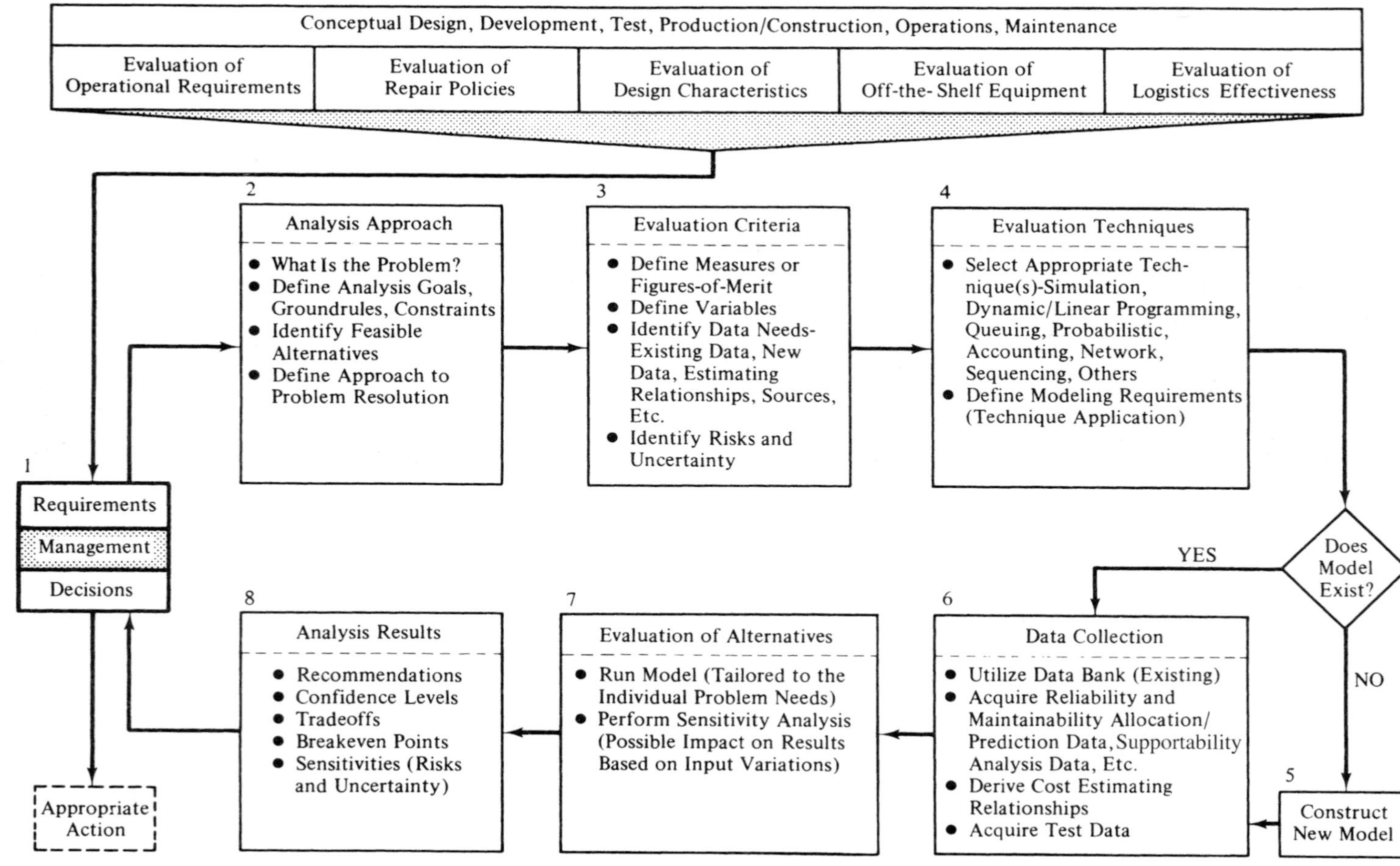

Figure 4.2 Supportability analysis approach.

that should be inherent within the design configuration itself, and the criteria against which the system should be evaluated.

The metrics associated with the maintenance and logistic support infrastructure (and its elements) are discussed in Chapter 2, and some of these factors are highlighted in Figure 4.3. The objective is to select those measures that are applicable for the system in question. Further, many of these measures are closely interrelated and some may be of a higher level than others. For example, in certain instances the measure of *availability (Ao)* may be a better measure for the system overall than *reliability (R)*. In any event, the appropriate measures must be initially established at the system level and then allocated to the subsystem level and below (refer to Section 3.7). Ultimately, there must be a top-down/bottom-up traceability of requirements as conveyed in Figure 3.18.

In the development of requirements, one often commences by expressing such in qualitative terms. Referring to the supportability application in Figure 4.4, the objective may be "to design and develop a system that can be supported effectively and efficiently." The question is: How does one respond to such a requirement and how does one measure the results for the purposes of validation?

In the absence of better guidance, the designer will need to interpret the specified requirements and make some assumptions relative to what is meant by "effectively" and "efficiently?" Although the objective is to design a system in response to customer requirements, it may not always happen unless there is a good communications link between the designer, logistics personnel, the customer, and others as necessary. Through a "team" effort, the development and use of an "objectives tree," as conveyed in Figure 4.4, can serve to help clarify requirements. Initially, it may be necessary to express design objectives in qualitative terms, showing their interrelationships in a top-down hierarchical manner. Subsequently, an attempt should be made to establish *quantitative* measures for each block in the figure, and to ensure that the appropriate "traceability" exists both downward and upward.

Referring to Figure 3.28 (Chapter 3), what measures should be applied and at what level in the overall hierarchical structure for the system? Given this, then the same question needs to be addressed for the various elements of the system to include the maintenance and support infrastructure illustrated in Figures 1.2 and 1.3. In other words, what measures, as described in Chapter 2 and identified in Figure 4.3, should be applied to the various blocks in Figure 4.4?

The application of such an approach should lead to a better definition of system requirements and greater visibility relative to selection of the appropriate evaluation criteria. Further, accomplishing this early in the design process, particularly with regard the maintenance and logistic support requirements, should result in a more cost-effective product output (refer to Figures 1.10 and 1.11).

Selection of a Model

Given the problem definition, the analysis objectives, and the selection of evaluation criteria, the analyst needs to identify the appropriate analytical techniques/methods that are available and can be used to facilitate the overall SA effort. Refernng to Figure 4.2 (block 4), there are a number of different tools that may be used in helping to

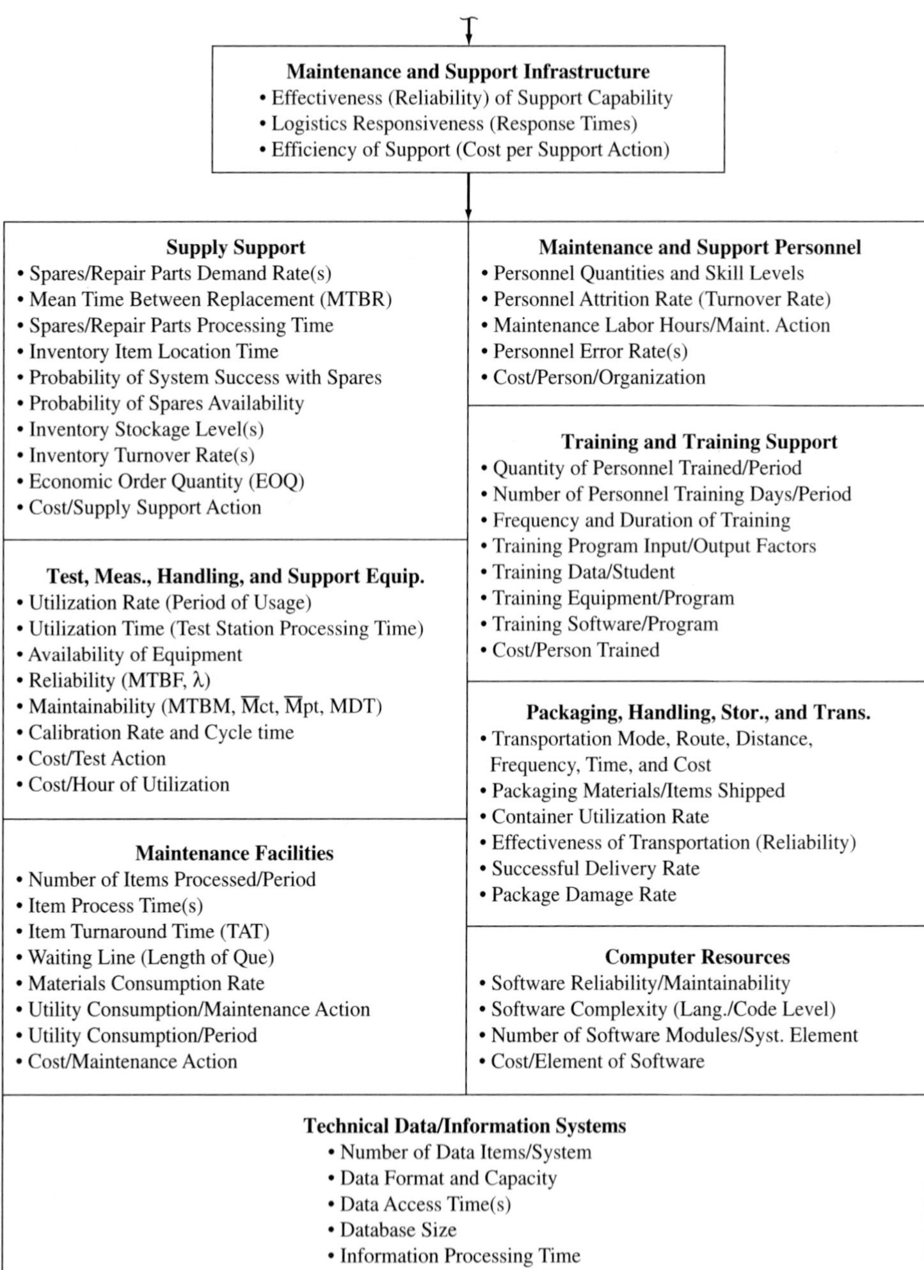

Figure 4.3 Selected technical performance measures for the support infrastructure.

solve specific problems of one type or another (i.e., the use of simulation methods, linear and dynamic programming, queuing theory, accounting methods, and related operations research tools). Knowledge of these methods and how they can be applied to

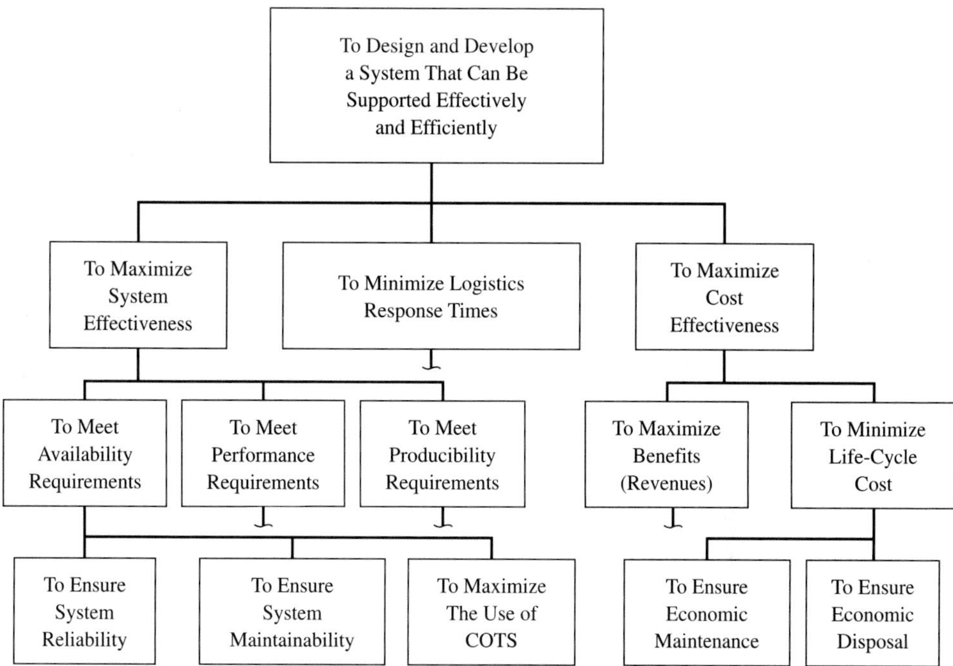

Figure 4.4 Partial design objectives tree.

the problem at hand is key to the subsequent selection or development of the appropriate computer-based model to be used in accomplishing supportability analysis requirements.[2]

A "model," presented in the context of Figure 4.2, may include either a single computer-based entity or an integrated set of tools as shown in Figure 3.35. The objective is select a model that incorporates the appropriate techniques (see block 4, Figure 4.2) and has the characteristics described in Section 3.8. The model must be sensitive to the problem at hand, and the analyst must be able to gain the necessary visibility required for evaluation of the system as a whole or any of its elements on an individual basis. More specifically, the selection of a good model may offer the following benefits:

1. In terms of system application, a number of considerations exist—operational considerations, design considerations, production/construction considerations, testing considerations, and logistic support considerations. There are many interrelated elements that must be integrated as a system and not treated on an individual basis. The model makes it possible to deal with the problem as an entity

[2] The development and application of various analytical techniques are covered in most textbooks on operations research. Two good references are (1) Hillier, F. S., and G. J. Lieberman, *Introduction to Operations Research*, 6th Ed., McGraw-Hill, New York, N.Y., 1995; and (2) Taha, H. A., *Operations Research: An Introduction*, 5th Ed., Macmillan, New York, N.Y., 1992.

and allows for consideration of all major variables of the problem on a simultaneous basis. Quite often the model will uncover relationships between the various aspects of a problem which are not apparent in the verbal description.

2. The model enables a comparison of *many* possible solutions and aids in selecting the best among them rapidly and efficiently.
3. The model often explains situations that have been left unexplained in the past by indicating cause-and-effect relationships.
4. The model readily indicates the type of data that should be collected to deal with the problem in a quantitative manner.
5. The model facilitates the prediction of future events, such as effectiveness factors, reliability and maintainability parameters, logistics requirements, and so on. In addition, the model aids in identifying areas of risk and uncertainty.

When analyzing a problem in terms of selecting a mathematical model for evaluation purposes, it is desirable to first investigate the tools that are currently available.[3] If a model already exists and is proven, then it may be feasible to adopt that model. However, extreme care must be exercised to relate the right technique with the problem being addressed and to apply it to the depth necessary to provide the sensitivity required in arriving at a solution. Improper application may not provide the results desired and may be costly.

Conversely, it might be necessary to construct a new model. In accomplishing such, one should generate a comprehensive list of system/equipment parameters that will describe the situation being simulated. Next, one should develop a matrix showing parameter relationships, each parameter being analyzed with respect to every other parameter to determine the magnitude of relationship. Model input/output factors and parameter feedback relationships must be established. The model is constructed by combining the various factors and then testing for validity. Testing is difficult to do since the problems addressed primarily deal with actions in the future which are impossible to verify. However, it may be possible to select a known system or equipment item which has been in the inventory for a number of years and exercise the model using established parameters. Data and relationships are known and can be compared with historical experience. In any event, the analyst might attempt to answer the following questions:

1. Can the model describe known facts and situations sufficiently well?
2. When major input parameters are varied, do the results remain consistent and are they realistic?
3. Relative to system application, is the model sensitive to changes in operational requirements, production/construction, and logistic support?
4. Can cause-and-effect relationships be established?

[3] A representative sample of some currently available models and their application is included in Appendix D.

Model development is an art and not a science, and is often an experimental process. Sometimes the analyst requires several iterations prior to accomplishing his (or her) objectives of providing a satisfactory analytical tool.

Data Generation and Application

One of the most important steps in the supportability analysis (SA) process is to assemble the appropriate input data (see Figure 4.2, block 6). The right type of data must be collected in a timely manner and must be presented in the proper format. Referring to Figure 1.7 (and the introduction to this chapter), there are input data requirements in the conceptual design phase, in the preliminary design phase, and so on, and the analyst needs to understand the depth and extent of the data required to solve the problem at hand. Quite often there is the tendency to generate too much data and too early, which can be counterproductive and very costly. Once again, the proper "tailoring" of requirements is critical.

As one proceeds in the system evaluation effort, there are a number of individual analytical techniques that may be applied within the overall SA spectrum and implemented with the objective of developing the required data. Referring to Figure 4.5, a necessary input constitutes a description of system operational requirements and the maintenance concept, the identification of operational and maintenance functions, and the identification of those factors that have been allocated to the various elements of the system (refer to Sections 3.2 through 3.7). This information provides a "baseline" and serves as an input for a life-cycle cost analysis, reliability and maintainability predictions, a level of repair analysis, and so on. Any one, or all, of the individual analysis activities shown in the figure may be implemented at any stage in the program. For example, a life-cycle cost analysis may be accomplished at the system-level early in conceptual design with a minimum amount of input data and utilizing the experience of the analyst. Later, a more indepth life-cycle cost analysis may be accomplished as the design definition becomes more complete and detailed. The objective is to understand the requirements for evaluation, the model, its inputs and outputs, and how it can be applied as one proceeds through the system acquisition process illustrated in Figure 1.6. Then, given this basic understanding, the analyst may wish to combine and integrate a set of models into the SA spectrum, as shown in Figure 4.6, where the input/ouput interrelationships are noted.[4]

In completing the individual analysis efforts identified in Figure 4.5, there are a variety of specific data input requirements, the depth and scope of which depends on the type of evaluation being performed and the program phase during which the evaluation is accomplished. In the early phases of system development, available data may be limited; thus, the analyst must depend on the use of various estimating relationships, projections based on past experiences associated with similar types of systems, and on the use of intuition. As the system development effort progresses, improved design

[4] A few selected of the tools shown in Figure 4.5 are discussed in Section 4.2. They include the life-cycle cost analysis (LCCA), FMECA, FTA, MTA, RCM, and the LORA.

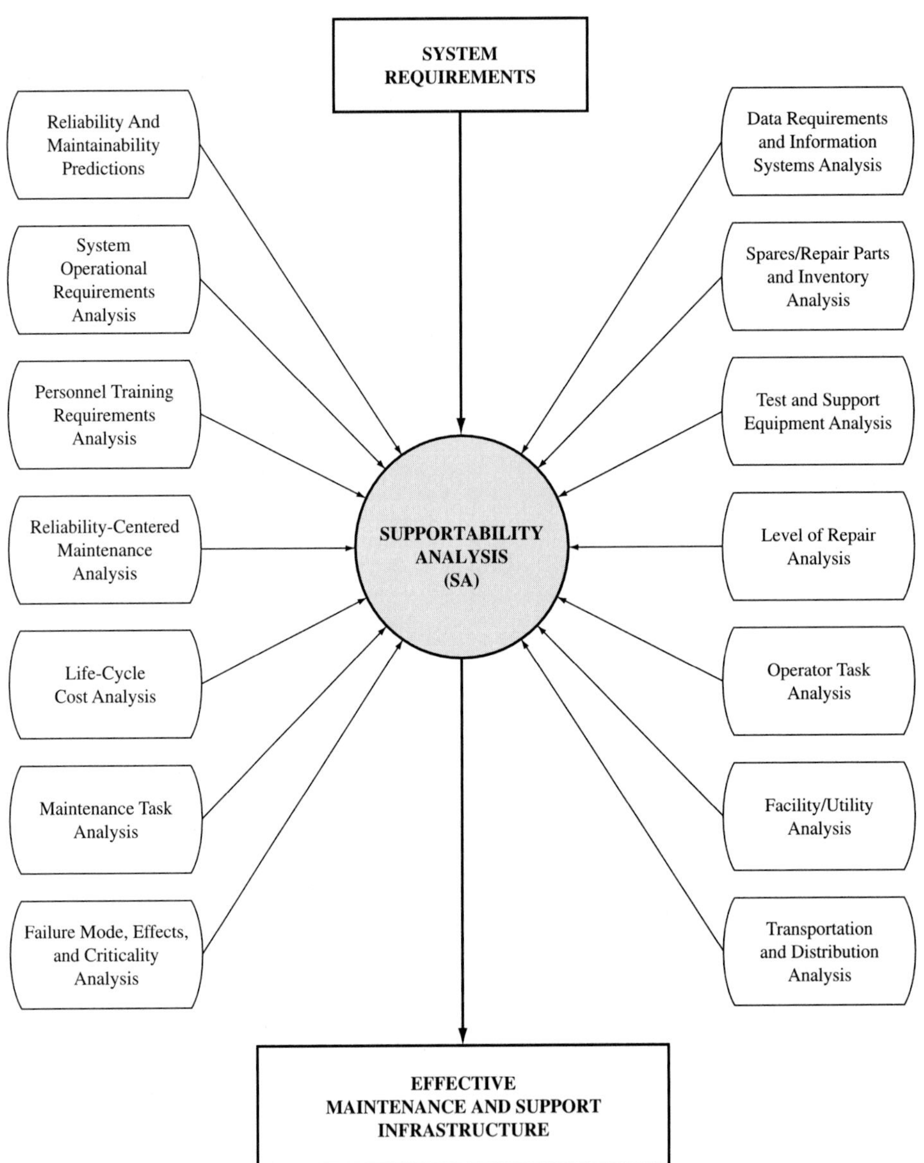

Figure 4.5 Supportability analysis and supplemental analysis efforts.

data are available as an input to the analysis effort. Ultimately, when the various elements of the system (hardware, software, facilities, elements of support, etc.) are available and in the field, testing and customer field data are available for the purposes of assessment. Thus, the analyst starts with a rough estimation and works toward a valid assessment through the system life cycle. The sources of data may be summarized in the following categories:

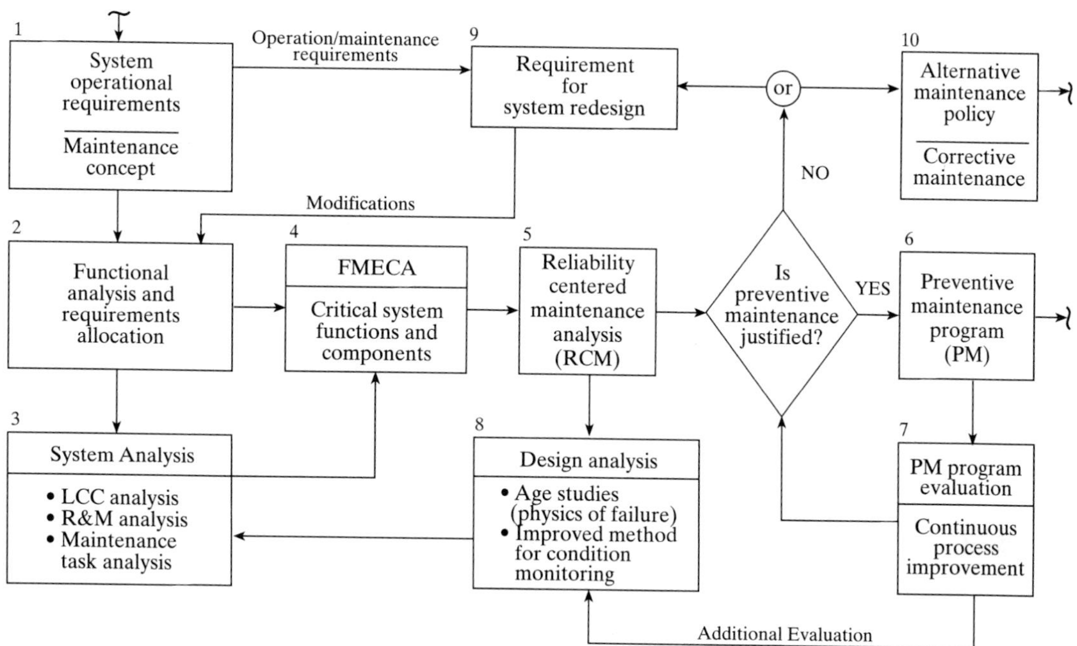

Figure 4.6 The integration of a set of analysis tools (example).

1. Current data banks, which provide historical information (field data) on existing systems in operational use and similar in configuration and function to the item being developed. Often, it is feasible to use such data and apply complexity factors to compensate for differences in technology, configuration, environment, and the time frame.
2. Estimating relationships, which relate one parameter in terms of another, provide rules of thumb or simple analogies from which specific factors are derived. For instance, in the cost area various cost categories are related to cost generating or explanatory variables (e.g., dollars per pound of weight, dollars per mile, dollars per part). Quite often, these explanatory variables represent characteristics of performance, physical configuration, logistics policy, and/or operational concept. Given one parameter, the analyst can estimate the second parameter. These relationships are developed by collecting and accumulating data on similar systems and correlating various individual factors to the appropriate characteristics of the new system. Forecasting techniques are used to facilitate the development process (e.g., use of Delphi estimates, multiple-regression techniques, etc.).
3. Advance system planning, preliminary system specifications, functional analyses, allocations, reliability and maintainability predictions, maintenance analyses, and related project reports provide much of the required input data when design information is first available.
4. Engineering test data and field data on equipment/software in the operational inventory are of course the best sources of data for actual assessment purposes.

Such data are usually employed when the SA is applied in evaluating the impact of modifications on prime equipment or the elements of logistic support.

The analyst relies on one or a combination of the sources listed above. The data utilized should be as accurate as possible, represent the operational situation, reflect current system conditions, and be used in sufficient quantity to provide a significant sample size covering the various system parameters being studied.

Analysis Results

As conveyed earlier, the supportability analysis (SA) constitutes the early application and integration of different analytical techniques (refer to Figure 4.5) applied to a variety of problem situations. Further, it is accomplished to provide a database identifying the specific requirements leading to the development of the maintenance and support infrastructure (refer to Figures 1.2 and 1.3). Referring to Figure 4.7, the SA leads to the definition of design criteria for each of the logistics elements identified in Figure 1.3. Associated with each of these elements are the applicable technical performance measures (TPMs) identified in Figure 4.3.

Regarding data formatting and packaging, the requirements may vary depending on the program, its organizational structure, the needs for specific information (type and amount of information, number of locations where the data are to be distributed, format, frequency, timeliness, etc.), and programing reporting requirements. The objective is, of course, to provide the right information, to the right location, and at the right time. This, in turn, requires the development of a database structure which is easily accessible and where the appropriate data are made available to all members of the design team concurrently. In the past, the development and processing of logistics-related data has been a major problem with too much data being generated that is not timely or accessible and which has been costly. As a result, the concept of *continuous acquisition and life-cycle support* (CALS) was implemented in the mid 1980s. CALS pertains to the application of computerized technology in the development and processing of data, primarily in a digital format, with the objectives of reducing preparation and processing times, eliminating redundancies, shortening the system acquisition process, and reducing the overall program costs. Specific applications have included the automation of technical publications, the preparation of a data package for the provisioning and procurement of spares/repair parts, and the development of design data defining products in a digital format.[5]

Referring to Figure 4.8, an objective is to develop some form of an integrated systems database structure that can (1) serve as a repository for all logistics and related data evolving from the supportability analysis, and (2) provide the necessary

[5] CALS, which may also be referred to as "computer-aided logistics support" or "computer-aided acquisition and logistics support," is described in detail in MIL-HDBK-59, "Computer-Aided Acquisition and Logistics Support (CALS) Program Implementation Guide," Department of Defense, Washington, D.C. The concept of CALS, which is currently quite popular and being implemented internationally (Asia, Australia, Europe, and North America), is closely associated with the activities dealing with *electronic commerce* (EC) and *Electronic Data Interchange* (EDI).

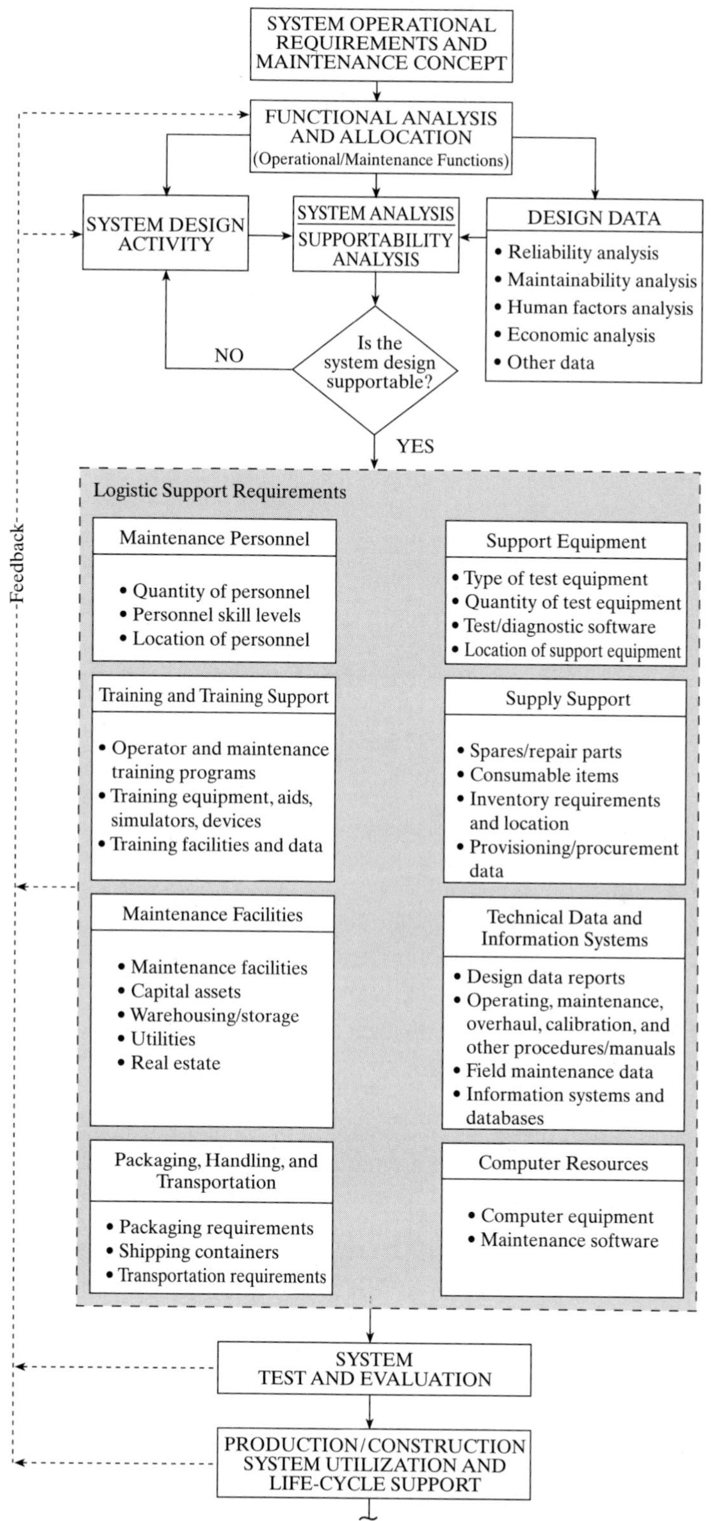

Figure 4.7 Development of logistics requirements through the supportability analysis.

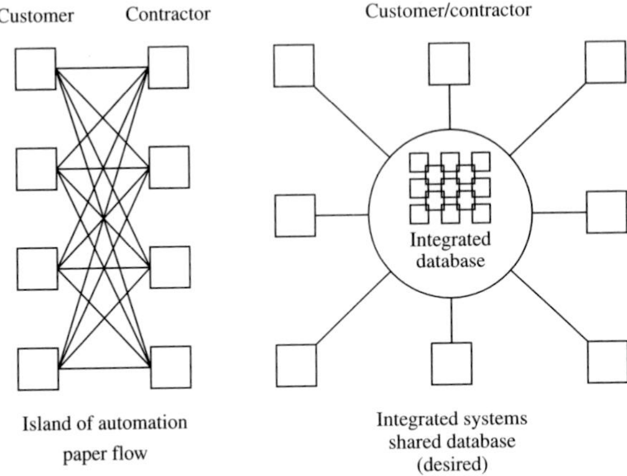

Figure 4.8 The data environment.

information at the right time and in the proper format in response to the reporting requirements for specific programs. For example, the current logistics reporting requirements for defense systems requires the preparation of "Supportability Analysis Summaries (SASs)" to include the following individual reports:[6]

1. Maintenance planning
2. Repair analysis
3. Support and test equipment
4. Supply support
5. Manpower, personnel, and training
6. Facilities
7. Packaging, handling, storage, and transportation
8. Post production support

Included within these summaries are approximately 160 individual data elements, and the goal is to be able to gain immediate access to the necessary information in the database and to print out the respective reports as required. Such "logistics management information (LMI)" must be properly integrated and directly accessible to all program personnel, in both the contractor and customer organizations. An integrated shared systems database is desired in lieu of following past practices where there have been many different reports/data packages flowing in many different directions and at different times.

[6] MIL-PRF-49506, "Performance Specification for Logistics Management Information," Department of Defense, Washington, D.C., November 1996. This specification was developed to replace MIL-STD-1388-2B, "Department of Defense Requirements for a Logistics Support Analysis Record (LSAR)." The objective is to state program requirements in terms of *performance* factors versus emphasis on the "how-to" issues.

4.2 ANALYSIS METHODS AND TOOLS

Referring to Figure 4.5, there are a number of individual analyses, using various analytical techniques/methods, that are inherent with the overall supportability analysis (SA) process. For the purposes of illustration, seven abbreviated problem exercises have been selected and presented in a summary manner. For each of the selected examples, identified in Figure 4.9, there is a problem statement, a description of the analysis approach or process, and the subsequent analysis results.

4.2.1 Life-Cycle Cost Analysis (LCC)

In addressing the *economic* side of the balance in Figure 4.10, one must deal with both revenues and costs. The emphasis here is on the element of *cost,* although the loss of revenues can constitute a "cost." In any event, costs must be addressed from a *life-cycle* perspective if one is to properly assess the risks associated with the day-to-day engineering and management decision-making processes. Although individual decisions may be based on some smaller aspect of cost (e.g., product procurement price), the analyst is remiss unless the consequences of those decisions are viewed in terms of *total* cost.

As described in Section 2.11, life-cycle costing addresses the *total cost* of the system and its supporting activities throughout its planned life cycle. It includes all future costs associated with research, design and development, construction and/or production, system utilization, maintenance and support, and system retirement and material recyling or disposal activities. In Figure 1.9, it can be seen that a major proportion of the projected life-cycle cost for a given system is due the maintenance and logistic support activities associated with that system throughout its period of utilization. Additionally, from Figure 2.25 it can be seen that much of this downstream cost is the consequence of design and management decisions made during the early stages of conceptual and preliminary design. Thus, the supportability analysis objectives described herein are critical in system design and development (or reengineering), and the use of life-cycle cost analysis (LCCA) methods is essential if one is to assess whether or not the system can be operated and supported in an effective and efficient manner.

In accomplishing a typical life-cycle cost analysis for any system (or element thereof), the analyst may wish to follow the steps presented in Figure 4.11. This approach must, however, be "tailored" to the problem at hand. In the abbreviated example described below, these steps were followed although the information is presented in the form of a summary.

Definition of the Problem. A ground vehicle currently in the development phase requires the incorporation of a new communications system. There is a need to design and develop a new system configuration, to be installed in the vehicle and operational 2 years from this point, and the per unit life-cycle cost of that system shall not exceed \$20,000. Two different supplier proposals have been received, and the objective is to evaluate each of the proposed design configurations and to select a preferred approach.

ANALYSIS TOOLS	DESCRIPTION OF APPLICATION
1. Life-Cycle Analysis (LCCA)	Determination of the system/product/process life-cycle cost (design and development, production and/or construction, system utilization, maintenance and support, and retirement/disposal costs) ; high-cost contributors; cause-and-effect relationships; potential areas of risk; and identification of areas for improvement (i.e. , cost reduction).
2. Failure Mode, Effects, and Criticality Analysis (FMECA)	Identification of potential product and/or process failures, the expected modes of failure and causes, failure effects and mechanisms, anticipated frequency, criticality, and the steps required for compensation (i.e., the requirement for redesign and/or the accomplishment of preventive maintenance). An Ishikawa "cause-and-effect" diagram may be used to facilitate the identification of causes, and a Pareto analysis may help in identifying those areas requiring immediate attention.
3. Fault-Tree Analysis (FTA)	A deductive approach involving the graphical enumeration and analysis of different ways in which a particular system failure can occur, and the probability of its occurrence. A separate fault tree may be developed for every critical failure mode, or undesired top-level event. Attention is focused on this top-level event and the first-tier causes associated with it. Each of these causes is next investigated for its causes, and so on. The FTA is narrower in focus than the FMECA and does not require as much input data.
4. Maintenance Task Analysis (MTA)	Evaluation of those <u>maintenance</u> functions that are to be allocated to the human. Identification of maintenance functions/tasks in terms of task times and sequences, personnel quantities and skill levels, and supporting resources requirements (i.e., spares/repair parts and associated inventories, tools and test equipment, facilities, transportation and handling requirements, technical data, training, and computer software). Identification of high resource-consumption areas.
5. Reliability-Centered Maintenance (RCM)	Evaluation of the system/process, in terms of the life cycle, to determine the best overall program for preventive (scheduled) maintenance. Emphasis is on the establishment of a cost-effective preventive maintenance program based on reliability information derived from the FMECA (i. e., failure modes, effects, frequency, criticality, and compensation through preventive maintenance).
6. Level-of-Repair Analysis (LORA)	Evaluation of maintenance policies in terms of levels of repair (i.e., should a component be repaired in the event of a failure or discarded, and, given the repair option, should the repair be accomplished at the intermediate level of maintenance, at the supplier's factory, or at some other level?). Decision factors include economic, technical, social, environmental, and political considerations. The emphasis here is based on life-cycle cost factors.
7. Evaluation of Design Alternatives	Evaluation of alternative design configurations using multiple criteria . Weighting factors are established to specify levels of importance.

Figure 4.9 A selection of supportability analysis tools.

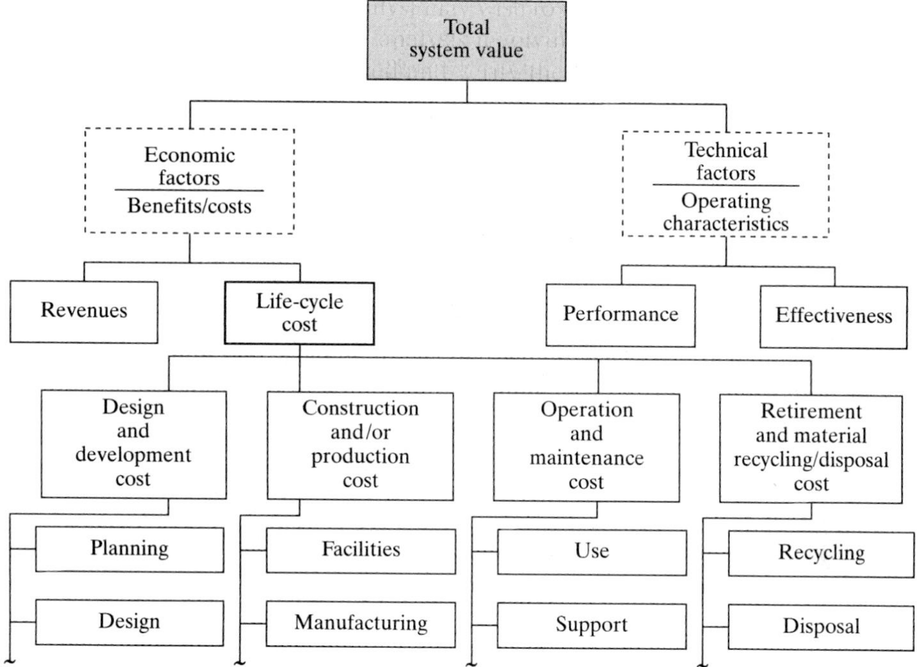

Figure 4.10 Total system value.

Analysis Approach.[7] Referring to Figure 4.11, the first step is to expand the problem definition by defining the system operational requirements and the maintenance and support concept (refer to Sections 3.3 and 3.4, respectively). Figure 4.12 illustrates these basic concepts.

The communications system is to be installed in a light vehicle, with 65 vehicles being distributed in three different geographical locations, and it shall be utilized on an average of 4 hours per day. It shall enable the day-to-day communications with other vehicles up to a range of 200 miles, overhead aircraft at an altitude of up to 10,000 feet, and with a centralized communications facility located within 250 miles. The system shall have a reliability MTBF of at least 450 hours, a $\overline{M}ct$ not to exceed 30 minutes, and a MLH/OH requirement of 0.2 or less. The system shall be operational 2 years hence, and it shall be retired from the inventory in 12 years.

The next step is to develop a cost breakdown structure (CBS) similar to the one shown in Figure 2.26 but "tailored" to this particular problem. The life-cycle plan, shown at the bottom of Figure 4.12, leads to the identification of the pertinent activities that must be covered by the CBS. Costs are estimated for each activity in the life cycle, on a year-to-year basis, and are reflected in the form of a cost profile as illustrated in Figure 4.13. Individual cost profiles are developed for each of the two configurations being evaluated. The costs for each configuration, Configurations *A* and *B*,

[7] Refer to Appendix E for a more indepth coverage of the LCCA, Appendix F, for the interest tables used in the comparison of alternatives in terms of economic equivalence, and the bibliography in Appendix H for some additional references on the subject.

1. *Define system requirements.* Define system operational requirements and the maintenance concept, identify the applicable technical performance measures (TPMs), and describe the system in functional terms (functional analysis at the system level).
2. *Describe the system life cycle and identify the activities in each phase.* Establish a "baseline" for the development of a cost breakdown structure and the estimation of cost for each year of the projected life cycle.
3. *Develop a cost breakdown structure* (CBS). Provide a top-down/bottom-up structure, to include all categories of cost, for the purposes of the initial allocation of costs (top-down) and the subsequent collection and summarization of costs (bottom-up). All life-cycle activities must be covered within.
4. *Identify data input requirements.* Identify the input data requirements and the possible sources from where such can be acquired. The type and amount of data will depend on the nature of problem being addressed, the phase of the system life cycle, and on the depth of analysis.
5. *Establish the costs for each category in the CBS.* Develop the appropriate cost-estimating relationships and estimate the costs for each category in the CBS on a year-by-year basis.
6. *Select a cost model for the purposes of analysis and evaluation.* Select (or develop) a computer-based model to facilitate the life-cycle cost analysis process. The model must be sensitive to the specific system being evaluated.
7. *Develop a cost profile and summary.* Construct a cost profile (i.e., cost stream) showing the "flow" of costs over the entire life cycle, and provide a summary identifying the cost for each category in the CBS and the percentage contribution in terms of the total.
8. *Identify the high-cost contributors and establish cause-and-effect relationships.* Highlight those functions, elements of the system, or segments of a process that should be investigated for possible areas of design improvement.
9. *Conduct a sensitivity analysis.* Evaluate the model, input-output data relationships, and the results of the "baseline" analysis to ensure that (1) the overall life-cycle cost analysis (LCCA) approach is valid; and (2) the model itself is well constructed and sensitive to the problem at hand. Given this, the sensitivity analysis can aid in identifying major areas of risk (as part of a risk analysis).
10. *Construct a Pareto diagram and identify priorities for problem resolution.* Conduct a Pareto analysis, construct a Pareto diagram, and identify priorities for problem resolution (i.e., those problems that require the most management attention).
11. *Identify feasible alternatives for design evaluation.* Having developed an approach for the LCC evaluation of a given single design configuration, it is now appropriate to extend the LCC analysis through the evaluation of multiple design alternatives.
12. *Evaluate feasible alternatives and select a preferred approach.* Develop a cost profile for each of the alternatives being evaluated, compare the alternatives considering the time value of money, construct a break-even analysis, and select a preferred design approach.

Figure 4.11 The basic steps in a life-cycle cost analysis (LCCA).

are summarized in Figure 4.14, indicating the cost for each significant category in the CBS. Note that the costs are presented in terms of *present value* for the purposes of comparing the two alternatives on an equivalent basis.[8]

Referring to Figure 4.14, note that the acquisition cost (R&D and Investment) is higher for Configuration *A* ($478,033 versus $384,131). This is due partially to a better design using more reliable components. Although the initial cost is higher, the overall life-cycle cost is lower due to a reduction in maintenance actions resulting in lower O&M costs. These characteristics in equipment design have a tremendous effect on life-cycle cost. In any event, it appears that Configuration *A* is preferred based on total

[8] The development of a cost breakdown structure (CBS), cost-estimating methods, the aspect of inflation and the application of learning curves, and the principles associated with the time value of money, are covered in detail in Appendix E.

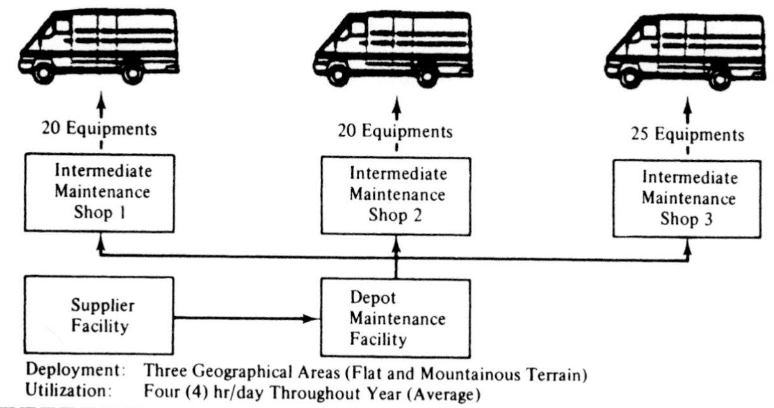

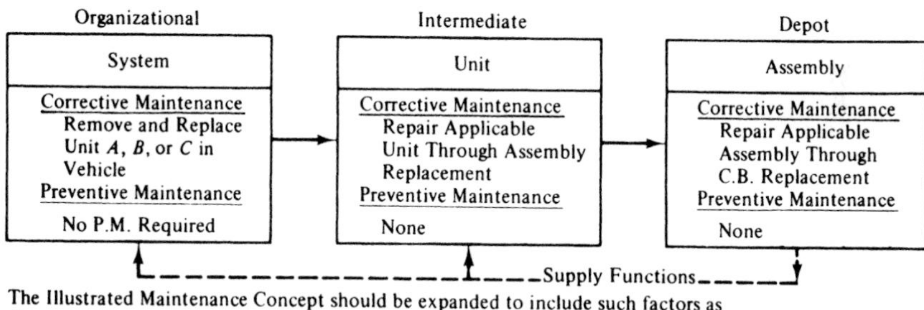

Research and Development

Full Complement of 65 Equipments Attained

0 Equipments in Inventory

40 Equipments in Inventory

65 Equipments in Inventory

Investment (Production)

Operations and Maintenance

0 1 2 3 4 5 6 7 8 9 10 11 12

Program Span, Years

Figure 4.12 Basic system concepts.

life-cycle cost. Before a final decision, however, the analyst should perform a break-even analysis to determine the point where Configuration *A* becomes more effective than Configuration *B*. Figure 4.15 illustrates a break-even point that is 6 years and 5

Category	Program Year												Total
	1	2	3	4	5	6	7	8	9	10	11	12	
Research and Development	32,119	38,100	—	—	—	—	—	—	—	—	—	—	70,219
Investment	—	94,110	156,852	156,852	—	—	—	—	—	—	—	—	407,814
Operations and Maintenance	—	—	12,180	32,480	60,492	57,472	53,480	50,484	50,470	50,494	37,480	17,185	422,217
Total	32,119	132,210	169,032	189,332	60,492	57,472	53,480	50,484	50,470	50,494	37,480	17,185	$900,250

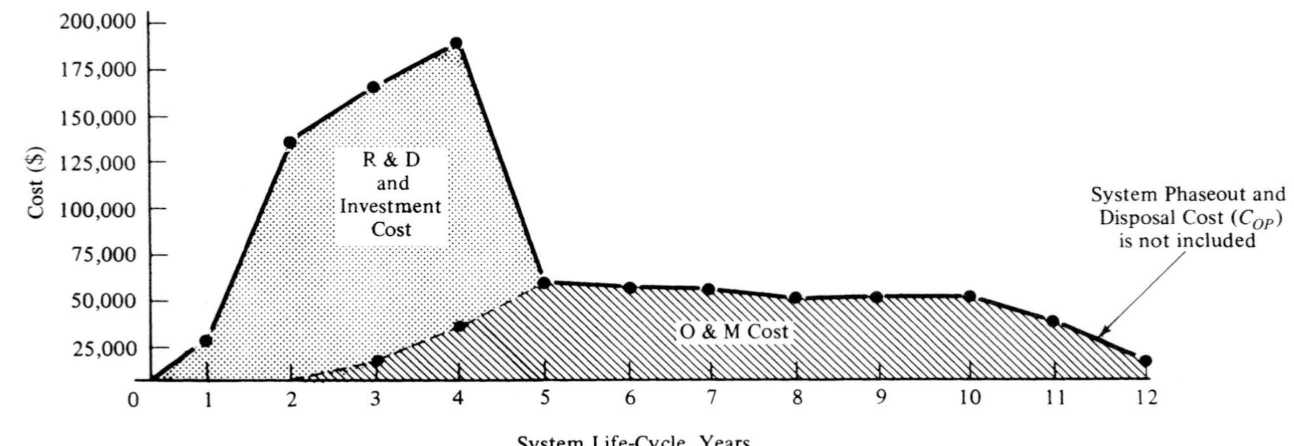

Figure 4.13 Cost profile for configuration *A*.

Cost Category	*Configuration A*		*Configuration B*	
	P.V. Cost ($)	*% of Total*	*P.V. Cost* ($)	*% of Total*
1. Research and development (C_R)	70,219	7.8	53,246	4.2
(a) Program management (C_{RM})	9,374	1.1	9,252	0.8
(b) Advanced R & D (C_{RR})	4,152	0.5	4,150	0.4
(c) Engineering design (C_{RE})	41,400	4.5	24,581	1.9
(d) Equipment development and test (C_{RT})	12,176	1.4	12,153	0.9
(e) Engineering data (C_{RD})	3,117	0.3	3,110	0.2
2. Investment (C_I)	407,814	45.3	330,885	26.1
(a) Manufacturing (C_{IM})	333,994	37.1	262,504	20.8
(b) Construction (C_{IC})	45,553	5.1	43,227	3.4
(c) Initial logistic support (C_{IL})	28,267	3.1	25,154	1.9
3. Operations and maintenance (C_O)	422,217	46.9	883,629	69.7
(a) Operations (C_{OO})	37,811	4.2	39,301	3.1
(b) Maintenance (C_{OM})	384,406	42.7	844,328	66.6
• Maintenance personnel and support (C_{OMM})	210,659	23.4	407,219	32.2
• Spare/repair parts (C_{OMX})	103,520	11.5	228,926	18.1
• Test and support equipment maintenance (C_{OMS})	47,713	5.3	131,747	10.4
• Transportation and handling (C_{OMT})	14,404	1.6	51,838	4.1
• Maintenance training (C_{OMP})	1,808	0.2	2,125	Neg.
• Maintenance facilities (C_{OMF})	900	0.1	1,021	Neg.
• Technical data (C_{OMD})	5,402	0.6	21,452	1.7
(c) System/equipment modifications (C_{ON})	...	...	...	...
(d) System phaseout and disposal (C_{OP})	...	...	...	...
Grand total*	$900,250	100%	$1,267,760	100%

*A 10% discount factor was used to convert to present value (PV).

Figure 4.14 Life-cycle cost analysis summary.

months, or a little more than 2 years after a full complement has been acquired in the operational inventory. This point is early enough in the life cycle to support the decision. On the other hand, if the break-even point were much further out in time, the decision might be questioned.

Referring to Figure 4.14, the analyst can readily pick out the high contributors (those that contribute more than 10% of the total cost). These are the areas where a more refined analysis is required and greater emphasis is needed in providing valid input data. For instance, maintenance personnel and support cost (C_{OMM}) and spare/repair parts cost (C_{COMX}) contribute 23.4% and 11.5%, respectively, of the total cost for Configuration *A*. This leads the analyst to reevaluate the design in terms of impact on personnel support and spares; the prediction methods used in determining maintenance frequencies and inventory requirements; the analytical model to ensure

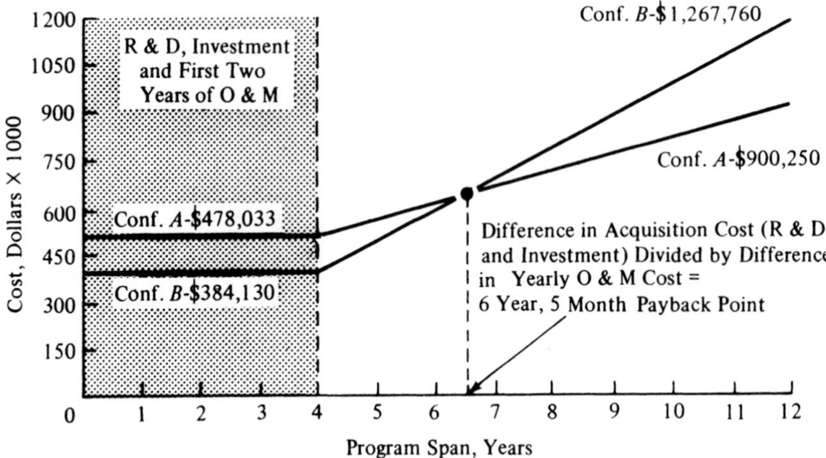

Figure 4.15 Investment payback (breakeven analysis).

that the proper parameter relationships are established; and cost factors such as personnel labor cost, spares material costs, inventory holding cost; and so on. If the analyst wishes to determine the sensitivity of these areas to input variations, he or she may perform a sensitivity analysis. In this instance, it is appropriate to vary MTBF as a function of maintenance personnel and support cost (C_{OMM}) and spare/repair parts cost (C_{OMX}). Figure 4.16 presents the results.

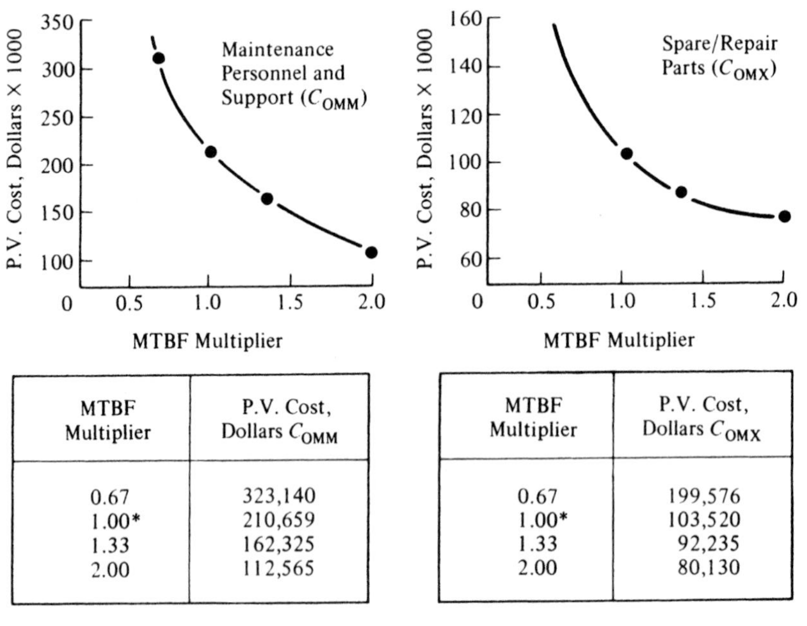

MTBF Multiplier	P.V. Cost, Dollars C_{OMM}
0.67	323,140
1.00*	210,659
1.33	162,325
2.00	112,565

*Baseline Configuration *A*

MTBF Multiplier	P.V. Cost, Dollars C_{OMX}
0.67	199,576
1.00*	103,520
1.33	92,235
2.00	80,130

*Baseline Configuration *A*

Figure 4.16 Sensitivity analysis.

The analyst or decision maker should review the break-even analysis in Figure 4.15 and determine how far out in time he or she is willing to go and remain with Configuration A. Assuming that the selected maximum break-even point is 7 years, the difference in alternatives is equivalent to approximately \$65,000 (the present value difference between the two configurations at the 7-year point). This indicates the range of input variations allowed. For instance, if the design of Configuration A changes or if the reliability prediction were in error (resulting in an MTBF of the required 450 hours versus a MTBF of 675 which was predicted by the supplier and was the basis for determining the reported LCC values), the maintenance personnel and support cost (Comm) will increase to approximately \$324,000, an increase of about \$113,340 above the baseline value. Thus, although the system reliability is within the specified requirements, the cost increase due to the input MTBF variation causes a decision shift in favor of Configuration B. The analyst must assess the sensitivity of significant input parameters and determine their impact on the ultimate decision.

Analysis Results. The descision in this case is to select Configuration A based on the delta cost, the results of the break-even analysis, and the relative relationships as conveyed in Figure 4.17. Given that both configurations will meet the specified reliability requirements, the one with the lowest predicted life-cycle cost is selected.

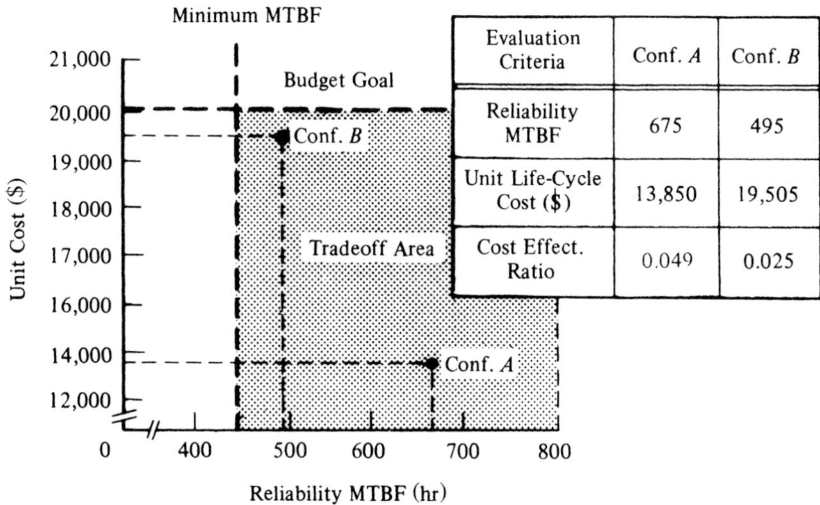

Figure 4.17 Reliability versus unit cost.

4.2.2 Failure Mode, Effects, and Criticality Analysis (FMECA)[9]

Definition of the Problem. Company ABC, a manufacturer of gaskets for automobiles, was experiencing problems related to declining productivity and increased product costs. At the same time, competition was increasing, and the company was losing

[9] This case study was taken in part from Blanchard, B. S., D. Verma, and E. L. Peterson, *Maintainability: A Key to Effective Serviceability and Maintenance Management*, John Wiley & Sons, Inc., New York, N.Y., 1995.

its share of the market. As a result, the company decided to implement a *continuous process improvement program* with the objective of identifying potential problem areas, and their impact and criticality on both internal company operations and on the product being delivered to the customer. To aid in facilitating this objective, the company's manufacturing operations were evaluated using the failure mode, effects, and criticality analysis (FMECA).

Analysis Approach. An initial step included the identification of the the major functions performed in the overall gasket manufacturing process by completing a functional flow diagram in accordance with the procedures described in Section 3.6 (Chapter 3). In this instance, there are 13 major functions that were subject to evaluation. For each function, required input factors and expected outputs were identified, along with the appropriate metrics. This led to the initial selection of one of the 13 functions, based on a perception by company personnel as to the area causing the most problems. Given this, the sequence of steps conveyed in Figure 4.18 was followed in completing an FMECA on the selected function.

Figure 4.19 represents the function, or portion the overall manufacturing process, that was selected for evaluation. Note that, although the emphasis is on the manufac-

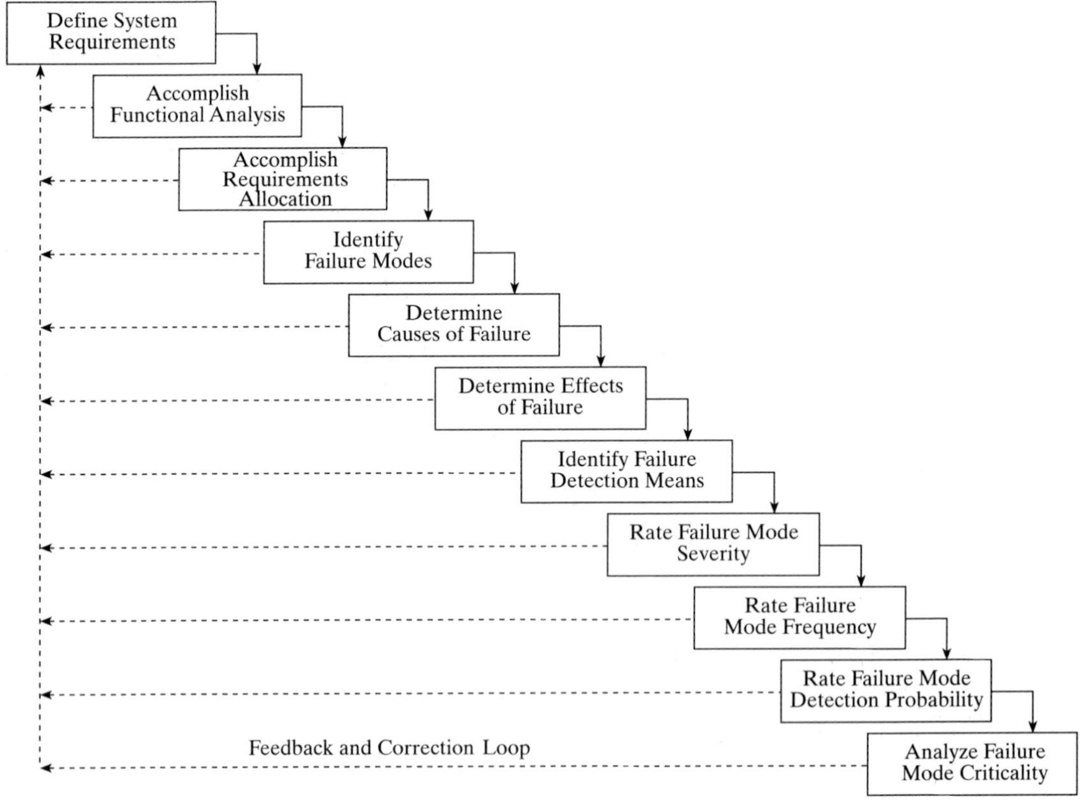

Figure 4.18 General approach to conducting a FMECA.

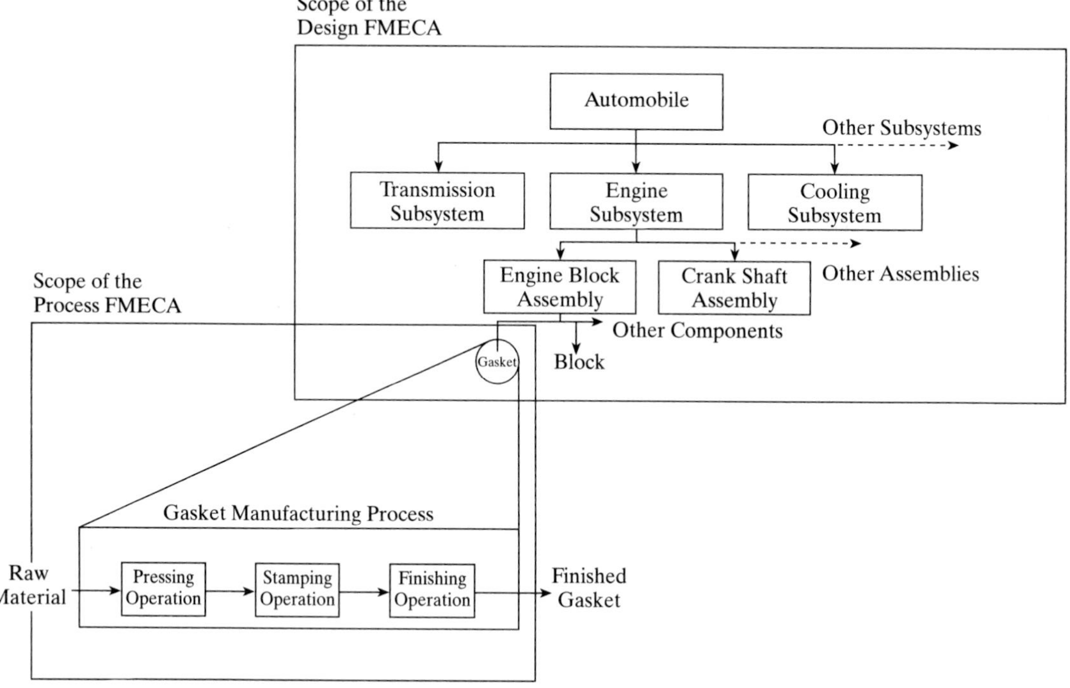

Figure 4.19 Design and process FMECA focus and scope.

turing process and its impact on the gasket, one must also consider the impact of a faulty gasket on the automobile. Thus, the FMECA needs to address both the *process* and the *product*.

In Figure 4.18, the approach selected for conducting the FMECA was in accordance with the practices followed in the automotive industry. This included the following:[10]

1. Identifying the different failure modes; that is, the manner in which a system element fails to accomplish its function.
2. Determining the cause(s) of failure; that is, the cause(s) responsible for the occurrence of each failure. An Ishikawa *cause-and-effect*, or *fishbone*, diagram, as illustrated in Figure 4.20, was utilized to help establish the relationships between failures and their possible causes.[11]
3. Determining the effects of failure; that is, the effects on subsequent functions/processes, on the next higher-level functional entity, and on the overall system.

[10] Several references were used to include (1) Instruction Manual, *Potential Failure Mode and Effects Analysis*, Ford Motor Company, 1988; (2) Instruction Manual, *Failure Mode and Effects Analysis*, Saturn Quality System, Saturn Corporation, 1990; and (3) *Potential Failure Mode and Effects Analysis (FMEA)*, Reference Manual FMEA-1, developed by FMEA teams at Ford Motor Company, General Motors, Chrysler, Goodyear, Bosch, and Kelsey-Hayes, under the auspices of ASQC and AIAG.

[11] K. Ishikawa, *Introduction to Quality Control*, Chapman and Hall, London, 1991.

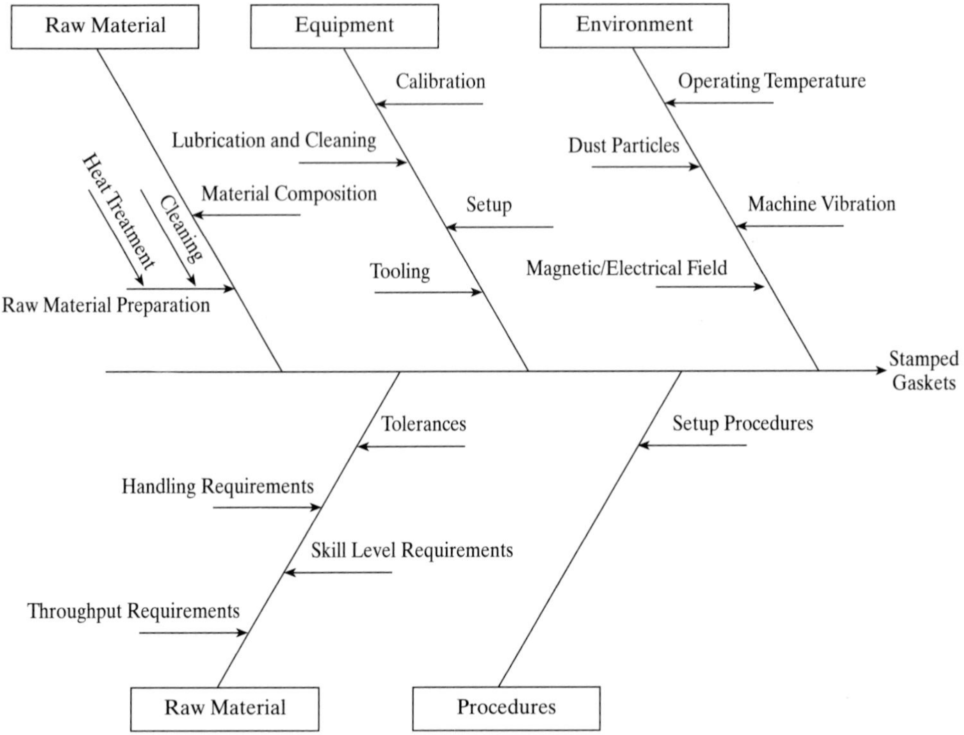

Figure 4.20 The Ishikawa cause-and-effect (fishbone) diagram.

4. Identifying failure detection means; that is, the current controls, design features, or verification procedures that will result in the detection of potential failure modes.
5. Determining the severity of a failure mode; that is, the seriousness of the effect or impact of a particular failure mode. The degree of severity was converted quantitatively on a scale of 1 to 10 with *minor* effects being 1, *low* effects being 2 to 3, *moderate* effects being 4 to 6, *high* effects being 7 to 8, and *very high* effects being 9 to 10. The level of severity was related to issues pertaining to safety and the degree of customer dissatisfaction.
6. Determining the frequency of occurrence; that is, the frequency of occurrence of each individual failure mode or the probability of failure. A scale of 1 to 10 was applied with *remote* (failure is unlikely) being 1, *low* (relatively few failures) being 2 to 3, *moderate* (occasional failures) being 4 to 6, *high* (repeated failures) being 7 to 8, and *very high* (failure is almost inevitable) being 9 to 10. These rating factors were based on the number of failures per segment of operating time.
7. Determining the probability that a failure will be detected; that is, the probability that the design features/aids or verification procedures will detect potential failure modes in time to prevent a system-level failure. For a process application, this refers to the probability that a set of process controls currently in place will be in a position to detect and isolate a failure before it gets transferred to the subsequent processes or to the ultimate product output. This probability is once

again rated on a scale of 1 to 10 with *very high* being 1 to 2, *high* being 3 to 4, *moderate* being 5 to 6, *low* being 7 to 8, *very low* being 9, and *absolute certainty of nondetection* being 10.

8. Analyzing failure mode criticality; that is, a function of severity (item 5), the frequency of occurrence of a failure mode (item 6), and the probability that it will be detected in time to preclude its impact at the system level (item 7). This resulted in the determination of the *risk priority number (RPN)* as a metric for evaluation. RPN can be expressed as

$$\text{RPN} = (\text{severity rating})(\text{frequency rating})(\text{probability of detection rating}) \quad (4.1)$$

The RPN reflects failure-mode criticality. On inspection, one can see that a failure mode with a high frequency of occurrence, with significant impact on system performance, and that is difficult to detect is likely to have a very high RPN.

9. Identifying critical areas and recommending modifications for improvement; that is, the iterative process of identifying areas with high RPNs, evaluating the causes, and initiating recommendations for process/product improvement.

Figure 4.21 shows a partial example of the format used for recording the results of the FMECA. The information was derived from the functional flow diagram and expanded to include the results from the steps presented in Figure 4.18. Figure 4.22 lists the resulting RPNs in order of priority (relative to requiring attention), and Figure 4.23 presents the results in the form of a Pareto analysis.

Analysis Results. After having completed the FMECA on the function identified in Figure 4.19, Company ABC proceeded to evaluate each of its other 12 major functions/processes in a similar manner, utilizing a *team* approach. The activity was very beneficial overall, the individuals participating in the effort learned more about their own activities, and numerous changes were initiated for the purposes of improvement.

4.2.3 Fault-Tree Analysis (FTA)

Definition of the Problem. During the very early stages of the system design process, and in the absence of the information required to complete a FMECA (discussed in Section 4.2.2), a fault-tree analysis (FTA) was conducted to gain insight into critical aspects of system design. The fault-tree analysis is a deductive approach involving the graphical enumeration and analysis of the different ways in which a particular system failure can occur, and the probability of its occurrence. A separate fault tree is developed for every critical failure mode, or undesired top-level event. Emphasis is on this top-level event and the first-tier causes associated with it. Each of these causes is next investigated for *its* causes, and so on. This top-down hierarchy, illustrated in Figure 4.24, and the associated probabilities, is called a *fault tree*. Figure 4.25 presents some of the symbology used in the development of such a structure.

Analysis Approach. One of the outputs from the FTA is the probability of occurrence of the top-level event or failure. If the probability factor is unacceptable, the causal hierarchy developed provides engineers with insight into aspects of the system

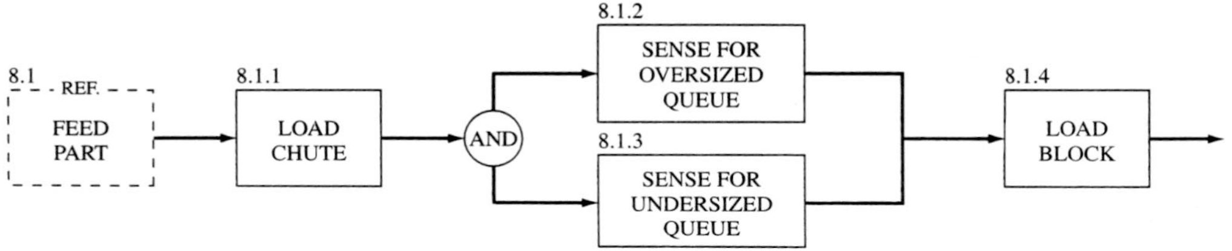

PROCESS FAILURE MODE AND EFFECTS ANALYSIS

Reference Number	Process Description	Potential Failure Mode	Potential Cause of Failure	Potential Effect(s) of Failure (company)	Potential Effect(s) of Failure (customer)	Current Controls	OCCURENCE	SEVERITY CO.	SEVERITY CU.	DETECTION	RPN	Recommended Action(s) and Status	Responsible Activity
8.1.2	Sense for Oversized Queue	Change in Freespread	Sensor Falls	Lip Process Jams		Machine Stops	1	1		1	1		
				Height Variation		1 Pc/5 Min	1	7		3	21		
					Bearing Loose When Installed in Engine	2 Pcs/Half Hr	1		7	5	35		
			Sensor Dirty	Lip Process Jams		Machine Stops	1	1		1	1		
				Height Variation		1 Pc/5 Min	1	7		3	21		
					Bearing Loose When Installed in Engine	2 Pcs/Half Hr	1		7	5	35		
			Improper Setup	Lip Process Jams		Machine Stops	1	1		1	1		
				Height Variation		1 Pc/5 Min	1	7		3	21		
					Bearing Loose When Installed in Engine	2 Pcs/Half Hr	1		7	5	35		
8.1.3	Sense for Undersized Queue	Lip Mislocated	Sensor Fails	Lip Process Jams		100% Visual	2	1		1	2		
					Fillet Ride	5 Pcs/Half Hr	2		3	7	42		
			Sensor Dirty	Lip Process Jams		100% Visual	2	1		1	2		
					Fillet Ride	5 Pcs/Half Hr	2		3	7	42		
			Improper Setup	Lip Process Jams		100% Visual	2	1		1	2		
					Fillet Ride	5 Pcs/Half Hr	2		3	7	42		
		Facing/Back Damage	Sensor Fails		Rejected at Assembly	100% Visual	3		5	4	60		
			Sensor Dirty		Rejected at Assembly	100% Visual	3		5	4	60		
			Improper Setup		Rejected at Assembly	100% Visual	3		5	4	60		
8.1.4	Load Block	Lip Mislocated	"Hold Down" Not Set Property	Lip Smashed in Broache		5 Pcs/Half Hr	3	7		7	147		
				Lip Process Jams		100% Visual	3	1		7	21		
					Fillet Ride	5 Pcs/Half Hr	3		3	7	63		
			Loose Load Block	Lip Smashed in Broache		5 Pcs/Half Hr	2	7		7	98		
				Lip Process Jams		100% Visual	2	1		7	14		
					Fillet Ride	5 Pcs/Half Hr	2		3	7	42		
		Facing/Back Damage	Misaligned Pusher		Rejected at Assembly	100% Visual	4		5	4	60		

Figure 4.21 Sample FMECA worksheet.

Causes	Risk Priority Numbers
Chip breaker angle ground incorrectly	273
Hold-down not set correctly	210
Undersize sensor fails	200
Undersize sensor dirty	200
Undersize sensor not positioned properly	200
Loose load block	161
Sharp die edge	120
Improper projection angle/resharpening of punch	108
Oversize sensor fails	105
Oversize sensor dirty	105
Oversize sensor not positioned properly	105
Improper sharpening of insert	93
Misaligned pusher	80
Worn tooling	72
Adapter reground to wrong dimension	60
Insert loose	60
Slivers in adapter	60
Insert off location	60
Worn/loose insert	60
Burrs from punch process caught	40
Ram stroke too long	36
Ram stroke too short	21
Broken/loose punch	12
Set screw fault	12
Insufficient stroke by ram/punch	12
Broken pressure spring	10
Total	2475

Figure 4.22 Risk priority numbers.

to which redesign efforts may be directed or compensatory provisions provided. The logic used in developing and analyzing a fault tree has its foundations in Boolean algebra. Axioms from Boolean algebra are used to collapse the initial version of the fault tree to an equivalent reduced tree with the objective of deriving *minimum cut sets*. Minimum cut sets are unique combinations of basic failure events that can cause the undesired top-level event to occur. These minimum cut sets are necessary to evaluate a fault tree from a qualitative and quantitative perspective. The basic steps in conducting a FTA are as follows:

1. Identify the top-level event. It is essential that the analyst be quite specific in defining this event. For example, it could be delineated as the "system catches fire" rather than the "system fails." Further, the top-level event should be clearly

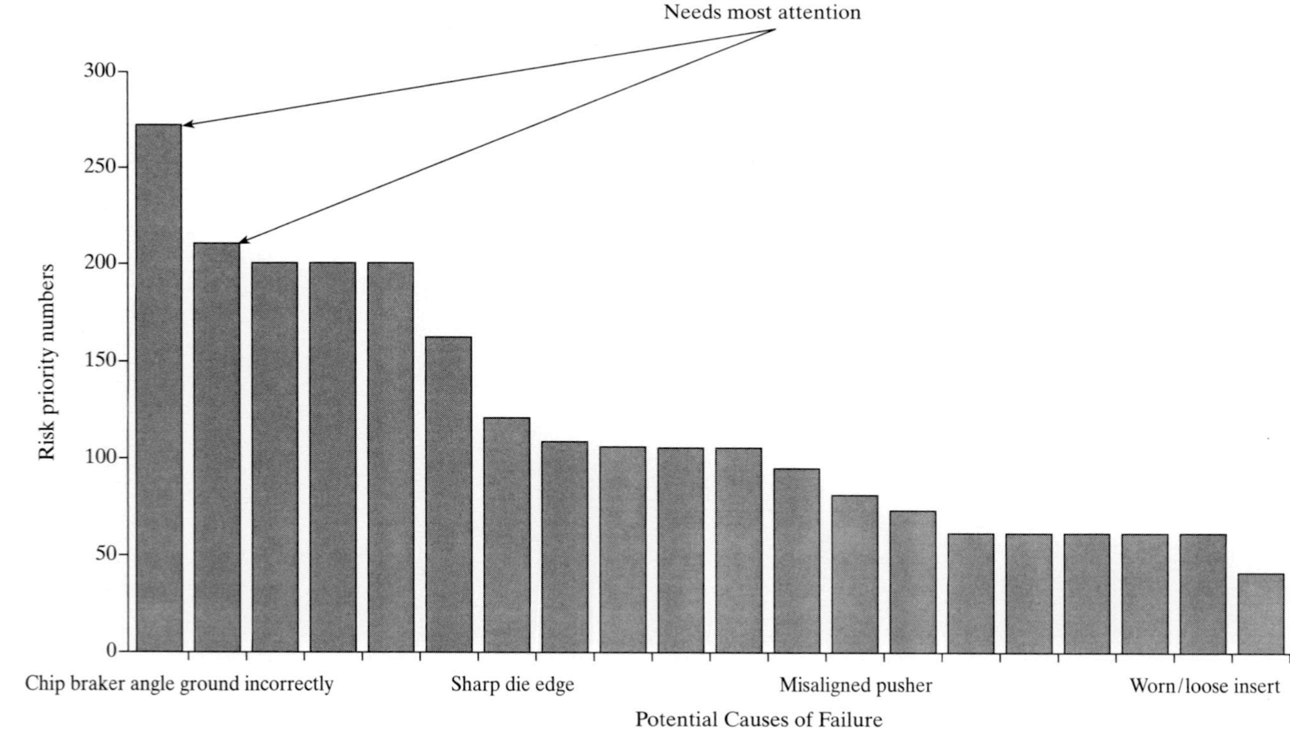

Figure 4.23 Pareto analysis (partial).

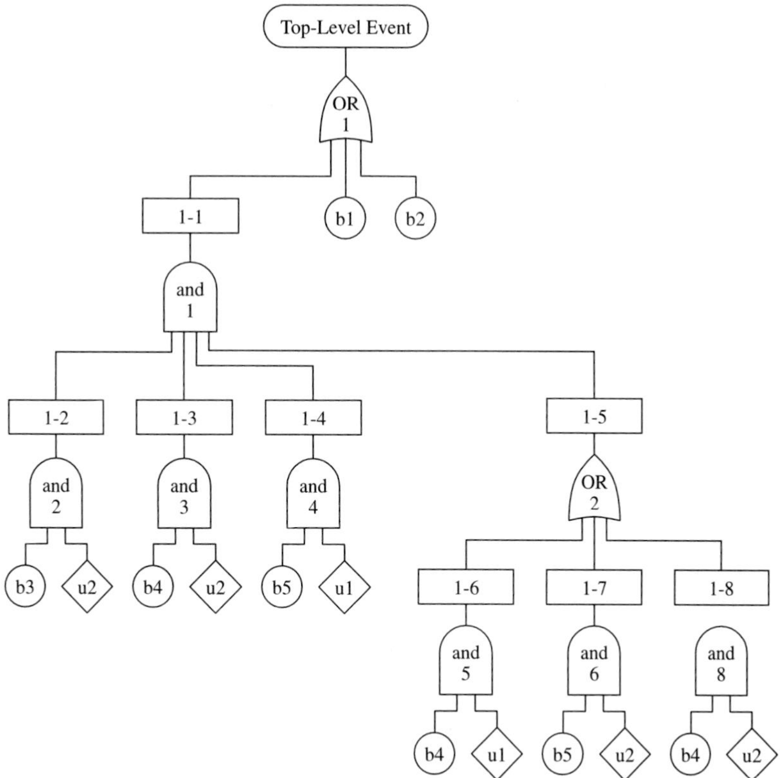

Figure 4.24 An illustrative fault tree.

observable, and unambiguously definable and measurable. A generic and non-specific definition is likely to result in a broad-based fault tree with too wide a scope and lacking in focus.

2. Develop the fault tree. Once the top-level event has been satisfactorily defined, the next step is to construct the initial causal hierarchy in the form of a fault tree. Once again, a technique such as Ishikawa's *cause-and-effect* diagram can prove to be beneficial (refer to Figure 4.20). While developing the fault tree, all hidden failures must be considered and incorporated.

   For the sake of consistency and communication, a standard symbology to develop the fault tree is recommended. Figure 4.25 depicts and defines the symbology to comprehensively represent the causal hierarchy and interconnects associated with a particular top-level event. In Figure 4.24, the symbols OR1 and OR2 represent the two OR logic gates, and 1 through 8 represent eight AND logic gates, I-1 through I-8 represent eight intermediate fault events, b1 through b5 represent five basic events, and u1 and u2 represent two undeveloped failure events. While constructing a fault tree, it is important to break every branch down to a reasonable and consistent level of detail.

3. Analyze the fault tree. The third step in conducting the FTA is to analyze the initial fault tree developed. A comprehensive analysis of a fault tree involves both

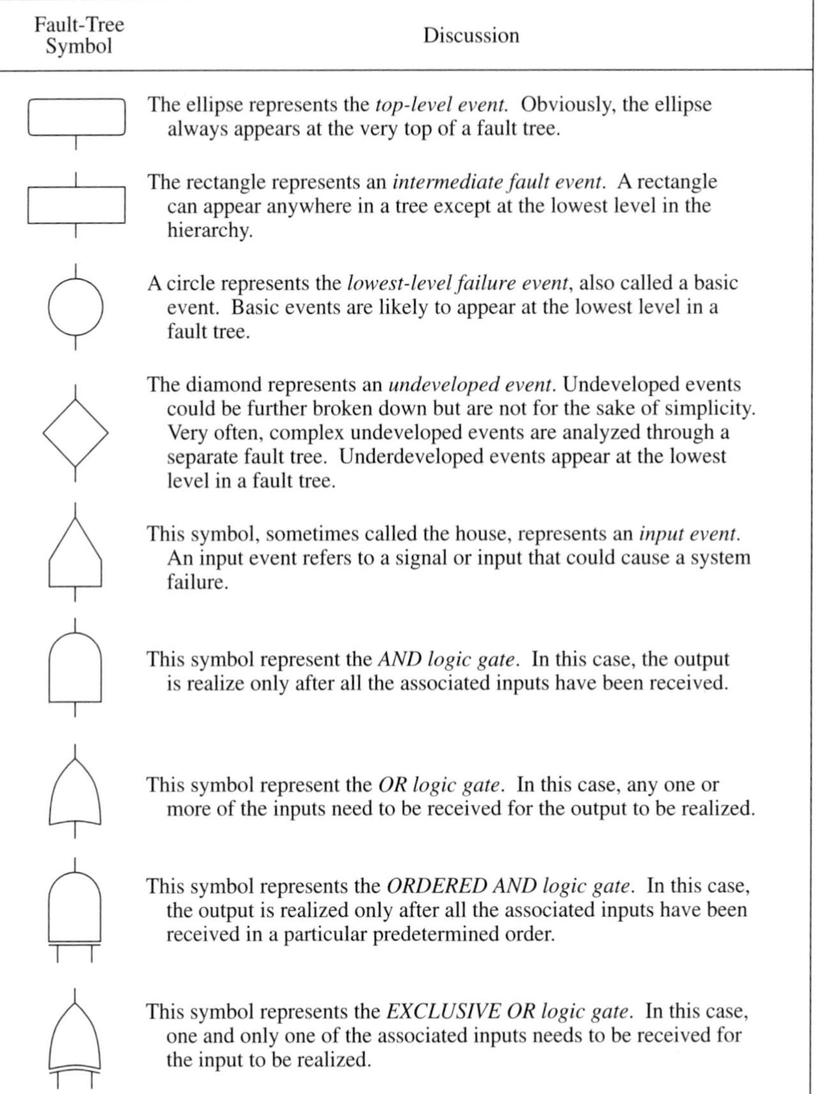

Figure 4.25 Fault-tree constructive symbology.

a quantitative and a qualitative perspective. The important steps in completing the analysis of a fault tree are as follows:

(a) *Delineate the minimum cut sets.* As part of the analysis process, the minimum cut sets in the initial fault tree are first delineated. These are necessary to evaluate a fault tree from a qualitative or quantitative perspective. The objective of this step is to reduce the initial tree to a simpler equivalent reduced fault tree. The minimum cut sets can be derived using two different approaches. The first approach involves a graphical analysis of the initial tree, an enumeration of all the cut sets, and the subsequent delineation of the minimal cut sets. The second approach, conversely, involves translating the graphical fault tree into an equivalent Boolean expression. This Boolean expression is then reduced to a simpler equivalent expression by eliminating all the redundancies and so on. As an example, the fault tree depicted in Figure 4.24 can be translated into a simpler and equivalent fault tree, through Boolean reduction, as depicted in Figure 4.26.

(b) *Determine the reliability of the top-level event.* This is accomplished by first determining the probabilities of all relevant input events, and the subsequent consolidation of these probabilities in accordance with the underlying logic of the tree. The reliability of the top-level event is computed by taking the product of the reliabilities of the individual minimum cut sets.

(c) *Review analysis output.* If the derived top-level probability is unacceptable, necessary redesign or compensation efforts will need to be initiated. The development of the fault tree and subsequent delineation of minimum cut sets provides engineers and analysts with the land of foundation needed for making sound decisions.

Analysis Results. The FTA can be effectively applied in the early phases of design to specific areas where potential problems are suspected. It is narrow in focus and easier to accomplish than the FMECA, requiring less input data to complete. For large and complex systems, which are highly software-intensive and where there are many interfaces, the use of the FTA is often preferred in lieu of the FMECA. The FTA is most beneficial if conducted, not in isolation, but as part of an overall supportability analysis process.[12]

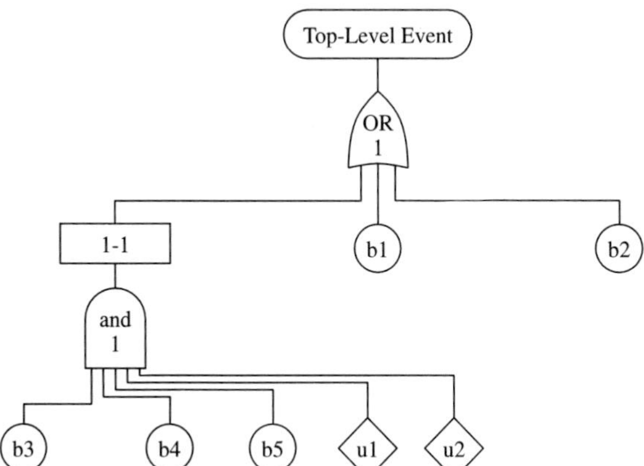

Figure 4.26 A reduced equivalent fault tree (reference: Figure 4.24).

[12] *Fault Tree Analysis Application Guide*, prepared by the Reliability Analysis Center (RAC), Rome Air Development Center, is an excellent "how-to" source for the application of FTA depicting numerous case studies.

4.2.4 Maintenance Task Analysis (MTA)[13]

Definition of the Problem. Company DEF has been manufacturing Product 12345 for the past few years. The costs have been higher than anticipated and international competition has been increasing! As a result, company management has decided to conduct an evaluation of the overall production capability, identify "high-cost" contributors through the accomplishment of a life-cycle cost analysis, and identify possible problem areas where improvement can be realized. One area for possible improvement is the manufacturing test function, shown in Figure 4.27, where frequent failures have occurred during Product 12345 test. By reducing maintenance costs, it is likely that one can reduce the overall cost of the product and improve the company's competitive position in the marketplace. With the objective of identifying some

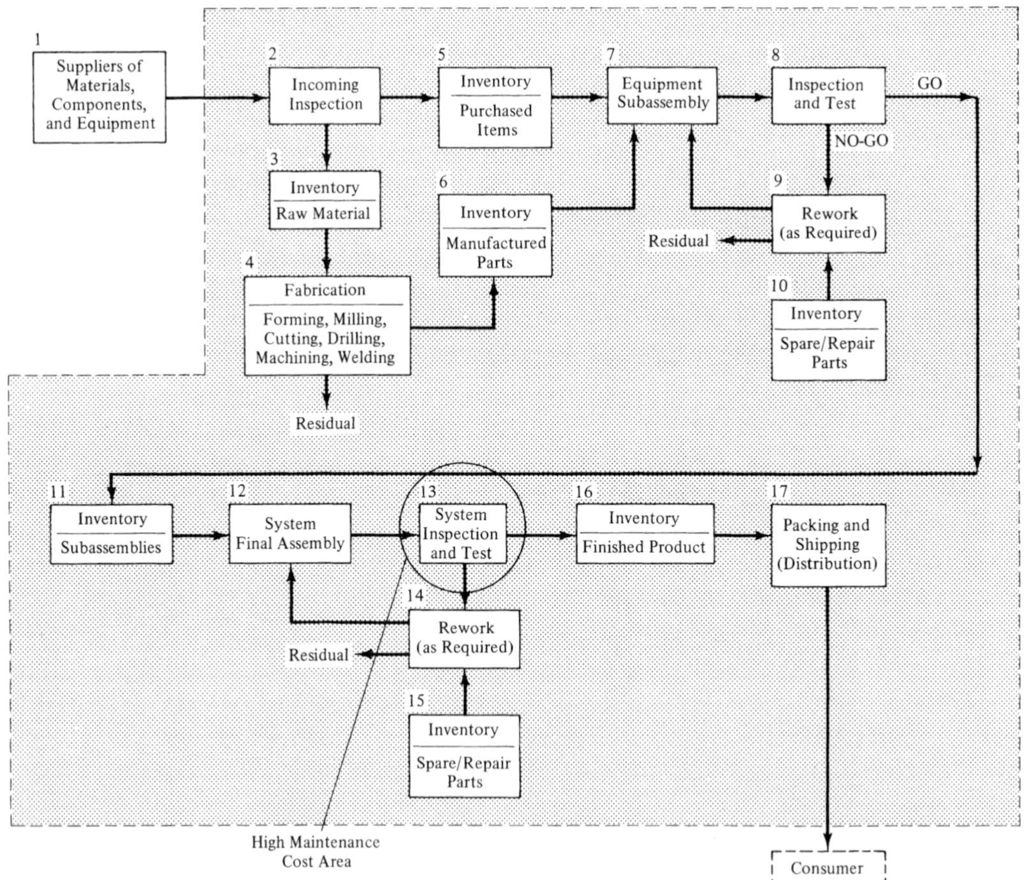

Figure 4.27 Production operation functional flow diagram.

[13] Refer to Appendix C for a more indepth coverage of the maintenance task analysis (MTA)

"specifics," a detailed maintenance task analysis of the manufacturing test function is accomplished. Specific recommendations for improvement are being solicited.

Analysis Approach. In response, a detailed maintenance task analysis is accomplished using the format included in Appendix C. The format, as adapted for the purpose of this evaluation, includes the following general steps:

1. Review of historical information covering the performance of the manufacturing test capability indicated the frequent loss of power during the final testing of Product 12345. From this, a typical "symptom of failure" was identified, and a sample logic troubleshooting flow diagram was developed as shown in Figure 4.28.
2. The applicable "go/no-go" functions, identified in Figure 4.28, are converted to the task analysis format in Figures 4.29 and 4.30. The functions are analyzed on the basis of determining task requirements (task durations, parallel-series relationships, sequences), personnel quantity and skill-level requirements, spare/repair part requirements, test and support equipment requirements, special facility requirements, technical data requirements, and so on. The intent of Figure 4.29 is to lay out the applicable maintenance tasks required, determine the anticipated frequency of occurrence, and identify the logistic support resources that are likely to be necessary for the performance of the required maintenance. This information, in turn, can be evaluated on the basis of cost.
3. Given the preliminary results of the analysis in terms of the layout of expected maintenance functions/tasks, the next step is evaluate the information presented in Figures 4.29 and 4.30, and suggest possible areas where improvement can be made.

Analysis Results. Review of the information presented in Figures 4.29 and 4.30 suggests that the following areas be investigated further:

1. With the extensive resources required for the repair of Assembly A-7 (e.g., the variety of special test and support equipment, the necessity for a "cleanroom" facility for maintenance, the extensive amount of time required for the removal and replacement of CB-1A5, etc.), it may be feasible to identify Assembly A-7 as being nonrepairable! In other words, investigate the feasibility of whether the assemblies of Unit B should be classified as "repairable" or "discard at failure" (refer to Section 4.2.6).

2. From Tasks 01 and 02, a "built-in test" capability exists at the organizational level for fault isolation to the subsystem. However, fault isolation to the unit requires a Special System Tester (0-2310B), and it takes 25 minutes of testing plus a highly skilled (supervisory skill) individual to accomplish the function. In essence, one should investigate the feasibility of extending the built-in test down to the unit level and eliminate the need for the special system tester and the high-skill level individual.

3. The physical removal and replacement of Unit B from the system take 15 minutes, which seems rather extensive. Although perhaps not a major item, it would be worthwhile investigating whether the removal/replacement time can be reduced (to less than 5 minutes, for example).

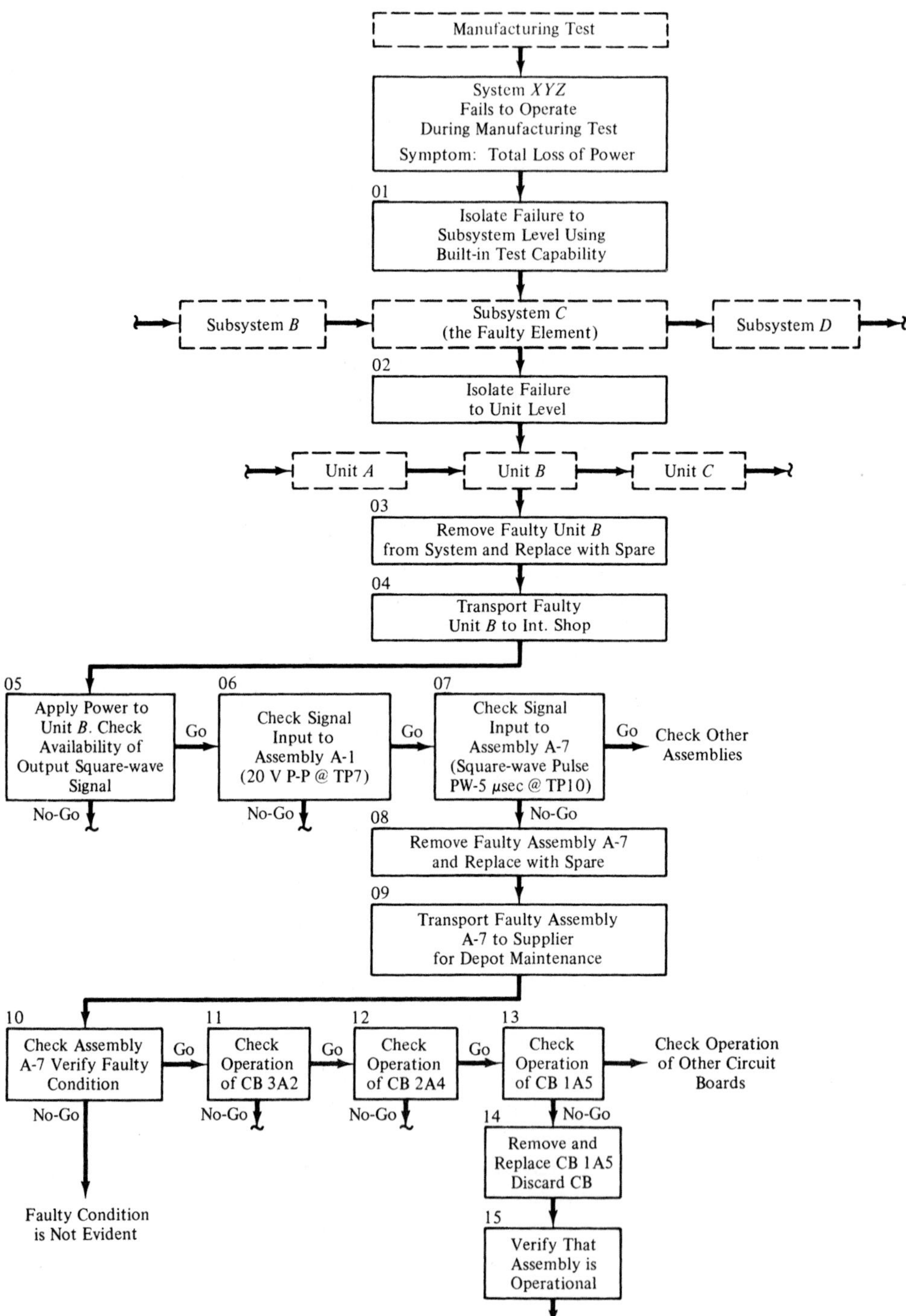

Figure 4.28 Abbreviated logic troubleshooting flow diagram.

1. System: *XYZ*	2. Item name/part no: Manufacturing Test/A4321		3. Next higher Assy.: Assembly and test		4. Description of requirement: During manufacturing and test of Product 12345 (Serial No. 654), System *XYZ* failed to operate. The symptom of failure was "loss of total power output" Requirement: Troubleshoot and repair system.
5. Req. No.: 01	6. Requirement: Diag./Repair	7. Reg. Freq.: 0.00450	8. Maint. Level Org/Inter.	9. Ma. Cont. No.: A12B100	

10. Task Number	11. Task Description	12. Elapsed time - minutes (2 4 6 8 10 12 14 16 18 20 22 24 26 28 30 32 34 36 38)	13. Total elap time	14. Task Freq.	Personnel Man-min 15 B	16 I	17 S	18 Total
01	Isolate failure to subsystem level		5	0.00450	5	-	-	5
	(Subsystem *C* is faulty)							
02	Isolate failure to unit level	(1)	25		-	25	-	25
	(Unit *B* is faulty)	(2)			-	-	25	25
03	Remove Unit *B* from system and	(2nd cycle)	15		15	-	-	15
	replace with a spare Unit *B*							
04	Transport faulty Unit to int. shop		30		30	-	-	30
05	Apply power to faulty unit. Check	(3rd cycle)	20		-	20	-	20
	for output squarewave signal							
06	Check signal input to Assembly A-1		15		-	15	-	15
	(20v P P @ T P.7)							
07	Check signal input to Assembly A-7	(4th cycle)	20		-	20	-	20
	(squarewave, PW-5 µsec @ T. P. 2)							
08	Remove faulty A-7 and replace		10		10	-	-	10
09	Transport faulty Assembly A-7 to	14 calendar days in transit						
	supplier for depot maintenance							
10	Check A-7 and verify faulty condition	(5th cycle)	25		-	-	25	25
11	Check operation of CB-3A2		15		-	-	15	15
12	Check operation of CB-2A4		10		-	-	10	10
13	Check operation of CB-1A5	(6th cycle)	20		-	-	20	20
14	Remove and replace faulty CB-1A5	(7th cycle)	40		-	40	-	40
	Discard faulty circuit board							
15	Verify that assembly is operational		15		-	-	15	15
	and return to inventory							
		Total	265	0.00450	60	120	110	290

Figure 4.29 Maintenance task analysis (Part 1).

1. Item name/Part No.: Manufacturing Test/A4321			2. Req. No.: 01	3. Requirement: Diagnostic Troubleshooting and repair		4. Req. Freq: 0.00450	5. Maint level: Organization, Intermediate, Depot	6. Ma. Cont. No.: A12B100
7. Task No.	Replacement Parts			Test and support/Handling equipment			16. Description of Facility Requirements	17. Special Technical Data Instructions
	8. Qty. per Assy.	9. Part nomenclature / 11. Part number	10. Rep Freq	12. Qty	13. Item part nomenclature / 15. Item part number	14. Use time (min)		
01	-	-	-	1	Built-in test equip. A123456	5	-	Organizational Maintenance
02				1	Special system tester 0-2310B	25	-	
03	1	Unit *B* B180265X	0.01866	1	Standard tool kit STK-100-B	15	-	
04	-	-	-	1	Standard cart (M-10)	30	-	Intermediate maintenance
05	-	-	-	1	Special system tester 1-8891011-A	20	-	
06	-	-	-	1	Special system tester 1-8891011-A	15	-	
07	-	-	-	1	Special system tester 1-8891011-A	20	-	
08	1	Assembly A-7 MO-2378A	0.00995	1	Special extractor tool EX20003-4	10	-	Refer to special removal instructions
09	-	-	-	1	Container, special handling T-300A	14 days	-	Normal trans. environment
10	-	-	-	1	Special system tester 1-8891011-B	25	Clean room environment	Supplier (depot) maintenance
11	-	-	-	1	C.B. test set D-2252-A	15		
12	-	-	-	1	C.B. test set D-2252-A	10		
13	-	-	-	1	C. B. test set D-2252-A	20		
14	1	CB-1A5 GDA-2210565C	0.00450	1	Special extractor tool/EX45112-63 Standard tool kit STK-200	40		
15	-	-	-	1	Special system tester 1-8891011-B	15		Return operating assy to inventory

Figure 4.30 Maintenance task analysis (Part 2).

4. From Tasks 10 to 15, a special cleanroom facility is required for maintenance. Assuming that the various assemblies of Unit B are repaired (versus being classified as "discard at failure"), then it would be worthwhile to investigate changing the design of these assemblies such that a cleanroom environment is not required for maintenance. In other words, can the expensive maintenance facility requirement be eliminated?

5. There is an apparent requirement for a number of new "special" test equipment/tool items; that is, special system tester (0-2310B), special system tester (I-8891011-A), special system tester (I-8891011-B), CB test set (D-2252-A), special extractor tool (EX20003-4), and special extractor tool (EX45112-63). Usually, these *special* items are limited as to general application for other systems, and are expensive to acquire and maintain. Initially, one should investigate whether or not these items can be eliminated; if test equipment/tools are required, can *standard* items be utilized (in lieu of special items)? Also, if the various special testers are required, can they be integrated into a "single" requirement? In other words, can a single item be designed to replace the three special testers and the CB test set? Reducing the overall requirements for special test and support equipment is a major objective.

6. From Task 09, there is a special handling container for the transportation of Assembly A-7. This may impose a problem in terms of the availability of the container at the time and place of need. It would be preferable if normal packaging and handling methods could be utilized.

7. From Task 14, the removal and replacement of CB-1A5 takes 40 minutes and requires a highly skilled individual to accomplish the maintenance task. Assuming that Assembly A-7 is repairable, it would be appropriate to simplify the circuit board removal/replacement procedure by incorporating plug-in components, or at least simplify the task to allow one with a basic skill level to accomplish.

4.2.5 Reliability-Centered Maintenance (RCM)[14]

Definition of the Problem. Reliability-centered maintenance (RCM) is a systematic approach to develop a focused, effective, and cost-efficient preventive maintenance program and control plan for a system or product. This technique is best initiated during the early system design process and evolves as the system is developed, produced, and deployed. However, this technique also can be used to evaluate preventive maintenance programs for existing systems, with the objective of continuous product/process improvement.

The RCM technique was developed in the 1960s primarily through the efforts of the commercial airline industry. The approach is through a structured decision tree that leads the analyst through a "tailored" logic to delineate the most applicable preventive maintenance tasks (their nature and frequency). The overall process involved in implementing the RCM technique is illustrated in Figure 4.31. Note that the functional

[14] This case study was taken in part from Blanchard, B. S., D. Verma, and E. L. Peterson, *Maintainability: A Key to Effective Serviceability and Maintenance Management*, John Wiley & Sons, Inc., New York, N.Y., 1995.

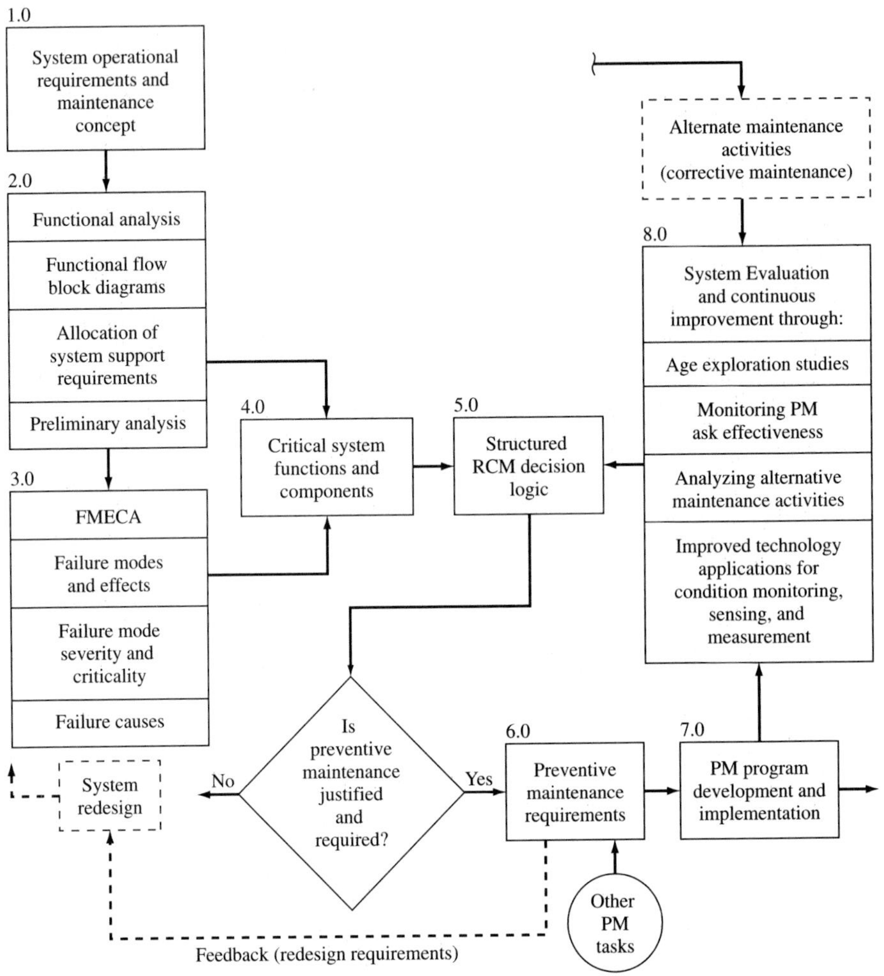

Figure 4.31 Reliability-centered maintenance analysis process.

analysis and the FMECA are necessary inputs to the RCM, and that there are trade-offs resulting in a balance between preventive maintenance and the accomplishment of corrective maintenance. Figure 4.32 presents a simplified RCM decision logic, where system safety is a prime consideration along with performance and cost.[15]

[15] A maintenance steering group (MSG) was formed in the 1960s which undertook the development of this technique. The result was a document called "747 Maintenance Steering Group Handbook: Maintenance Evaluation and Program Development (MSG-1)" published in 1968. This effort, focused toward a particular aircraft, was next generalized and published in 1970 as "Airline/Manufacturer Maintenance Program Planning Document-MSG2." The MSG-2 approach was further developed and published in 1978 as "Reliability Centered Maintenance." Report Number A066-579, prepared by United Airlines, and in 1980 as "Airline/Manufacturer Maintenance Program Planning Document-MSG3." The MSG-3 report has been revised and is currently available as "Airline/Manufacturer Maintenance Program Development Document (MSG-3), 1993." These reports are available from the Air Transport Association.

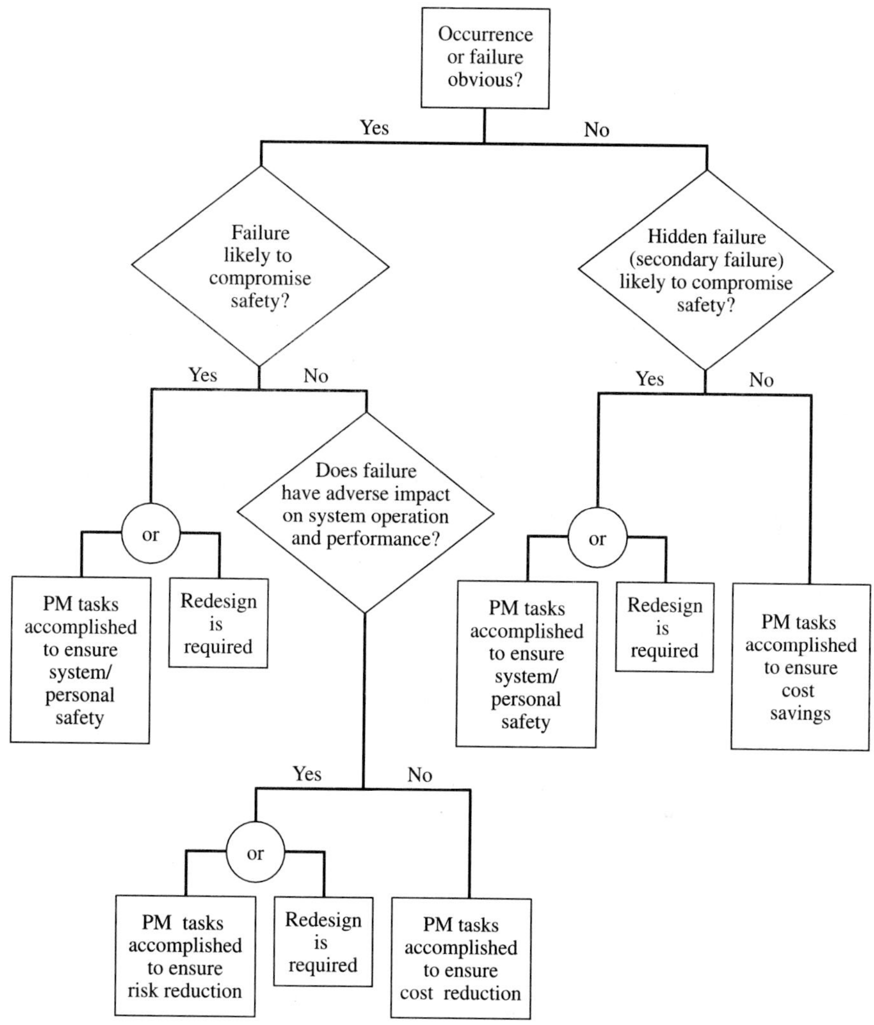

Figure 4.32 Simplified RCM decision logic.

Analysis Approach. The major steps in accomplishing an RCM analysis follow:

1. Identify the critical system functions or components. The first step in this analysis is to identify critical system functions or components; for example, airplane wings, car engine, printer head, video head, and so on. Criticality in terms of this analysis is a function of the failure frequency, the failure effect severity, and the probability of detection of the relevant failure modes. The concept of criticality is discussed in more detail in Section 4.2.2. This step is facilitated through outputs from the system functional analysis (see Section 3.6) and the failure mode, effects, and criticality analysis (FMECA). This is also depicted in Figure 4.31, Blocks 1.0 to 4.0.

2. Apply the RCM decision logic and PM program development approach. The critical system elements are next subjected to the tailored RCM decision logic. The objective here is to better understand the nature of failures associated with the critical system functions or components. In each case, and whenever feasible, this knowledge is translated into a set of preventive maintenance tasks, or a set of redesign requirements. A simplified illustrative RCM decision logic is depicted in Figure 4.32. Numerous decision logics, with slight variations to the original MSG-3 logic and tailored to better address certain types of systems, have been developed and are currently being utilized.[16]

These slight variations notwithstanding (as illustrated in Figure 4.32), the first concern is whether a *failure is evident or hidden.* A failure could become evident through the aid of certain color-coded, visual gauges, or alarms. It may also become evident if it has a perceptible impact on system operation and performance. Conversely, a failure may not be evident (i.e., hidden) in the absence of an appropriate alarm, and even more so if it does not have an immediate or direct impact on system performance. For example, a leaking engine gasket is not likely to reflect an immediate and evident change in the automobile's operation, but it may in time and, after most of the engine oil has leaked, cause engine seizure. In the event that a failure is not immediately evident, it may be necessary to either institute a specific fault-finding task as part of the overall PM program or design in an alarm that signals a failure (or pending failure).

The next concern is whether the failure is likely to compromise personal safety or system functionality. Queries exist in the decision logic to clarify this and other likely impacts of failures. This step in the overall process can be facilitated by the results of the FMECA (Section 4.22). The objective is to better understand the basic nature of the failure being studied. Is the failure likely to compromise the system or personnel safety? Does it have an operational or economic impact? For example, a failure of an aircraft wing may be safety-related, whereas a certain failure in the case of an automobile engine may result in increased oil consumption without any operational degradation, and will therefore have an economic impact. In another case, a failed printer head may result in a complete loss of printing capability and is said to have an operational impact, and so on.

Once the failure has been identified as a certain type, it is then subjected to another set of questions. However, to answer this next set of questions adequately, the analyst must thoroughly understand the nature of the failure from a *physics-of-failure* perspective. For example, in the event of a crack in the airplane wing, how fast is this crack likely to propagate? How long before such a crack causes a functional failure?

These questions have an underlying objective of delineating a feasible set of compensatory provisions or preventive maintenance tasks. Is a lubrication or servicing task applicable and effective, and, if so, what is the most cost-effective and efficient fre-

[16] RCM decision logics, with some variations, have also been proposed in: (a) MIL-STD-2173(AS)—*Reliability-Centered Maintenance Requirements for Naval Aircraft, Weapons Systems, and Support Equipment;* (b) AMC-P-750-2—*Guide to Reliability-Centered Maintenance:* and (c) Moubray, John, *Reliability-Centered Maintenance*, Butterworth-Heinemann, London, United Kingdom, 1991.

quency? Will a periodic check help preclude the failure, and at what frequency? Periodic inspections or checkouts are likely to be most applicable in situations where a failure is unlikely to occur immediately, but is likely to develop at a certain rate over a period of time. The frequency of inspections can vary from very infrequently to continuously, as in the case of condition monitoring. Some of the more specific queries are presented in Figure 4.32. In each case, the analyst must not only respond with a "yes" or "no," but should also give specific reasons for each response. Why would lubrication either make, or not make, any difference? Why would periodic inspection be a *value-added* task? It could be that the component's wearout characteristics have a predictable trend, in which case inspections at predetermined intervals could preclude corrective maintenance. Would it be effective to discard and replace certain system elements in order to upgrade the overall inherent reliability? And, if so, at what intervals or after how many hours of system operation (e.g., changing the engine oil after 3000 miles of driving)? Further, in each case a trade-off study, in terms of the benefit/cost and overall impact on the system, needs to be accomplished between performing a task and not performing it.

In the event that a set of applicable and effective preventive maintenance requirements are delineated, they are input to the preventive maintenance program development process and subsequently implemented, as shown in Figure 4.31, Blocks 5.0 to 7.0. If no feasible and cost-effective provisions or preventive maintenance tasks can be identified, a redesign effort may need to be initiated.

3. Accomplish PM program implementation and evaluation. Very often, the PM program initially delineated and implemented is likely to have failed to consider certain aspects of the system, delineated a very conservative set of PM tasks, or both. Continuous monitoring and evaluation of preventive maintenance tasks along with all other (corrective) maintenance actions is imperative in order to realize a cost-effective preventive maintenance program. This is depicted in Figure 4.31, Block 8.0. Further, given the continuously improving technology applications in the field of condition monitoring, sensing, and measurement. PM tasks need to be reevaluated and modified whenever necessary.

Often, when the RCM technique is conducted in the early phases of the system design and development process, decisions are made in the absence of ample data. These decisions may need to be verified and modified, whenever justified, as part of the overall PM evaluation and continuous improvement program. Age exploration studies are often conducted to facilitate this process. Tests are conducted on samples of unique system elements or components with the objective of better understanding their reliability and wear-out characteristics under actual operating conditions. Such studies can aid the evaluation of applicable PM tasks, and help delineate any dominant failure modes associated with the component being monitored or any correlation between component age and reliability characteristics. If any significant correlation between age and reliability is noticed and verified, the associated PM tasks and their frequency may be modified and adapted for greater effectiveness. Also, redesign efforts may be initiated to account for some, if any, of the dominant component failure modes.

Analysis Results. Quite often in the early design process, as system components are being selected, the issue of maintenance is ignored altogether. If addressed, however, the designer may tend to specify components requiring some preventive maintenance (usually recommended by the manufacturer). By doing so, the perception is that such PM recommendations are based on actual knowledge of the component in terms of its physical characteristics, expected modes of failure, and so on. It is also believed that the more preventive maintenance required, the better the reliability. In any event, there is often a tendency to overspecify the need for PM because of the reliability issue, particularly if the component *physics-of-failure* characteristics are not known and the designer assumes a conservative approach just in case.

Experience has indicated that although the accomplishment of some selective preventive maintenance is essential, the overspecification of PM activities can actually cause a degradation of system reliability and can be quite costly! The objective is to specify the correct amount of PM, to the depth required, and at the proper frequency;—that is, *not too much or too little!* Further, as systems age, the required amount of PM may shift from one level to another. The application of RCM methods on a continuing basis is highly recommended, particularly when evaluating systems from a life-cycle cost perspective.

4.2.6 Level-Of-Repair Analysis (LORA)[17]

Definition of the Problem. In defining the detailed maintenance concept and establishing criteria for system design, it is necessary to determine whether items should be repaired at the intermediate level, repaired at the depot/supplier facility, or discarded in the event of failure. This example evaluates these alternatives through a level of repair analysis.

A computer subsystem is to be distributed in quantities of 65 throughout three major geographical areas. The subsystem will be utilized to support both scientific and management functions within various industrial firms and government agencies. Although the actual system utilization will vary from one consumer organization to the next, an average utilization of 4 hours per day (for a 360-day year) is assumed.

The computer subsystem is currently in the early development stage, should be in production in 18 months, and will be operational in 2 years. The full complement of 65 computer subsystems is expected to be in use in 4 years, and will be available through the eighth year of the program before subsystem phaseout begins. The life cycle, for the purposes of the analysis, is 10 years.

[17] Repair-level analysis is one of the key supportability analysis inputs to system design, initially in the development of the maintenance concept and in the establishment of design criteria at the subsystem-level (leading to the identification of COTS applications), and later in the evaluation of an item to determine the most efficient approach (i.e., repair at the organizational level, repair at the intermediate level, repair at the depot or supplier level, repair by a third-party maintenenance organization, discard at failure, and the extent of recylcing and level of disposal in the event of a discard decision). Refer to MIL-HDBK-502, "Department of Defense Handbook Acquisition Logistics," Department of Defense, Washington, D.C., May 1997, for applications related to defense systems.

Based on early design data, the computer subsystem will be packaged in five major units, with a built-in test capability that will isolate faults to the unit level. Faulty units will be removed and replaced at the organizational level (i.e., customer's facility), and sent to the intermediate maintenance shop for repair. Unit repair will be accomplished through assembly replacement, and assemblies will be either repaired or discarded. There is a total of 15 assemblies being considered, and the requirement is to justify the assembly repair or discard decision on the basis of life-cycle cost criteria. The operational requirements, maintenance concept, and program plan are illustrated in Figure 4.33.

Analysis Approach. The stated problem definition pertains primarily to the analysis of 15 major assemblies of the given computer subsystem configuration to determine whether the assemblies should be repaired or discarded when failures occur. In other words, the various assemblies will be individually evaluated in terms of (1) assembly repair at the intermediate level of maintenance, (2) assembly repair at the supplier or depot level of maintenance, and (3) disposing of the assembly. Lifecycle costs, as applicable to the assembly level, shall be developed and employed in the alternative selection process. Total overall computer subsystem costs have been determined at a higher level and are not included in this case study.

Given the information in the problem statement, the next step is to develop a cost-breakdown structure (CBS) and to establish evaluation criteria. The CBS employed in this analysis and specific cost categories are similar to the structure presented in Figure 2.26 and in Appendix E. Not all cost categories in the CBS are applicable in this case; however, the structure is used as a starting point, and those categories that are applicable are identified accordingly.

The evaluation criteria include consideration of all costs in each applicable category of the CBS, but the emphasis is on operation and support (O&S) costs as a function of acquisition cost. Thus, the research and development cost and the production cost are presented as one element, whereas various segments of O&S costs are identified individually. Figure 4.34 presents evaluation criteria, cost data, and a brief description and justification supporting each category. The information shown in the figure covers only one of the 15 assemblies but is typical for each case.

In determining these costs, the analyst must follow an approach similar to that conveyed in the process illustrated in Figure 4.11 and discussed in Section 4.2.1. That is, operational requirements and a basic maintenance concept must be defined; a program plan must be established; an inventory profile must be identified; a CBS must be developed; reliability, maintainability, and logistic support factors must be identified; cost-estimating relationships must be developed; and so on.

The next step is to employ the same criteria presented in Figure 4.34 to determine the recommended repair-level decision for each of the other 14 assemblies (i.e., assemblies 2 through 15). Although acquisition costs, reliability and maintainability factors, and certain logistics requirements are different for each assembly, many of the cost-estimating relationships are the same. The objective is to be *consistent* in analysis approach and in the use of input cost factors to the maximum extent possible and where appropriate. The summary results for all 15 assemblies are presented in Figure 4.35.

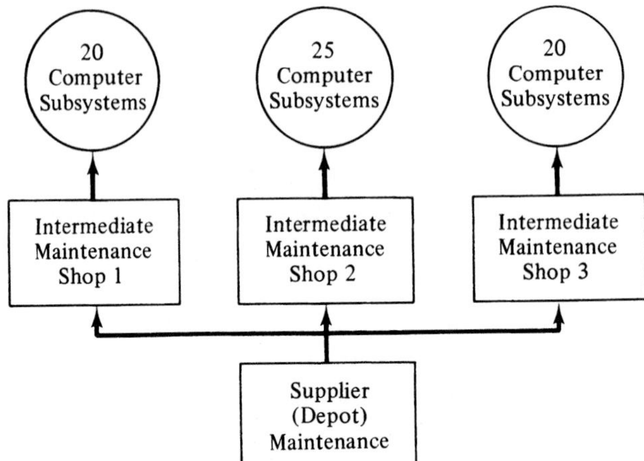

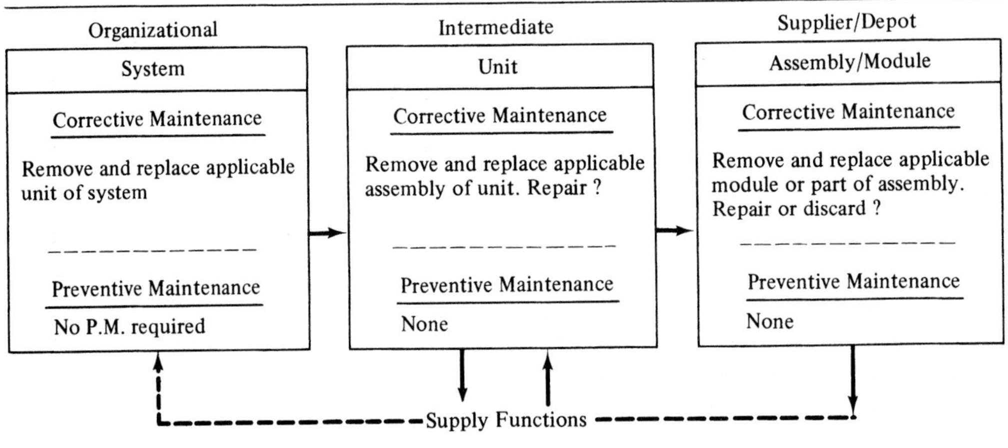

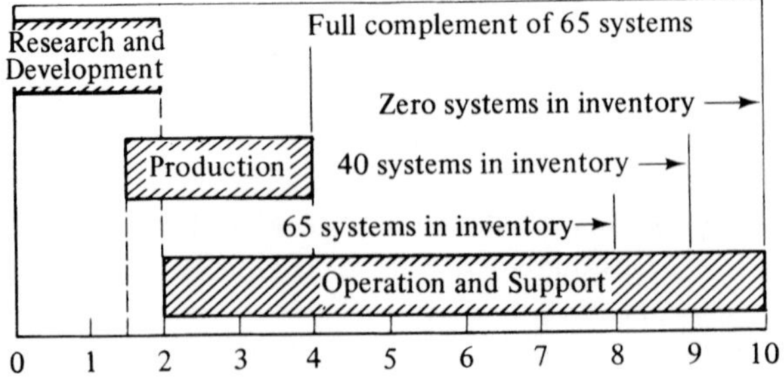

Figure 4.33 Basic subsystem concepts and plan.

Evaluation Criteria	Repair at Intermediate Cost ($)	Repair at Supplier Cost ($)	Discard at Failure Cost ($)	Description and Justification
1. Estimated acquisition costs for Assembly A-1 (to include R&D cost and production cost)	550/assy. or 35,750	550/assy. or 35,750	475/assy. or 30,875	Acquisition cost includes all applicable costs in categories C_R, C_{PI}, C_{PM}, and C_{PQ} allocated to each Assembly A-1, based on a requirement of 65 systems. Assembly design and production are simplified in the discard area.
2. Unscheduled maintenance cost (C_{OLA})	6,480	8,100	Not applicable	Based on the 8-year useful system life, 65 systems, a utilization of 4 hours/day, a failure rate (λ) of 0.00045 for Assembly A-1, and a $\overline{M}_{ct}$ of 2 hours, the expected number of maintenance actions is 270. When repair is accomplished, one technician is required on a full-time basis. The labor rates are $12/hour for intermediate maintenance and $15/hour for supplier maintenance.
3. Supply support—spare assemblies (C_{PLS} and C_{OLS})	3,300	4,950	128,250	For intermediate maintenance 6 spare assemblies are required to compensate for transportation time, the maintenance queue, TAT, etc. For supplier/depot maintenance, 9 spare assemblies are required. 100% spares are required in the discard case.
4. Supply support—spare modules or parts for assembly repair (C_{PLS} and C_{OLS})	6,750	6,750	Not applicable	Assume $25 for materials per repair action.
5. Supply support—inventory management (C_{PLS} and C_{OLS})	2,010	2,340	25,650	Assume 20% of the inventory value (spare assemblies, modules, and parts).
6. Test and support equipment (C_{PLT} and C_{OLE})	5,001	1,667	Not applicable	Special test equipment is required in the repair case. The acquisition and support cost is $25,000 per installation. The allocation for Assembly A-1 per installation is $1,667. No special test equipment is required in the discard case.
7. Transportation and handling (C_{OLH})	Not applicable	4,725	Not applicable	Transportation costs at the intermediate level are negligible. For supplier maintenance, assume 540 one-way trips at $175/100 pounds. One assembly weighs 5 pounds.
8. Maintenance training (C_{OLT})	260	90	Not applicable	Delta training cost to cover maintenance of the assembly is based on the following: Intermediate—26 students, 2 hours each, $200/student week, Supplier—9 students, 2 hours each, $200/student week.
9. Maintenance facilities (C_{OLM})	594	810	Not applicable	From experience, a cost-estimating relationship of $0.55 per direct maintenance labor hour is assumed for the intermediate level and $0.75 is assumed for the supplier level.
10. Technical data (C_{OLD})	1,250	1,250	Not applicable	Assume 5 pages for diagrams and text covering assembly repair at $250/page.
11. Disposal (C_{DIS})	270	270	2,700	Assume $10/assembly and $1/module or part as the cost of disposal
Total estimated cost	61,665	66,702	187,475	

*The cost breakdown structure (CBS) used as a basis for the data presented herein was taken from Blanchard, B.S., *Design and Manage to Life Cycle Cost*, M/A Press (Dilithium Press, Inc.), Beaverton, Oreg., 1978.

Figure 4.34 Repair versus discard evaluation (Assembly "A-1").

Referring to the figure, note that the decision for Assembly A-1 favors repair at the intermediate level; the decision for Assembly A-2 is repair at the supplier or depot level; the decision for Assembly A-3 is not to accomplish repair at all but to discard the assembly when a failure occurs; and so on. The table reflects recommended policies for each individual assembly. The cost of this mixed approach is reflected in the fifth column ($983,984). On the other hand, an overall policy, addressing all 15 assemblies as an integral package considering repair at a designated location, favors repair at the supplier ($ 1,037,362).

Prior to arriving at a final conclusion, the analyst should reevaluate each situation where the decision is close. Referring to Figure 4.34, it is clearly uneconomical to accept the discard decision; however, the two repair alternatives are relatively close.

Assembly Number	*Maintenance Status*				*Decision*
	Repair at Intermediate Cost ($)	*Repair at Supplier Cost ($)*	*Discard at Failure Cost ($)*	*Optimum Policy Cost ($)*	
A-1	61,665	66,702	187,475	61,665	Repair—intermediate
A-2	58,149	51,341	122,611	51,341	Repair—supplier
A-3	85,115	81,544	73,932	73,932	Discard
A-4	85,778	78,972	65,072	65.072	Discard
A-5	66,679	61,724	95,108	61,724	Repair—supplier
A-6	65,101	72,988	89,216	65,101	Repair—intermediate
A-7	72,223	75,591	92,114	72,223	Repair—intermediate
A-8	89,348	78,204	76,222	76,222	Discard
A-9	78,762	71,444	89,875	71,444	Repair—supplier
A-10	63,915	67,805	97,212	63,915	Repair—intermediate
A-11	67,001	66,158	64,229	64,229	Discard
A-12	69,212	71,575	82,109	69,212	Repair—intermediate
A-13	77,101	65,555	83,219	65,555	Repair—supplier
A-14	59,299	62,515	62,005	59,299	Repair—intermediate
A-15	71,919	65,244	63,050	63,050	Discard
Policy cost	1,071,267	1,037,362	1,343,449	983,984	

Figure 4.35 Summary of repair-level costs for 15 assemblies within a "unit".

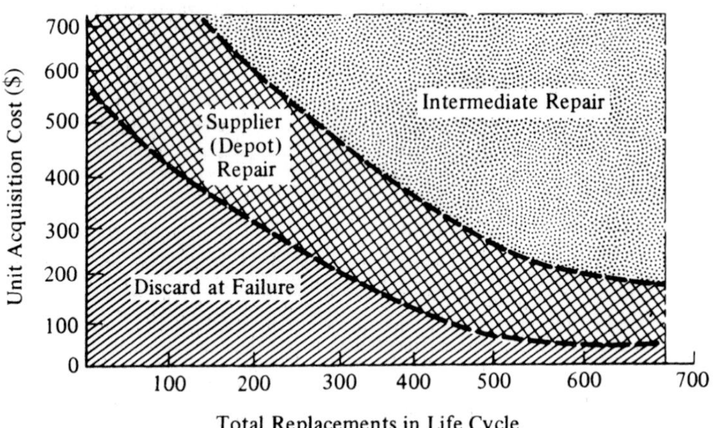

Figure 4.36 Economic screening criteria.

Based on the results of the various individual analyses, the analyst knows that repair-level decisions are highly dependent on the unit acquisition cost of each assembly and the total estimated number of replacements over the expected life cycle (i.e., maintenance actions based on assembly reliability). The trends are illustrated in Figure 4.36, where the decision tends to shift from discard to repair at the intermediate level as the unit acquisition cost increases and the number of replacements increases (or the reliability decreases).[18] In instances where the individual analysis result lies close to the

[18] The curves projected in Figure 4.36 are characteristic for this particular life-cycle cost analysis and will vary with changes in operational requirements, system utilization, the maintenance concept, production requirements, and so on.

crossover lines in the figure, the analyst may wish to review the input data, the assumptions, and accomplish a sensitivity analysis involving the high-cost contributors. The purpose is to assess the risk involved and verify the decision. This is the situation for Assembly A-1, where the decision is close relative to repair at the intermediate level versus repair at the supplier's facility.

After reviewing the individual analyses of the 15 assemblies to ensure that the best possible decision is reached, the results in Figure 4.35 are updated as required. Assuming that the decisions remain basically as indicated, the analyst may proceed in either of two ways. First, the decisions in Figure 4.35 may be accepted without change, supporting a *mixed* policy with some assemblies being repaired at each level of maintenance and other assemblies being discarded at failure. With this approach, the analyst should review the interaction effects that could occur (i.e., the effects on spares, utilization of test and support equipment, maintenance personnel utilization, etc.). In essence, each assembly is evaluated individually based on certain assumptions; the results are reviewed in the context of the whole; and possible feedback effects are assessed to ensure that there is no significant impact on the decision.

A second approach is to select the overall least-cost policy for all 15 assemblies maintained in one area as an entity (i.e., assembly repair at the supplier or depot level of maintenance). In this case, all assemblies are designated as being repaired at the supplier's facility and each individual analysis is reviewed in terms of the criteria in Figure 4.34 to determine the possible interaction effects associated with the single policy. The result may indicate some changes to the values in Figure 4.35.

Finally, the output of the repair-level analysis must be reviewed to ensure compatibility with the initially specified system maintenance concept. The analysis data may either directly support and be an expansion of the maintenance concept, or the maintenance concept will require change as a consequence of the analysis. If the latter occurs, other facets of system design may be significantly impacted. The consequences of such maintenance concept changes must be thoroughly evaluated prior to arriving at a final repair-level decision. The overall process employed in level of repair analyses is illustrated in Figure 4.37.

Analysis Results. Based on the results of this particular analysis, the decision favors the "mixed" policy identified in Figure 4.35. This is based on an evaluation in terms of *economic* criteria only. However, as conveyed in Figure 4.37, there may be other influencing noneconomic factors that will influence the final decision. Such may include *technical* factors (e.g., those related to the complexity of the technology where maintenance can not be effectively accomplished at the intermediate level), *proprietary* (e.g., those uniquely protected items where maintenance can only be accomplished at the supplier/producer level and where there are warranties/guarantees in place), *safety* or *security* (e.g., those items where safety is of concern and where higher-skilled personnel or better facilities are required to perform the necessary maintenance), *policy* issues and/or *political* factors (e.g., where a policy exists in favor of "outsourcing" to industries in selected countries around the world or in helping to maintain critical industries in selected states in the United States). These issues may ultimately influence many of the decisions in this area. However, the *economic screening* approach described through this case study constitutes a good place to begin.

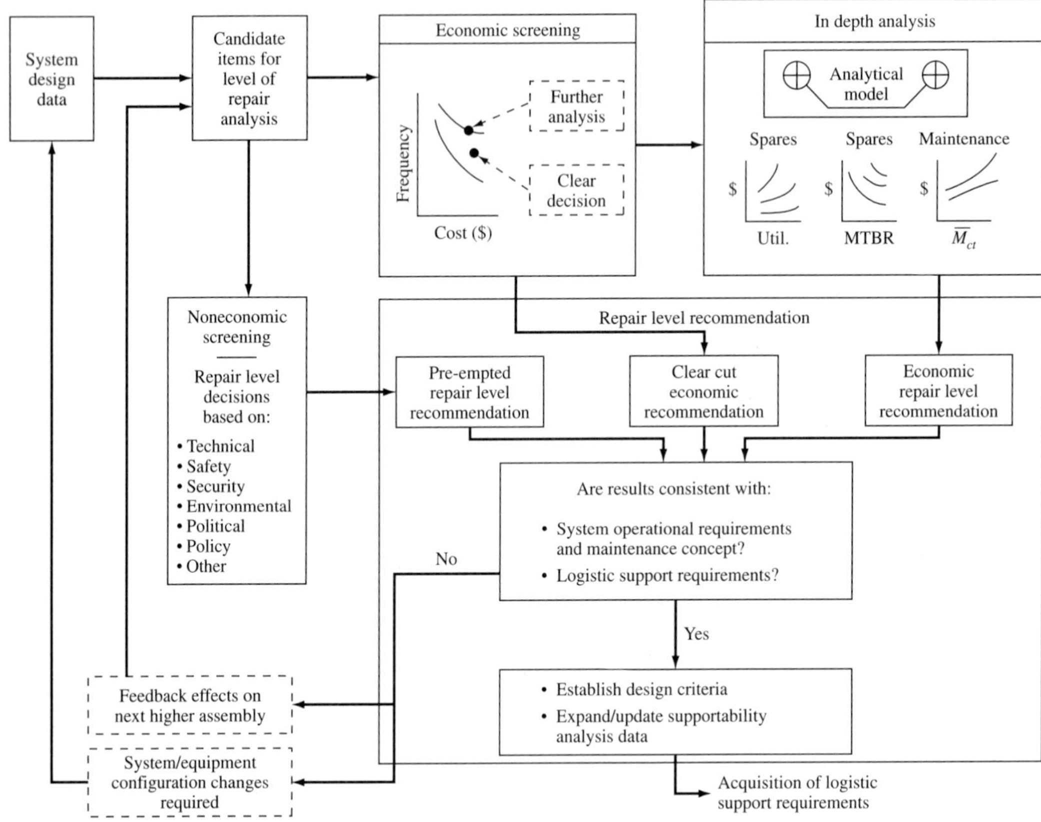

Figure 4.37 Level of repair analysis process.

4.2.7 Evaluation of Design Alternatives

Definition of the Problem. Company DEF is responsible for the design and development of a major system, which, in turn, is comprised of a number of large subsystems. Subsystem XYZ is to be procured from an outside supplier, and there are three different configurations being evaluated for selection. Each of the configurations represents an existing design, with some redesign and additional development necessary to be compatible with the requirements for the new system. The evaluation criteria include different parameters such as performance, operability, effectiveness, design characteristics, schedule, and cost. Both qualitative and quantitative considerations are covered in the evaluation process.

Analysis Approach. The analyst commences with the development of a list of evaluation parameters, as depicted in Figure 4.38. In this instance, there is no single parameter (or figure of merit) that is appropriate by itself, but there are 11 factors that must be considered on an integrated basis. Given the evaluation parameters, the next step is to determine the level of importance of each. Quantitative weighting factors from

Item	Evaluation Parameter	Weighting Factor	Conf. A		Conf. B		Conf. C	
			Base Rate	Score	Base Rate	Score	Base Rate	Score
1	Performance—compatibility, maintenance capability, maintenance load, method/mode of testing, versatility of testing	14	6	84	9	126	3	42
2	Operability—simplicity of operation	4	10	40	7	28	4	16
3	Effectiveness—A_o, MTBM, $\overline{M}_{ct}$, $\overline{\text{M}}$pt, MLH/OH	12	5	60	8	96	7	84
4	Design characteristics—reliability, maintainability, human factors, safety, transportability, producibility, interchangeability	9	8	72	6	54	3	27
5	Design data—design drawings, specifications, logistics provisioning data, technical manuals, reports	2	6	12	8	16	5	10
6	Test aids—general test equipment, calibration standards, maintenance tapes	3	5	15	8	24	3	9
7	Facilities and utilities—space, weight, volume, environment, power, heat, water, air conditioning	5	7	35	8	40	4	20
8	Spare/repair parts—part type and quantity, standard parts, procurement time	6	9	54	7	42	5	30
9	Flexibility/growth potential—test accuracies, performance range, space, reconfiguration, design change acceptability	3	4	12	8	24	6	18
10	Schedule—R&D, production	17	7	119	8	136	9	153
11	Cost—life cycle (R&D, investment, O&M)	25	10	250	9	225	5	125
Subtotal				753		811		534
Derating factor (development risk)				113 15%		81 10%		107 20%
Grand total		100		640		730		427

Figure 4.38 Evaluation summary (three design configurations).

zero to 100 are assigned to each parameter in accordance with the degree of importance. The QFD, Delphi method, or some equivalent evaluation technique, may be used to establish the weighting factors. The sum of all weighting factors is 100.[19]

For each of the 11 parameters identified in Figures 4.38, the analyst may wish to develop a special checklist including criteria against which to evaluate the three proposed configurations. For instance, the parameter "PERFORMANCE" may be described in terms of degrees of desirability; that is, "highly desirable," "desirable," or "less desirable." Although each configuration must comply with a minimum set of requirements, one may be more desirable than the next when looking at the proposed performance characteristics. In other words, the analyst should break down each evaluation parameter into "levels of goodness"!

Each of the three proposed configurations of Subsystem *XYZ* is evaluated independently using the special checklist criteria. Base rating values from zero to 10 are applied according to the degree of compatibility with the desired goals. If a "highly desirable" evaluation is realized, a rating of 10 is assigned.

The base-rate values are multiplied by the weighting factors to obtain a score. The total score is then determined by adding the individual scores for each configuration. Because some redesign is required in each instance, a special derating factor is applied to cover the risk associated with the failure to meet a given requirement. The resultant values from the evaluation are summarized in Figure 4.38.

Analysis Results. From Figure 4.38, Configuration *B* represents the preferred approach based on the highest total score of 730 points. This configuration is recommended on the basis of its inherent features relating to performance, operability, effectiveness, design characteristics, design data, and so on.

4.3 SUPPORTABILITY ANALYSIS APPLICATIONS

The examples presented herein represent a very small sample of the type of problems facing the designer at the early stages of system development. Figures 4.39 and 4.40 illustrate some additional trade-off areas. In addition, the problems as defined are addressed individually. Although these examples are realistic, quite often one is required to evaluate many different facets of a system on a relatively concurrent basis. For instance, for a given design configuration it may be feasible to accomplish a system availability analysis, a shop turnaround-time analysis, a level-of-repair analysis, and a spares inventory policy evaluation as part of verifying design adequacy in terms of compliance with system requirements. One may also wish to determine the impact of one system parameter on another or the interaction effects between two elements of logistic support. In accomplishing such, problem resolution may require the utilization of a number of different models combined in such a manner to provide a variety of output factors.[20]

[19] The selection of evaluation criteria should reflect the results from the selection and prioritization of TPMs (refer to Section 3.5).

[20] Appendix D provides some examples of experience in this area.

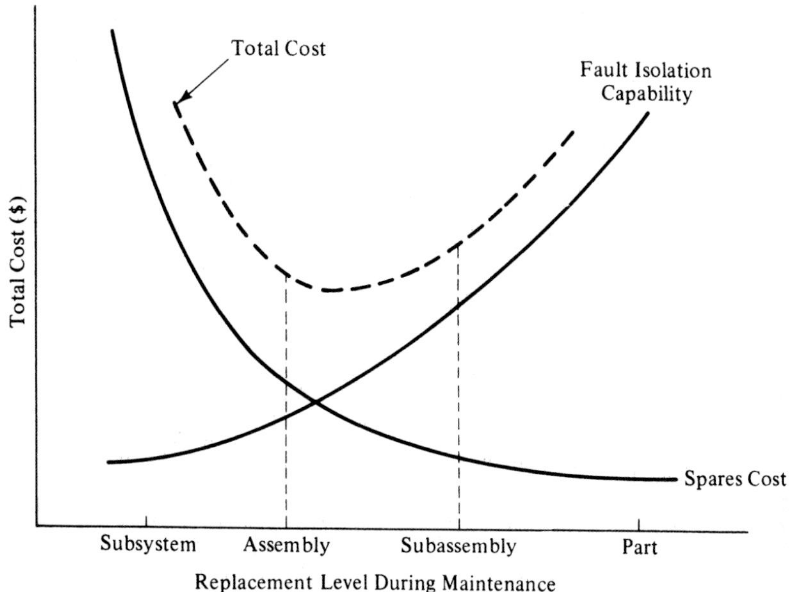

Figure 4.39 Cost-effectiveness of diagnostic capability.

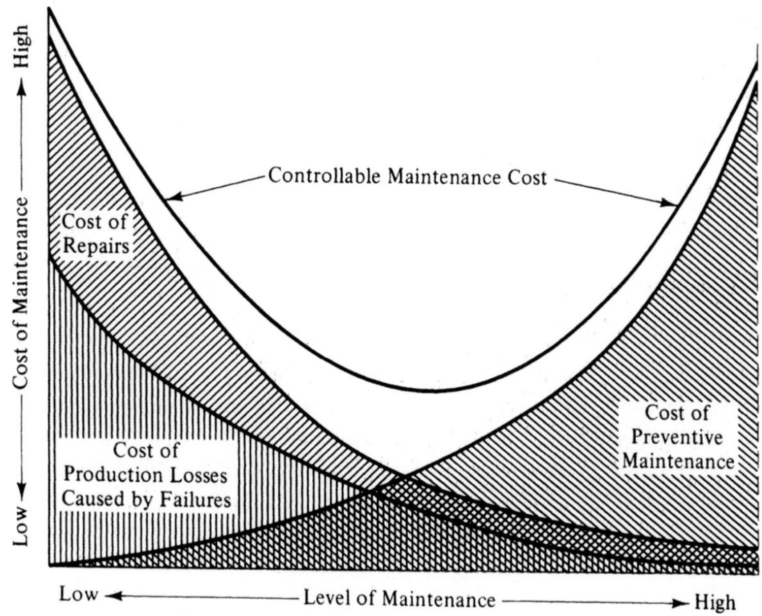

Figure 4.40 Corrective versus preventive maintenance cost.

The analysis of a system (to include the prime equipment, software, facilities, and its associated elements of logistic support) is sometimes fairly complex; however, this process can be simplified by applying the right techniques and developing the proper analytical tools. The combining of techniques to facilitate the analysis task can be accomplished by following the steps in Section 4.1 (i.e., defining the problem, identifying feasible alternatives, selecting evaluation criteria, etc.). The output may appear as illustrated in Figure 4.41.

Referring to the figure, the analyst may wish to develop a series of individual models as illustrated. Each model may be used separately to solve a specific detailed problem, or the models may be combined to solve a higher-level problem. The models may be used to varying degrees depending on the depth of analysis required. In any event, the overall analytical task must be approached in an organized methodical manner and applied judiciously to prove the results desired.

The supportability analysis process shown in Figure 4.2 can be implemented, as part of the overall system engineering analysis effort, both in the development of new systems and in the evaluation and reengineering (i.e., upgrading) of systems already in use. Referring to Figure 4.42, the system engineering process can be applied as shown. Every time that there is a *newly defined consumer need* or a *new requirement*, the steps identified in blocks 7 through 11 are applicable. This, of course, equates to the process described in Chapter 3. Whenever, there is a desire to *reengineer* or *upgrade* a system that is currently in use, a revised set of goals should be established in the form of "new benchmarks," the current configuration should be evaluated relative to these benchmarks, and modifications for improvement should be incorporated in order to comply with the new set of requirements. These steps are represented by blocks 2 through 6, and 11, of Figure 4.42. Figure 4.43 represents an expansion of these blocks covering the system evaluation and improvement process. The supportabilty analysis activity, shown in Figure 1.7, addresses the requirements analysis, functional analysis and reallocation, system evaluation, identification of high-cost or low-effectiveness contributors, and the evaluation of alternatives as illustrated by blocks 2 through 6 of Figure 4.43. Note that in block 5, the life-cycle cost analysis (LCCA) can be applied in the identification of the *high-cost contributors*.

4.4 CONTINUOUS ACQUISITION AND LIFE-CYCLE SUPPORT (CALS)[21]

CALS is a combined Department of Defense and industry strategy that enables a transition from paper-intensive acquisition and logistic processes to a highly automated and integrated mode of operation. It focuses on the generation, access, management, maintenance, distribution, and use of technical data. Such data include engineering drawings, product definition data, logistics management information (LMI), the results from the supportability analysis summaries (SASs), technical manuals and reports, training materials, and so on. The objective is to provide a framework for the integration and processing of large amounts of data in a timely manner and with accuracies greater than experienced in the past.

[21] Refer to the information provided on CALS in Section 4.1, Figure 4.8, and in footnote 5.

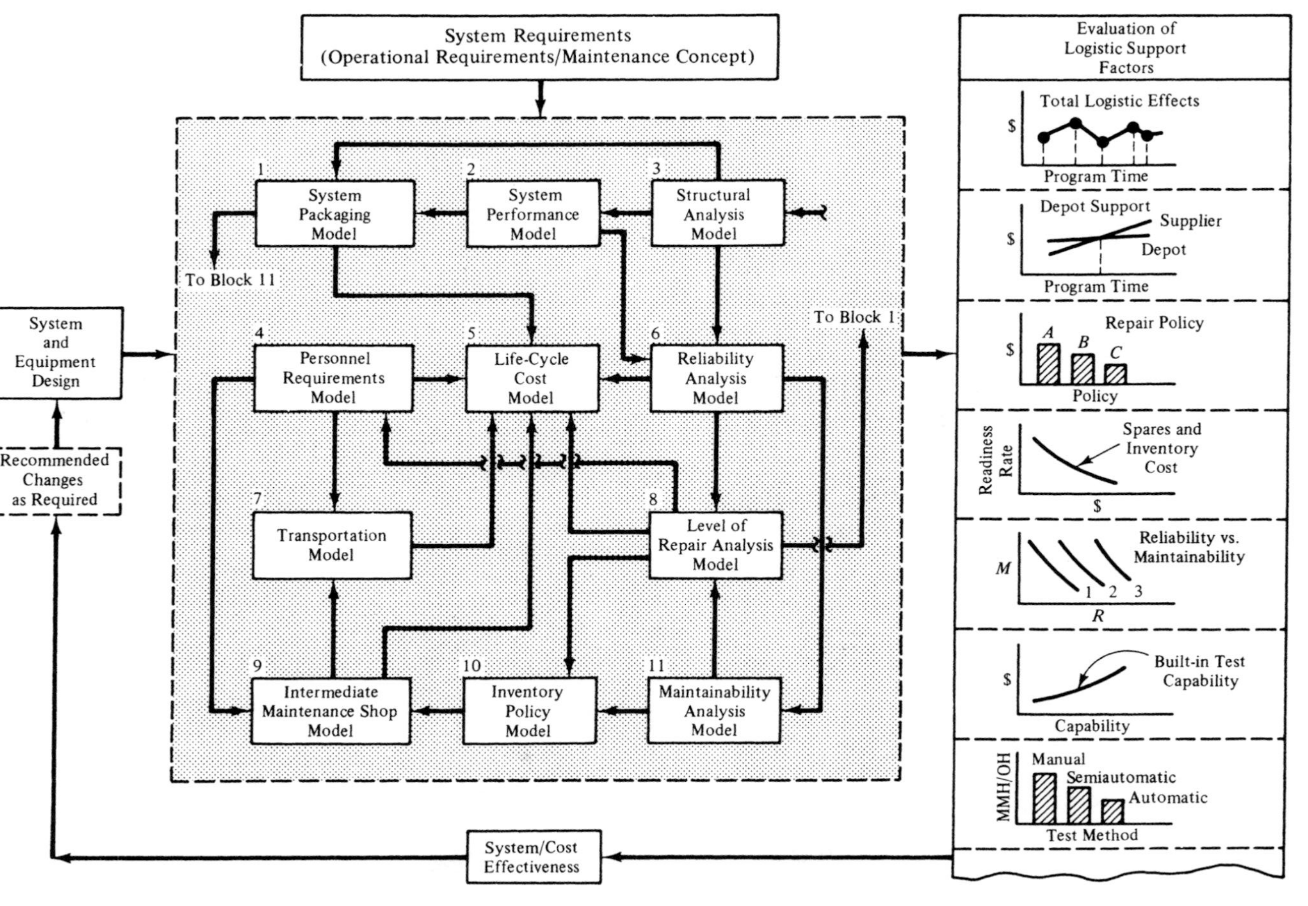

Figure 4.41 Application of support models.

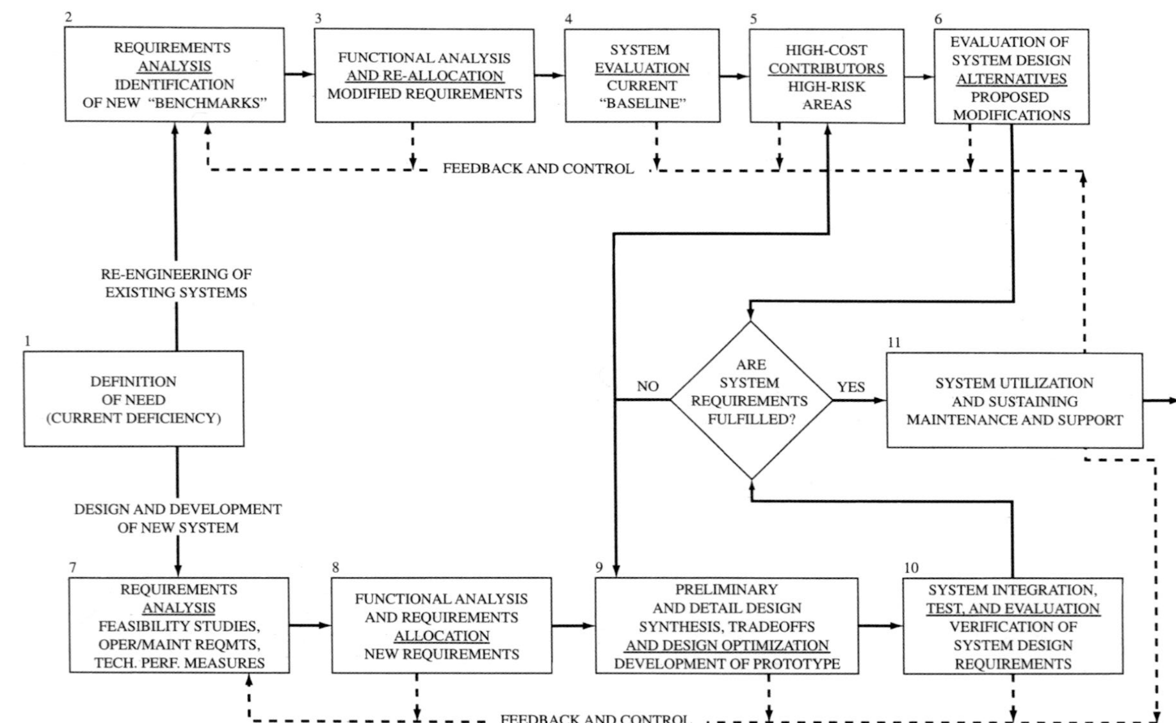

Figure 4.42 Application of the system engineering process with supportability analysis (reference: Figure 1.7).

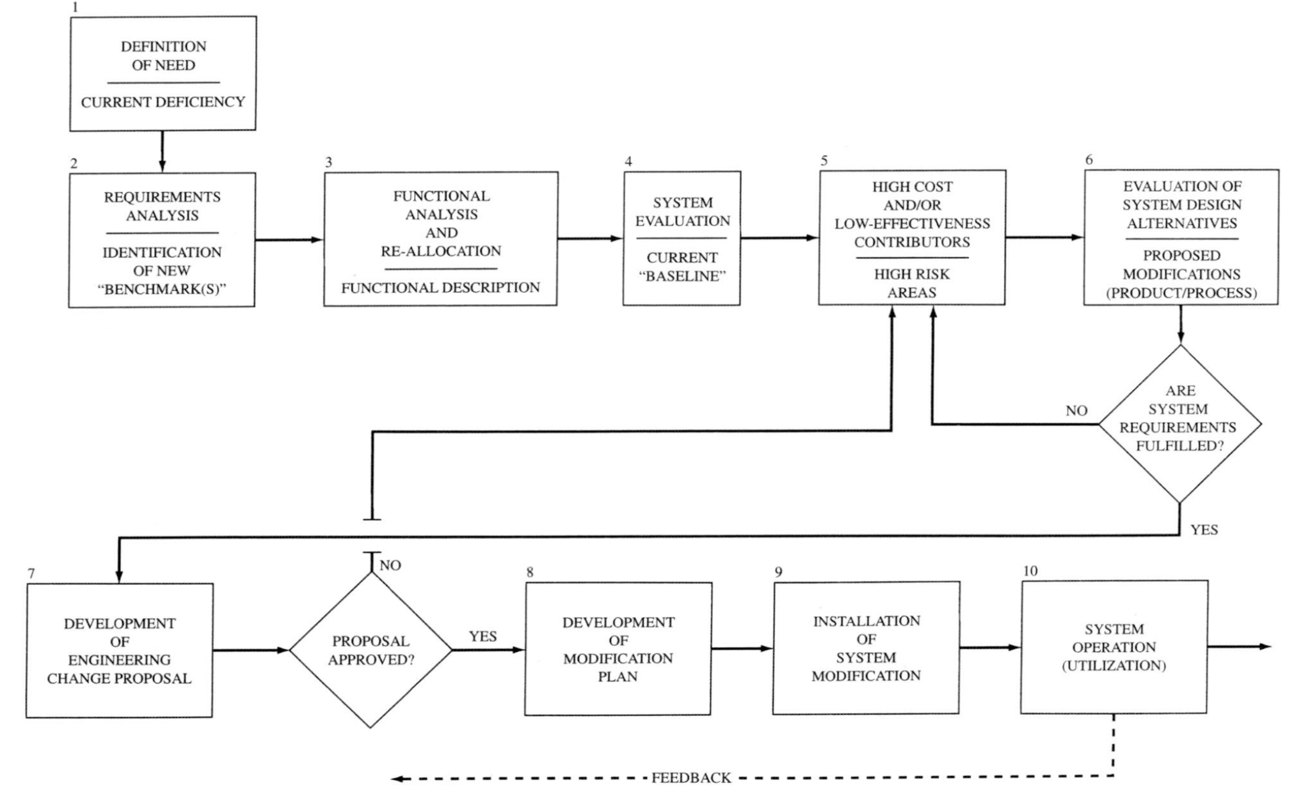

Figure 4.43 System evaluation and improvement process.

The CALS effort was initiated in 1985 primarily because of the inefficiencies and high cost of data preparation and processing experienced in the past. In the design and development of large-scale systems, there are numerous design drawings, materials and parts lists, change notices, and technical reports describing the prime elements of the system. In the logistics area, data requirements have included maintenance and supportability analyses, logistic support analysis records (LSARs), spare/repair parts provisioning data, procurement documentation, test and support equipment requirements reports and design data, technical manuals, calibration requirements data, training data, modification instructions, field data and reports, and so on. The sources of these data may include suppliers located throughout the world, some involved in the design and development of new products and others providing common and standard off-the-shelf components. Because of the many categories and types of data required in system acquisition, combined with the multiple quantity of widely dispersed suppliers providing such data, there have been numerous problems related to (1) the excessive data preparation and processing times; (2) the inability to provide the appropriate data to support design, manufacturing, and/or logistics decisions in a timely manner; (3) the redundancies or the duplication of efforts in data preparation; and (4) the inability to "track" changes and modifications to the system through timely data/documentation revisions. The net result has been costly.

The objectives of the CALS program are to improve the timeliness of data preparation and processing, improve the quality relative to data content and consistency, and to reduce cost. The approach, in the short term, is to convert existing data from a paper product to a digital format using an electronic database means. With the advent of computer-aided technologies, these digital data files can be easily transmitted both throughout a project and between the customer, contractor, and supplier operations. At this stage, the most appropriate applications involve the processing of system design data (or *product definition data*), the generation and processing of logistics management information (LMI) and SAS reparts, the development of training materials, and the development of technical publications (i.e., operating and maintenance instructions, calibration procedures, etc.). In the long term, the objective is to provide a fully integrated technical information system, appropriately tied in with the computer-aided design (CAD) and computer-aided manufacturing (CAM) processes.

With the continued implementation of the CALS approach, it is anticipated that the following benefits will be realized:

1. The preparation of technical data in a digital format, using an appropriate integrated electronic database configuration, can be accomplished in less time, should be more accurate and of higher quality, and should eliminate unnecessary redundancies and the extensive amount of paper required in the past.
2. The use of a shared database (refer to Figure 4.8), with information distributed through local area networks and to suppliers located worldwide, should provide a common baseline definition of the system (and components thereof) to many different organizational groups *concurrently*. Not only should the communications improve, but the time(s) allotted for the transfer of information should be reduced significantly.

3. The processing of logistics provisioning data (and associated procurement information) can be accomplished in an expedited manner. This, in turn, will result in shorter lead times for the acquisition of system components. Additionally, the costs of provisioning should be reduced significantly.

In summary, the CALS initiative is designed to revolutionize the capabilities within the logistics field in terms of data development, access, processing, and utilizations. With the implementation of computer-aided technologies, it should be possible not only to improve the accuracy and quality of the data generated, but to process such data in a timely manner. While the CALS activity is basically directed toward Department of Defense program applications, the concepts are applicable to commercial systems as well.

QUESTIONS AND PROBLEMS

1. What is the *supportability analysis?* What is its purpose and what is included? When is it accomplished in the system life cycle?
2. How does the supportability analysis relate to system engineering and the system engineering process?
3. If you were to implement a supportability analysis effort on a system of your choice what steps would you take in accomplishing your objective(s)?
4. In accomplishing system evaluation, what *criteria* would you use for the purpose? How would you establish the proper quantitative and qualitative "design-to" factors for evaluation?
5. What is a *model*? Identify the desired characteristics of such. What are some of the precautions that should be addressed in selecting a model?
6. How would you determine the type of data requirements needed for accomplishing a supportability analysis (refer to Figure 4.2)?
7. Describe what is meant by *logistics management information* (LMI)? What is included? How does it relate to the supportability analysis?
8. What are some of the objectives in designing a logistics database?
9. What is meant by *continuous acquisition and life-cycle support* (CALS)? What is included? How does it relate to the supportability analysis?
10. What is a *sensitivity analysis*? What are the benefits of such an analysis? How does this type of analysis relate to a *risk analysis*?
11. What is the purpose of a *life cycle cost analysis* (LCCA)? When in the system life cycle can it be accomplished? What are some of the benefits derived from a LCCA?
12. Identify and describe the basic steps of a LCCA.
13. What is the CBS? What factors should be considered in the development of a CBS?
14. Describe some of the methods used in cost estimating? What is meant by *activity-based costing* (ABC)? What is a cost estimating relationship? Give several examples.
15. What is the purpose of a break-even analysis?
16. How can learning curves affect life-cycle cost analysis results? What factors influence the development of learning curves?
17. In the comparison of alternatives, why must the *time value of money* be considered? What steps would you take in comparing alternatives on an equivalent basis?

18. Develop a life-cycle cost model for a system of your choice. Identify model input/output factors, develop the model, accomplish the necessary analysis, and evaluate the results. Describe how you would use the results to improve the design.
19. What is the FMECA? Its purpose(s)? Its benefits? What does the RPN tell you? Describe a situation where the FMECA can be used to facilitate the accomplishment of a LCCA. How can the FMECA be used to improve the design (give several examples)?
20. What is the FTA? How does it compare with the FMECA?
21. What is the MTA? Its purpose(s)? Its benefits? Describe a situation where the MTA can be used to facilitate the accomplishment of a LCCA. Describe how the MTA can be used to improve the design (give several examples)?
22. Select a system of your choice and identify a particular element or component of that system. Assume that the component selected has failed. Develop a MTA covering the necessary repair action (to inlcude the maintenance tasks identified in Figure 2.11).
23. What is the RCM? Its purpose(s)? Its benefits? How do the results of the RCM relate to the FMECA, the MTA, and system safety? How can the RCM be used to improve the design (give several examples)?
24. What is the LORA? Its purpose(s)? Its benefits? When can it be applied in the system life cycle? Describe how the results of the LORA and the MTA can impact each other (give an example).
25. What is the Ishikawa "Fishbone" diagram? Its purpose? Under what circumstances would you apply it?
26. Refer to Figure 4.14. What do the figures presented tell you? What actions would you take to improve the design?
27. What is a *Pareto analysis*? How can it be applied?
28. Refer to Figure 4.34. If the reliability of Assembly A-1 is four times the value indicated, what would be the results of the evaluation? What would be the results if the reliability was one-tenth (0.1) of the value indicated?
29. Based on the information provided below, compute the life-cycle cost in terms of present value using a 10% discount factor for System *XYZ*. Indicate the total value at the start of the program (decision point) and plot the cost stream or profile.

System *XYZ* is installed in an aircraft which will be deployed at five operational bases. Each base will have a maximum force level of 12 aircraft with the bases being activated in series (e.g., base 1 at the end of year 3, base 2 at the end of year 4, etc.). The total number of System *XYZ*'s in operation are

Year Number									
1	2	3	4	5	6	7	8	9	10
0	0	0	10	20	40	60	55	35	25

System *XYZ* is a newly designed configuration packaged in three units (Unit *A*, Unit *B*, and Unit *C*). The specified requirements for each of the units are shown in Figure 4.44.

The average System *XYZ* utilization is 4 hours per day and Units *A* to *C* are operating 100% of the time when System *XYZ* is on. One of the aircraft crew members will be assigned to operate several different systems throughout flight and it is assumed that 10% of his time is allocated to System *XYZ*.

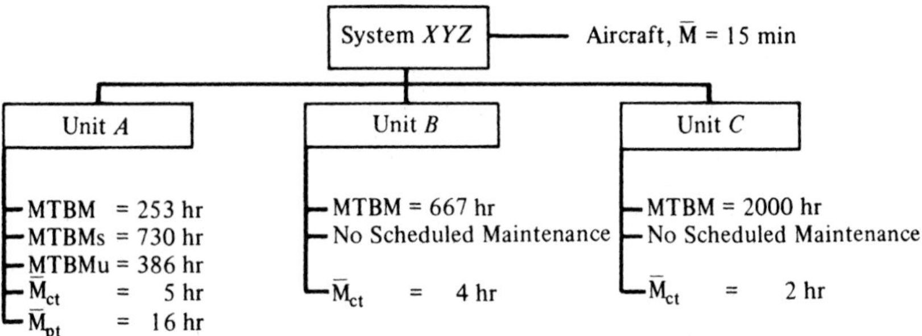

Figure 4.44 System requirements.

Relative to the maintenance concept, System XYZ incorporates a built-in self-test that enables rapid system checkout and fault isolation to the unit level. No external support equipment is required at the aircraft. In the event of a no-go condition, fault isolation is accomplished to the unit and the applicable unit is removed, replaced with a spare, and sent to the intermediate-level maintenance shop (located at the operational base) for corrective maintenance. Unit repair is accomplished through module replacement with the modules being discarded at failure. Scheduled (preventive) maintenance is accomplished on Unit A in the intermediate shop every 6 months. No depot maintenance is required; however, the depot does provide backup supply and support functions as required.

The requirements for System XYZ dictate the program profile shown in Figure 4.45. Assume that life-cycle costs are broken down into three categories, represented by the blocks in the program profile.

Year Number

1	2	3	4	5	6	7	8	9	10

Research, Design, Test, Evaluation.

Production

Operations and Maintenance

Figure 4.45 Life-cycle activities.

In an attempt to simplify the problem, the following additional factors are provided:

(1) RDT&E costs for System XYZ (to include labor and material) are \$100,000 for year 1, \$200,000 for year 2, and \$250,000 for year 3.

(2) RDT&E costs for special support equipment are \$50,000 in year 2 and \$10,000 in year 3.

(3) System XYZ operational models are produced, delivered, and purchased in the year prior to the operational deployment need. System unit costs are Unit A = \$6000, Unit B = \$3000, and Unit C = \$1000. Recurring and nonrecurring manufacturing costs are amortized on a unit basis.

(4) Support equipment is required at each intermediate maintenance shop at the start of the year when System *XYZ* operational models are deployed. In addition, a backup support equipment set is required at the depot when the first operational base is activated. The cost per set of support equipment is $20,000. Support equipment maintenance is based on a burden rate of $0.50 per direct maintenance labor-hour for the prime equipment.

(5) Spares units are required at each intermediate maintenance shop at the time of base activation. Assume that 2 Unit *A*'s, 1 Unit *B*, and 1 Unit *C* are provided at each shop as safety stock. Also, assume that one spare System *XYZ* is stocked at the depot for backup support.

Additional spares constitute modules. Assume that material costs are $100 per corrective maintenance action and $50 per preventive maintenance action. This includes inventory maintenance costs.

In the interests of simplicity, the effects of the total logistics pipeline and shop turnaround time on spares are ignored in this problem.

(6) For each maintenance action at the system level, one low-skilled technician at $9 per direct maintenance labor-hour is required on a full-time basis. $\overline{\text{M}}$ is 15 minutes. For each corrective maintenance action involving Unit *A*, Unit *B*, or Unit *C*, two technicians are required on a full-time basis. One technician is low-skilled at $9 per hour and one technician is high-skilled at $11 per hour. Direct and indirect costs are included in these rates. For each preventive maintenance action, one high-skilled technician at $11 per hour is required on a full-time basis.

(7) System operator personnel costs are $12 per hour.

(8) Facility costs are based on a burden rate of $0.20 per direct maintenance labor-hour associated with the prime equipment.

(9) Maintenance data costs are assumed to be $20 per maintenance action. Assume that the design-to-unit-acquisition cost (i.e., unit flyaway cost) requirement is $15,000. Has this requirement been met? Assume that the design-to-unit-O&M cost requirement is $20,000. Has this requirement been met? In solving this problem, be sure to state all assumptions in a clear and concise manner.

30. A need has been identified which will require the addition of a new performance capability to an existing aircraft. In response, System *XYZ* is to be developed to fulfill this need, and there are two different configurations of System *XYZ* being considered for procurement.

Based on the information provided below: (**a**) compute the life-cycle cost for each of the two configurations; (**b**) select a preferred approach; (**c**) plot the applicable cost stream (undiscounted and discounted); and (**d**) accomplish a break-even analysis. In computing present value costs, assume a 15% discount factor.

System *XYZ* is to be installed in an aircraft that will be deployed at five operational bases. Each base will have a maximum force level of 12 aircraft, with the bases being activated in series (e.g., base 1 at the end of year 2, base 2 at the end of year 3, etc.). The total number of System *XYZ*s in operation are noted subsequently.

Year Number									
1	2	3	4	5	6	7	8	9	10
0	0	10	20	40	60	60	60	35	25

In addition to fulfilling mission requirements, System *XYZ* will be utilized on the average of 4 hours per day, 365 days per year. One of the aircraft crew members will be assigned to operate several different systems throughout flight, and it is assumed that 1% of his time is allocated to System *XYZ*. The required system MTBM is 175 hours, MTBF is 250 hours, and $\overline{\text{M}}$ is 30 minutes. System *XYZ* is a newly designed entity, and each of the two candidate configurations is packaged in three units: Unit *A* to *C*, as illustrated.

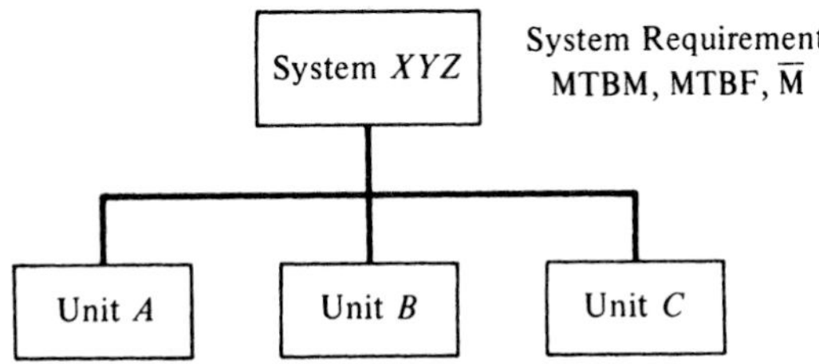

The predicted reliability and maintainability factors associated with each of the two candidate configurations are noted in the following table.

Parameter*	Configuration A	Configuration B
System Level (Organizational Maintenance)		
MTBM	195 hr	249 hr
MTBM_u (or MTBF)	267 hr	377 hr
$\overline{\text{M}}$	30 min	30 min
Unit Level (Intermediate Maintenance)		
Unit *A*		
MTBM	382 hr	800 hr
MTBM_u	800 hr	800 hr
MTBM_s	730 hr	–
$\overline{\text{Mct}}$	5 hr	5 hr
$\overline{\text{Mpt}}$	16 hr	–
Unit *B*		
MTBM	500 hr	422 hr
MTBM_u	500 hr	1,000 hr
MTBM_s		730 hr
$\overline{\text{Mct}}$	4 hr	5 hr
$\overline{\text{Mpt}}$	–	12 hr
Unit *c*		
MTBM	2,000 hr	2,500 hr
$\overline{\text{Mct}}$	2 hr	3 hr

* Assume that MTBM_u = MTBF. When there is no scheduled maintenance, MTBM_u = MTBM.

Relative to the maintenance concept, the two configurations of System *XYZ* incorporate a built-in self-test capability that enables rapid system checkout and fault isolation to the unit level. No external support equipment is required at the aircraft. In the event of a no-go condition, fault isolation is accomplished to the unit and the applicable unit is removed, replaced with a spare, and sent to the intermediate-level maintenance shop (located at the operational base) for corrective maintenance. Unit repair is accomplished through module replacement, with the modules being discarded-at-failure (i.e., modules are assumed as being nonrepairable). Scheduled (preventive) maintenance is accomplished for configuration *A* (Unit *A*) and configuration *B* (Unit *B*), as noted in the table, in the intermediate shop every six (6) months. No depot maintenance is required; however, the depot does provide backup supply and support functions as required.

The requirements for System *XYZ* dictate the program profile shown in Figure 4.46.

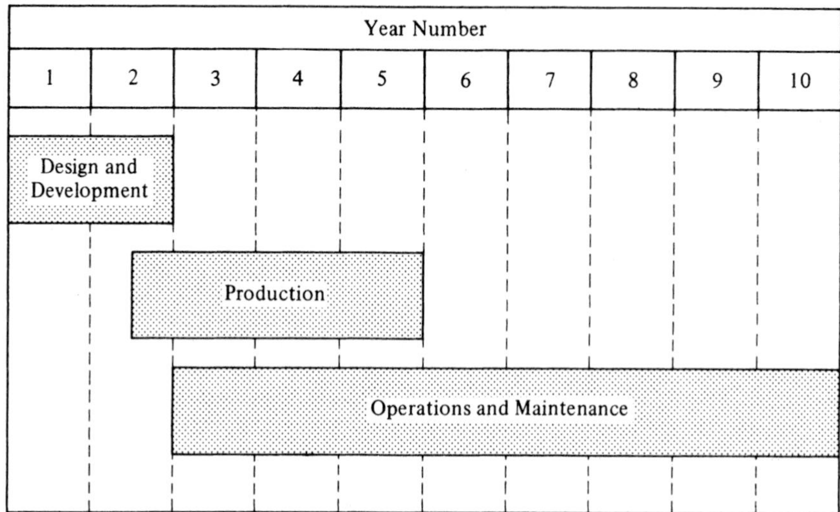

Figure 4.46 Life-cycle activities.

Assume that life-cycle costs are broken down into the three categories represented by the blocks in the program profile (i.e., Design and Development, Production, and Operations and Maintenance).

In an attempt to simplify the problem, the following additional factors are assumed:

(1) Design and development costs for System *XYZ* (to include labor and material) are:

Configuration *A*: $80,000 ($50,000 in year 1 and $30,000 in year 2)

Configuration *B*: $100,000 ($70,000 in year 1 and $30,000 in year 2)

(2) Design and development costs for special support equipment at the intermediate level of maintenance are

Configuration *A*: $30,000 ($20,000 in year 1 and $10,000 in year 2)

Configuration *B*: $23,000 ($17,000 in year 1 and $6,000 in year 2)

(3) System *XYZ* operational models are produced and delivered in the year prior to the operational deployment need (i.e., 10 models are produced and delivered in year 2, etc.). Production costs for each System *XYZ* are

Configuration *A*: $21,000/system

Configuration *B*: $23,000/system

(4) Special support equipment is required at each intermediate maintenance shop (for corrective maintenance of units) at the start of the year when System *XYZ* operational models are deployed (i.e., base 1 at the beginning of year 3). In addition, a backup special support equipment set is required at the depot when the first operational base is activated. Special support equipment is produced and delivered at a cost of (recurring and amortized nonrecurring costs are included).

Configuration *A* support equipment: $13,000

Configuration *B* support equipment: $12,000

(5) Spare units are required at each intermediate maintenance shop at the time of base activation. Assume that one (1) Unit *A*, one (1) Unit *B*, and one (1) Unit *C* constitute a set of spares, and that the cost of a set is equivalent to the cost of a production system (i.e., $21,000 for configuration *A* and $23,000 for configuration *B*). Also, assume that one set of spares is stocked at the depot when the first operational base is activated.

Additional spares constitute components (i.e., modules, subassemblies, parts, etc.). Assume that the material costs are $250 per corrective maintenance action, and $100 per preventive maintenance action. These cost factors include amortized inventory maintenance costs. In the interest of simplicity, the effects of the total logistics pipeline and maintenance shop turnaround time on spares are ignored in this problem.

(6) Maintenance facilities (as defined here) includes resources required for System *XYZ* maintenance and support, above and beyond spares, personnel, and data. This includes the use of intermediate-level maintenance facilities and the sustaining maintenance support of special support equipment. A burden rate of $1 per direct maintenance manhour associated with the prime equipment is assumed.

(7) Maintenance data include the preparation and distribution of maintenance reports, failure reports, and related data associated with each maintenance action. Maintenance data costs are assumed to be based on a rate of $25 per maintenance action.

(8) For each maintenance action at the system level, one low-skilled technician at $20 per direct maintenance manhour is required on a full-time basis. For the purposes of simplicity, it is assumed that this rate is an average value, applied throughout the life cycle, and includes direct, indirect, and inflationary factors. $\overline{\mathrm{M}}$ is 30 minutes for each of the two configurations.

(9) For each corrective maintenance action involving Unit *A*, Unit *B*, or Unit *C*, two technicians are required on a full-time basis (i.e., duration of the $\overline{\mathrm{M}}\mathrm{ct}$ value). One (1) lowskilled technician at $20 per hour and one (1) high-skilled technician at $30 per hour are required. Direct, indirect, and inflationary factors are considered in these average values.

(10) For each preventive maintenance action involving units (configurations *A* and *B*), one high-skilled technician at $30 per hour is required on a full-time basis (i.e., duration of the $\overline{\mathrm{M}}\mathrm{pt}$ value).

(11) For operation of System *XYZ*, the allocated crew operator cost is $40 per hour.

In solving this problem, be sure to state all assumptions in a clear and concise manner.

31. The following information is known for a given system:

(1) The total system operating time for a designated period is 10,000 hours.

(2) The corrective maintenance rate is 0.004 maintenance action per hour of system operation.

(3) The preventive maintenance rate is 0.001 maintenance action per hour of system operation.

(4) The average downtime per maintenance action is 50 hours, with a mean downtime of 5 hours for corrective maintenance.

(a) Calculate the following: MTBF, MTBM, A_i, and A_o.

(b) Assume that the value for operational availability calculated above is sufficient to meet mission requirements. Also, assume that the actual demonstrated MTBM is

100 hours, which is less than the calculated value and the requirement. What corrective action needs to be taken to meet mission requirements?

32. In defining the detailed maintenance plan and establishing criteria for equipment design (and system support), it is necessary to determine whether it is economically feasible to repair certain assemblies or to discard them when failures occur. If the decision is to accomplish repair, it is appropriate to determine the maintenance level at which the repair should be accomplished (i.e., intermediate maintenance or depot maintenance). For the purposes of this exercise, Assembly A-1 (one of 15 assemblies in Unit *B* of System *XYZ*) is to be evaluated in terms of the three options. The following information is provided to facilitate the evaluation task:

(1) System *XYZ* is installed in each of 60 aircraft that are deployed at five operational bases over an 8-year period. System *XYZ* will be utilized on the average of 4 hours per day, 365 days per year, and the total system operating time is 452,600 hours.

(2) System *XYZ* is packaged in three units: Unit *A*, Unit *B*, and Unit *C*. When corrective maintenance is required, a built-in self-test capability enables rapid checkout and fault isolation to the unit level. In the event of a no-go condition, the applicable unit is removed, replaced with a spare, and the faulty unit is sent to the intermediate-level maintenance shop (located at the operational base) for corrective maintenance. Unit repair is accomplished through fault isolation to the applicable assembly (e.g., Assembly A-1), removal of the faulty assembly and replacement with a spare, and checkout of the unit to verify satisfactory operation. The faulty assembly must now be processed in accordance with the designated maintenance plan.

(3) Pertinent data associated with Assembly A-1 are noted below.

(a) The estimated acquisition cost for Assembly A-1 (to include design and development cost and production cost) is \$1700 if the assembly is designed to be repairable, and \$1600 if the assembly is designed to be discarded at failure.

(b) The estimated failure rate (or corrective maintenance rate) is 0.00045 failure per hour of system operation.

(c) When failures occur, repair is accomplished by one technician assigned on a full-time basis (i.e., for the duration of the predicted elapsed active repair time). The estimated $\overline{\text{M}}\text{ct}$ is 3 hours. The labor rate is \$20 per labor-hour for intermediate maintenance and \$30 per labor-hour for depot maintenance.

(d) Supply support involves three categories of cost: (1) the cost of spare assemblies in the pipeline, (2) the cost of spare components to enable the repair of faulty assemblies, and (3) the cost of inventory maintenance. Assume that five spare assemblies will be required in the pipeline when maintenance is accomplished at the intermediate level, and that 10 spares will be required when maintenance is accomplished at the depot level. For component spares, assume a material cost of \$50 per maintenance action. The estimated cost of inventory maintenance will be 20% of the inventory value.

(e) When assembly repair is accomplished, special test and support equipment is required for fault diagnosis and checkout. The cost per set is \$12,000, which includes acquisition cost and amortized maintenance cost.

(f) Transportation and handling cost is considered to be negligible when maintenance is accomplished at the intermediate level. However, assembly maintenance accomplished at the depot level will involve an extensive amount of transportation. For depot maintenance, assume \$150 per 100 pounds per one-way trip (independent of the distance), and that the packaged assembly weighs 20 pounds.

(g) The allocation for Assembly A-1 relative to maintenance facility cost is categorized in terms of (1) an initial fixed cost, and (2) a sustaining recurring cost proportional to facility use requirements. The initial fixed cost is $1000 per installation, and the assumed usage cost allocation is $1.00 per direct maintenance labor-hour at the intermediate level and $1.50 per direct maintenance labor-hour at the depot level.

(h) Technical data requirements will constitute (1) the maintenance instructions to be included in the technical manual to support assembly repair activities, and (2) the failure reporting and maintenance data covering each maintenance action in the field. Assume that the cost for preparing and distributing maintenance instructions is $1000 and that the cost for field maintenance data is $25 per maintenance action.

(i) There will be some initial formal training costs associated with maintenance personnel when considering the assembly repair option. Assume 30 student-days of formal training for intermediate-level maintenance and 6 student-days for depot-level maintenance. The cost of training is $150 per student-day.

(j) As a result of maintenance actions, there will be a requirement for the disposal of either faulty assemblies or faulty components. The assumed disposal cost is $20 per assembly and $2 per component.

The object of this exercise is to evaluate Assembly A-1 (based on the information provided above) and make a recommendation. Should Assembly A-1 be (1) *repaired at the intermediate level of maintenance*, (2) *repaired at the depot level of maintenance*, or (3) *discarded at failure*?

33. Operational and maintenance facility requirements are based on what factors?

34. Personnel requirements for the operation and maintenance of a system are based on what factors?

35. Personnel training requirements are based on what factors?

36. Test equipment requirements are based on what factors? Describe the steps leading to the selection of test and support equipment.

37. How are transportation requirements determined? Identify the steps leading to the selection of transportation and packaging resources.

CHAPTER 5

LOGISTICS IN THE DESIGN AND DEVELOPMENT PHASE

System design begins with the identification of a consumer need and extends through a series of steps to include conceptual design, preliminary design, and detail design and development (refer to Figure 3.1). Design is an evolutionary, top-down, process leading to a functional entity that can be produced, or constructed, with the objective of fulfilling a customer requirement in an effective and efficient manner. Inherent within this process is the integration of many different design disciplines, as well as the application of various design methods, tools, and technologies. The selection of technologies may include the utilization of existing or standard approaches, may evolve as a result of directed research, or may include a combination thereof.

Effective system design can best be realized through implementation of the system engineering process described in Chapter 3. An objective is to bring the system into being through the iterative process of requirements analysis, functional analysis and allocation, synthesis, optimization, design definition, test and evaluation. Fulfillment of this objective is accomplished through the proper planning and integration of the appropriate resources. This integration pertains not only to the classical design disciplines of electrical engineering, mechanical engineering, and so on, but to those design support disciplines such as reliability engineering, maintainability engineering, human factors engineering, and logistics engineering. Thus, the function of logistics, as it pertains to design, is an integral part of the system engineering process.

The purpose of this chapter is to briefly discuss the major phases of the system design process, and to describe some of the critical activities of those disciplines that are closely aligned with logistics and whose output is required in order to perform logistics functions properly. The disciplines selected include reliability, maintainability, human factors and safety, producibility, disposability, and quality engineering. An understanding of the requirements in these areas is essential, and many of the products produced become an integral part of logistics products. Finally, it is appropriate to address the requirements associated with the *design for supportability*, and to cover the functions of design review and test and evaluation with these supportability requirements in mind.

5.1 THE DESIGN PROCESS

The design process, as defined within the context of this book includes the phases of conceptual design, preliminary system design, and detail design and development as shown in Figure 1.4. This process is amplified in Figure 1.7 to reflect some of the steps that are inherent within the mainstream activities of design, and to show a relationship with a few of the more significant logistics activities. These relationships are discussed further in the paragraphs to follow. The activites emphasized throughout this chapter are highlighted in Figure 5.1.

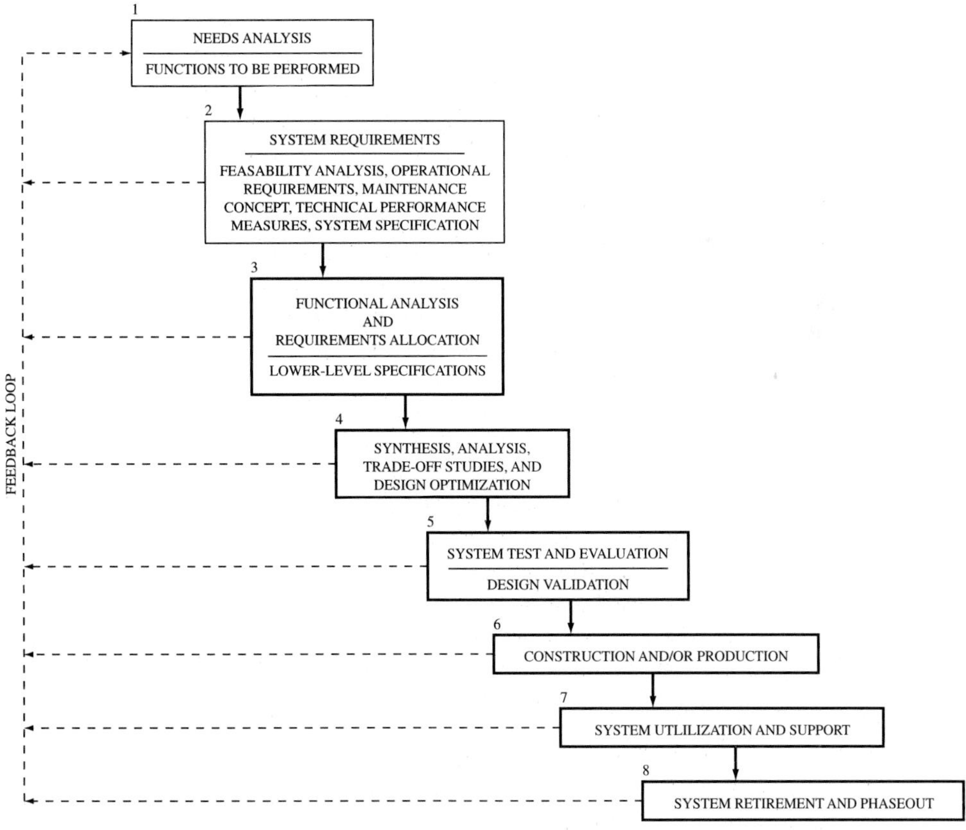

Figure 5.1 Steps in the life cycle (reference: Figure 3.1).

5.1.1 Conceptual Design

Conceptual design constitutes the first step in the overall design process and is initiated in response to an identified customer need. Feasibility studies are accomplished with the objective of identifying alternative technical approaches in meeting design requirements, and trade-off studies are accomplished to support early design decisions at the *system* level. System operational requirements and the maintenance concept are defined, a top-level functional analysis for the system may be completed, and the System Specification (Type "A") is prepared to describe the design requirements for the system. This stage of requirements analysis and system definition is represented by blocks 1 and 2 in Figure 5.1 and discussed in Sections 3.1 through 3.5.

Logistics requirements for the system are initially defined through the criteria established from the feasibility analysis, the operational requirements, the maintenance concept, and through the selection and prioritization of the appropriate technical performance measures (TPMs). The selection of technology applications in system design does have supportability implications; the operational requirements for the system lead to the identification of geographical locations where the support will be required and for how long; the maintenance concept defines the levels, responsibilites, and the effectiveness of the support required; and the prioritized TPMs indicate the critical areas that need to be addressed in the design. The extent of logistics activity in this phase, represented by blocks 1 and 2 in Figure 1.7, has been discussed in Chapter 3. Logistics engineering representation, as part of the design team, is essential in the requirements analysis definition process.

5.1.2 Preliminary System Design

Preliminary system design starts with the "functional" baseline system configuration described in the System Specification (Type "A") prepared during the conceptual design phase and proceeds toward the translation of the established system-level requirements into detailed qualitative and quantitative design characteristics. Preliminary design, illustrated by blocks 3 through 8 in Figure 1.7, includes the process of functional analysis and the allocation of requirements to various system elements, the accomplishment of trade-off studies and design optimization, system synthesis, the accomplishment of supportability analysis, and the ultimate "allocated" configuration baseline definition in the form of Development (Type "B"), Product (Type "C"), and Process (Type "D") Specifications, as shown in Figure 3.1.

Referring to Figure 5.2, the functional analysis and allocation process leads to the identification of specific resource requirements to include equipment, software, people, facilities, data, and so on (refer to Sections 3.6 and 3.7, and Figure 3.25). In other words, the process evolves from the functional analysis activity at the system-level (Figure 3.1, block 0.2) to the identification of the hardware, software, and human requirements identified in Figure 5.2 (blocks 0.251, 0.252, and 0.253, respectively). These specific requirements are then developed through a series of steps that represent their particular cycles of evolution. It should be noted that the example in the fig-

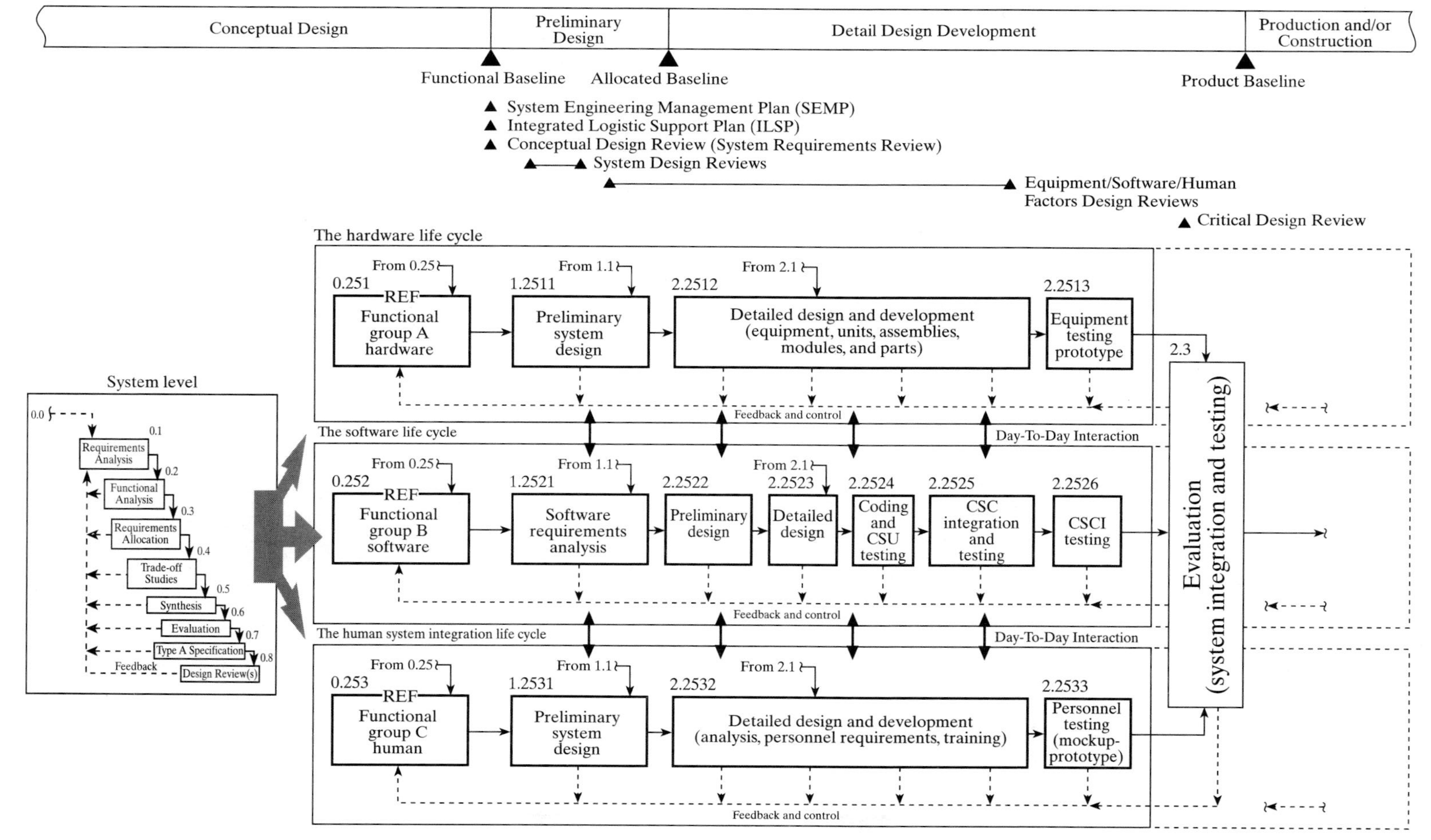

Figure 5.2 The integration of the hardware, software, and human life cycles. (Source: Blanchard, B. S., "The System Engineering Process: An Application for the Identification of Resource Requirements," *Systems Engineering,* Journal of the International Council on Systems Engineering, Volume 1, Number 1, July 1994.)

ure only addresses hardware, software, and the human (as an example); however other requirements to include facilities, data, etc., are identified in a similar manner. A major objective in system engineering is to ensure that these individual design "cycles" are well integrated as one proceeds toward system integration and testing (Figure 5.2, block 2.3).

Referring to Figure 5.3, as the level of definition evolves from the system-level to the subsystem-level and on down, the basic design requirements (and "design-to" criteria) are specified and applied at each level as appropriate. As described in Chapter 3, there must be a top-down/bottom-up *traceability* of requirements which can be facilitated through application of the QFD (and related processes) described in Section 3.5. Further, from a system engineering perspective, the appropriate design tasks/methods/tools must be selected and integrated into the mainstream design activity described in Section 3.8.

As is the case in conceptual design, logistics requirements must be considered as an integral part of the preliminary design process. The functional analysis includes the development of maintenance and support functions (Section 3.6); the allocation of requirements includes supportability factors and the establishment of supportability criteria for design (Section 3.7); the accomplishment of trade-off studies and the design optimization includes logistics and the application of supportability criteria in the evaluation of alternatives (Section 3.8); and the supportability analysis described in Chapter 4 continues to be a major activity relative to ensuring that logistics is adequately addressed in the design process.

5.1.3 Detail Design and Development

The *detail design and development* phase begins with the concept and configuration derived through preliminary system design: that is, a system configuration with performance, effectiveness, logistic support, cost, and other requirements described through the *A*, *B*, *C*, and other specifications as applicable. An overall system design configuration has been established, and now it is necessary to convert that configuration to the definition and subsequent realization of hardware, software, data, and specific items of support. The process from here on includes:

1. *Definition of system elements.* Subsystems, units, assemblies, lower-level component parts, software, data, and the elements of logistic support (spare/repair parts, test and support equipment, facilities, personnel, technical data).
2. *Preparation of design data.* Specifications, drawings, databases and electronic data files, trade-off study reports, analysis results, predictions, and so on, describing all facets of the system.
3. *Development of physical models of the system or major system components.* Engineering laboratory models, service test models, mock-ups, and prototype models for the purposes of test and evaluation.
4. *Conductance of system integration and test.* The verification of system characteristics and that the requirements have been met. Deficiencies are noted and corrected through redesign, system modification, and retest as necessary.

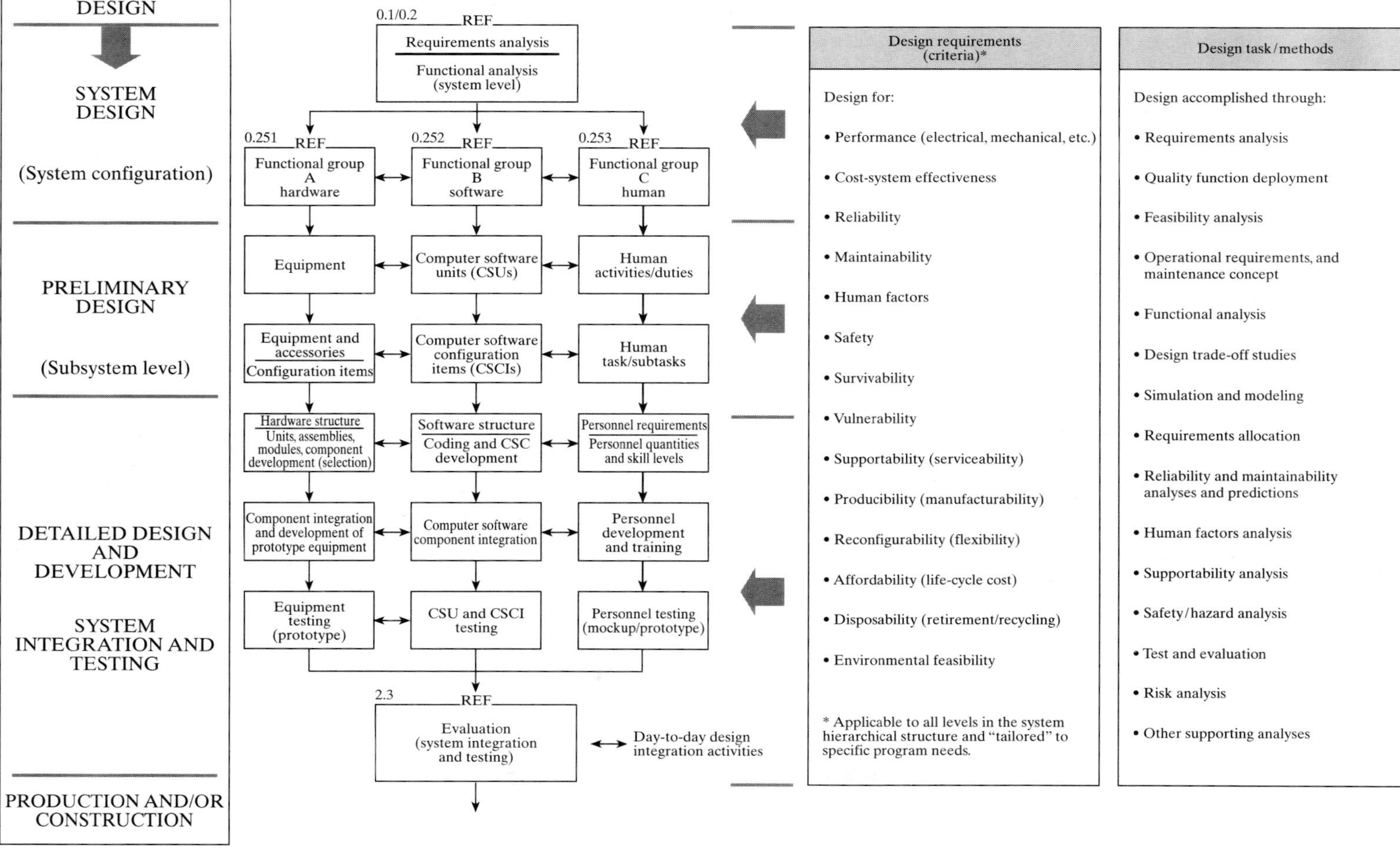

Figure 5.3 A top-down application of requirements. (Source: Blanchard, B. S., "The System Engineering Process: An Application for the Identification of Resource Requirements," *Systems Engineering*, Journal of the International Council on Systems Engineering, Volume 1, Number 1, July 1994.)

These requirements must, of course, be tailored to the system being developed. Systems requiring the application of new technologies and a large amount of new design will receive a greater degree of coverage through documentation and may be more extensively tested than systems comprising large amounts of standard off-the-shelf components.

Application of design criteria. With the preliminary design stage completed for the top system-level configuration, it is now necessary to define the subsystems, units, assemblies, and on down to the part level. This definition process results in the identification of specific characteristics which have a considerable impact on logistic support. To facilitate the design task, appropriate criteria are developed to provide guidelines covering areas such as accessibility, packaging, mobility, transportability, human factors, standardization, and many others. These criteria are directed toward incorporating the necessary characteristics compatible with the system goals for optimum logistic support, and are derived from the prioritized TPMs (refer to Section 3.5).

Design criteria can be classified as *general* or *specific*. Within the overall guidelines established through functional analysis and allocation, there may be a number of options available to the designer. It is important that supportability considerations be inherent in the decision process. To ensure that this is accomplished, illustrated qualitative and quantitative data, checklists, and related material are made available. Figures 5.4 through 5.7 represent examples of general criteria. These and other examples are supported by documentation covering the results of studies, engineering experiments, and field tests.

As a supplement to the general criteria appropriate checklists may be developed which serve to remind the designer of areas of particular concern. An example is presented in Figure 5.8. Referring to the figure, the designer may quickly review the appropriate factors, determine applicability, and assess the extent to which a design reflects consideration of these factors. If the designer desires to investigate further the meaning of certain checklist items, he or she may call on a specialist for an interpretation or refer to a more indepth coverage as presented in Appendix A.[1] Appendix B includes some questions pertaining to the evaluation and selection of a supplier.

Conversely, as design progresses, the designer may be faced with certain problems which require specific guidance. Data, consistent with overall system design objectives for supportability and compatible with the general criteria referenced above, may be developed in response to a particular need. Quite often, several alternative approaches may be feasible, and in such instances, the designer formalizes the decision through the accomplishment of trade-off studies.

Analysis and trade-off studies. Throughout the design process, analyses and trade-off studies are accomplished in the evaluation of alternatives. Early in design, these trade-off evaluations are conducted at a relatively high level, as discussed in the previous chapters. As design progresses, evaluations are accomplished at a lower level in the system hierarchy. For instance, it may be necessary to

[1] The design review questions presented in Appendix A directly support the checklist items included in Figure 5.8. These questions are intended to highlight the characteristics in design that are considered desirable.

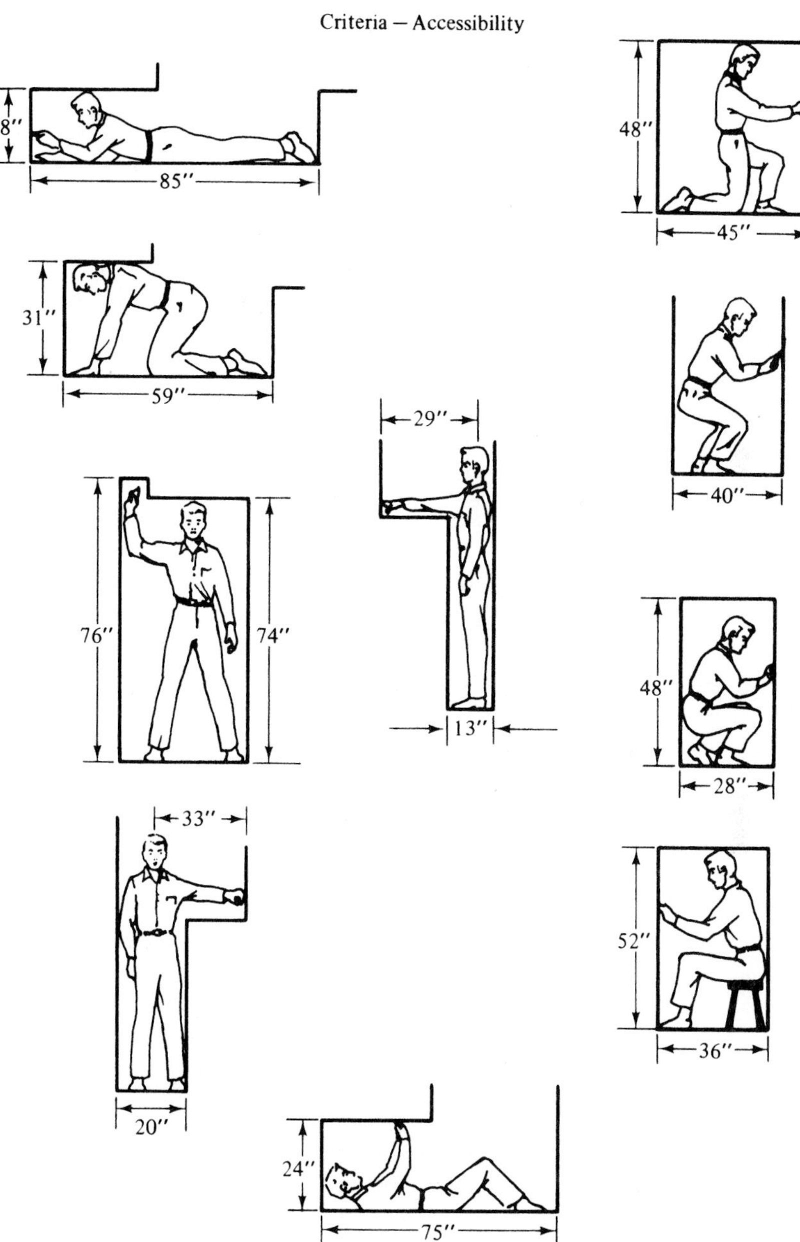

Figure 5.4 Limiting clearances required for various body positions. (NAVSHIPS 94324, *Maintainability Design Criteria Handbook for Designers of Shipboard Electronic Equipment*, U.S. Navy, Washington, D.C.)

Criteria — Accessibility

Minimum Openings for Using Common Hand Tools

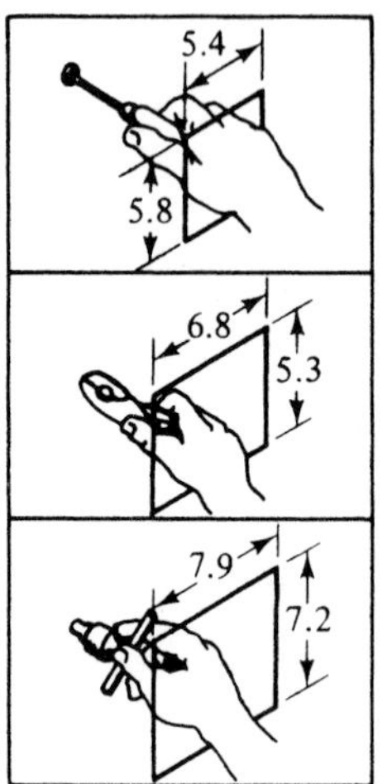

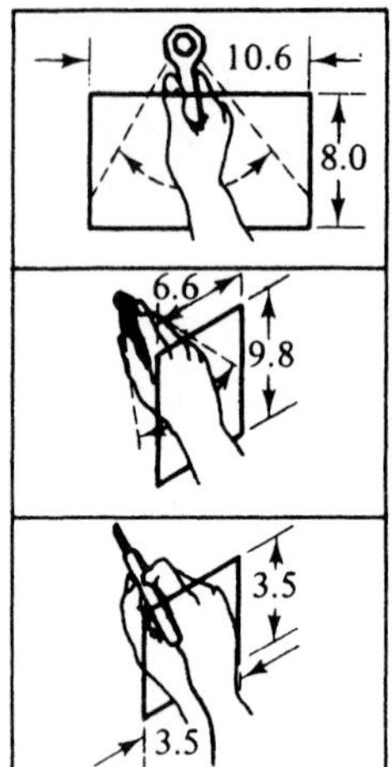

Space Required for Using Common Hand Tools

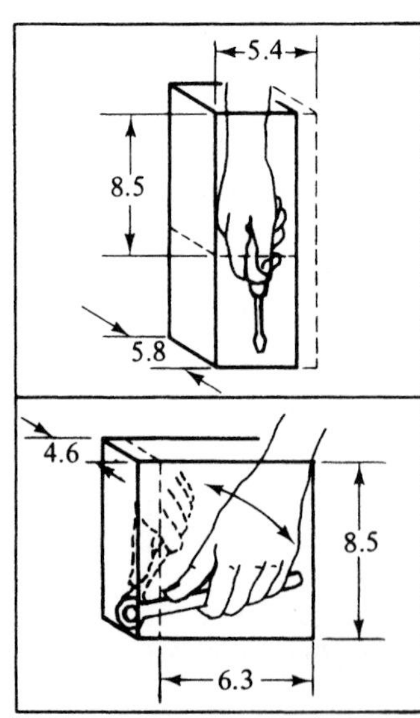

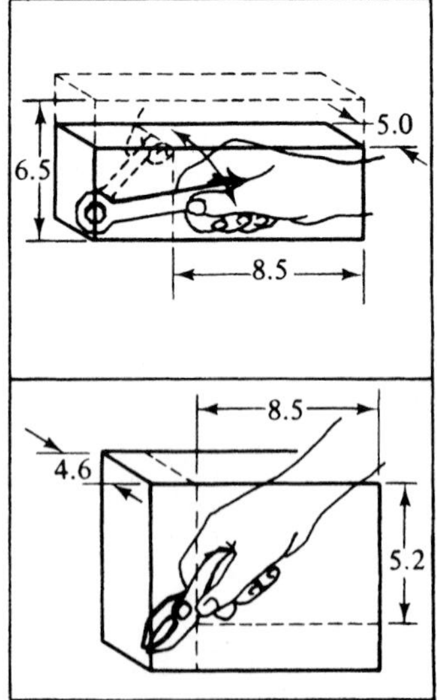

Figure 5.5 General criteria (accessibility). (NAVSHIPS 94324, *Maintainability Design Criteria Handbook for Designers of Shipboard Electronic Equipment*, U.S. Navy, Washington, D.C.)

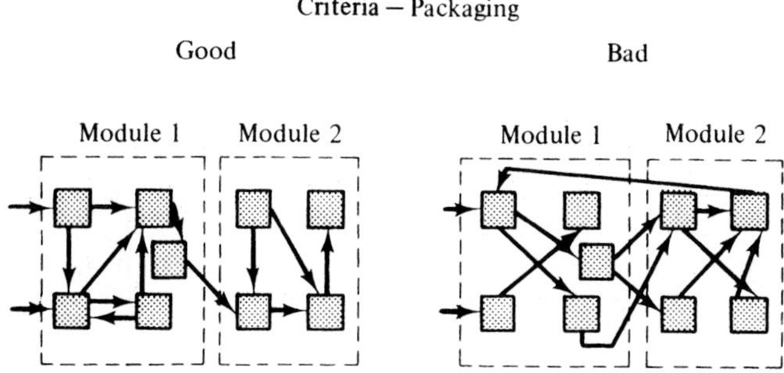

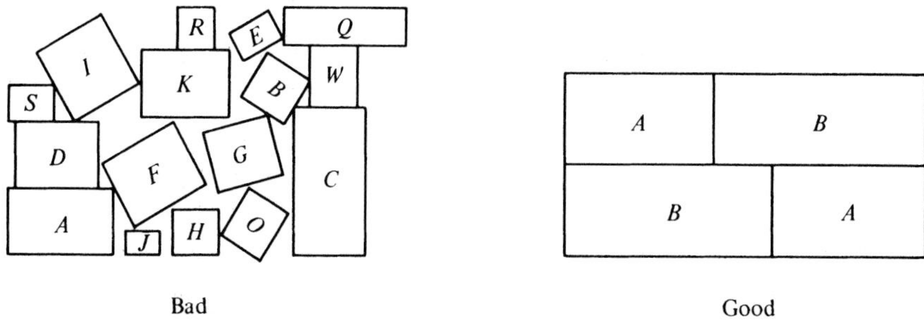

Figure 5.6 General criteria. (AMCP 706-134, *Engineering Design Handbook: Maintainability Guide for Design*, U.S. Army, Washington, D.C.).

1. Determine alternative methods for mounting components in an assembly or on a designated surface.
2. Determine whether it is desirable to use a light indicator or a meter on a front operator panel to provide certain information.
3. Determine whether it is more feasible to design a repairable assembly internally within the organization or to purchase a comparable item from an outside supplier (i.e., *make* or *buy*).
4. Determine the feasibility of repairing a given subassembly when a failure occurs or discarding it (this constitutes an extension of the level-of-repair analysis described in Sections 4.2.6).
5. Determine whether to use standard components in a given application or to use new nonstandard components with higher reliability.
6. Determine alternative inventory stock levels for a given spare part consumption.
7. Determine whether new test equipment should be developed or whether existing items should be used.
8. Determine the extent to which built-in test should be incorporated versus the use of external support equipment.

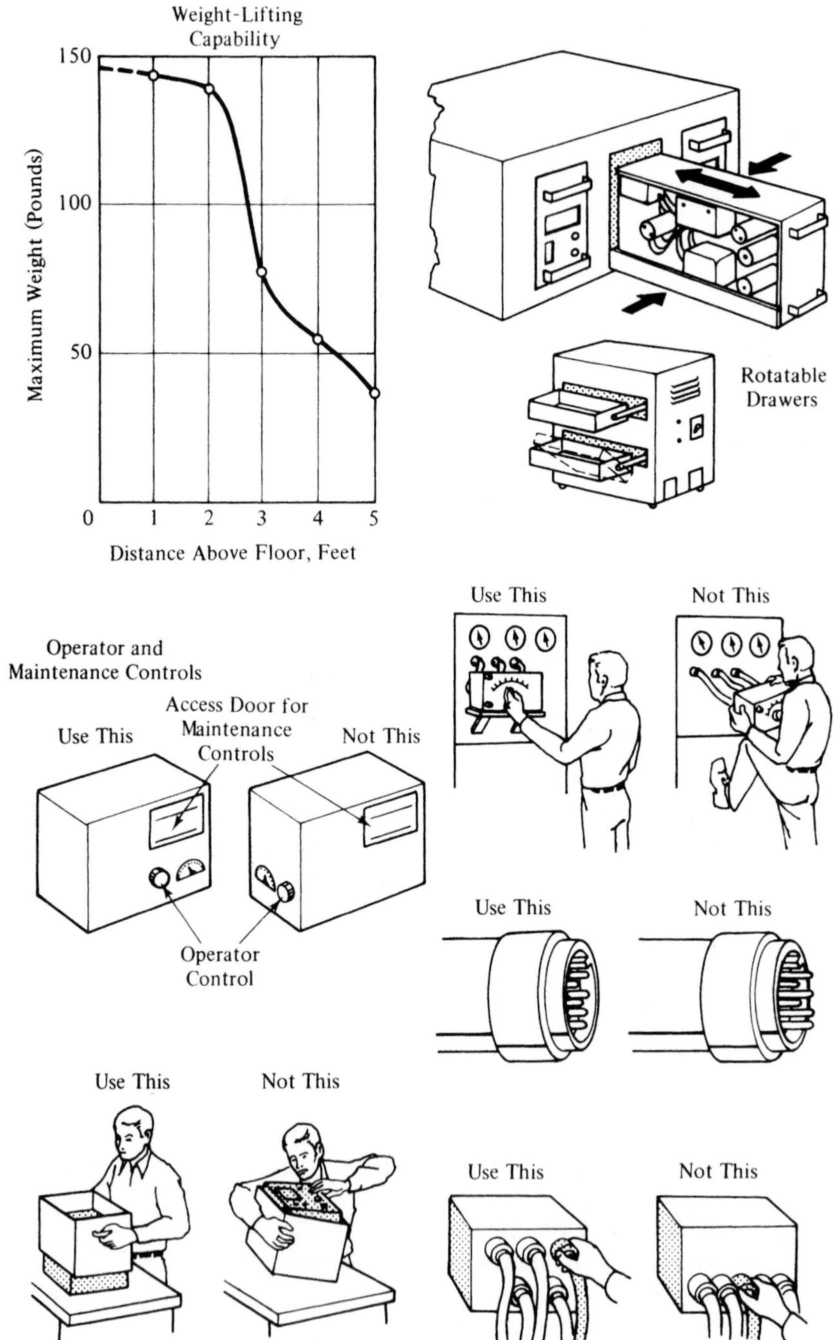

Figure 5.7 General criteria. (DH 1-3, *Design Handbook: Personnel Subsystems*, AFSC, U.S. Air Force, Washington, D.C.).

System Design Review Checklist

<u>General</u>

1. System Operational Requirements Defined
2. Effectiveness Factors Established
3. System Maintenance Concept Defined
4. Functional Analysis and Allocation Accomplished
5. Supportability Analysis Accomplished
6. Logistic Support Operation Plan Complete

<u>Logistic Support Elements</u>

Requirements Known and Optimized for

1. Test and Support Equipment
2. Supply Support (Spare/Repair Parts)
3. Personnel and Training
4. Technical Data (Procedures)
5. Facilities and Storage
6. Transportation and Handling

<u>Design Features</u>

Does the Design Reflect Adequate Consideration Toward

1. Selection of Parts
2. Standardization
3. Test Provisions
4. Packaging and Mounting
5. Interchangeability
6. Accessibility
7. Handling
8. Fasteners
9. Panel Displays and Controls
10. Adjustments and Alignments
11. Cables and Connectors
12. Servicing and Lubrication
13. Calibration
14. Environment
15. Storage
16. Transportability
17. Producibility
18. Safety
19. Reliability
20. Software
21. Disposability

When reviewing design (layouts, drawings, part lists, engineering reports), this checklist may prove beneficial in covering various program functions and design features applicable to logistic support. The items listed are supported with more detailed criteria in Appendix A. The response to each item listed should be "YES."

Figure 5.8 Sample design review checklist.

9. Determine whether automation is desired versus a manual approach in the accomplishment of maintenance actions.

The concepts and techniques employed in the evaluation of alternative design approaches are the same here as described in Chapter 4. However, one must ensure that the depth of application is tailored to the problem at hand.

Utilization of design aids. During the early stages of the design process, the design engineer is confronted with the task of defining the basic system configuration, developing equipment layout and packaging schemes, performing various types of design analyses, selecting components and identifying suppliers, incorporating reliability and maintainability provisions, evaluating human-machine interfaces, and the like. He or she has to solve many problems and evaluate numerous alternatives in a relatively limited period of time, yet turn out an effective product. With the complexity of systems continually increasing, combined with the need to shorten the time span of design projects, the design engineer needs all the assistance that he or she can attain to reach the ultimate design objective. This ever-increasing challenge has led to the development of various categories of design aids ranging from design standards documentation to the use of computers in design, the use of three-dimensional mock-ups or physical models, and so on. The advent of these new design aids, particularly with regard to computer applications, is already having and will continue to have a significant impact on the accomplishment of logistics and related functions in the design process.

Computers have been used rather extensively in (1) the performance of analytical modeling and design analyses (e.g., system requirements analysis, the evaluation of alternatives, predictions, finite element analysis, stress-strength analysis, analysis of design variables and margins, statistical and mathematical analyses, optimization studies); (2) the transmitting of design concepts into drawings, three-dimensional displays, graphic illustrations, material and component parts lists, and related product definition data; (3) the storage and retrieval of specific types of design data in varying configurations; and (4) in the generation of reports and documentation using automated means (e.g., technical manuals, maintenance instructions, manufacturing procedures, checklists).

More recently, there has been a great deal of emphasis on the application of computer-aided design (CAD), computer-aided manufacturing (CAM), and the integration of these technologies to provide a smooth flow of activity from design to production. Additionally, the emphasis on continuous acquisition and life-cycle support (CALS) has become prominent, as discussed in Section 4.4. The relationship between these three areas of activity is shown in Figure 5.9, and some of the major activity interfaces are illustrated in Figure 5.10.

In evaluating the design environment further, particularly for large projects, one often finds that design engineers are located in different buildings, if not geographically separated as in the case for various suppliers distributed throughout a large area. With the design process changing due to the advent of greater concurrency and less time for acquisition, there is a need for a highly efficient and responsive design communications network as conveyed by the illustration in Figure 5.11. The design process in the future will include more activities accomplished on a concurrent basis, in lieu of normal serial approach followed in the past.

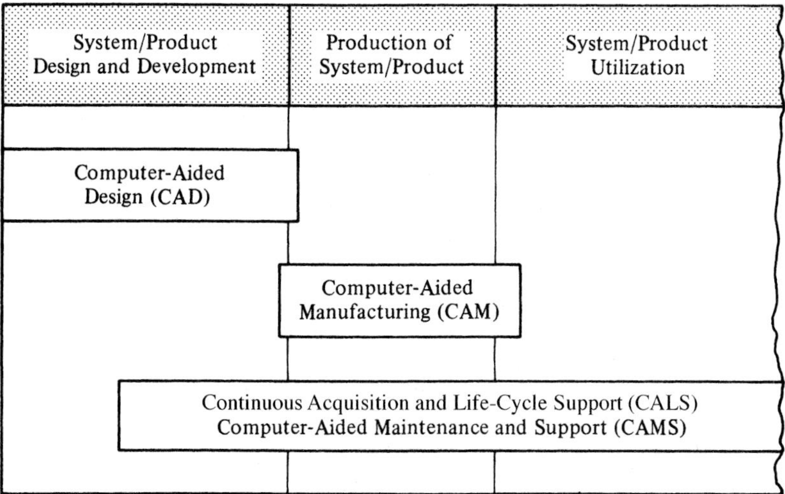

Figure 5.9 Application of analytical methods in the life cycle.

What does this all mean to the logistics engineer? As indicated, the methods/techniques being utilized in the design process are rapidly changing, and those who should be involved as members of the design team must adapt accordingly or they will fall behind. Because it is important that supportability be considered as a design parameter, it is necessary for those who represent the field to be conversant with the latest design practices. One must be able to understand and utilize computer-aided methods, interpret database presentations and evaluate graphic illustrations reflecting various alternative configurations, and understand the methods associated with data storage, retrieval, and the replication of various bits of information.

As with the design engineer, the challenges for the logistics engineer are ever present. Not only will he or she be accountable for understanding the design process, but the logistics engineer will require the appropriate tools to help in design evaluation and decision making. Inherent in many of the design workstations is the capability of being able to accomplish a structural analysis, or a finite element analysis, or a thermal analysis. However, current workstation capabilities do not cover logistics areas where analytical models are utilized to support design decisions (e.g., level-of-repair analysis, reliability and maintainability predictions and analyses, life-cycle cost analysis). These activities are currently accomplished on a *stand-alone basis*, with the results being fed into the design. This, in turn, often takes too long and is inefficient. Thus, an objective for the future is to integrate these analytical models into the CAD workstation capability. Conceptually speaking, the models described in Chapter 4 (and in Appendix D as applicable) should be tied to the design process in such a way as to enable the proper communications in a timely manner. This concept is illustrated in Figures 3.35 and 4.41.

While the use of computer models and simulation exercises is highly beneficial to the design engineer, particularly in the early stages of system design and development, there are occasions where it is feasible to develop three-dimensional scale models or physical mock-ups to provide a realistic simulation of a proposed final equipment configuration. These models or mock-ups, usually constructed during the detail design

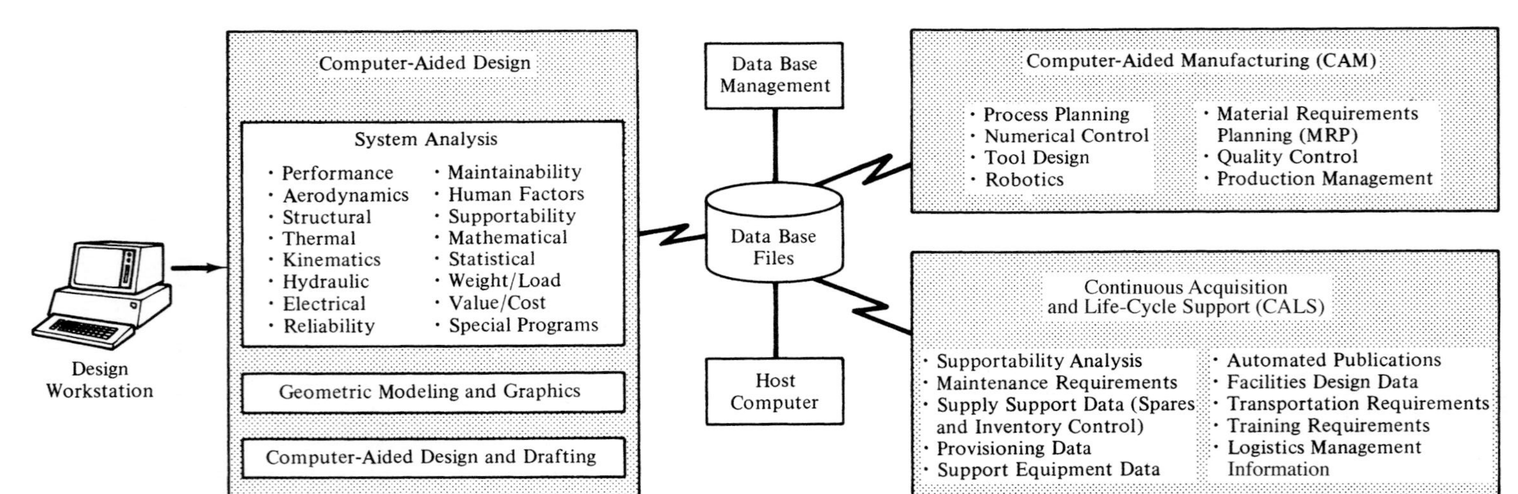

Figure 5.10 Major CAD/CAM/CALS interfaces.

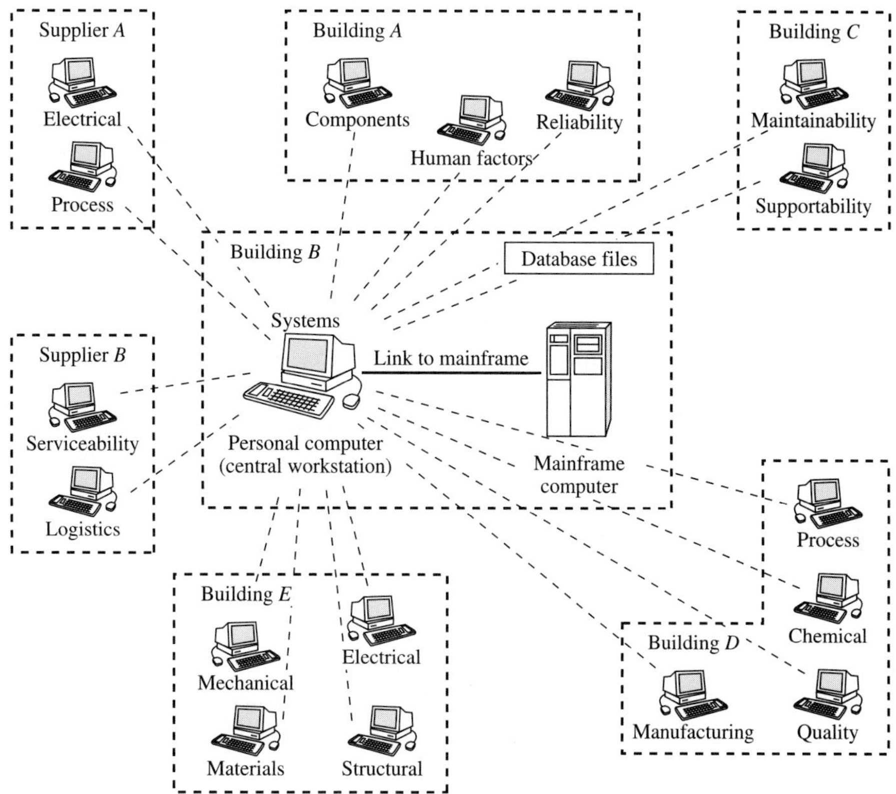

Figure 5.11 Example project design communications network.

and development phase, can be produced to any desired scale and to varying degrees of detail depending on the level of emphasis required. Mock-ups may be constructed of heavy cardboard, wood, metal, or a combination of materials. Mock-ups can be developed on a relatively inexpensive basis and in a short period of time when utilizing the right materials and personnel services (industrial design, human factors, or model-shop personnel are usually well oriented to this area and should be utilized to the greatest extent possible). The uses and values of a mock-up are numerous.

1. It provides the design engineer with the opportunity of experimenting with different facility layouts, packaging schemes, panel displays, etc., prior to the preparation of formal design data.
2. It provides the reliability/maintainability/human factors/logistics engineer with the opportunity to accomplish a more effective review of a proposed design configuration for the incorporation of supportability characteristics. Problem areas readily become evident.
3. It provides the maintainability/human factors engineer with a tool for use in the accomplishment of predictions and detailed task analyses. It is often possible to simulate operator and maintenance tasks to acquire task sequence and time data.

4. It provides the design engineer with an excellent tool for conveying the final design approach during a formal design review.
5. It serves as an excellent marketing tool.
6. It can be employed to facilitate the training of system operator and maintenance personnel.
7. It is utilized by production and industrial engineering personnel in developing fabrication and assembly procedures and in the design of factory tooling and associated test fixtures.
8. At a later stage in the system life cycle, it may serve as a tool for the verification of a modification kit design prior to the preparation of formal data and the development of kit hardware.

In general, the mock-up is extremely beneficial. It has been used effectively in facility design, aircraft design, and the design of smaller systems/equipments.

5.2 RELATED DESIGN DISCIPLINES

The accomplishment of system design and development is a team effort and there are many disciplines involved in the process. Of particular significance, when considering the objectives pertaining to *design for supportability,* are the disciplines of reliability, maintainability, human factors, system safety, producibility, quality, disposability and economic feasibility. These areas are covered herein to the extent required to show their relationships with logistics support.

5.2.1 Design for Reliability

Reliability engineering includes those design-related activities that are accomplished to fulfill the objectives described in Chapter 1 (Section 1.5). The intent is to design and develop (or reengineer) a system that will meet the specified operational requirements in an effective and efficient manner. As the design progresses, there are a number of methods/techniques/tools that may be utilized to facilitate the *design for reliability*. The purpose here is to briefly describe a select number of those methods that have the greatest impact on logistic support.[2]

Reliability functional analysis. As conveyed in Section 3.6, the system functional analysis forms the baseline upon which many subsequent design analysis activities build. From the top-level functional analysis developed in the conceptual design phase, second-, third-, and lower-level functional diagrams are developed. Operational func-

[2]The intent herein is to provide an introductory overview of reliability engineering, both in terms of definitions and program requirements, and not to cover the subject in depth! However, it is highly recommended that the subject area be pursued further. Three good references are (1) D. Kececiogly, *Reliability Engineering Handbook*, 2 vols., Prentice Hall, Upper Saddle River, NJ, 1991; (2) W. G. Ireson, and C. F. Coombs (Eds.), *Handbook of Reliability Engineering and Management*, McGraw-Hill, New York, 1988; and (3) J. Knezevic, *Reliability, Maintainability, and Supportability*, McGraw-Hill, New York, 1933. Additional references on reliability are included in the bibliography in Appendix H.

tions are identified, which, in turn, lead into the description of maintenance functions. When evaluating the various possible design approaches for meeting a given functional requirement, trade-off studies are conducted and a preferred configuration is selected. The objective here is to extend the functional analysis, described in Section 3.6, and to utilize this baseline as a starting point for the accomplishment of different reliability tasks. Referring to Figure 5.12, the first step is to define the specific reliability requirements for the individual functions that are to be accomplished.

Reliability allocation. Once that the reliability requirements have been established for the system overall, these requirements must then be allocated (or apportioned) to the various subsystems, units, assemblies, and so on, as described in Section 3.7 (refer to Figures 3.28 and 3.31). These allocated factors are included in the specifications for the applicable elements of the system and serve as *criteria* for the designer. The objective is to ensure that all design requirements are traceable from the system-level down to the depth necessary for the purposes of influencing the ultimate product output.

Reliability modeling. Early in the design, a reliability model is usually developed for the purposes allocation, prediction, the accomplishment of stress/strength analysis, tolerance analysis, and so on. The model constitutes a block-diagram description of the system, as illustrated in Figure 5.13, and reflects the baseline described through the functional analysis in Section 3.6. Referring to the figure (which is an extension of Figure 3.30), the appropriate reliability metrics may be estimated for each of the blocks, combined, variances may be introduced, and a predicted MTBF for the overall system may be accomplished. The reliability model serves as a prime tool in the initial development of reliability requirements and later in the prediction and assessment task as the design configuration becomes better defined. The results from reliability analyses and predictions constitute a major input to the supportability analysis (SA) described in Chapter 4.

Failure mode, effects, and criticality analysis (FMECA). Given a description, both in *functional* and *physical* terms, the designer needs to be able to evaluate a system relative to possible failures, the anticipated modes and expected frequency of failure, their causes, the consequences and impact(s) of such on the system overall, and where preventive measures can be initiated to preclude such in the future. The objective is to identify the weak links in the system, the criticality and impact of failures on the system in completing its mission, potential safety problems, and areas of high risk. The FMECA is an excellent design tool, and it can be applied in the development or assessment of any product or process. Figure 5.14 shows an application of the FMECA to a package handling system and a case study illustration pertaining to a manufacturing system is included in Section 4.2.2 (Chapter 4).

Fault-tree analysis (FTA). Fault-tree analysis (FTA) is a deductive approach involving a graphical enumeration and analysis of different ways in which a particular system failure can occur, and the probability of its occurrence. A separate fault tree may be developed for every critical failure mode or undesired event. Events must be defined in rather specific terms, as a generic and nonspecific definition is likely to result in a rather broad-based fault tree with too wide a scope and lacking in focus. Attention is

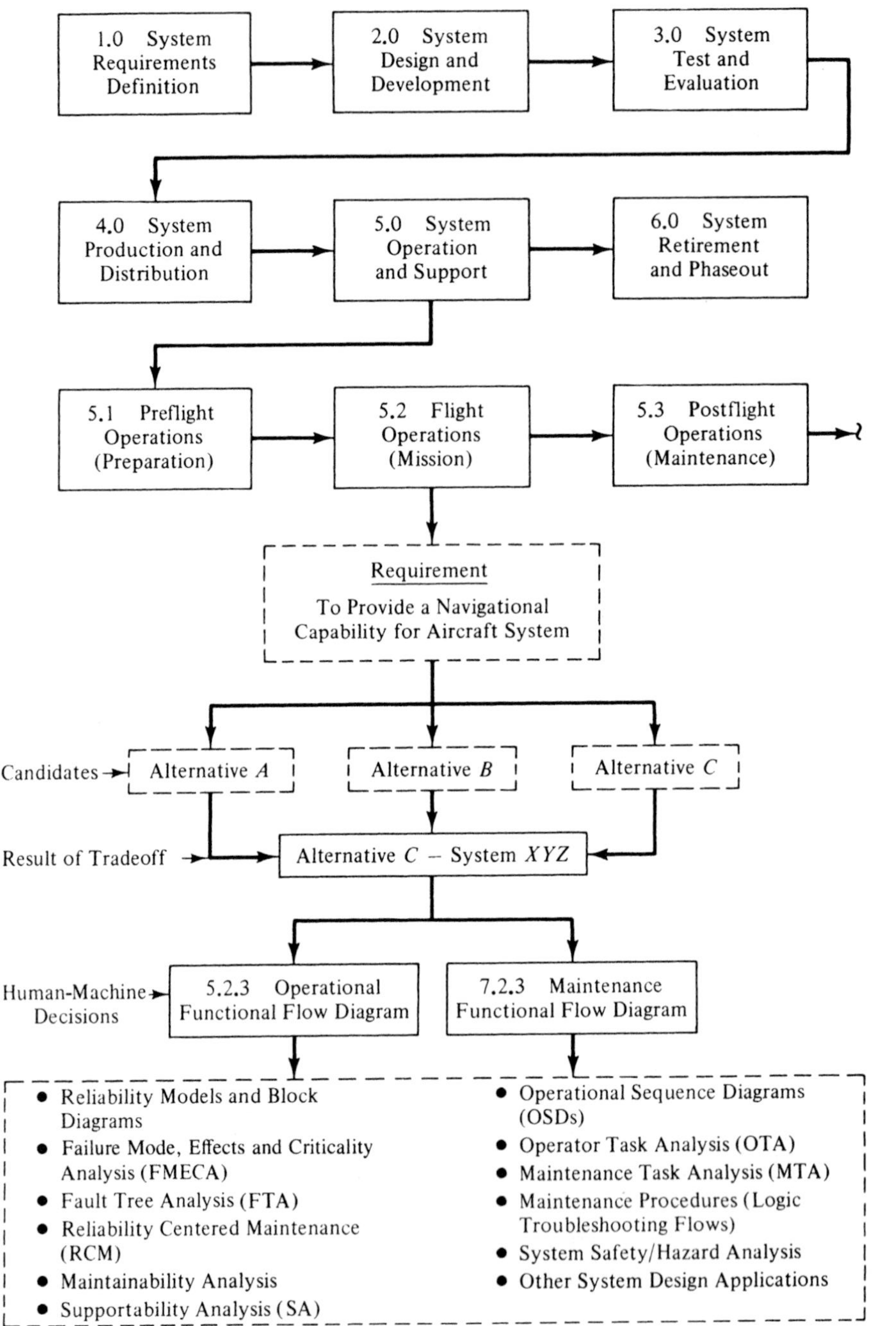

Figure 5.12 Evolution of functional design (partial application).

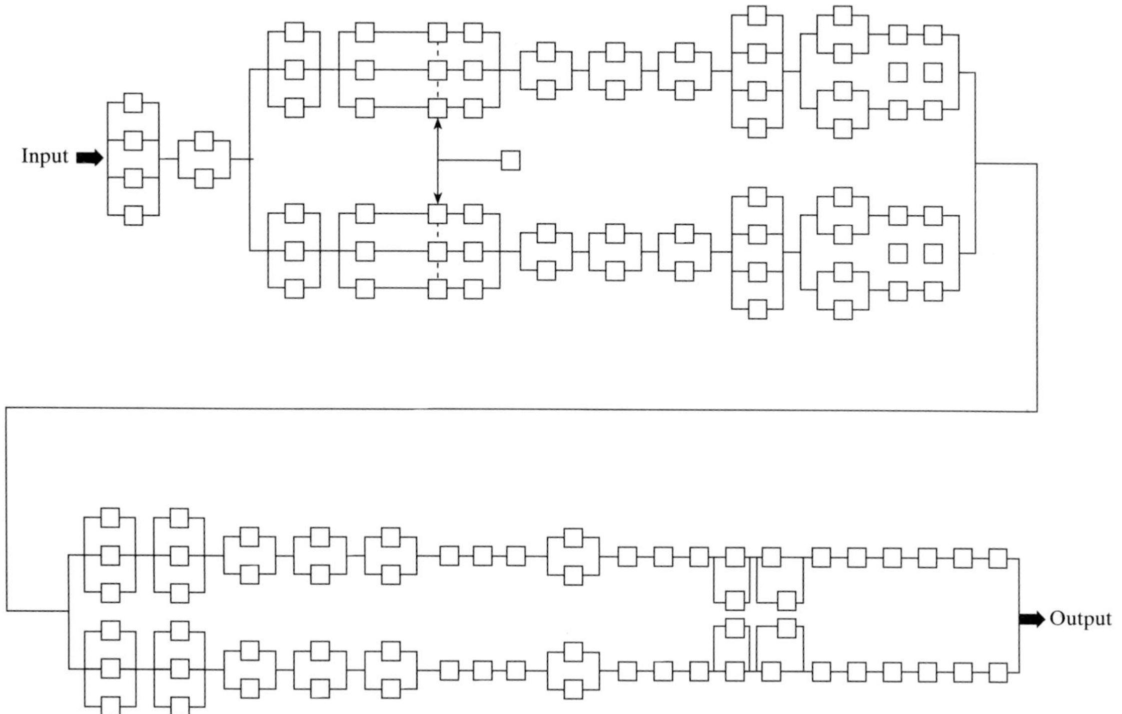

Figure 5.13 Expanded reliability block diagram of a system.

directed to the top-level event and the first-tier causes associated with it. Each of these causes is next investigated for *its* causes, and so on.

This top-down causal hierarchy, and associated probabilities, is called a fault tree. A sample fault tree is illustrated in Figure 4.24, and the FTA analysis process is described through the case study in Section 4.2.3. The FTA, which is narrower in focus than the FMECA, is often used in the early phases of conceptual and preliminary design when available design data are limited and when the analyst needs to gain insight relative to the feasibility of a selected technology application or to investigate potential safety issues. While the FMECA provides more information, the FTA can be applied faster and easier.

Reliabilty-centered maintenance (RCM). Reliability-centered maintenance (RCM) is a systematic approach to develop a focused and cost-effective preventive maintenance program and control plan for a product or a process. In the past, preventive maintenance requirements have often been established without the benefit of any specific knowledge of the system in terms of its reliability characteristics and when selected critical components need to be replaced to sustain system life. This, in turn, has led to the accomplishment of either too little or too much preventive maintenance, the results of which have been costly. The RCM technique, which was intoduced in the

Functional Flow Diagrams (Top Level)

A1.0 Incoming package unloading → A2.0 Package processing (labeling) → A3.0 Package processing (sorting) → A4.0 Package processing (distribution) → A5.0 Outgoing package loading

Functional Flow Diagrams (Second Level)

A1.1 Ref. Unload package to incoming vehicle (voice induction) → A1.2 Transfer package from unloader to rollers → A1.3 Position package (sled photo cell) → A2.0

A2.0 → A2.1 Ref. Align package to side pan for label application → A2.2 Determine height of package and transmit information to label applicator → A2.3 Apply label to package (label applicator) → A3.0

A3.0 → A3.1 Ref. Merge package flow from three lines to single belt → A3.2 Align packages to side pan for proper diversion → A3.3 Adjust distances between packages for proper diversion

A3.4 Determine package height and transmit information to camera → A3.5 Focus camera based on information from height detector and decodes label → A3.6 Check label and apply zip code/destination to package (PC) → A4.0

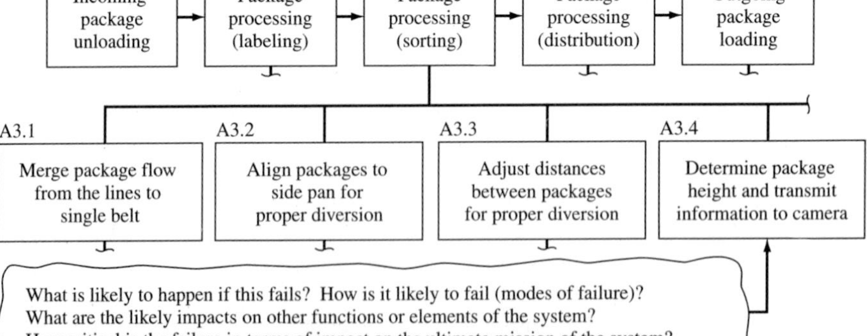

Figure 5.14 Application of the FMECA to a package handling system.

1960s and popularized by the airline industry in the 1970s, utilizes the FMECA in identifying areas of criticality where either redesign is required or scheduled item replacements are necessary in order to maintain operations. The RCM analysis process is described in Figure 4.31, an illustrative RCM decision logic is depicted in Figure 4.32, and the case study in Section 4.2.5 provides an example of the approach. This task may also be accomplished as a maintainability program requirement and is a inherent part of the supportability analysis (refer to Figure 4.5).

Reliability prediction. As engineering data become available, reliability prediction is accomplished as a check on design in terms of the system requirement and the factors specified through allocation. The predicted values of MTBM, MTBF, or failure rate (λ) are compared against the requirement, and areas of incompatibility are evaluated for possible design improvement.

Prediction is accomplished at different times in the design process, and will vary somewhat depending on the type of data available. Reliability block diagrams, models, and computer methods are employed to varying degrees depending on the problem at hand. Basic prediction techniques are summarized as follows:

1. Prediction may be based on the analysis of similar items. This technique should only be used when the lack of data prohibits the use of more sophisticated techniques. The prediction uses MTBF values for similar equipments of similar degrees of complexity performing similar functions and having similar reliability characteristics. The reliability of the new equipment is assumed to be equal to that of the equipment which is most comparable in terms of performance and complexity. Part quantity and type, stresses, and environmental factors are not considered. This technique is easy to perform, but not very accurate.

2. Prediction may be based on an estimate of *active element groups* (AEG). The AEG is the smallest functional building block that controls or converts energy. An AEG includes one active element (e.g., relay, transistor, pump, machine) and a number of passive elements. By estimating the number of AEGs and using a complexity chart, one can predict MTBF.

3. Prediction may be accomplished from an equipment parts count. There are a variety of methods used that differ somewhat due to data source, the number of part-type categories, and assumed stress levels. Basically, a design parts list is used and parts are classified in certain designated categories. Failure rates are assigned and combined to provide a predicted MTBF at the system level. A representative approach is illustrated in Figure 5.15.

4. Prediction may be based on a stress analysis. When detailed design is relatively firm, the reliability prediction becomes more sophisticated. Part types and quantities are determined, failure rates are applied, and stress ratios and environmental factors are considered. The interaction effects between components are addressed. This approach is peculiar and varies somewhat with each particular system design. Computer methods are often used to facilitate the prediction process.

Component Part	*λ/Part (%/ 1000 Hours)*	*Quantity of Parts*	*(λ/Part) (Quantity)*
Part *A*	0.161	10	1.610
Part *B*	0.102	130	13.260
Part *C*	0.021	72	1.512
Part *D*	0.084	91	7.644
Part *E*	0.452	53	23.956
Part *F*	0.191	3	0.573
Part *G*	0.022	20	0.440
Failure rate (λ) = 48.995%/1000 hours			Σ = 48.995%
$\text{MTBF} = \dfrac{1000}{0.48995} = 2041$ hours			

Source: **Data from MIL-HDBK-217,** ***Military Standardization Handbook, Reliability and Failure Rate Data for Electronic Equipment.***

Figure 5.15 Reliability prediction data summary. (Source: Data from MIL-HDBK-217, *Military Standardization Handbook, Reliability and Failure Rate Data for Electronic Equipment.*) Department of Defense, Washington, D.C., latest edition.

The figures derived through reliability prediction constitute a direct input to maintainability prediction data, supportability analysis, and the determination of specific support requirements (e.g., test and support equipment, spare/repair parts, etc.). Reliability basically determines the frequency of maintenance and the quantity of maintenance actions anticipated throughout the life cycle; thus, it is imperative that reliability prediction results be as accurate as possible.

Reliability growth modeling. During the early phases of design, new systems/products often display a lower reliability than what is ultimately required. Experience has indicated that, by accomplishing certain "test, analyze, and fix (TAAF)" activities as one progresses through the overall development effort, some degree of reliability growth can be realized. In other words, system reliability can be improved by analyzing and fixing some of the failure modes experienced. For many programs, it may be appropriate to develop a reliability growth model for the purposes of monitoring progress toward meeting the ultimate reliability objective. Figure 5.16 depicts a sample reliability growth curve.[3]

Effects of storage, packaging, transportation, and handling. When considering reliability as a system metric, the design emphasis is usually on how well the system must be able to function during the performance of some mission. Thus, the designer concentrates on the environment and the stresses to which the system will be subjected during that mission. Conversely, experience has indicated that the various elements of the system may be subjected to greater stresses and a more rigorous environment during the process of packaging and handling, when being stored or sitting on a palet in

[3] The concept of reliability growth and the development of growth curves was initiated in the 1960s with the development of the *Duane Model* and the *Gompertz Model.* These two models are discussed in more detail in Kececioglu, D., *Reliability Engineering Handbook*, Vols. 1 and 2, Prentice Hall, Inc., Upper Saddle River, N.J., 1991.

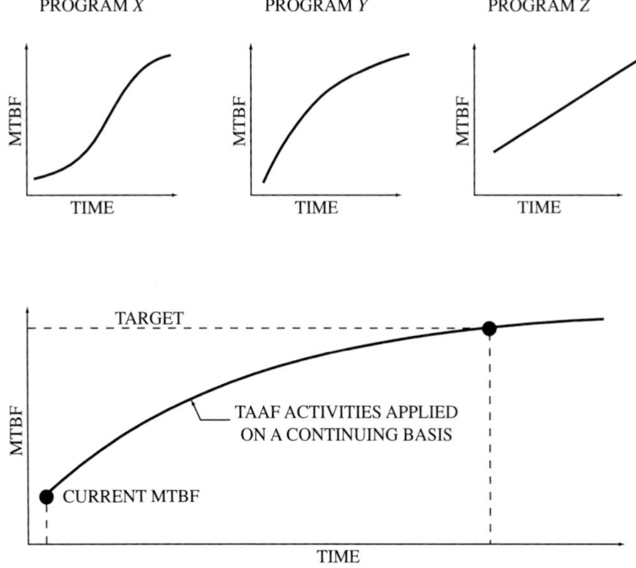

Figure 5.16 Development of reliability growth curve.

the rain or snow, or being transported from one location to another. Further, these environmental stresses may be significant when system components are undergoing maintenance.

Referring to Figures 1.1 and 1.2, the entire *outward* and *reverse* activity flows must be considered. Items will be packaged and shipped from the supplier or manufacturer to the user's operational site, some items may be stored as spares, and repairables may be returned for maintenance. Sometimes these environments will include extreme conditions of vibration, shock, high and low temperatures, rain, high humidity, sand, salt spray, and so on. The question is: How much degradation will occur when the system and its elements are being processed through the activities conveyed in the figures? System design requirements must address the entire spectrum as illustrated.

Failure reporting, analysis, and corrective-action system (FRACAS). When system failures occur, the objective is to accomplish the necessary repair actions and to return the system to full operational status in an expeditious manner. In accomplishing such, there is a tendency to ignore the associated data requirements and the recording of the steps involved in accomplishing the necessary corrective maintenance (i.e., the steps identified in Figure 2.11 commencing with a good description of the symptom of failure). Also, there is often a "disconnect" between the maintenance actions accomplished at the system level and the specific detailed repair actions that are necessary to correct the problem at hand. In other words, the data recording, analysis, reporting, and feedback capability for many systems has been inadequate. Yet, in the design and development of new systems (or the reengineering of existing systems), the designer is heavily dependent on having access to field data and an understanding of what has happened in the past. The validity of future reliability analyses and predictions is a function of good historical data feedback to the designer in a timely manner. Thus, the establishment of a good comprehensive failure analysis and reporting capability is critical, and the results constitute a major input for maintainability design and the supportability analysis.

Reliability design—summary. These and related tasks are often included as part of a formal reliability program, and are accomplished on a progressive basis as one proceeds through the steps illustrated in Figure 1.6. Because the reliability characteristics that are inherent within the system design actually dictate the requirements for the subsequent maintenance and support of that system throughout its life cycle, it is difficult to separate reliability from maintainability or supportability. In many instances, these same tasks may be accomplished within the context of a formal logistics program. In any event, independent of how one may "classify" a given requirement in this area, it is critical that one understand each of these tasks, the concepts and principles involved, and how it may be applied in meeting the objectives described throughout this text.

5.2.2 Design for Maintainability

Maintainability design involves those actions (or activities) that deal with the ease, accuracy, and economy in the performance of maintenance tasks (refer to Sections 1.5 and 2.2). Maintainability design includes those functions in the design process necessary to ensure that the ultimate product configuration is compatible with the top system-level objectives from the standpoint of the allocated MTBM, MDT, $\overline{\text{M}}$ct, $\overline{\text{M}}$pt, MLH/OH, cost/maintenance action, and related factors. Maintainability is concerned with maintenance times and frequencies, supportability factors in design, and projected maintenance cost over the life cycle.

Because of its objectives, maintainability is perhaps the largest contributor in the design relative to addressing logistics support from an optimum viewpoint. Much of logistics support stems from maintenance, and maintenance is a result of design. Maintainability is concerned with influencing design such that maintenance is optimized and life-cycle cost is minimized. The scope of coverage in the following paragraphs includes those activities that are considered to have the greatest impact on logistic support.[4]

Maintainability functional analysis. The basic requirements for system maintenance and support stem from the maintenance concept (Section 3.4) and from the maintenance functional flow diagrams (Section 3.6). Maintenance functions are identified from operating functions as shown in Figure 3.22, and the resources required for sys-

[4]The objective is to provide an introductory overview of maintainability engineering, including definitions and program requirements, and not to cover the subject in depth. However, for more information, two good references are (1) Patton, J. D., *Maintainability and Maintenance Management*, Instrument Society of America, Research Triangle Park, N.C., 1980; and (2) Blanchard, B. S., Verma, D., and Peterson, E., *Maintainability: A Key to Effective Serviceability and Maintenance Management*, John Wiley, New York, N.Y., 1995. Additional references are included in Appendix H.

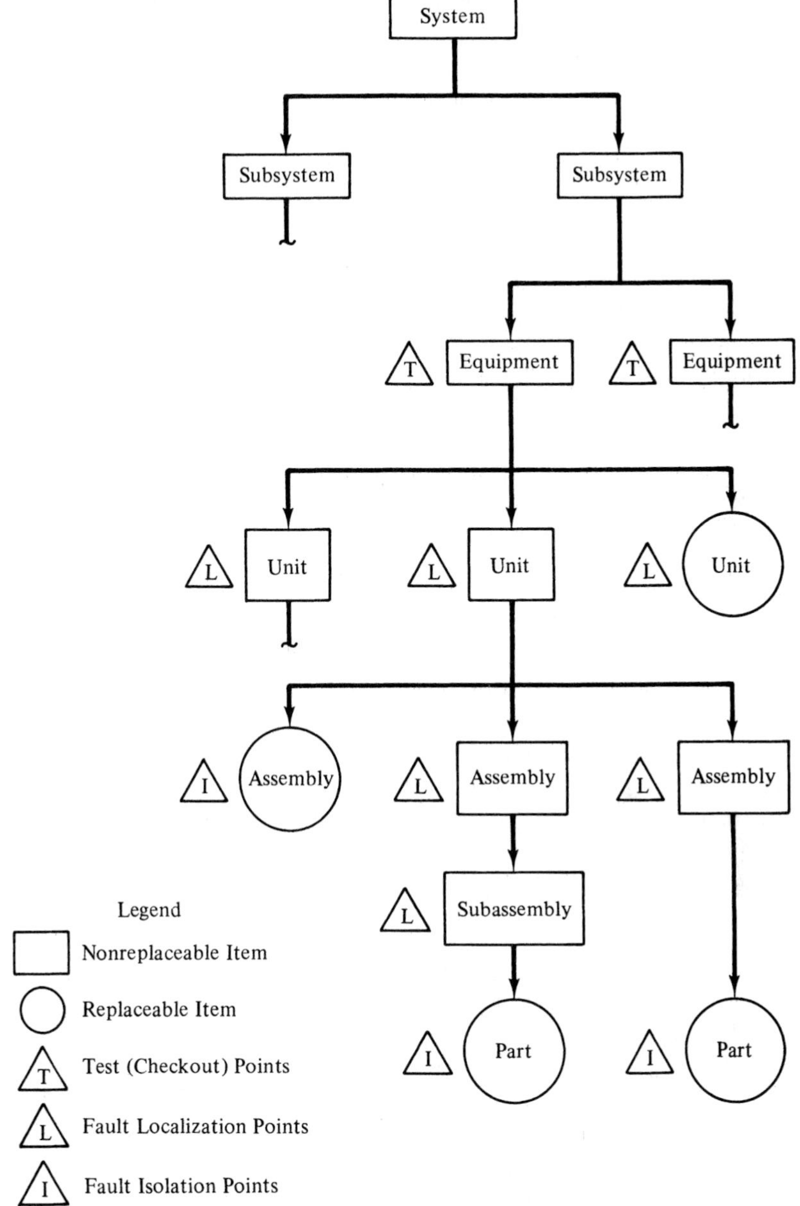

Figure 5.17 Functional-level diagram.

tem support evolve from the process conveyed in Figure 3.25. These requirements are iterated from the top down, level of repair analyses are accomplished, and the design progresses with the inclusion of specific maintainability characteristics as shown in Figure 5.17; e.g., the development of packaging approaches and inclusion of diagnostic provisions. This, in turn, leads to the identification of specific maintenance tasks and the completion of the maintenance task analysis (MTA).

Maintainability allocation. Maintainability allocation is accomplished, along with reliability allocation, as one of the initial steps in the design process. Requirements at the system level, stated both quantitatively and qualitatively, are allocated (or apportioned) among the various subsystems, units, and assemblies to provide guidelines for the designer. Referring to Section 3.7, the allocation process stems from the identification and prioritization of the applicable technical performance measures (TPMs) and evolves downward to the extent necessary for control of the design. Trade-offs are sometimes accomplished to attain the right balance of reliability and maintainability characteristics, with the results as shown in Figure 3.32.

Maintainability prediction. Maintainability prediction commences early in system design. The predicted values of MTBM, $\overline{\text{M}}\text{ct}$, $\overline{\text{M}}\text{pt}$, MLH/OH, $\overline{\text{MLH}}_C$, and so on, are compared with the allocated factors for compatibility with system requirements. Maintainability prediction is a design tool used to identify possible problem areas where redesign might be required to meet system requirements.

Several prediction techniques are available, and their particular application will vary somewhat depending on the definition of the design (supported by engineering data) at the time. These are summarized as follows:

1. Prediction of corrective maintenance time may be accomplished using a system functional-level breakdown and determining maintenance tasks and associated times in progressing from one function to another. The functional breakdown is an expansion of the illustration in Figure 5.17, and covers subsystems, units, assemblies, and parts. Maintainability characteristics such as localization, isolation, accessibility, repair, and checkout as incorporated in the design are evaluated and identified with one of the functional levels. Times applicable to each part (assuming that every part will fail at some point) are combined to provide factors for the next higher level. A sample data format for an assembly is presented in Table 5.1.

Similar data prepared on each assembly in the system are combined as illustrated in Table 5.2, and the factors are computed to arrive at the predicted $\overline{\text{M}}\text{ct}$ or $\overline{\text{MLH}}_C$, whichever is desired. Maintenance tasks and task times are estimated from experience data obtained on similar systems in the field. Failure rates are derived from reliability predictions.

2. Prediction of preventive maintenance time may be accomplished using a method similar to the corrective maintenance approach previously summarized. Preventive maintenance tasks are estimated along with frequency and task times. An example is presented in Table 5.3.

3. Prediction of corrective maintenance time may be accomplished using a checklist developed from experience on similar systems. The checklist provides scoring criteria for desired maintainability characteristics in the design. A random sample of parts reflected in the new equipment is identified and the characteristics of design as related to each part are evaluated against the checklist criteria. Scores are noted, and a predicted $\overline{\text{M}}\text{ct}$ is derived using a regression equation that supports the checklist.

The figures derived through maintainability prediction are a direct input to the supportability analysis (particularly the life-cycle cost and maintenance task analyses),

TABLE 5.1 Maintainability Prediction Worksheet

Item: Assembly 4/Part Number: 12345/Sheet No: 4

Part Category	λ	N	$(N)(\lambda)$	Maintenance Times (Hours)							$(N)(\lambda)(\text{Mct}_i)$
				Loc	Iso	Acc	Ali	Che	Int	Mct_i	
Part *A*	0.161	2	0.322	0.02	0.08	0.14	0.01	0.01	0.11	0.370	0.119
Part *B*	0.102	4	0.408	0.01	0.05	0.12	0.01	0.02	0.12	0.330	0.134
Part *C*	0.021	5	0.105	0.03	0.04	0.11	—	0.01	0.14	0.330	0.034
Part *D*	0.084	1	0.084	0.01	0.03	0.10	0.02	0.03	0.11	0.300	0.025
Part *E*	0.452	9	4.060	0.02	0.04	0.13	0.02	0.03	0.08	0.320	1.299
Part *F*	0.191	8	1.520	0.01	0.02	0.11	0.01	0.02	0.07	0.240	0.364
Part *G*	0.022	7	0.154	0.02	0.05	0.15	—	0.05	0.15	0.420	0.064
Total			6.653							Total	2.039

N = quantity of parts; Iso = isolation; Che = Checkout
λ = failure rate; Acc = access; Int = interchange
Loc = localization; Ali = alignment; Mct_i = maintenance cycle time
For determination of $\overline{\text{MMHc}}$, enter manhours for maintenance times.

TABLE 5.2 Maintainability Prediction Data Summary

Work Sheet No.	Item Designation	Work Sheet Factors	
		$\Sigma(N)(\lambda)$	$\Sigma(N)(\lambda)(\text{Mct}_i)$
1	Assembly 1	7.776	3.021
2	Assembly 2	5.328	1.928
3	Assembly 3	8.411	2.891
4	Assembly 4	6.653	2.039
5	Assembly 5	5.112	2.576
13	Assembly 13	4.798	3.112
Grand total		86.476	33.118

$$\overline{\text{Mct}} = \frac{\Sigma(N)(\lambda)(\text{Mct}_i)}{\Sigma(N)(\lambda)} = \frac{33.118}{86.486} = 0.382 \text{ hour}$$

and form the basis for determining logistic resource requirements for a given design configuration.

Level of repair analysis (LORA). The initial definition of repair policies is accomplished in the development of the maintenance concept during the conceptual design phase (refer to Section 3.4). A repair policy may dictate that an item should be

TABLE 5.3 Preventive Maintenance Data Summary

Description of Preventive Maintenance Task	Task Frequency $(fpt_i)(N)$	Task Time (Mpt_i)	Product $(fpt_i)(N)(Mpt_i)$
1. Lubricate...	0.115	5.511	0.060
2. Calibrate...	0.542	4.234	0.220
31. Service...	0.321	3.315	0.106
Grand Total	13.260		31.115

$$\overline{Mpt} = \frac{\Sigma(fpt_i)(N)(Mpt_i)}{\Sigma(fpt_i)(N)} = \frac{31.115}{13.260} = 2.346 \text{ hour}$$

designed to be fully repairable, partially repairable, or nonrepairable. In the event that a repair decision is made, at what level should the repair be accomplished? Should the item be repaired at the organizational level of maintenance, at the intermediate level, or at the manufacturer/supplier level? Further, to what depth should the repair be accomplished? Such decisions are often based on the results of a level of repair analysis (LORA), which considers a combination of economic, technological, political, social, and environmental factors. The approach in conducting a level of repair analysis is illustrated through the case study presented in Section 4.2.6.

The results from a repair-level analysis are significant in that they will lead to the identification of the functions and maintenance tasks that will be accomplished at each level of maintenance (refer to Figure 3.14), which, in turn, will influence the resource requirements (e.g., spares/repair parts, test equipment). Further, they will influence whether or not a commercial off-the-shelf (COTS) item will be selected, item procurement policies, the level at which warranties are established, the degree and level at which configuration management controls are established, the level at which interchangeability requirements must be specified, and so on. As the system life cycle evolves and producer/supplier relationships change, repair policies may change. This is particularly true for systems that are still being utilized while the initial production capabilities have been discontinued (i.e., the *postproduction support* phase). In essence, the LORA can have a major impact on the design of the overall maintenance and support infrastructure.

Maintenance task analysis (MTA). Having identified the major scheduled and unscheduled maintenance functions that must be accomplished, the next step is to break these functions down into *job operations*, *duties*, *tasks*, *subtasks*, and *task elements*. A maintenance task analysis (MTA) may be accomplished following the steps illustrated through the case study described in Section 4.2.4 and amplified though the procedures included in Appendix C. Referring to Figures 4.29 and 4.30, note that the tasks have been identified for a given corrective maintenance cycle, and that the specific resource requirements have been identified for each of the tasks listed. By accomplishing such a task analysis for each repairable item within the overall system structure, the analyst can combine and integrate the resources specified into an overall package for the system as an entity.

The maintenance task analysis (MTA) can be utilized for several purposes. First, it allows for an excellent evaluation of a given design configuration relative to the incorporation (or lack thereof) of maintainability characteristics in the design. Referring to Section 4.2.4, the results from the illustrated case study indicate where design improvements can be made to improve system maintainability. Second, it leads to the identification of maintenance personnel quantities and skill levels, spares/repair parts, test and support equipment, maintenance facilities, software, technical data, and so on. This, in turn, provides an input in the development of the system maintenance and support infrastructure. High-resource "drivers" can be identified, leading to recommendations for improving the overall system support capability.

Maintainability design—summary. The level of repair analysis (LORA) and the maintenance task analysis (MTA), in particular, may be included within the context of a formal maintainability program effort or within the accomplishment of a supportability analysis (SA). The objective here is not to "classify" these activities as to where (and by whom) they may be accomplished, but to emphasize that the results are a necessary input for the SA.

5.2.3 Design for Human Factors and Safety

Until fairly recently, the function of human factors in the design has received little priority in relation to performance, schedule, cost, and even reliability and maintainability. However, it has been realized that for the system design to be complete, one also needs to address the human element and the interface(s) between the human being and the other elements of the system. Optimum hardware (and software) design alone will not guarantee effective results. Consideration must be given to anthropometric factors (e.g., human physical dimensions), human sensory factors (e.g., sight, hearing, feel), human physiological factors (e.g., reaction to environment), psychological factors (e.g., need, expectation, attitude, motivation), and their interrelationships. Human factors in design deal with these considerations, and the results affect not only system operation (i.e., the operator) but the human being in the performance of maintenance and support activities. Human physical and psychic behavior is a major consideration in determining operational and maintenance functions, personnel and training requirements, procedural data requirements, and facilities.

The gross-level operational and maintenance functions of the system (defined through the functional analysis described in Section 3.6 and iterated further in Figure 5.18) should be allocated between human being and equipment so as to utilize the capabilities of each in the most efficient manner possible. It is obvious that human beings and equipment are not directly comparable. Equipment, when operating, performs in a consistent manner but is relatively inflexible. Human beings, on the other hand, are flexible but do not always perform in a consistent manner.

The human factors effort in the design process is directed toward providing an optimum human interface with equipment and software. Where manual functions are

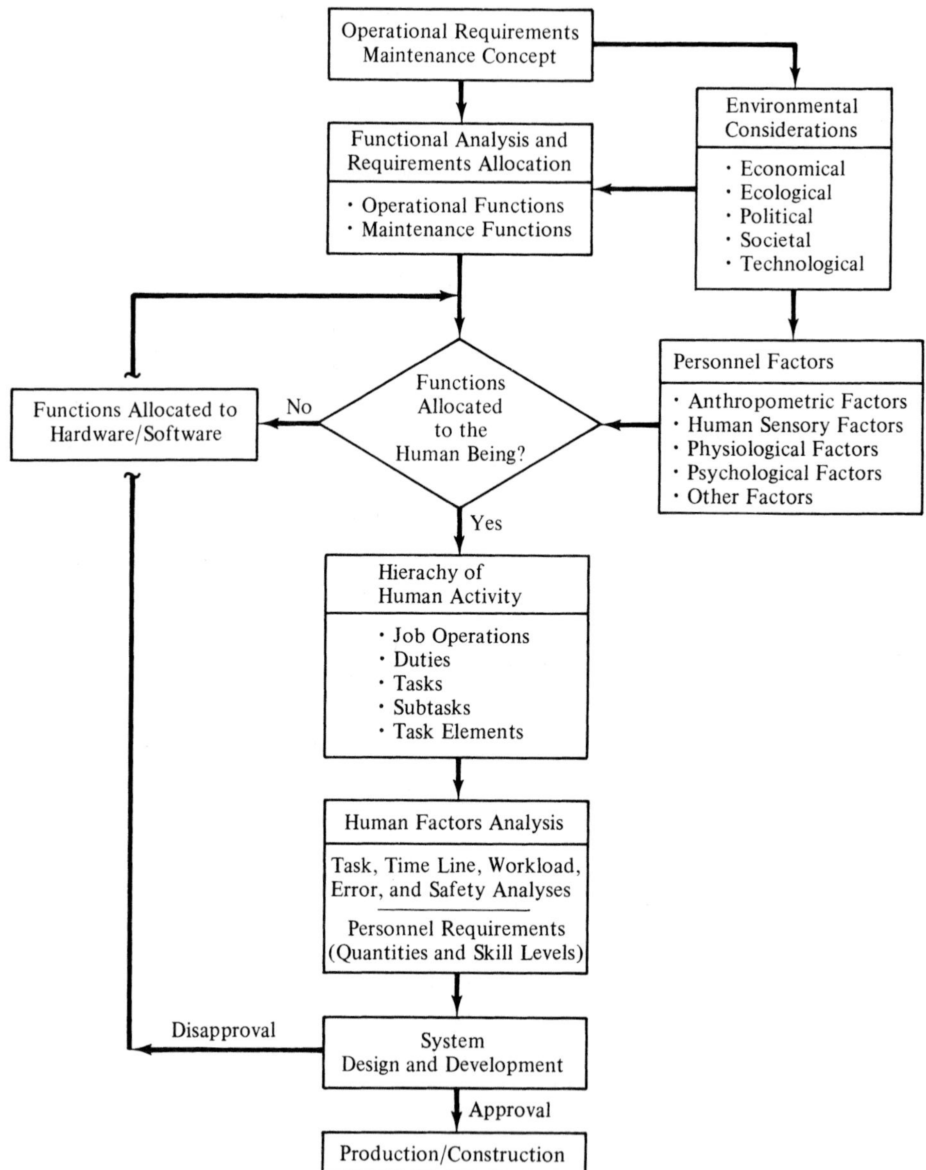

Figure 5.18 Human factors requirements.

performed (requiring information handling, communications, decision making, and coordination), it is necessary to ensure that personnel performance and labor utilization are maximized, and that personnel attrition and training costs are held to a minimum. Further, the personnel errors in the operation and maintenance of equipment must be eliminated if possible. The inclusion of features in the design that are simple to understand, facilitate task accomplishment, and result in clear-cut decisions is necessary.

The relationship between human factors and logistic support is rather pronounced since personnel and training requirements constitute a large factor in the support picture, and these requirements are a direct result of human factor considerations in the design. In addition, the aspect of human factors is very closely allied and integrated with reliability and maintainability, and in some areas an overlap exists.[5]

Human factors analysis. Throughout system design, a human factors analysis is performed as an integral part of the overall system analysis effort. The human factors analysis constitutes a composite of individual program activities directed toward (1) the initial establishment of human factors requirements for system design, (2) the evaluation of system design to ensure that an optimum interface exists between the human being and other elements of the system, and (3) the assessment of personnel quantity and skill-level requirements for a given system design configuration. The analysis effort employs a number of the analytical techniques and is closely related to the reliability analysis, maintainability analysis, supportability analysis, and life-cycle cost analysis.

The human factors analysis begins with conceptual design when functions are identified and trade-off studies are accomplished to determine whether these functions are to be performed manually using human resources, automatically with equipment, or by a combination thereof (refer to Chapter 3). Given the requirements for human resources, one must then ensure that these resources are utilized as efficiently as possible. Thus, the analysis continues through an iterative process of evaluation, system modifications for improvement, reevaluation, and so on. In support of this latter phase of the overall analysis process, there are a number of methods and techniques that can be employed for evaluation purposes. These include the accomplishment of the operator task analysis (OTA), the generation of operational sequence diagrams (OSDs), the performance of an error analysis and a safety analysis, the preparation of duty and task worksheets, and so on.

1. *Operator task analysis (OTA).* This facet of analysis involves a systematic study of the human behavior characteristics associated with the completion of a system task(s). It provides data basic to human engineering design and to the determination of personnel types and skill-level requirements. Tasks may be classified as being discrete or continuous. Further, there are operator tasks and maintenance tasks. Thus, one may wish to divide the analysis effort into the *operator task analysis* and the *maintenance task analysis*. The portion of the analysis covering maintenance tasks may evolve directly from the maintenance task analysis (MTA).

In accomplishing a task analysis, there are varying degrees of emphasis and the type of format used. However, the following general steps apply in most instances:

[5] The objective is to provide an introduction to human factors (or human engineering), and not to cover the subject in depth. However, for more information, three good references are (1) Meister, D., *Behavioral Analysis and Measurement Methods*, John Wiley, New York, N.Y., 1985; (2) Sanders M. S., and McCormick, E. J., *Human Factors in Engineering Design*, 7th Ed., McGraw-Hill, New York, N.Y., 1992; and (3) Woodson, W. E., Tillman, B., and Tillman, P., *Human Factors Design*, 2nd Ed., McGraw-Hill, New York, N.Y., 1991. Additional references are included in Appendix H.

(a) Identify system operator and maintenance functions and establish a hierarchy of these functions in terms of job operations, duties, tasks, subtasks, and task elements.

(b) Identify those functions (or duties, tasks, etc.) that are controlled by the human being and those functions that are automated.

(c) For each function involving the human element, describe the specific information necessary for operator or maintenance personnel decisions. Such decisions may lead to the actuation of a control, the monitoring of a system condition, or the equivalent. Information required for decision making may be presented in the form of a visual display, or an audio signal of some type.

(d) For each action, determine the adequacy of the information fed back to the human being as a result of control activations, operational and maintenance sequences, and so on.

(e) Determine the impact of the environmental and personnel factors and constraints on the human activities identified.

(f) Determine the time requirements, frequency of occurrence, accuracy requirements, and criticality of each action (or series of actions) accomplished by the human being.

(g) Determine the human skill-level requirements for all operator and maintenance personnel actions.

(h) Integrate and group these skill-level requirements into specific position descriptions and describe individual workstation requirements.

A task analysis is generated to ensure that each stimulus is tied to a response, and that each response is directly related to a stimulus. Further, individual human motions are analyzed on the basis of dexterity, mental and motor skill requirements, stress and strain characteristics of the human being performing the task, and so on. The purpose is (1) to identify those areas of system design where potential human-machine problems exist, (2) to identify the necessary personnel skill-level requirements for operating and maintaining the system in the future, and (3) to describe personnel quantities and skill-level requirements for supporting organizations.

2. *Operational sequence diagrams (OSDs).* As part of the human factors analysis activity, one of the major tasks is the evaluation of the flow of information from the point in time when the operator first becomes involved with the system to completion of the mission. Information flow in this instance pertains to human decisions, human control activities, and the transmission of data.

There are a number of different techniques that can be employed to show information flow. The use of operational sequence diagrams is one. Operational sequence diagrams are decision-action flow devices that integrate operational functions and equipment design. More specifically, these diagrams project different sequences of operation showing:

(a) Manual operations

(b) Automatic operations

(c) Operator decision points

(d) Operator control actuations or movements

(e) Transmitted information
(f) Received information using indicator displays, meter readouts, and so on

Operational sequence diagrams are similar to industrial engineering work-flow process charts and time line analyses, and are used to evaluate decision-action sequences and human-machine interfaces. The evaluation of operator control panel layouts and workspace design configurations are good examples of where these diagrams may be profitably used. Figure 5.19 illustrates a sample operational sequence diagram.

3. *Error analysis.* An error occurs when a human action exceeds some limit of acceptability, where the limits of acceptable performance have been defined. Errors may be broken down into errors of omission when a human fails to perform a necessary task and errors of commission when a task is performed incorrectly (i.e., selection, sequence, or time errors). The possible causes of error are due to the following:

(a) Inadequate work space and work layout—poor workstation design relative to seating, available space, activity sequences, and accessibility to system elements
(b) Inadequate design of facilities, equipment, and contol panels for human factors—inadequate displays and readout devices, poor layout of controls, and lack of proper labeling
(c) Poor environmental conditions—inadequate lighting, high or low temperatures, and high noise level
(d) Inadequate training, job aids, and procedures—lack of proper training, and poorly written operating and maintenance procedures
(e) Poor supervision—lack of communications, no feedback, and lack of good planning resulting in overtime

The error analysis can be accomplished in conjunction with the OTA, MTA, during the development of OSDs, and physically as part of the system test and evaluation effort discussed in Section 5.7. The objective is to select a human with the appropriate skills and training, simulate the operation of the system by undertaking a series of functions/tasks, and measure the number of errors that occur in the process. Using an Ishikawa "cause-and-effect" diagram approach, as conveyed in Figure 4.20, can aid in determining the causes of the errors (i.e., how was the error introduced in the system)? Additionally, it is important to note the effects of the error on other elements of the system and on the system as an entity. Thus, it is important that the error analysis be closely integrated with both the FMECA (Section 4.2.2) and the System Safety/Hazard Analysis discussed subsequently.

4. *Safety/hazard analysis.* The safety/hazard analysis is closely aligned with the FMECA described in Section 4.2.2. Safety pertains to both personnel and the other elements of the system with personnel being emphasized herein. The safety/hazard analysis generally includes the following basic information.[6]

[6] The objective is to provide an introduction to safety engineering. For a more in-depth coverage, two good references are (1) Hammer, W., *Occupational Safety Management and Engineering*, 4th Ed., Prentice Hall, Upper Saddle River, NJ, 1989; and (2) Roland, H. E., and Moriarty, B., *System Safety Engineering and Management*, 2nd Ed., John Wiley, New York, N.Y., 1990.

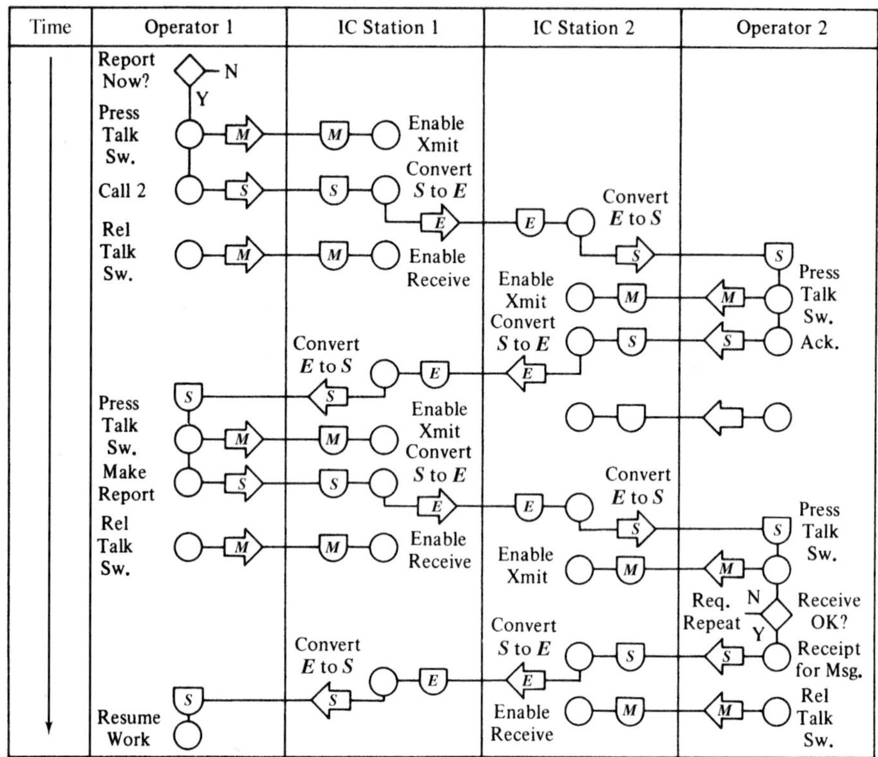

Notes on Operational Sequence Diagram

Symbols		Links	
◇	Decision	*M*	Mechanical or Manual
○	Operation	*E*	Electrical
⇨	Transmission	*V*	Visual
◡	Receipt	*S*	Sound
D	Delay	etc.	
□	Inspect, Monitor		
▽	Store		

Stations or subsystems are shown by columns; sequential time progresses down the page.

(Source: MIL-H-46855, Military Specification, "Human Engineering Requirements for Military Systems, Equipment and Facilities," Department of Defense, Washington, D.C.)

Figure 5.19 Operational sequence diagram (example). (Source: MIL-H-46855, Military Specification, "Human Engineering Requirements for Military Systems, Equipment and Facilities," Department of Defense, Washington, D.C.)

(a) *Description of hazard.* System operational requirements and the system maintenance concept are reviewed to identity possible hazardous conditions. Past experience on similar systems or products utilized in comparable environments serves as a good starting point. Hazardous conditions may include acceleration and motion, electrical shock, chemical reactions, explosion and fire, heat and temperature, radiation, pressure, moisture, vibration and noise, and toxicity.

(b) *Cause of hazard.* Possible causes should be described for each identitied hazard. In other words, what events are likely to occur in creating the hazard?

(c) *Identification of hazard effects.* Describe the effects of each identified hazard on both personnel and equipment. Personnel effects may include injuries, such as cuts, bruises, broken bones, punctures, heat exhaustion, asphyxiation, trauma, and respiratory or circulatory damage.

(d) *Hazard classification.* Hazards may be categorized according to their impact on personnel and equipment as follows:

(1) *Negligible hazard (category I).* Such conditions as environment, personnel error, characteristics in design, errors in procedures, or equipment failures that will not result in significant personnel injury or equipment damage.

(2) *Marginal hazard (category II).* Such conditions as environment, personnel errors, characteristics in design, errors in procedures, or equipment failures that can be controlled without personnel injury or major system damage.

(3) *Critical hazard (category III).* Such conditions as environment, personnel error, characteristics in desiring errors in procedures, or equipment failures that will cause personnel injury or major system damage, or that will require immediate corrective action for personnel or system survival.

(4) *Catastrophic hazard (category IV).* Such conditions as environment, personnel error, characteristics in design, errors in procedures, or equipment failures that will cause death or severe injury, or complete system loss.

(e) *Anticipated probability of hazard occurrence.* Through statistical means, estimate the probability of occurrence of the anticipated hazard frequency in terms of calendar time, system operation cycles, equipment operating hours, or equivalent.

(f) *Corrective action or preventive measures.* Describe the action(s) that can be taken to eliminate or minimize (through control) the hazard. It is hoped that all hazardous conditions will be eliminated; however, in some instances it may only be possible to reduce the hazard level from category IV to one of the lesser critical categories.

The safety/hazard analysis serves as an aid in initially establishing design criteria and as an evaluation tool for the subsequent assessment of design for safety. Although the format may vary somewhat, a safety analysis may be applied in support of requirements for both industrial safety and system/product safety.

Personnel and training requirements. Personnel skill levels and quantities are identified by evaluating the complexity and frequency of tasks in the detailed task analysis. Job proficiency levels are established for each location where prime elements of the system are operated and where maintenance support is performed. These requirements are compared with the personnel goals initially specified for the system.

Given the requirements for personnel as dictated by the system design, one must determine the personnel resources that actually will be assigned to operate and maintain the system in the field. The difference in skills between the specified requirements and the personnel that will be assigned is the basis for a formalized training program (i.e., that effort required to upgrade personnel to the desired proficiency level). Training needs are defined in terms of program content, duration of training, training data, software, and training equipment requirements.

Human factors engineers are interested in the personnel and training requirements to ensure that these requirements are realistic for the system. If skill level requirements are high and a large amount of training is anticipated, then the system design should be reevaluated to see if changes can be made to simplify the situation. In addition, the need for high personnel skills significantly limits the market in terms of finding qualified people.

Human factors—summary. The ultimate result of this effort, accomplished during the design process, not only is directed toward assessing the human-equipment interface, but forms the basis for the personnel and training element of total logistic support.

5.2.4 Design for Producibility

Producibility is a measure of the relative ease and economy of producing a system or a product. The characteristics of design must be such that an item can be produced easily and economically using conventional and flexible manufacturing methods and processes without sacrificing function, performance, effectiveness, or quality. Some major objectives in designing a system for producibility are noted as follows:

1. The quantity and variety of components utilized in system design should be held to a minimum. Common and standard items should be selected where possible, and there should be a number of different supplier sources available throughout the planned life cycle of the system.
2. The materials selected for constructing the system should be standard, available in the quantities desired and at the appropriate times, and should possess the characteristics for easy fabrication and processing. The design should preclude the specification of peculiar shapes requiring extensive machining and/or the application of special manufacturing methods.
3. The design configuration should allow for the easy assembly (and disassembly as required) of system elements (i.e., equipment, units, assemblies, modules, etc.). Assembly methods should be simple, repeatable, economical, and should not require the utilization of special tools and devices or high personnel skill levels.
4. The design configuration should be simplistic to the extent that the system (or product) can be produced by more than one supplier, using a given data package and conventional manufacturing methods/processes. The design should be compatible with the application of CAD/CAM technology where appropriate.

The basic underlying objectives are simplicity and flexibility in design. More specifically, it is the goal to minimize the use of critical materials and critical processes, the use of proprietary items, the use of special production tooling, the application of unrealistic tolerances in fabrication and assembly, the use of special test systems, the use of high personnel skills in manufacturing, and the production/procurement lead times.

As one can see, the objectives of producibility are directly in line with supportability goals. The use of common and standard components simplifies the spare/repair parts provisioning and procurement processes; the ease of assembly and disassembly of system components facilitates the accomplishment of repair activities in response to corrective maintenance requirements; and the availability of multiple supplier sources provides for the necessary postproduction support. Producibility, as a characteristic of system design, is a significant contributor to the design for supportability. Figure 5.20 conveys some of the key considerations that need to be included in the design process.

5.2.5 Design for Quality

In the past, the fulfillment of quality objectives has been accomplished primarily in the production/construction phase of the life cycle through the implementation of a formal quality control (QC) or quality assurance (QA) program. Statistical process control (SPC) techniques, incoming and in-process inspection activities, closely monitored supplier control programs, periodic audits, and selected problem-solving methods have been implemented with the objective of attaining a designated level of system quality. The efforts have basically been accomplished after-the-fact, and the overall results have been questionable.

Recently, the aspect of *quality* has been viewed more from a top-down, life-cycle perspective, and the concept of *total quality management* (TQM) has evolved.[7] TQM, as defined in Chapter 1 (Section 1.5), represents a total integrated management approach that addresses system/product quality during all phases of the life cycle and at each level in the overall system hierarchy. It provides a before-the-fact orientation to quality, and it focuses on system design and development activities as well as production, manufacturing, assembly, construction, product support, and related functions.

Included within the function of system design and development is the consideration for (1) the inherent characteristics of the various system components, and (2) the characteristics of the processes that will be utilized to produce the system components and its elements of support. Quality in design pertains to many of the issues discussed throughout the earlier sections of this chapter. Emphasis is directed toward design simplicity, flexibility, standardization, and so on. Of a more specific nature are the concerns for *variability*, where a reduction in the variation of the dimensions for specific component designs, or tolerances in process designs, might result in an overall improvement. Taguchi's general approach to *robust design* is to provide a design that is insensitive to the variations normally encountered in production or in operational use. The more robust the design, the less the support requirements, the lower the life-cycle cost, and the higher the degree of effectiveness. Overall design improvement is anticipated through a combination of careful component evaluation and selection, the

[7] The Department of Defense, in particular, has been advocating TQM, and there are many references covering the subject from different perspectives. DOD requirements are covered by DODD 5000.51G, *Total Quality Management: A Guide for Implementation*, Department of Defense, Washington, D.C.

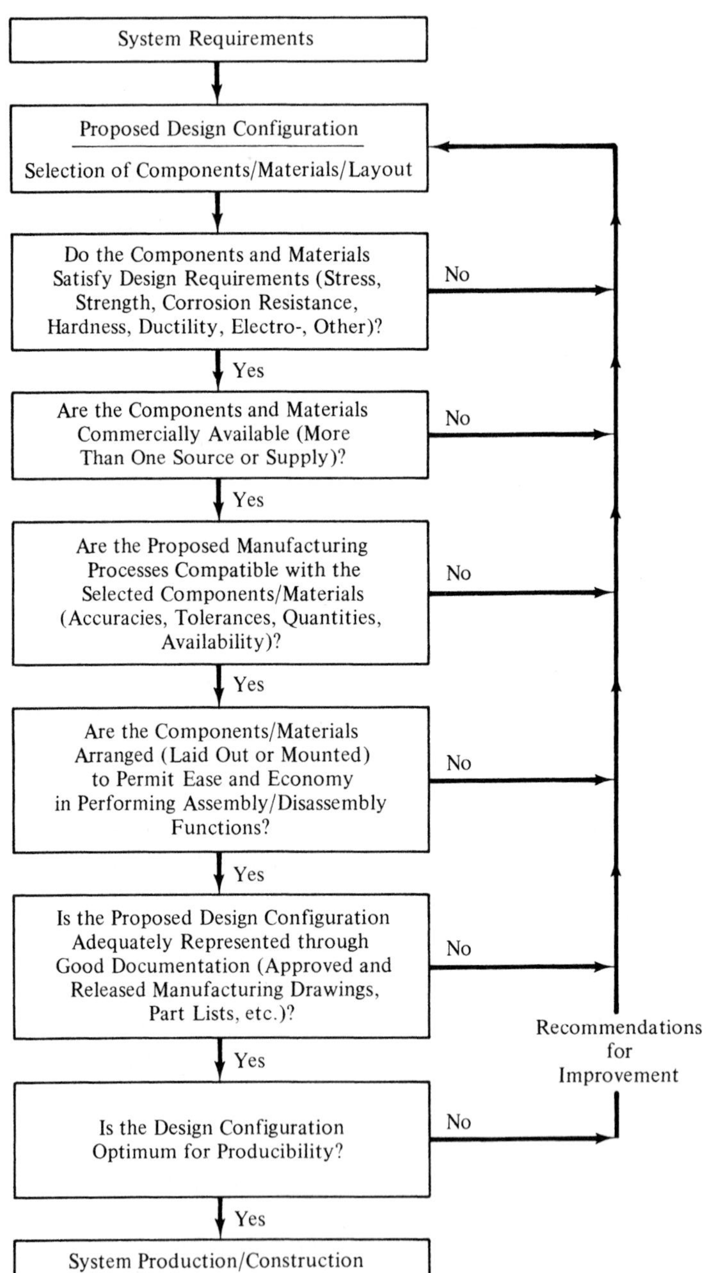

Figure 5.20 Producibility considerations.

appropriate use of statistical process control (SPC) methods, and application of experimental testing approaches, applied on a continuous basis.[8]

The subject of quality pertains to both the technical characteristics of design and the humanistic aspects in the accomplishment of design activities. Not only is there a concern relative to the selection and application of components, but the successful fulfillment of quality objectives is highly dependent on the behavioral characteristics of those involved in the design process. A thorough understanding of customer requirements, good communications, a team approach, the willingness to accept the basic principles of TQM, and so on, are all necessary. In this respect, the objectives of quality engineering are inherent within the scope of system engineering and are directly applicable to logistic support.

5.2.6 Design for Economic Feasibility

The significance of addressing life-cycle economics in the system design and development process has been emphasized throughout the earlier chapters of this book. Although it is often popular to consider only certain segments of cost in making design decisions (i.e., initial price or acquisition cost), there are life-cycle implications associated with almost all decisions. Therefore, one must view cost from the total life-cycle perspective as presented in 1.5 and 4.2.1.

When considering economics in system design (i.e., *design for economic feasibility*), the steps are similar to those followed in addressing reliability requirements, maintainability requirements, and so on, and are illustrated in Figure 5.21. More specifically:

1. Cost targets, reflecting life-cycle cost considerations, should initially be established during conceptual design. Such a target (or goal) may be expressed as *design to cost* (DTC), *design to unit life-cycle cost*, and so on. This factor may be further broken down into *design to unit acquisition cost, design to unit production cost, design to unit operational and support cost*, or any combination of these. These target values should be specified in the definition of system technical performance measures (Section 3.5).
2. The specified quantitative economic factors applicable at the system level should be allocated to the appropriate elements of the system, as necessary to ensure that economics is reflected in the design of these elements (refer to Section 3.7).
3. Life-cycle cost analyses are accomplished throughout the design process, in the evaluation of alternatives, to ensure that the ultimate approach selected reflects economic considerations. Alternative technology applications, operational concepts, support policies, equipment packaging schemes, levels of repair, diagnostic routines, and so on, are evaluated through a life-cycle cost analysis (refer to Section 4.2.1).
4. Throughout production and during system operational use, life-cycle cost analyses are accomplished to identify high-cost contributors and to aid in determining

[8] Dr. Genichi Taguchi has developed mathematical techniques for application relative to the evaluation of design variables. Refer to Rose, P. J., *Taguchi Techniques for Quality Engineering*, McGraw-Hill Book Company, New York, N.Y., 1988.

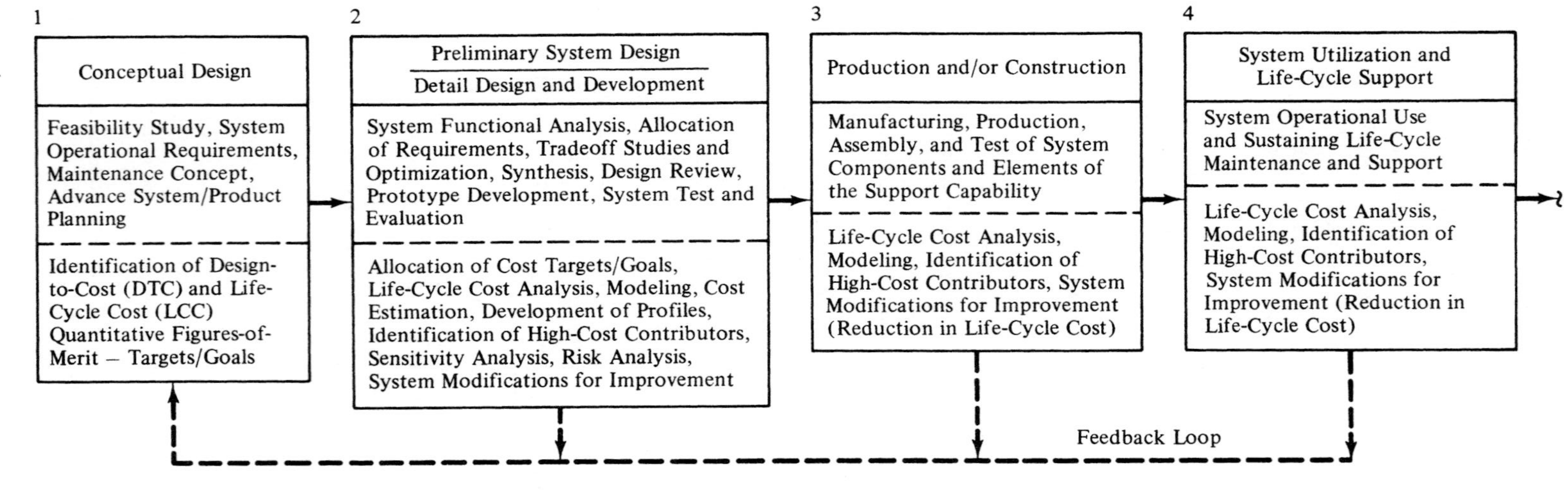

Figure 5.21 Cost considerations in the systems life cycle.

cause-and-effect relationships, leading to the recommendations for system/product improvement. Proposals may be initiated with the objective of reducing life-cycle cost.

Referring to Figure 5.21, life-cycle cost analyses (in one form or another) are accomplished throughout system design and development, during production, and while the system is in operational use. The completion of such an effort generally requires that one follow certain steps such as those presented in Figure 4.11, and amplified in Appendix E. Referring to the figure, one needs to progress from the problem definition stage, through the definition of system requirements, the development of a cost breakdown structure (CBS), cost profiles, summaries, and so on. Although this process is characteristic of a typical system analysis approach, the important consideration is to ensure that economic issues are addressed in terms of the entire system life cycle.

5.2.7 Design for the Environment

Although the previous sections in this chapter dealt primarily with some of the more tangible considerations in design, it is essential that one also consider the aspect of "design for the environment." *Environment*, in this context, refers to the numerous external factors that must be addressed in the overall system development process. In addition to the *technological* and *economic* factors discussed previously, one must deal with *ecological*, *political*, and *social* considerations as well. The system being developed must be compatible with, acceptable, and ultimately must exist within an environment that addresses the many factors illustrated in Figure 5.22. A requirement within the

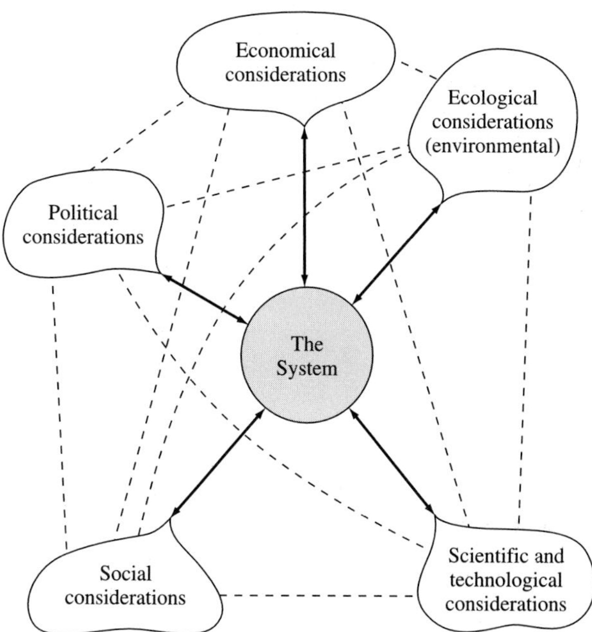

Figure 5.22 Environmental influences on system design and development.

spectrum of systems engineering is to ensure that the system being developed will be socially acceptable, is compatible with the political structure, is technically and economically feasible, and will not cause a degradation to the environment overall.

Of particular interest here are the ecological considerations. Ecology generally pertains to the study of the relationships between various organisms and their environment. This includes consideration of plant, animal, and human populations in terms of rate of population growth, food habits, reproductive habits, and ultimate death. In other words, one is addressing the conventional biological process as viewed in the broad context.

In recent decades, the world population growth, combined with the technological changes associated with our living standards, has created a greater consumption of our resources, resulting in potential shortages, which, in turn, has stimulated shifts toward establishing other means for accomplishing objectives. Concurrently, the amount of waste has increased significantly. The net effects of this have caused alterations to the basic biological process, and to some extent these alterations have been harmful. Of particular concern are those problems dealing with the following:

1. *Air pollution and control.* Any gaseous, liquid, or solid material suspended in air that could result in health hazards to humans. Air pollutants may fit into categories to include particulate matter (small substances in air resulting from fuel combustion, incineration of waste materials, or industrial processes), sulfur oxides, carbon monoxide, nitrogen oxides, and hydrocarbons.
2. *Water pollution and control.* Any contaminating influence on a body of water brought about by the introduction of materials that will adversely affect the organisms living in that body of water (measure of dissolved oxygen content).
3. *Noise pollution and control.* The introduction of industrial noise, community noise, and/or domestic noise that will result in harmful effects on the humans (e.g., loss of hearing).
4. *Radiation.* Any "natural" or "human-made" energy transmitted through space that will result in harmful effects on the humans.
5. *Solid waste.* Any garbage or refuse (e.g., paper, wood, cloth, metals, plastics, etc.) that cannot be decomposed and will result in a health hazard. Roadside dumps, piles of industrial debris, junk car yards, and so on, are good examples of solid waste. Improper solid-waste disposal may be a significant problem in view of the fact that flies, rats, and other disease carrying rodents are attracted to areas where there are solid wastes. In addition, there may be a significant impact on air pollution if windy conditions prevail or on water pollution if the solid waste is located near a lake, river, or stream.

In the design of systems, all phases of the life cycle must be addressed, including the "retirement and material disposal" block in Figures 1.4 and 1.5. When the system and its components are retired from the inventory, either because of obsolescence and there is no longer a need or for the purposes of maintenance when items are removed in order to accomplish repair, those items must be of such a makeup that they can be disposed of without causing any negative impacts on the environment. More specifically, a prime objective is to design components such that they can be *reused* in other similar applications. If there are no opportunities for "reuse," then the component should be designed such that it can be decomposed, with the residual elements being *recycled* and converted into materials that can be remanufactured for other purposes. Further, the *recycling process* itself should not create any detrimental effects on the environment.

Thus, in the development of systems and in the selection of components, the designer needs to be sure that the materials selected can be reused if possible, will not cause any toxicity problems, and can be decomposed without adding to the solid-waste inventory that currently exists in many areas. Care must also be taken to ensure that the product characteristics do not generate the need for a nonreusable container or packing materials for transportation that will cause problems. Figure 5.23 conveys a decision-making logic approach that may be applied and that will be helpful in the design and development of systems.

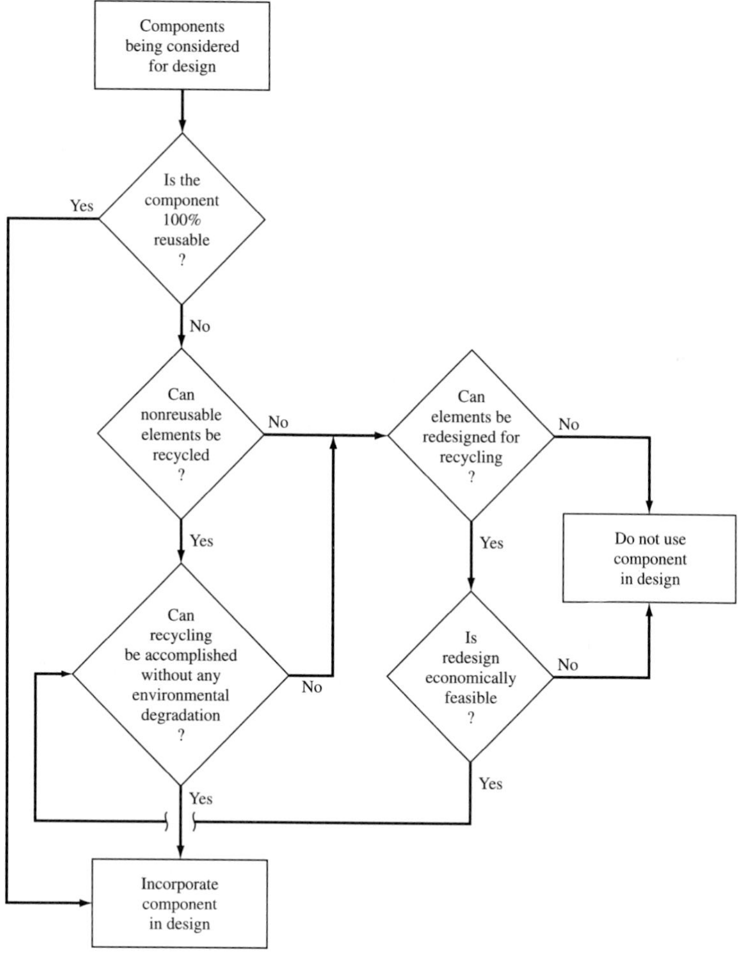

Figure 5.23 An abbreviated component-selection process.

5.3 SUPPLIER DESIGN ACTIVITIES

The term *supplier* refers to a broad class of external organizations that provide products, components, materials, or services to a producer, a prime contractor, or directly to the ultimate consumer. This may range from the delivery of a major subsystem or configuration item down to a small component part. Suppliers may provide services to include (1) the design, development, and manufacture of a major element of the system; (2) the production and distribution of items already designed, providing a manufacturing source; (3) the distribution of commercial and standard components from an established inventory, serving as a warehouse and providing parts from various sources of supply; and/or (4) the implementation of a process in response to some functional requirement, providing a service.

For many systems, suppliers provide a large number of elements that make up the system (i.e., more than 50% of the components in some cases), as well the spares and repair parts that are required to support maintenance activities. Further, with the current trends toward globalization and more "outsourcing," the percentage of supplier activity for a given program is likely to increase in the future. For large programs, there may be a *layering* of suppliers, as shown in Figure 5.24, with one or more suppliers providing services to the supplier of a major subsystem or configuration item. Additionally, these suppliers may be located in designated geographical centers throughout the world.[9]

Of particular interest in this section are those suppliers who are involved in the design and development of new system elements. Through accomplishment of the functional analysis and allocation process (Sections 3.6 and 3.7), proposed technical solutions for design are identified and evaluated, and "make or buy" decisions are made. Should the design and manufacture of an item of equipment, the development

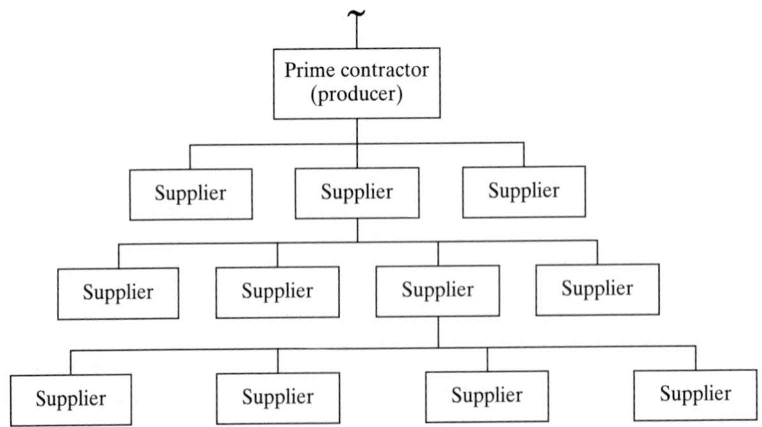

Figure 5.24 Typical structure involving the layering of suppliers.

[9] The term *outsourcing* refers to the practice of soliciting the support of component suppliers to accomplish selected packages of work externally from the producer or prime contractor. Experience indicates that there is a greater use of external suppliers today than in the past.

of a software package, or the completion of a process be accomplished *in house* by the producer or prime contractor (i.e., the "make" decision), or should an *external* source of supply be selected (i.e., the "buy" decision)? Trade-off studies are conducted, and the requirements for items selected for external design are described in the appropriate Development (Type "B") Specification(s).

In preparing these specifications, it is essential that the "traceability" of requirements be maintained as one progresses downward through hierarchy shown in Figure 5.3. The appropriate TPMs must be identified and prioritized, design criteria must be established and included in the appropriate specification(s), and the suppliers selected must comply with these requirements. In other words, the supplier in question must address such considerations as reliability, maintainability, human factors and safety, producibility, disposability, and related factors as they apply to the system element being developed. Further, the analytical techniques/methods/tools described in Chapter 4 may be applied as necessary to enhance the design of the item in question. The supportability analysis must be extended and "tailored" to include the supplier design activity, particularly for the larger elements of the system. In such cases, the supplier should be an integral part of the *design team*.

5.4 DESIGN INTEGRATION

Design integration activities commence during the early stages of conceptual design and extend through system development, production or construction, distribution, system operational use and sustaining support, and ultimate retirement and the disposal (or recycling) of materials. Initially, such activities are more associated with the development of a new system, whereas, subsequently, these activities may pertain to the evaluation, modification, and improvement of systems already in being. As requirements are established, the *design team* is formed.

In the early stages of conceptual design, the emphasis is on the *system* as an entity and development of the requirements for such! The design team may include only a small number of selected qualified individuals, with the objective of preparing a comprehensive System Specification (Type "A"). The requirements conveyed in Figure 5.25 must be properly integrated into this top-level specification. It is important that personnel with the appropriate backgrounds and experience be selected, and these individuals must be able to work together and effectively communicate on a day-to-day basis. The assignment of of a large number of individual domain specialists, whose expertise lie in given technical fields, is not appropriate at this stage.

As system development progresses, the appropriate design specialists are added to the team. Additionally, supplier activities are introduced, and the team "makeup" may include representation located in various parts of the world. There may be activities that must be accomplished in parallel and where the principles of *concurrent engineering* should be implemented. The objective is to ensure that the right specialists are available at the time required and that their individual contributions are properly integrated into the whole. The selection of domain specialists is highly dependent on the requirements developed through the functional analysis and allocated process.

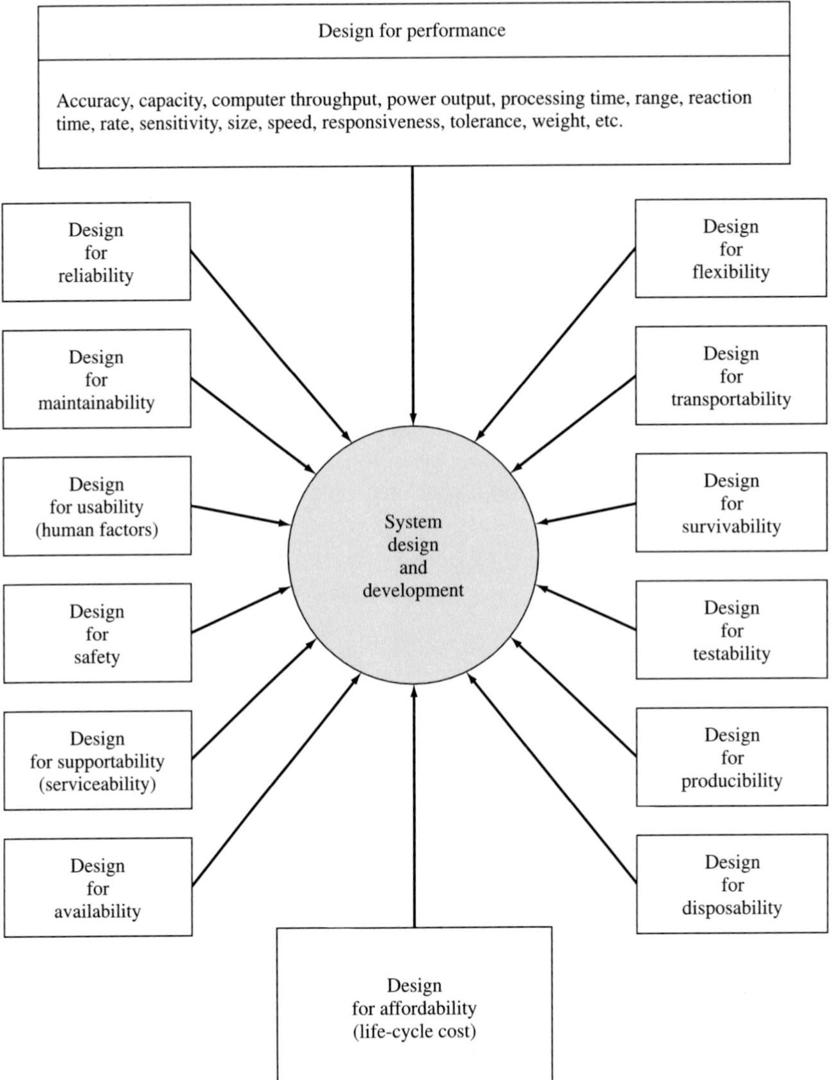

Figure 5.25 System design integration requirements.

As the criteria for design will vary with the type of system and its mission, the emphasis in assigning the proper level of expertise to the team will differ from one project to the next.

During the latter phases of the life cycle, the emphasis is on *evaluation* and the introduction and processing of design changes as necessary. The requirement(s) for *change* may stem from some deficiency (i.e., the failure to meet an initially specified requirement), or for the purposes of *continuous product/process improvement*. Each "engineering change proposal" (ECP) must be evaluated in terms of not just performance alone, but in terms of reliability, maintainability, human factors and safety, producibility, supportability, disposability, and economic feasibility as well.

Inherent within the established design team activity is the requirement for good communications on a day-to-day basis. Although the colocation of personnel in one geographical area is preferred, the trends toward "outsourcing" and decentralization often result in the introduction of many different suppliers and design activities being conducted concurrently and at remote locations. Thus, the design team becomes heavily dependent on the utilization of computer-aided tools, operating in a network such as illustrated in Figure 5.11. Success in this area is, in turn, dependent on the structure of the design database. Such a database may include design drawings and layouts, the presentation of three-dimensional visual models, parts and material lists, prediction and analysis results, supplier data, and whatever else is necessary to describe the system configuration as designed. The designer must be able to gain access to the database and provide input easily, and the results must be transmitted to other members of the design team accurately and in a timely manner. The data, usually presented in a digital format, must be available to all members of the design team concurrently. Instead of many different data items "flowing" back and forth between different members of the design team, between the producer (contractor) and consumer (customer), and so on, an integrated shared database structure is necessary, as illustrated in Figure 4.8. This, of course, should facilitate the process of communications, with every member of the design team having access to the same system description.[10]

5.5 DESIGN REVIEW

Design is a progression from an abstract notion to something that has form and function, is fixed, and can be reproduced in designated quantities to satisfy a need. The designer produces a model that is used as a template for the replication of additional models. In the course of production, an error made in any one model will result in a single rejection. However, an error in the design, repeated in all subsequent models, may lead to a serious problem (e.g., complete recycling of all equipment for a major modification). Thus, the designer's responsibility is significantly large. The basic philosophy or evolution of design is illustrated in Figure 5.26.

Initially, a requirement or need is specified. From this point, design evolves through a series of phases (i.e., conceptual design, preliminary system design, detail design and development). In each major phase of the design process, an evaluative function is accomplished to ensure that the design is correct at that point prior to proceeding with the next phase. The evaluative function includes both the informal day-to-day project coordination and data review, and the formal design review. A more detailed procedure illustrating the evaluative function is presented in Figure 5.27.

[10] With the advent of new technologies on an almost continuing basis, it is anticipated that the nature of the *data environment* will be changing almost constantly. The objective here is to emphasize the need for good communications through the integration and transfer of design data among members of the design team, supporting organizations, and management.

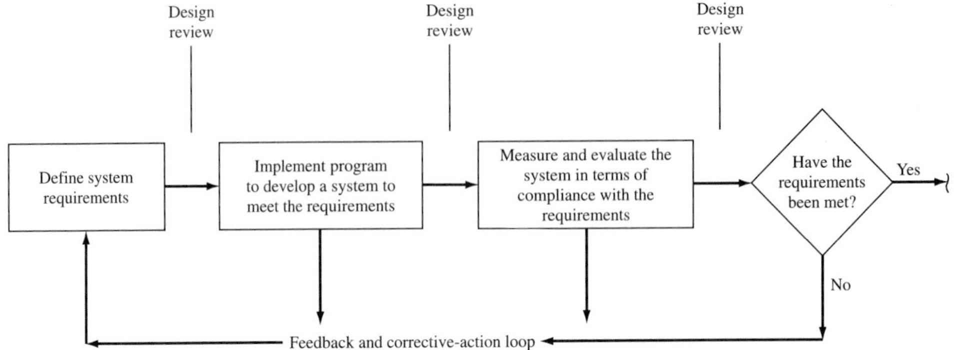

Figure 5.26 The basic system requirements, evaluation, and review process.

Referring to Figure 5.27, design information is released and reviewed for compliance with the basic system/equipment requirements (i.e., performance, reliability, maintainability, human factors). If the requirements are satisfied, the design is approved as is. If not, recommendations for corrective action are prepared and submitted to the designer for action. If no action is taken as a result of the day-to-day liaison activity, the recommendations are presented and discussed as part of the next formal design review. The informal daily liaison activity is basically accomplished through the ongoing communications processes (review of individual drawings, review of a configuration presented through an electronic database, review of a supplier's technical proposal). This informal review process is supported by the checklist criteria identified in Figure 5.8 and amplified in Appendix A.

The formal design review constitutes a coordinated activity (including a meeting or series of meetings) directed to satisfy the interests of the design engineer, the technical discipline support areas (reliability, maintainability, human factors), logistics, manufacturing, industrial engineering, quality control, program management, and so on. The purpose of the design review is to formally and logically cover the proposed design from the "total system standpoint" in the most effective and economical manner through a combined integrated review effort. The formal design review serves a number of purposes.

1. It provides a formalized check (audit) of the proposed system/equipment design with respect to contractual and specification requirements. Major problem areas are discussed and corrective action is taken.
2. It provides a common baseline for all project personnel. The design engineer is provided the opportunity to explain and justify his or her design approach, and representatives from the various supporting organizations (e.g., maintainability, logistics support) are provided the opportunity to hear of the design engineer's problems. This serves as a tremendous communication medium and creates a better understanding among design and support personnel.
3. It provides a means for solving interface problems, and promotes the assurance that all system elements will be compatible. For instance, major interface problems between engineering and manufacturing, relative to lack of producibility,

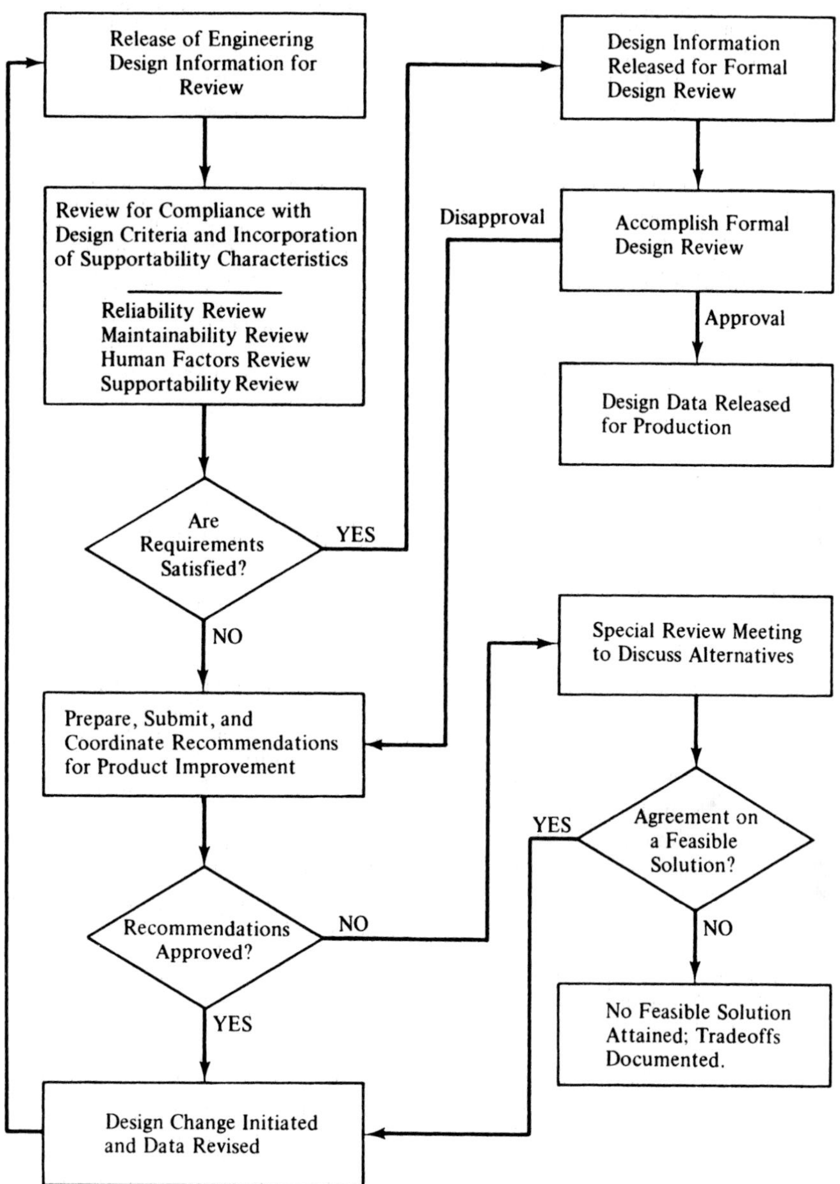

Figure 5.27 Design engineering data review.

are often not detected until after design data are released and production is underway. The results of major problems discovered at that time are quite costly. Another major problem (which seems to be a common occurrence) is the lack of compatibility between the different elements of logistic support with the prime equipment and the elements of logistic support with each other. Such problems are often undetected at an early point in time because of a wide variance of organizational interests and activity, the rush to get the hardware/software into production, the physical separation of members of the design team, or some other related reason. A formal design review is intended to prevent the occurrence of these problems.

4. It provides a formalized record of what design decisions were made and the reasons for making them. Analyses, predictions, and trade-off study reports are noted and are available to support design decisions. Compromises to reliability, maintainability, human factors, and supportability are documented and included in the tradeoff study reports.
5. It promotes a greater probability of mature design as well as the incorporation of the latest techniques (where appropriate). Group review may identify new ideas, possibly resulting in simplified processes and ultimate cost savings.

In summary, the formal design review, when appropriately scheduled and conducted in an effective manner, causes a reduction in the producer's risk relative to meeting contract and specification requirements, and often results in improvement of the producer's methods of operation.

Scheduling of design reviews. Design reviews are generally scheduled prior to each major evolutionary step in the design process. In some instances, this may entail a single review toward the end of each phase (i.e., conceptual, preliminary system design, detail design and development). For other projects, where a large system is involved and the amount of new design is extensive, a series of formal reviews may be conducted on designated elements of the system. This may be desirable to allow for the early processing of some items while concentrating on the more complex high risk items.[11]

Although the quantity and type of design reviews scheduled may vary from program to program, four basic types are readily identifiable and common to most programs. They include the conceptual design review (i.e., system requirements review), the system design review, the equipment software design review, and the critical design review. The time phasing of these reviews is illustrated in Figure 5.28.

1. *Conceptual design review.* The conceptual design review may be scheduled during the early part of a program (preferably not more than 4 to 8 weeks after program start) when operational requirements and the maintenance concept have been defined. Feasibility studies justifying preliminary design concepts should be reviewed. Logistic support requirements at this point are generally included in the specification of supportability constraints and goals and in the maintenance concept definition, generally contained in the System Specification (Type "A").

[11] Items that are procured as off-the-shelf or items where the design is basic and proven may be processed expeditiously to suit both engineering and production schedules. Complex newly designed items (those pushing the state of the art) will require more in-depth reviews followed by some modifications and possibly a second review before being released for production.

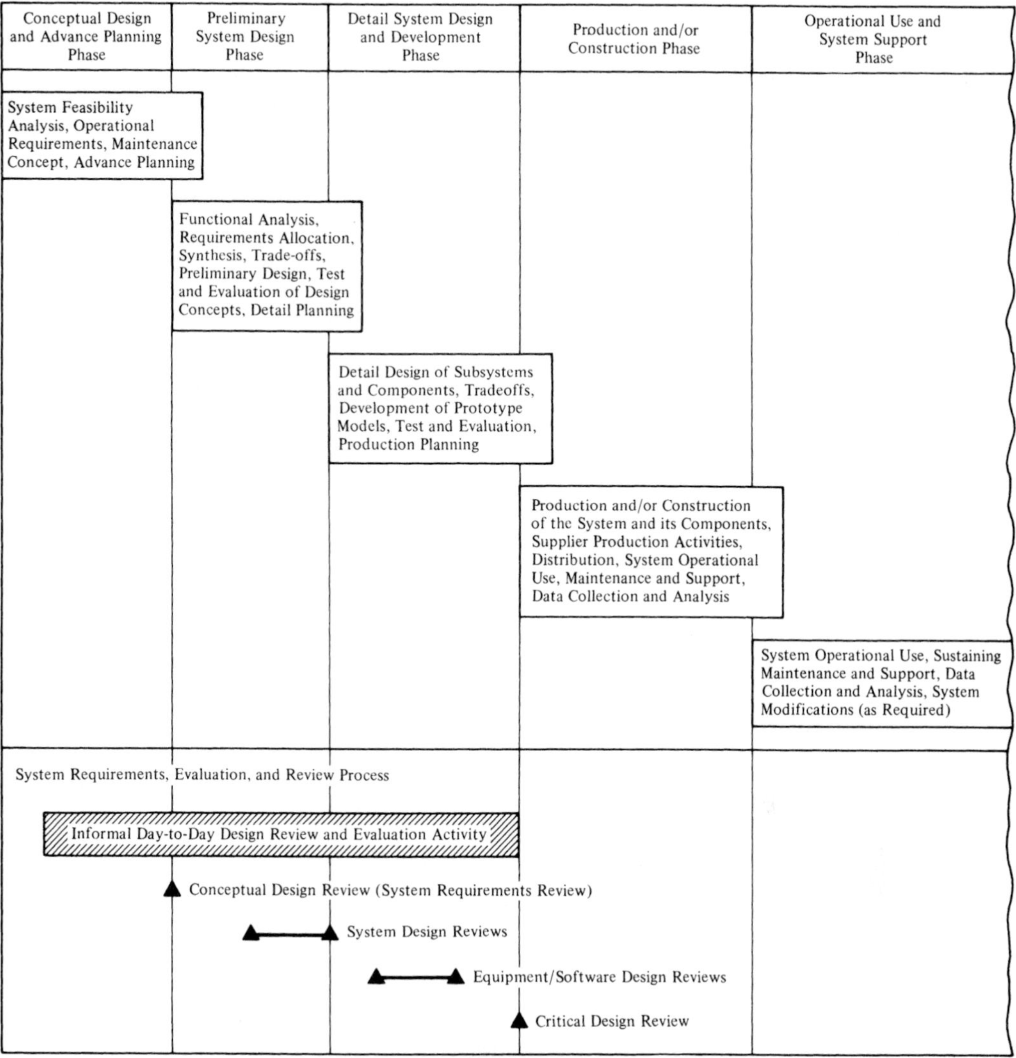

Figure 5.28 Design review schedule in relationship to program phases.

2. *System design review.* System design reviews are generally scheduled during the preliminary system design phase when preliminary system layouts and specifications have been prepared (before their formal release). These reviews are oriented to the overall system configuration in lieu of individual equipment items. Supporting data may include functional analyses and allocations, preliminary supportability analyses, and trade-off study reports. There may be one or more formal reviews scheduled depending on the size of the system and the extent and complexity of the new design. The purpose of the review is to determine whether the design is compatible with all system requirements and whether the documentation supports the design.

3. *Equipment/software design review.* Equipment/software design reviews are scheduled during the detail design and development phase when layouts, preliminary mechanical and electrical drawings, functional and logic diagrams, design data bases, and component part lists are available. In addition, these reviews cover engineering breadboards (hardware), software models or mock-ups, and prototypes. Supporting the design are reliability analyses (FMECA, critical-item data, etc.) and predictions, maintainability analyses and predictions, human factors analyses (system analysis and detailed task analysis), and logistic support analyses. The design process at this point has identified specific design constraints, additional or new requirements, and major problem areas. Such reviews are conducted prior to proceeding with finalization of the detail design.

4. *Critical design review.* The critical design review is scheduled after detail design has been completed but prior to the release of firm design data to production. Such a review is conducted to verify the adequacy and producibility of the design. Design is essentially *frozen* at this point, and manufacturing methods, schedules, and costs are reevaluated for final approval.

The critical design review covers all design efforts accomplished subsequent to the completion of the equipment/software review. This includes changes resulting from recommendations for corrective action stemming from the equipment/software design review. Data requirements include manufacturing drawings and material lists, a production management plan, final reliability and maintainability predictions, engineering test reports, a firm supportability analysis (i.e., maintenance task analysis), and a formal integrated logistic support plan (ILSP).

Design review requirements. The success of a formal design review is dependent on the depth of planning, organization, and data preparation prior to the review itself. A tremendous amount of coordination is required relative to the definition of

1. The item(s) to be reviewed.
2. A selected date for the review.
3. The location or facility where the review is to be conducted.
4. An agenda for the review (including a definition of the basic objectives).
5. A design review board representing the organizational elements and disciplines affected by the review. Reliability, maintainability, human factors, and logistics representation are included. Individual organization responsibilities should be identified. Depending on the type of review, the user or individual component suppliers may be included.
6. Equipment (hardware) and software requirements for the review. Engineering breadboards, prototypes, or mock-ups may be required to facilitate the review process.
7. Design data requirements for the review. This may include all applicable specifications, material/parts lists, drawings, data files, predictions and analyses, logistics data, and special reports.
8. Funding requirements. Planning is necessary in identifying sources and a means for providing the funds for conducting the review.

9. Reporting requirements and the mechanism for accomplishing the necessary follow-up action(s) stemming from design review recommendations. Responsibilities and action item time limits must be established.

Conducting the design review. As indicated earlier, the design review involves a number of different discipline areas and covers a wide variety of design data and, in some instances, hardware and software. In order to fulfill its objective expeditiously (i.e., review the design to ensure that all system requirements are met in an optimum manner), the design review must be well organized and firmly controlled by the design review board chairman. Design review meetings should be brief and to the point and must not be allowed to drift away from the topics on the agenda. Attendance should be limited to those having a direct interest and who can contribute to the subject matter being presented. Specialists who participate should be authorized to speak and make decisions concerning their area of specialty. Finally, the design review must make provisions for the identification, recording, scheduling, and monitoring of corrective actions. Specific responsibility for follow-up action must be designated by the design review board chairman.

5.6 CONFIGURATION CHANGE CONTROL[12]

Quite often in the design and development process (after a baseline has been established as a result of a formal design review), or in the production process, changes are initiated to correct a deficiency or to improve the product. A change may result from the redesign of a prime equipment item, a software modification, the revision of a production process, or a combination of these. In most instances, a change in one element of a system will have a direct impact on other elements. For instance, a change in the design configuration of prime equipment (e.g., change in size or weight, repackaging, added performance capability) will in all probability affect the design of test and support equipment, the type and quantity of spare/repair parts, technical data, facilities, the production process, and so on. A change in software will have a likely impact on equipment, which, in turn, will affect spare/repair parts, technical data, and test equipment. A change in a production process may have an impact on the quality or reliability of the product, which, in turn, affects the overall requirements for logistic support.

Each proposed system change must be thoroughly evaluated in terms of its impact on other elements of the system prior to a decision on whether or not to incorporate the change. Figure 5.29 describes the basic steps in the change procedure. The incorporation of a change after production begins may be quite costly, particularly if the change is accomplished in the latter stages of the production cycle. The change will require the acquisition of additional materials and services, and a given amount of the already expended product-input resources will have been wasted. For equipments that are produced in multiple quantities, the additional resources and resultant waste may

[12] By change control, the author is referring to the necessary management functions required to ensure that complete compatibility is maintained between *all* elements of a system whenever any single element is changed for any reason. This activity is also known as *configuration management*.

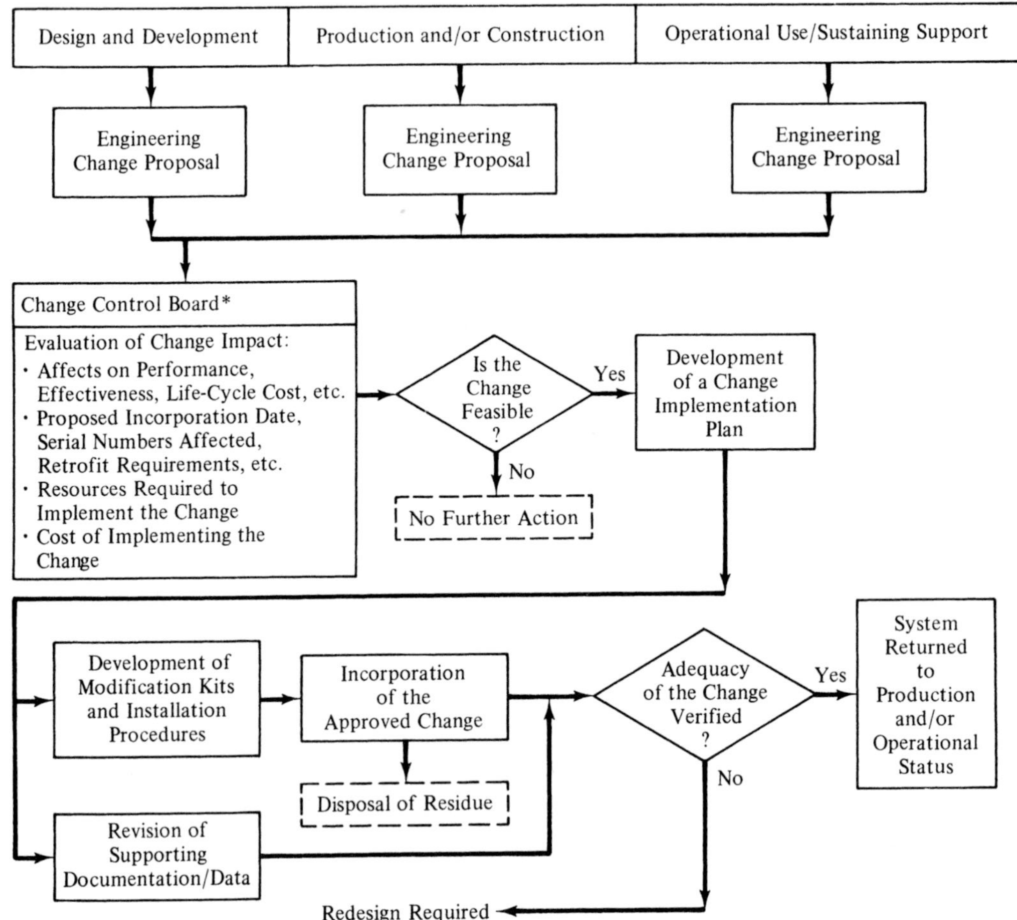

*This activity is also known as the "Configuration Control Board".

Figure 5.29 System change control procedure.

be multiplied on the basis of the number of equipments requiring incorporation of the change. If the change is included in certain equipments (i.e., equipment item serial numbers 24 and on) and not retrofitted on earlier models (i.e., equipment items with serial numbers 1 through 23), then each of the two configurations will require a different type of logistic support, which is highly undesirable.

Past experience with a variety of systems has shown that configuration changes late in the design and development phase and during the production phase are costly (refer to Figure 1.11). With some systems, many different changes have been initiated at various points. Not always have the required changes been incorporated in all applicable production models, nor have the effects of each change on other elements of the system been considered. This has resulted in the delivery of prime equipment models of different configurations and elements of logistic support that are not compatible with the prime equipment. In such situations, system effectiveness is seriously compromised, the right type of system support is not available, and much waste occurs. The causes related to these occurrences may vary.

1. The system or equipment may enter into the production phase without first establishing a *fixed* design configuration. In other words, the design has not been formally reviewed, verified analytically, or demonstrated through test.
2. The original need and operational requirements for the system may change causing a necessary equipment/software change downstream.
3. The technical state of the art may be advanced through the introduction of a new design technique or process, and the producer decided to initiate a change as a result.
4. The supplier of certain items may decide to discontinue his source of supply and no other suppliers of the same item are available. Thus, redesign is often required.

There may be other reasons for incorporating changes in production; however, in most instances, these causes can be avoided through proper planning and progressing systematically through the steps of system design and development described in earlier chapters of this text. Changes to equipment hardware, software, and manufacturing operations must be controlled and held to a minimum.

Conversely, if changes are required for corrective-action purposes, then it is essential that change control be implemented. That is, the impact of each change on the total system and its various elements must be thoroughly evaluated. When a change is incorporated in any one item, all other affected elements of the system must be modified for compatibility. As an example, it is assumed that Unit *B* of System *XYZ* (refer Figure 3.32) is modified to improve performance, and the system is to be produced in multiple quantities. The change involves the redesign of three assemblies, and the change is scheduled to be incorporated in system serial numbers 25 and on. A retrofit for serial numbers 1 through 24 is recommended.

First, the change will require some modification to the production processes (e.g., fabrication and subassembly methods) and material inventories to cover the future production of system serial numbers 25 and on. This, in turn, may affect manufacturing tools, jigs, fixtures, and so on. For the earlier models, modification kits and installation instructions must be developed. The systems are then pulled out of operation (assuming that the first 24 systems have been delivered to the user) while the change is implemented and verified. The change involves added production costs and operation costs in the field.

As a second consideration, the change will affect each element of logistic support to some extent. Reliability and the frequency of maintenance (anticipated number of maintenance actions) are affected, either upward or downward. Maintainability characteristics will change along with detailed maintenance tasks, task frequencies, and possibly the level of repair decisions made earlier. Test and support equipment must be modified to incorporate test provisions covering the added performance capability. Spare parts must be changed at the assembly level and below if the assemblies are designated repairable. Personnel training, training data, and operating and maintenance procedures for field use must be revised to reflect the change. Facilities may also require some modification.

Thus, what initially appears to be a simple system modification often has a tremendous impact on the prime equipment, associated software, the production capability, and logistic support. The incorporation of such a change must be identified with a specific equipment serial number. Each unit, assembly, and subassembly of that particular equipment must be marked in such a way as to be identified with that equipment. Spares with the same part number as items in the prime equipment must be treated in a like manner. A change incorporated in the system must be traceable through *all* affected elements (i.e., hardware, software, and data).

Configuration change control is particularly significant when the system undergoes a number of changes in the production phase. The results of the single change to System *XYZ* described above may be multiplied many times. Without the proper controls, there is no guarantee that the production output will provide effective results.

5.7 TEST AND EVALUATION

As the system design and development activity progresses, there needs to be an ongoing measurement and evaluation (or validation) effort, as indicated in Figure 5.26. In the true sense, a complete evaluation of the system, in terms of meeting the initially specified consumer requirements, cannot be accomplished until the system is produced and functioning in an operational environment. However, if problems occur and system modifications are necessary, the accomplishment of this so far downstream in the life cycle may turn out to be quite costly. In essence, the earlier that problems are detected and corrected, the better off one is in terms of both incorporating the required changes and the costs thereof.

When addressing the subject of evaluation, the objective is to acquire a high degree of confidence, as early in the life cycle as possible, that the system will ultimately perform as intended. The realization of this, through the accomplishment of laboratory and field testing involving a physical replica of the system (or its components), can be quite expensive. The resources required for testing are often quite extensive, and the necessary facilities, test equipment, personnel, and so on, may be difficult to schedule. Yet we know that a certain amount of formal testing is required to properly verify that system requirements have been met.

Conversely, with a more comprehensive analysis effort and the use of prototyping, it may be possible to verify certain design concepts during the early stages of preliminary and detail design. With the advent of three-dimensional databases and the application of simulation techniques, the designer can now accomplish a great deal relative to the evaluation of system layouts, component relationships and interferences, human-machine interfaces, and so on. There are many functions that can now be accomplished with computerized simulation that formerly required a physical mockup of the system, a preproduction prototype model, or both. The availability of computer-aided design (CAD), computer-aided manufacturing (CAM), continuous acquisition and life-cycle support (CALS) methods, and related technologies has made it possible to accomplish much in the area of system evaluation, relatively early in the system life cycle when the incorporation of changes can be accomplished with minimum cost.

In determining the needs for test and evaluation, one commences with the initial specification of system requirements in conceptual design. As specific technical performance measures (TPMs) are established, it is necessary to determine the methods by which compliance with these factors will be verified. How will these TPMs be measured and what resources are necessary to accomplish such? Response to this question may be in the form of using simulation and related analytical methods, using an engineering model for test and evaluation purposes, testing a production model, evaluating an operational configuration in the consumer's environment, or a combination of these. In essence, one needs to review the requirements for the system, determine the methods that can be used in the evaluation effort and the anticipated effectiveness of these methods, and develop a comprehensive plan for an overall integrated test and evaluation effort. As a point of reference, Figure 5.30 is presented to illustrate suggested categories of testing as they may apply in system evaluation.

Categories of test and evaluation[13]. In Figure 5.30, the first category is "analytical," which pertains to certain design evaluations that can be conducted early in the system life cycle using computerized techniques to include CAD, CAM, CALS, simulation,

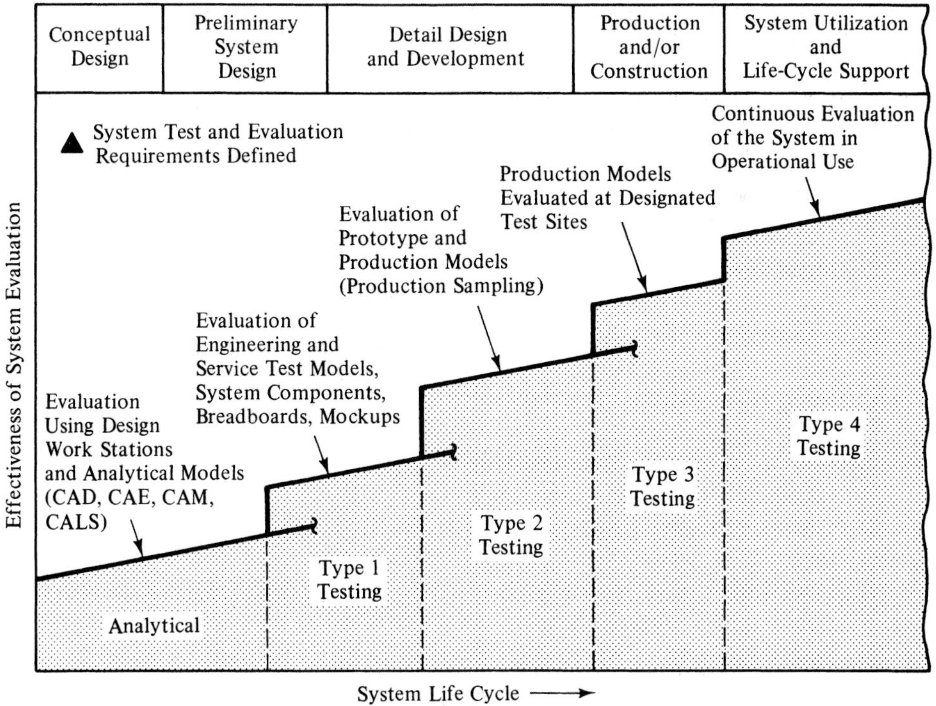

Figure 5.30 Stages of system evaluation during the life cycle.

[13] The categories of test and evaluation may vary by type of system or by functional organization. These categories have been selected as a point of reference for discussions throughout this text.

rapid prototyping, and related approaches. With the availability of a wide variety of models, three-dimensional databases, and so on, the design engineer is now able to simulate human-equipment interfaces, equipment packaging schemes, the hierarchical structures of systems, and activity/task sequences. In addition, through the utilization of these technologies, the design engineer is able to do a better job of predicting, forecasting, and the accomplishment of sensitivity/contingency analyses with the objective of reducing future risks. In other words, a great deal can be now accomplished in system evaluation that, in the past, could not be realized until equipment became available in the latter phases of detail design and development.

"Type 1 testing" refers primarily to the evaluation of system components in the laboratory using engineering breadboards, bench test models, service test models, rapid prototyping, and the like. These tests are designed primarily with the intent of verifying certain performance and physical characteristics, and are developmental by nature. The test models used operate functionally, but do not by any means present production equipment or software. Such testing is usually performed in producer/supplier's laboratory facility by engineering technicians using "jury-rigged" test fixtures and engineering notes for procedures. It is during this initial phase of testing that design concepts and technology applications are validated, and changes can be initiated on a minimum-cost basis.

"Type 2 testing" includes formal tests and demonstrations accomplished during the latter stages of the detail design and development phase when preproduction prototype equipment and software are available. Prototype equipment is similar to production equipment (that which will be delivered for operational use), but is not necessarily fully qualified at this point. A test program in this area may constitute a series of individual tests, tailored to the need, including the following:[14]

1. *Environmental qualification.* Temperature cycling. shock and vibration, humidity, sand and dust, salt spray, acoustic noise, explosion proofing, and electromagnetic interference.
2. *Reliability qualification.* Sequential testing, life testing, environmental stress screening (ESS), and test, analyze, and fix (TAAF).
3. *Maintainability demonstration.* Verification of maintenance tasks, task times and sequences, maintenance personnel quantities and skill levels, degree of testability and diagnostic provisions, prime equipment—test equipment interfaces, maintenance procedures, and maintenance facilities.
4. *Support equipment compatibility.* Verification of the compatibility among the prime equipment, test and support equipment, and ground handling equipment.
5. *Technical data verification.* The verification (and validation) of operating procedures, maintenance procedures, and supporting data.

[14] "Qualified" equipment refers to the production configuration that has been verified through the *successful completion* of environmental qualification tests (e.g., temperature cycling, shock and vibration), reliability qualification, maintainability demonstration, and supportability compatibility tests. Type 2 testing primarily refers that activity associated with the qualification of a system.

6. *Personnel test and evaluation.* Verification to ensure the compatibility among the human and equipment, the personnel quantities and skill levels required, and training needs.
7. *Software compatibility.* Verification that software meets the system requirements, the compatibility between software and hardware, and that the appropriate quality provisions have been incorporated. This includes computer software unit (CSU) and computer software configuration item (CSCI) testing, as reflected in Figure 5.2.

Another facet of testing in this category is production sampling tests, used when multiple quantities of an item are being produced. Although the system (and its components) may have successfully passed the initial qualification tests, there needs to be some assurance that the *same* level of quality has been maintained throughout the production process. The process is usually dynamic by nature, conditions change, and there is no guarantee that the characteristics that have been built into the design will be retained throughout production. Thus, sample systems/components may be selected (based on a percentage of the total produced), and qualification tests may be conducted on a recurring basis. The results are measured and evaluated in terms of whether improvement or degradation has occurred.

"Type 3 testing" includes the completion of formal tests at designated field test sites by user personnel over an extended period of time. These tests are usually conducted after initial system qualification and prior to the completion of the production/construction phase. Operating personnel, operational test and support equipment, operational spares, applicable computer software, and validated operating and maintenance procedures are used. This is the first time that *all* elements of the system (i.e., prime equipment, software, and the elements of support) are operated and evaluated on an integrated basis. A series of simulated operational exercises are usually conducted, and the system is evaluated in terms of performance, effectiveness, the compatibility between the prime mission-oriented segments of the system and the elements of support, and so on. Although Type 3 testing does not completely represent a fully operational situation, the tests can be designed to provide a close approximation.

"Type 4 testing," conducted during the system operational use and life-cycle support phase, includes formal tests that are sometimes conducted to acquire specific information relative to some area of operation or support. The purpose is to gain further insight of the system in the user environment, or of user operations in the field. It may be desirable to vary the mission profile or the system utilization rate to determine the impact on total system effectiveness, or it may be feasible to evaluate several alternative maintenance support policies to see whether system operational availability can be improved. Type 4 testing is accomplished at one or more user operational sites, in a realistic environment, by operator and maintenance personnel, and is supported through the normal maintenance and logistics capability. This is actually the first time that we will really know the true capability of the system.

Integrated test planning. Test planning starts in the conceptual design phase when system requirements are initially established. If a requirement is to be specified, there needs to be a way to evaluate and validate the system at a later point to ensure that the requirement has been met. Thus, considerations for test and evaluation are intuitive from the beginning.

Initial test planning is included in a Test and Evaluation Master Plan (TEMP), prepared in the conceptual design phase. The document includes the requirements for test and evaluation, the categories of test, the procedures for accomplishing testing, the resources required, and associated planning information (i.e., tasks, schedules, organizational responsibilities, and cost).[15]

One of the key objectives of this plan, and of particular significance for system engineering, is the *complete integration* of the various test requirements for the overall system. By referring to the content of Type 2 testing, individual requirements may be specified for environmental qualification, reliability qualification, maintainability demonstration, software functionality, and so on. These requirements, which often stem from a series of "stand-alone" specifications, may be overlapping in some instances, and conflicting in other cases. Further, not all system configurations should be subjected to the same test requirements. In situations where there are new design technology applications, more up-front evaluation may be desirable, and the requirements for Type I testing may be different than for a situation involving the use of well-known state-of-the-art design methods. In other words, in areas where the potential technical risks are high, the requirement for a more extensive evaluation effort early in the system life cycle may be feasible.

In any event, the TEMP represents a significant input relative to meeting the objectives of system engineering. Not only must one understand the system requirements overall, but knowledge of the functional relationships among the various components of the system is necessary. Also, those involved in test planning must be familiar with the objectives of each specific test requirement such as reliability qualification, maintainability demonstration, and so on. A total integrated approach to test and evaluation is essential, particularly when considering the costs associated with testing activities.

Preparation for system test and evaluation. Before the start of formal testing, an appropriate period of time is designated for the purposes of test preparation. During this time, the proper conditions must be established to ensure effective results. These conditions will vary depending on the category of testing being undertaken.

During the early phases of design and development, as analytical evaluations and Type I testing are accomplished, the extent of test preparation is minimal. Conversely, the accomplishment of Type 2 and Type 3 testing, where the conditions are designed to simulate realistic consumer operations to the maximum extent possible, will likely require a rather extensive preparation effort. In order to promote a realistic environment, the following factors need to be addressed:

[15] In the defense sector, the TEMP is required for most large programs and includes the planning and implementation of procedures for Development Test and Evaluation (DT&E) and Operational Test and Evaluation (OT&E). DT&E basically equates to the Analytical, Type 1, and Type 2 testing, and OT&E is equivalent to Type 3 and Type 4 testing.

1. *Selection of test item.* The system (and its components) selected for test should represent the most up-to-date design or production configuration, incorporating all of the latest approved engineering changes.
2. *Selection of test site.* The system should be tested in an environment that will be characteristic for user operations—that is, arctic or tropics, flat or mountainous terrain, airborne or ground. The test site selected should simulate these conditions to the maximum extent possible.
3. *Testing procedures.* The fulfillment of test objectives usually involves the accomplishment of both operator and maintenance tasks, and the completion of these tasks should follow formal approved procedures (e.g., validated technical manuals). The recommended task sequences must be followed to ensure proper system operation.
4. *Test personnel.* This includes (a) the individuals who will actually operate and maintain the system throughout the test, and (b) supporting engineers, technicians, data recorders, analysts, and administrators who provide assistance in conducting the overall test program. Personnel selected for the first category should be representative of user (or consumer) requirements in terms of the recommended quantities, skill levels, and supporting training needs.
5. *Test and support equipment/software.* The accomplishment of system operational and maintenance tasks may require the use of ground handling equipment, test equipment, software, and/or a combination thereof. Only those items that have been approved for operation should be used.
6. *Supply support.* This includes all spares, repair parts, consumables, and supporting inventories that are necessary for the completion of system test and evaluation. Again, a realistic configuration, projected in a real-world environment, is desired.
7. *Test facilities and resources.* The conductance of system testing may require the use of special facilities, test chambers, capital equipment, environmental controls, special instrumentation, and associated resources (e.g., heat, water, air conditioning, power, telephone). These facilities and resources must be properly identified and scheduled.

In summary, the nature of the test preparation function is highly dependent on the overall objectives of the test and evaluation effort. Whatever the requirements may dictate, these considerations are important to the successful completion of these objectives.

Test performance and evaluation. With the necessary preparations in place, the next step is to commence with the formal test and evaluation of the system. The system (or elements thereof) is operated and supported in a designated manner, as defined in the TEMP. Throughout this process, data are collected and analyzed, and the results are compared with the initially specified requirements. With the system in operational status (either "real" or "simulated"), the following questions arise:

1. How well did the system actually perform and did it accomplish its mission objective?
2. What is the *true* effectiveness of the system
3. What is the *true* effectiveness of the system support capability?
4. Does the system meet all of the requirements as covered through the specified technical performance measures (TPMs)?
5. Does the system meet all customer requirements?

A response to these questions requires a formalized data-information feedback capability with the appropriate output in a timely manner. A data subsystem must be developed and implemented with the goal of achieving certain objectives, and these objectives must relate to these questions.

The process associated with formal testing, data collection, analysis, and evaluation is presented in Figure 5.31. Testing is conducted, data are collected and evaluated, and decisions are made as to whether the system configuration (at this stage) meets the requirements. If not, problem areas are identified, and recommendations are initiated for corrective action.

The final step in this overall evaluation effort is the preparation of a final test report. The report should reference the initial test planning document (i.e., the TEMP), describe all test conditions and the procedures followed in conducting the test, identify data sources and the results of the analysis, and include any recommendations for corrective action or improvement. Because this phase of activity is rather extensive and represents a critical milestone in the life cycle, the generation of a good comprehensive test report is essential from the historical standpoint.

System modifications. The introduction of a change in an item of equipment, a software program, a procedure, or an element of support will likely affect many different components of the system. Equipment changes will likely affect software, spare parts, test equipment, technical data, and possibly certain production processes. Procedural changes will affect personnel and training requirements. Software changes may impact hardware and technical data. A change in any given component of the system will likely have an impact (of some kind) on most, if not all, of the other major components of that system.

Recommendations for changes, evolving from test and evaluation, must be dealt with on an individual basis. Each proposed change must be evaluated in terms of its impact on the other elements of the system, and on life-cycle cost, before a decision on whether or not to incorporate the change. The feasibility of incorporating the change will depend on the extensiveness of the change, its impact on the system in terms of its ability to perform the designated mission, and the cost of change implementation.

If a change is to be incorporated, the necessary change control procedures described in Section 5.6 must be implemented. This includes consideration of the time when the change is to be incorporated, the appropriate serial-numbered item(s) affected in a given production quantity, the requirements for retrofitting on earlier serial-numbered items, the development and "proofing" of the change modification kits, the geographic location where the modification kits are to be installed, and the requirements for system checkout and verification following the incorporation of the change. A plan should be developed for each approved change being implemented.

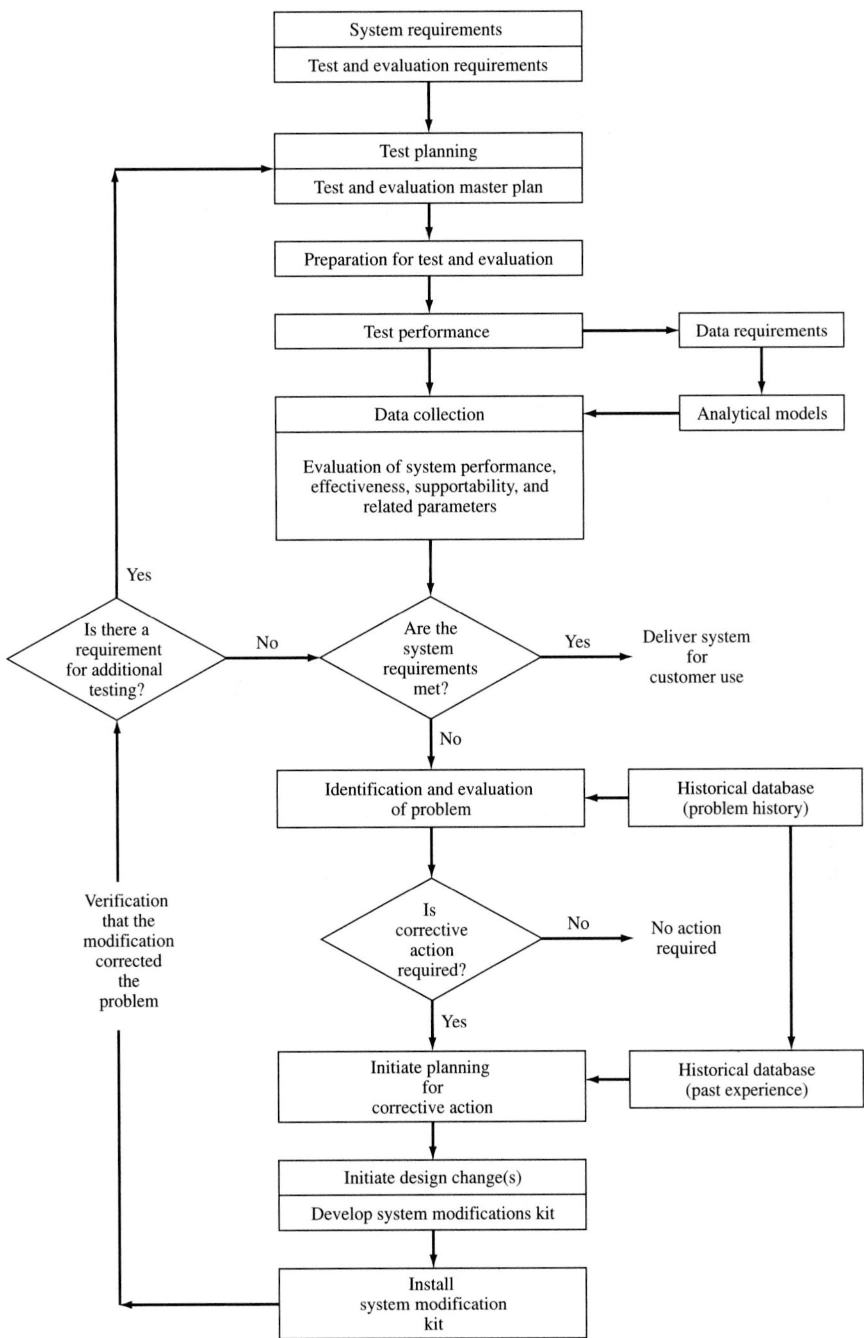

Figure 5.31 System evaluation and corrective-action loop.

QUESTIONS AND PROBLEMS

1. Briefly describe the design process and the major steps in system development.
2. How is logistics considered in the design process? What are the objectives?
3. Refer to Figure 5.2. How does the design for *supportability* fit into the various life cycles? How does *supportability analysis* fit in?
4. What is meant by *design criteria?* Provide some examples.
5. How does reliability in design affect logistic support?
6. For each of the following, briefly describe the tool and provide an example of how and when it can be applied in the system life cycle: reliability functional analysis, reliability allocation, reliability model, FMECA, FTA, RCM, reliability prediction, reliability growth model, and FRACAS.
7. How does maintainability affect logistic support?
8. For each of the following, briefly describe the tool and provide an example of how and when it can be applied in the system life cycle: maintainability functional analysis, maintainability allocation, LORA, MTA, and maintainability prediction.
9. When should the LORA be completed? Why is the completion of the LORA important? What information is provided?
10. How does the FMECA relate to the RCM? How does the FMECA relate to the MTA? How can the FMECA be utilized to enhance the LCCA? How does the FTA differ from the FMECA? How can the FMECA be used to aid in acquiring reliability growth during a program? Provide an example in each instance.
11. How does the RCM relate to the MTA? How does the LORA relate to to the MTA? How does the LORA relate to the LCCA? How does maintainability prediction relate to the MTA? How does the OTA (operator task analysis) relate to the MTA? Provide an example in each instance.
12. What is meant by CAD, CAM, and CALS? How do they interrelate (if at all)?
13. What criteria are considered in determining whether a function should be accomplished by human beings or through automation?
14. What is the purpose of an OSD? How does it relate to the OTA?
15. Briefly describe how training requirements are determined?
16. What is meant by *design for producibility? Design for disposability? Design for quality? Design for the environment? Design for economic feasibility?* Include a brief description of each, and identify how each is related to *supportability.*
17. How are supplier requirements determined? What criteria might you use in the evaluation and selection of suppliers?
18. Identify some of the key requisites for good *design integration.*
19. Refer to Figure 5.15. Which item(s) would you investigate for improvement? Why?
20. Select a system (or element thereof) of your choice and accomplish the following:
    (a) A reliability prediction.
    (b) A maintainability prediction.
    (c) A FMECA.
    (d) A FTA.
    (e) An operator task analysis (OTA).
    (f) A maintenance task analysis (MTA).

(g) A level of repair analysis (LORA).
(h) A life-cycle cost analysis (LCCA).

21. What are the objectives of a formal design review? When should you schedule such? What items should be selected for review (how determined)? What are some of the benefits that can be derived from conducting formal design reviews?
22. Refer to Figure 5.8. Develop an abbreviated design review checklist for a system of your choice. Describe the steps that you should follow in the development of such!
23. When considering a design change, what factors should be evaluated in the decision-making process?
24. How can a design change in a prime mission-related element of the system impact the various elements of the maintenance and support infrastructure? Provide some examples.
25. Why is configuration control important in logistics? What can happen in logistics when the proper level of configuration management is not maintained?
26. How are formal design changes initiated? Describe the steps involved.
27. How are system test requirements determined? What measures would you select and why? When should test planning be accomplished? What factors should be considered in test planning? How are the supporting resources for the completion of system testing determined?
28. How would you evaluate the various elements of logistics through formal testing?
29. How may reliability growth be realized through testing?
30. Why is the proper *integration* of test requirements important? What can happen if such requirements are not properly integrated?.
31. What steps would you take if the results from a given test indicate noncompliance with a specific system requirements?
32. Under what conditions would you specify a "retest" requirement?
33. The data output and resultant reporting of test results can provide what benefits (from a supportability perspective)?

CHAPTER 6

LOGISTICS IN THE PRODUCTION/ CONSTRUCTION PHASE

The earlier chapters deal with the design, development, test, and evaluation of a system. The system may take the form of a configuration that can be reproduced in multiple quantities through a production process, or a one-of-a-kind entity that is to be constructed at a designated location. This chapter addresses the phase of the life cycle that involves production or the construction of the system, with emphasis directed primarily toward the production activity.[1]

Logistics in the production/construction phase of the system life cycle includes several major facets of activity.

1. The aspects of support necessary in the production process itself and relating to the initial purchasing of items from various suppliers, the flow of materials through the manufacturing facility, the establishment and maintenance of inventories, packaging and transportation, warehousing, and the ultimate distribution of products for customer use. This includes the business-oriented activities described in Section 1.1 (refer to footnote 5).
2. The provisioning, procurement, production, and distribution of the elements of logistics, identified through the supportability analysis (SA), that are required for the sustaining maintenance and support of the system throughout its intended period of utilization. This includes the provisioning, procurement or production

[1] A system configuration has been defined, the design has been formally reviewed (through the critical design review), system test and evaluation have been accomplished, design changes for corrective-action purposes have been incorporated, and a fixed baseline for production/construction has been established. Any changes from here on, initiated for producibility or equivalent purposes, should be processed in accordance with the configuration management procedures described in Section 5.6.

of spares and repair parts, special test and support equipment, software, technical data, and so on.

In addressing the subject of production, one must consider the total flow of materials, beginning with the procurement of raw materials from suppliers and extending through the delivery of a finished product to the ultimate user or consumer. Referring to Figure 6.1 (which is an extension of Figure 1.1), these activities may be broken down into three basic phases (i.e., the *physical supply* or the delivery of items as an *input* to the production or manufacturing process, the *internal flow* of materials through the manufacturing process, and the *physical distribution* of products from the factory to the user's operational site.) This represents the *outward* flow of activities as reflected in Figure 1.1, whereas the reverse flow is addressed in Chapter 7.

6.1 PRODUCTION/CONSTRUCTION REQUIREMENTS

Referring to Figure 6.1, the objective is to design, develop, and implement a total *process flow* (i.e., the flow of materials from the supplier to the user's operational site) that will be completely responsive to customer/consumer requirements. More specifically, the intent is provide a high-quality product output, in minimum time, and at an

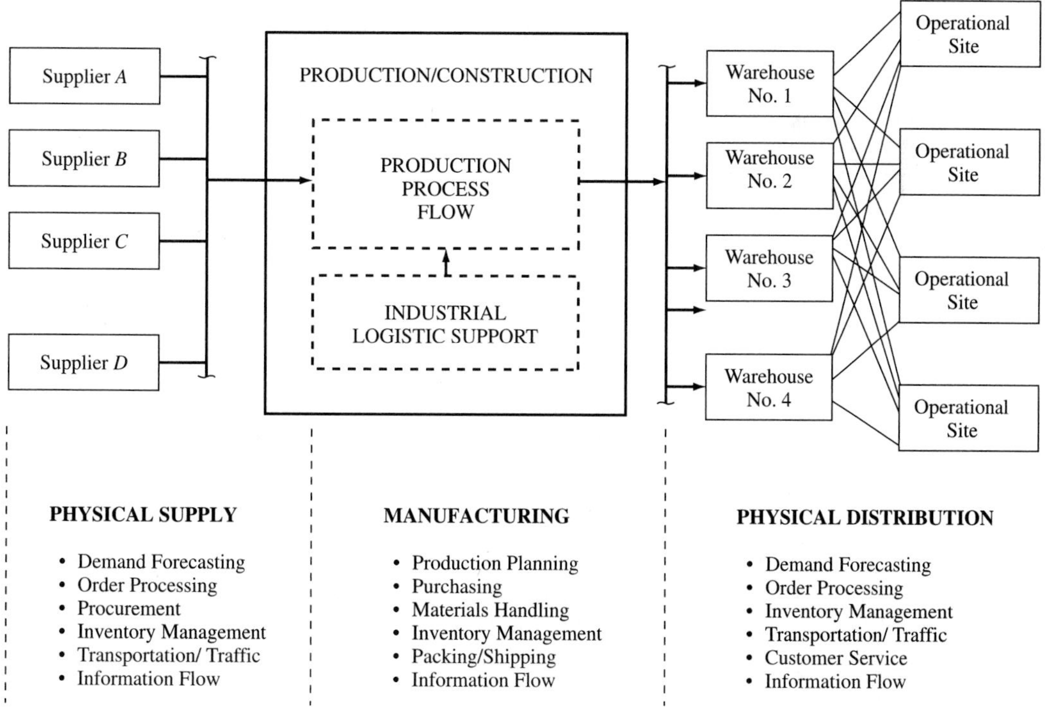

Figure 6.1 Logistics activities in production/construction (reference: Figure 1.1).

affordable price. This, in turn, requires that one address the total overall flow, its supporting products and subprocesses, as illustrated in the figure. In the past, the tendency has been to address only certain facets of the process, on an individual-by-individual basis, and not the overall process as an entity.[2]

With this in mind, one may wish to establish some specific quantitative requirements or technical performance measures that will lead toward (1) minimizing the overall response time from the point where a need is first identified to the delivery and installation of the item at the user' s site; (2) minimizing the number of steps in the decision-making loop; (3) increasing asset visibility and minimizing inventory requirements, the number of warehouses necessary, and storage space; (4) minimizing transportation times; and (5) minimizing costs from a life-cycle perspective. The realization of these objectives is, highly dependent on the availability of a good and highly effective information system and database capability (refer to Section 1.2 and Figure 1.2).

The development of the overall capability illustrated in Figure 6.1 is accomplished by following the same basic steps as described earlier for the design and development of the prime mission-oriented elements of the system. Specific quantitative and qualitative requirements are initially established, alternative approaches for meeting these requirements are evaluated, and a preferred configuration(s) is recommended. The evaluation criteria selected must address the system life cycle, and the supportability analysis tools/methods described in Chapter 4 may be applied as appropriate.

In responding to the specified design requirements and criteria, an initial step is to determine whether an existing commercial and standard process or material solution is available. For instance, it may be more appropriate to utilize an existing commercially available method of transportation (e.g., United Parcel Service, Federal Express) rather than establishing an "organic" capability within a given organizational structure. It may be preferable to select commercially available materials or components (e.g., COTS items) versus the development of new nonstandard items. Further, in the manufacture of items needed for future application, it may be more appropriate to select a manufacturing capability that incorporates flexibility, group technology, and agile-oriented processes than a facility with unique and highly complex processes. In any event, there are a series of steps, shown in Figure 6.2, that are appropriate in the decision-making process.

Production requirements initially stem from system operational data developed during the early planning and conceptual design stage (refer to Chapter 3) and are refined throughout the system development process. The basic information desired includes a detailed description of the prime mission-related item(s) to be produced, the quantity of items needed, the time of need and the place of delivery, and the general environmental conditions associated with the transportation of the finished product from the production facility to the user's operational site. The *product* to be produced is defined through the development (Type "B"), process (Type "C"), product (Type "D"), and/or material (Type "E") specifications as applicable.

Concurrently, the production requirements for the elements of logistics are defined through the supportability analysis (SA) described in Chapter 4. Referring to

[2] Many elements of the current logistics infrastructure are fragmented and characterized by subsystems and processes that are not interoperable, interrelated, or well integrated.

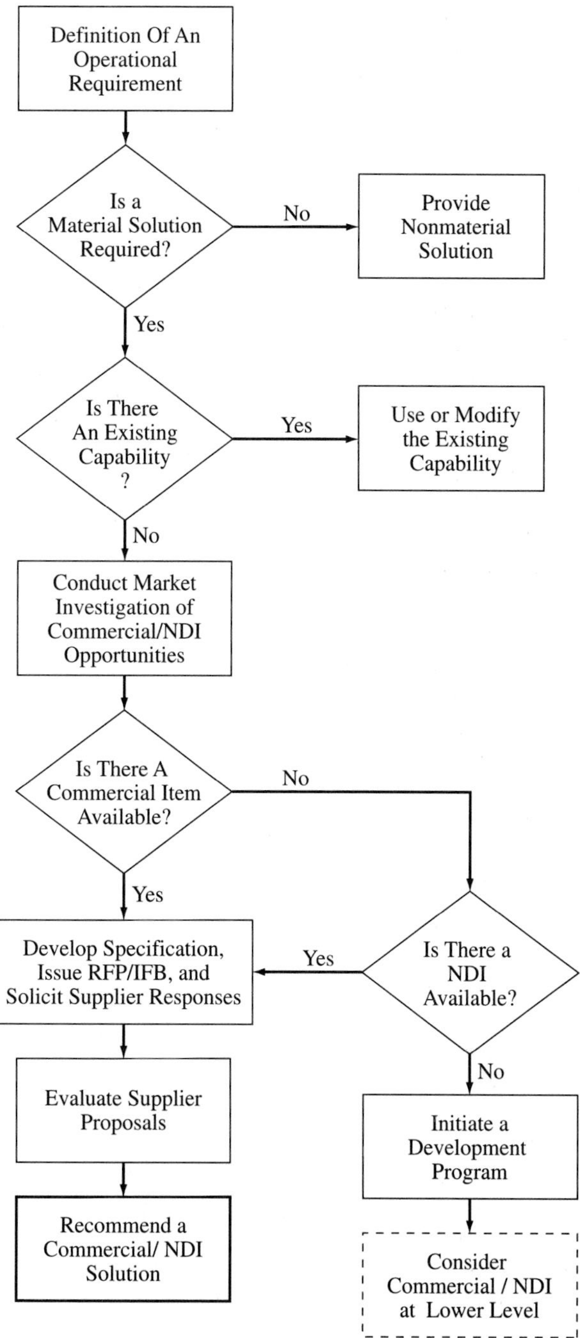

Figure 6.2 The commercial/ nondevelopmental item (NDI) decision process.

Figures 6.3 and 6.4, for example, the requirements for spares/repair parts and test and support equipment are determined by following the steps of the SA identified within the shaded areas respectively. Evolving from these activities is the development of

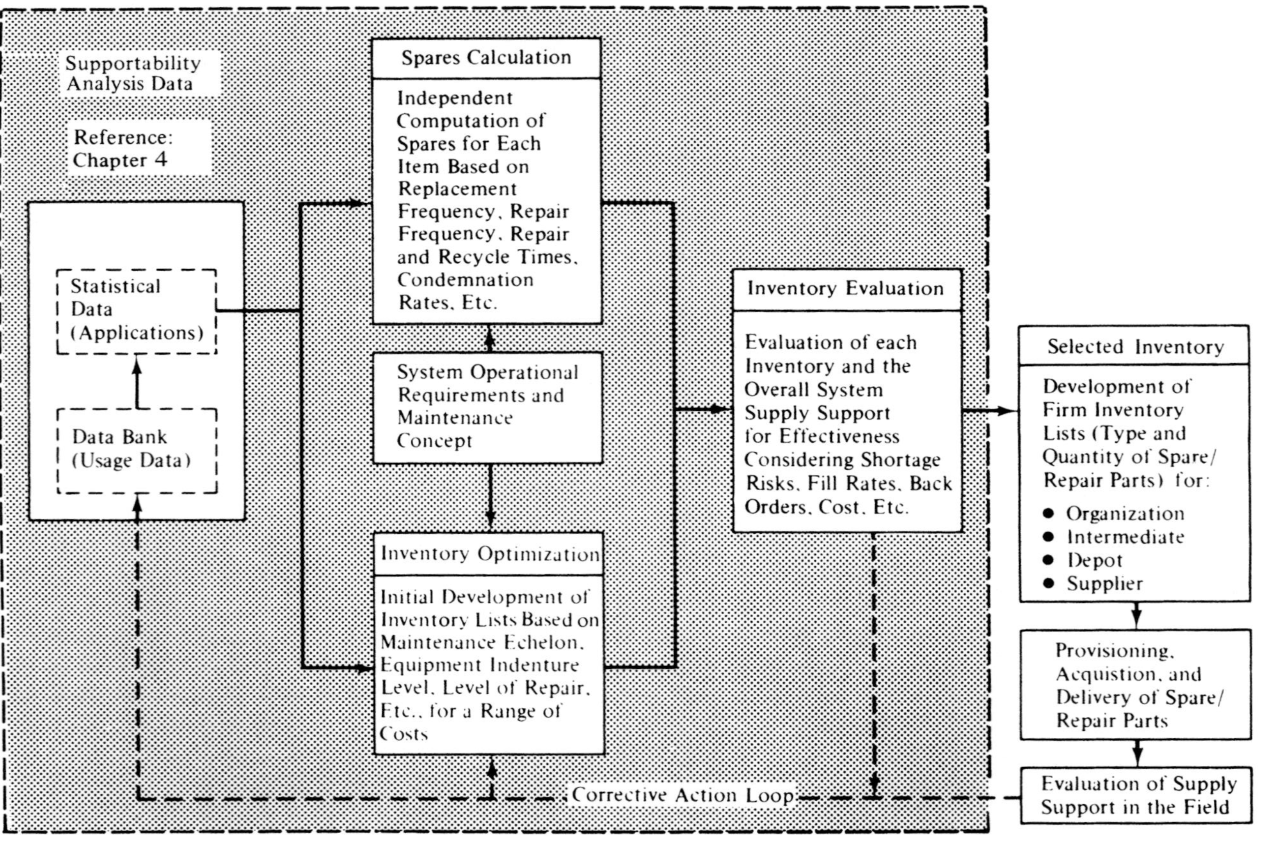

Figure 6.3 Spare/repair-parts development process.

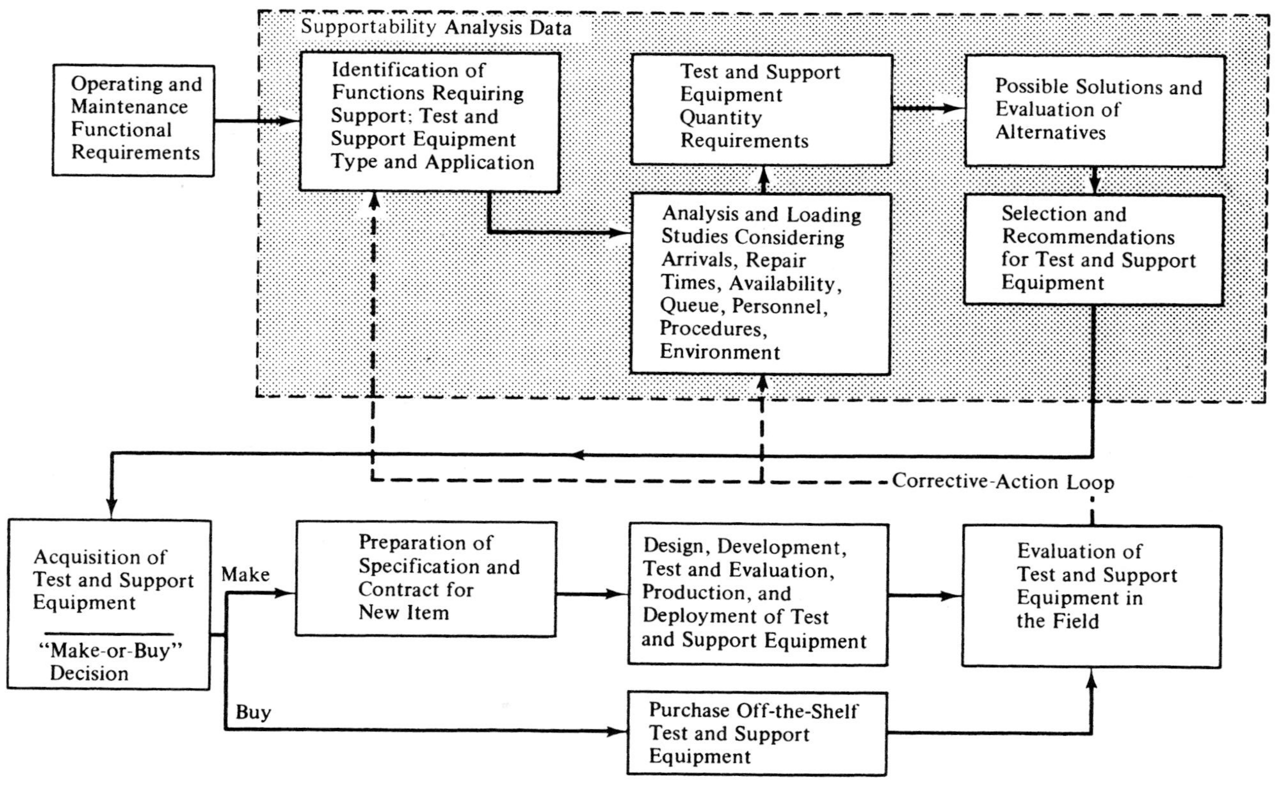

Figure 6.4 Test and support equipment development process.

logistics management information (LMI) and the identification of specific requirements for acquisition. *Make-or-buy* decisions are made, leading to the identification of products that are either to be produced internally within prime contractor's facility or procurred from an outside source.[3] Given the decision relative to source of supply, the requirements may specify multiple quantities of a single item, multiple quantities of many different items, or a single quantity of a wide variety of items. Further, when dealing with multiple quantities, production may be continuous or discontinuous as illustrated in Figure 6.5.

With the requirements established relative to the items to be produced, one can now identify design approaches for the overall production capability. The production of multiple quantities of an item assumes a *flow-shop* pattern where the output variety is limited and each kind of output follows the same basic path and sequence of processing steps. Production facility layout, inventory stockpoints, assembly sequences, inspection and test stations, and personnel functions are designed to handle a large volume (i.e., mass production). When production is continuous and at an approximate steady rate (Figure 6.5, example *A*), the associated logistics resources required are simplified, somewhat constant, and production costs are relatively easy to determine.[4]

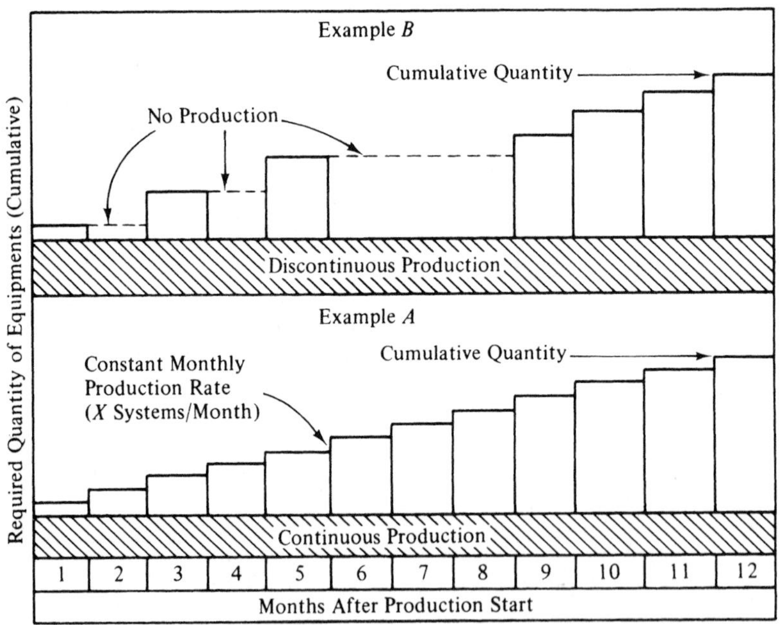

Figure 6.5 Continuous and discontinuous production runs.

[3] *Make-or-buy* decisions are discussed further in Chapter 9. The discussion in this chapter is directed primarily to the activities pertaining to the production of items within the prime contractor's facility.

[4] A major concern today is that of *postproduction support*—that is, the need for spare/repair parts to support systems that are in operational use long after the production line for the system has been shut down. Additional follow-on production sources for the resupply of spares are difficult (if not impossible) to find and, when such requirements are identified, the costs associated with the procurement of these items are prohibitive. Thus, an early decision involves evaluating the feasibility of producing a projected quantity of spares for the entire system life cycle at a time when the production line is in operation versus delaying such

Conversely, when production is discontinuous (including production line startups and shutdowns illustrated by example *B* in Figure 6.5), the logistics resources required are greater, somewhat variable, and the costs are usually higher. Discontinuities result in inventory variations, personnel retraining, and problems related to workmanship, product quality level, and production realization rates. The probability of items being rejected through inspection and test is generally high. This, in turn, results in a large number of items being recycled through certain steps in the production process, resulting in added resource requirements and possibly a shift in production schedules.

Given a large-quantity production requirement, there may be a number of approaches considered which will result in the same final overall output. It may be feasible to establish a faster buildup rate at the beginning; it may be appropriate to include varying degrees of automation through the application of numerical control (NC) or computer-aided manufacturing (CAM) methods; it may be feasible to consider single versus multiple production line capabilities; it may be appropriate to combine inventory stockage locations; and so on. Each variation will impact the logistics requirements in production, and must be evaluated in terms of effectiveness and total cost.

In contrast to the multiple-quantity situation, a requirement may constitute the production of a mix of different items or the construction of a single item such as the commercial airline facility example in Chapter 3. This assumes a *job-shop* pattern where the output varies from item to item and the activity includes a mix of jobs following different paths through a program network. In the pure sense, no two jobs are exactly alike, and each job requires a unique setup. For a small group of unique products, or for a single construction project, the logistics resources will, of course, be peculiar to the particular product output. As the product mix increases, there may be opportunities for the combining and integrating of logistics resources, with a resultant overall cost savings.

Production requirements are developed through an iterative process of planning and analysis. During the early evaluation of technology applications in conceptual design, one needs to consider the associated requirements for manufacturing, fabrication and assembly processes, and *design for producibility* is a major objective (refer to Section 5.2.4). At the same time, the implications for logistics resources need to be addressed.

The steps for designing and developing a production capability are similar to those for the prime elements of the system: that is, the definition of products to be produced, a functional analysis and allocation, the accomplishment of trade-off studies, and so on. A functional analysis may be developed as shown in Figure 3.2.4. The combining of requirements can lead to the preparation of a flow diagram such as the one illustrated in Figure 6.6. The specific functions identified in the figure are evaluated in terms of resource requirements, alternative methods for responding to these functional requirements are considered, and so on.[5]

a decision at the risk of either facing a stock-out condition or a production *start-up* problem later which may turn out to be quite costly.

[5] The flow process in Figure 6.6 serves as a basis for later discussions pertaining to production requirements. The author realizes that the figure does not reflect incorporation of some of the latest production technologies, methods, management concepts, and so on.

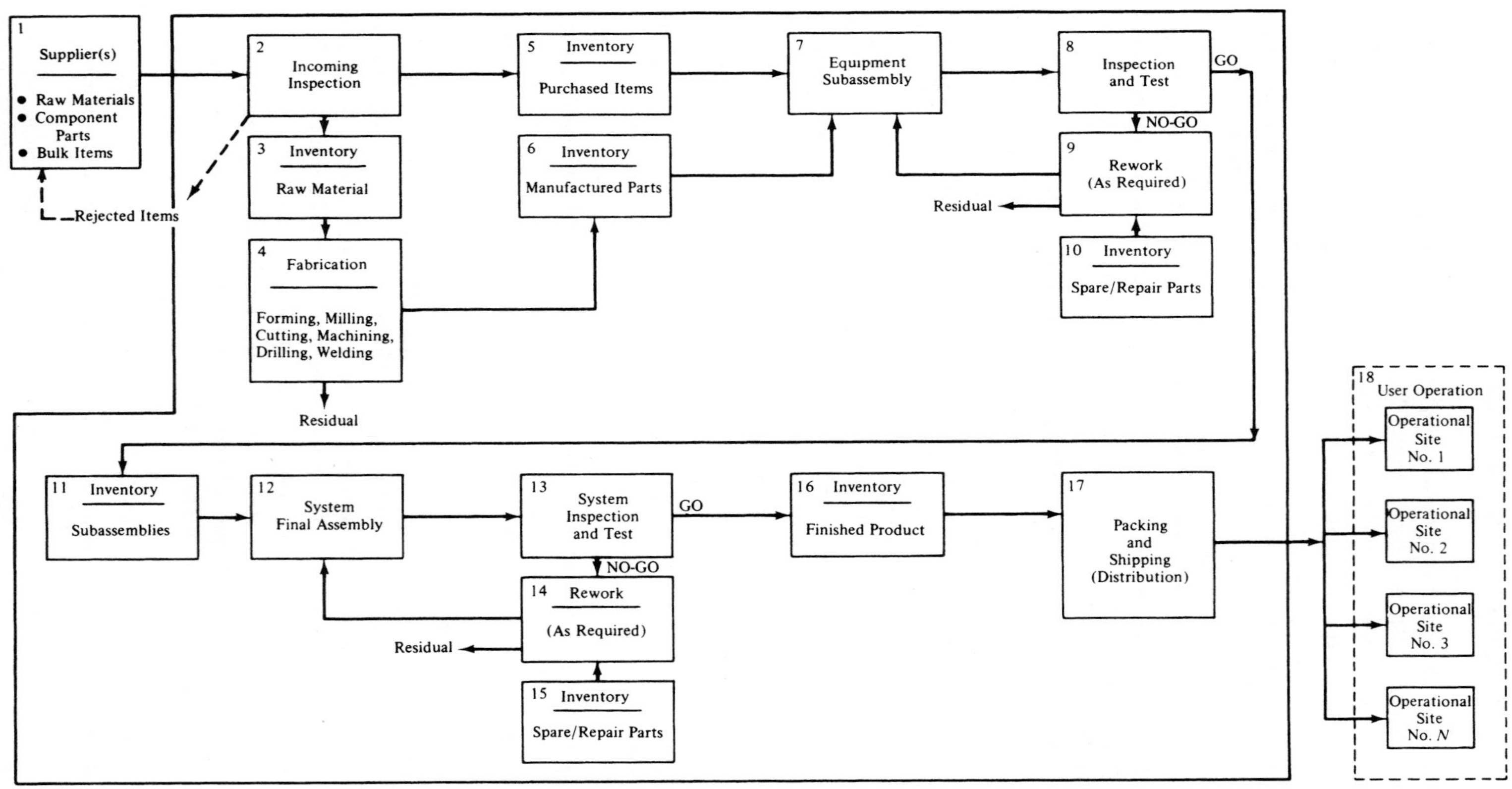

Figure 6.6 The production process (flow of activities).

6.2 INDUSTRIAL ENGINEERING AND OPERATIONS ANALYSIS

Industrial engineering in this instance refers to a composite of activities responsible for the design and development of a production capability. Given the basic requirements through the activities discussed in Section 6.1, it is necessary to develop a capability that will enable the production and delivery of the needed quantities of the system configuration that has been described and approved through the critical design review. Included within the context of industrial engineering are a number of engineering organizations which, in many companies, are involved in the design of production configurations. These are briefly discussed in the paragraphs to follow:[6]

Plant engineering. Plant engineering includes the general design, development, construction, operation, and maintenance of production facilities. Specific objectives are to:[7]

1. Determine the capacity and location of both production and storage facilities (fabrication, assembly, inspection, and test).
2. Determine capital equipment needs (machines, heavy equipment, computer-aided manufacturing equipment, large jigs and fixtures).
3. Determine material handling provisions (conveyor belts, hooks, and cranes).
4. Determine utility requirements (power, environmental controls, fuels, telephone).
5. Accomplish plant layout showing production lines, capital equipment locations, computer-aided manufacturing (CAM) applications, numerical control (NC) equipment and robotic applications, material handling provisions, and utility needs.
6. Establish an integrated plant maintenance capability for the life-cycle support of the production facility and the capital assets contained within.

Regarding plant maintenance, there is a need to establish a pro-active capability that will be responsive in the accomplishment of corrective and preventive maintenance actions as they occur throughout the life cycle of the production capability. This need, which is similar in approach to that pertaining to the prime mission-oriented segments of the system in the field, must be addressed if products are to be produced in a cost-effective manner. Breakdowns resulting from equipment failures, downtime due to setups and adjustments, slowdowns resulting from minor discrepancies and stoppages, defects resulting in rework, and so on, are costly. This, in turn, often results in higher product costs, an undesirable factor in today's highly competitive environment.[8]

[6] This definition is used to establish a frame of reference for further discussion. It is not intended that the definition be aligned to any particular organization structure.

[7] Three references dealing with production are (a) Buffa, E., *Modern Production and Operations Management*, 5th ed., John Wiley & Sons, Inc., New York, 1987; (b) Tompkins, J. A. and J. A. White, *Facilities Planning*, John Wiley & Sons, Inc., New York, 1984; and (c) Starr, M., *Operations Management*, Prentice Hall, Inc., Upper Saddle River, N.J., 1978. Refer to Appendix H for references pertaining to computer-aided manufacturing (CAM), computer-integrated manufacturing (CIM), and so on.

[8] A total intagrated maintenance management approach applied in a commercial factory environment is being implemented by the Japanese through their concept of *total productive maintenance* (TPM). Refer to (a) Nakajima, S., *Total Productive Maintenance: An Introduction*, Productivity Press, Inc., Cambridge,

Manufacturing engineering. Manufacturing engineering often assumes the lead role in the definition of specific production requirements, or the particular functions that are to be accomplished in producing the product. Like the system-level requirements defined through the Type *A* Specification, requirements associated with production are defined through-one or a series of process specifications. The results are used to support plant engineering and methods engineering requirements. Typical tasks are to

1. Evaluate the results of early system design and assist in determining *make-or-buy* decisions (whether an item should be produced within or purchased from an outside supplier).
2. Evaluate and help select the material(s) to be used in the fabrication of each "make" item.
3. Select the basic process (or processes) that is to be used in item fabrication. The process should consider degree of accuracy, precision, allowable tolerances, neatness, reliability (repeatability), or other characteristics that are required and are economically feasible. The process specification should include criteria related to tolerances and allowable process variations, and should be prepared with the support of the quality control and applicable engineering design organizations.[9]
4. Specify, in conjunction with methods engineering, the function/task sequences that are to be implemented in manufacturing operations. Develop flow diagrams describing these sequences in functional terms, and help in identifying the anticipated resource requirements for each function (refer to Figure 6.6).
5. Identify, in conjunction with human factors engineering, the human-machine interfaces. Apply human factors criteria in determining those functions that are to be performed manually versus those that can be automated through the use of computer-aided manufacturing (CAM), numerical control (NC), and so on. Decisions in this area must be made in conjunction with plant engineering and methods engineering.
6. Select the appropriate machines, tools, jigs, and fixtures for each applicable manufacturing operation.
7. In areas where the requirements dictate a peculiar need, design the appropriate tools, jigs, and fixtures as required. In some instances, a new item of test equipment or handling equipment may be required to support manufacturing operations. At the same time, a similar requirement has been identified in support of consumer operations in the field. Where feasible, the same item should be used in both applications. For instance, if an assembled item requires a test before being processed to the next manufacturing station and that test is comparable to

Mass., 1988; and (b) S. Nakajima, ed., *TPM Development Program: Implementing Total Productive Maintenance*, Productivity Press, Inc., Cambridge, Mass., 1989.

[9] In recent years, there has been much concern pertaining to the variability in both system design characteristics (the responsibility of the design engineer) and in manufacturing/fabrication processes (the responsibility of the manufacturing engineer). If this variability is significant, the results in terms of overall system quality can be costly. Refer to Ross, P., *Taguchi Techniques for Quality Engineering*, McGraw-Hill Book Company, New York, N.Y., 1988. Additional references in the area of quality control are included in Appendix H.

a test that is required for the system in the field, the operational test equipment should be used rather than designing a new item specifically for production purposes. This tends to ensure compatibility in system testing, helps to verify the adequacy of operational test equipment, and reduces the cost of production.

Methods engineering. The aspect of analyzing production operations on the basis of effectiveness and cost is accomplished through methods engineering. Specifically, this includes the following functions:

1. Establish work methods, and time and cost standards. Select jobs that can be standardized, analyze the jobs in terms of elements, synthesize and evaluate alternative job approaches, select a preferred approach, establish cost and time standards for each job, and apply the standards for all applicable manufacturing operations. Cost data generated here are used in the accomplishment of life-cycle cost analyses.
2. Estimate component part costs and the cost of each manufacturing operation.
3. Determine personnel job-skill requirements for all functions that are to be accomplished manually, and estimate manufacturing personnel quantities and labor grades.
4. Combine functions and design subassembly, assembly, and test operations.
5. Analyze overall production operations in terms of cost and effectiveness criteria, and initiate changes for improvement where appropriate.

Production control requirements. Associated with plant engineering, manufacturing engineering, and methods engineering are the aspects of production control (i.e., control of the production process illustrated in Figure 6.6). Initially, one needs to determine production requirements from the standpoint of the user or consumer. Given these requirements, it is necessary to design and develop a capability that will provide the needed product(s) in a timely and cost-effective manner. Finally, there is an ongoing evaluation effort implemented for the purposes of assessing the overall effectiveness of the production capability. In areas where corrective action is required, or where improvements can be made, changes may be initiated. Changes in the process, however, must be controlled in a manner similar to the configuration management practices described in earlier chapters (refer to Section 5.6).

To enable the necessary production control, the design must consider such questions as

1. Where should the production facility be located? Is a new facility required, or should an existing facility be modified?
2. How many production lines should be established, and what capital equipment should be assigned to each line?
3. What is the optimum production rate and output on a month-to-month basis?
4. How should the production facility and associated equipment be laid out for optimum output?

5. Where should inventory stockpoints be located, and what levels of inventory should be maintained? What economic order quantities and procurement cycles are appropriate to adequately support the production flow process?
6. What functions should be automated and what functions should be accomplished through manual means?
7. How should maintenance and rework of the item being produced be accomplished? Where should it be accomplished?
8. How should maintenance of production equipment, utilities, facilities, and so on, be accomplished? Where should it be accomplished? Has a maintenance concept/plan been developed for the production capability?
9. What methods for materials handling and modes of transportation should be provided in order to support product delivery or maintenance requirements?
10. How are changes to the production capability implemented? Changes may be initiated as a result of product changes or for the purposes of improving the production capability.

Answers to these and other questions require a rather indepth operations analysis. There are numerous alternatives to consider, and the input variable relationships may become quite complex. In many instances, the analysis effort relative to the evaluation of alternatives will require the use of analytical methods such as those described in Chapter 4. Requirements are identified, alternatives are proposed, and the industrial engineer must evaluate each feasible alternative and recommend a preferred solution.[10]

Referring to Figure 6.6, the objective is to show the major functional requirements associated with producing a specific product. An expansion of each function can be accomplished through a work process analysis illustrated in Figure 6.7. The functional flowchart and work-process analysis approach is similar to the maintenance task analysis procedure described in Section 4.2.4 (and Appendix C) for the prime elements of the system. That is, functions and tasks are identified, analyzed, and the results are used to determine the specific resource requirements for production operations. A life-cycle cost analysis can be accomplished, *high-cost contributors* and cause-and-effect relationships can be identified, and recommendations for improvement can be initiated as appropriate. The approach is similar to that described in earlier chapters except that the functional descriptions are different.

When defining production requirements, the industrial engineering activity is charged with the responsibility of seeking an optimum solution. The functions of plant engineering, manufacturing engineering, methods engineering, and production control are accomplished, and the results are considered in the decision-making process. The student, when analyzing the situation, might consider the production capability as a

[10]Typical problems may pertain to the evaluation of single versus multiple production alternatives, optimum equipment replacement plans, alternative inventory configurations, product sequencing and waiting-line effects, optimum machine assignments, alternative transportation options, optimum quality control levels, and so on. The operational analysis of a production capability can be a separate subject in itself, and the student is advised to review additional literature in order to understand the scope and complexities involved. Refer to Appendix H.

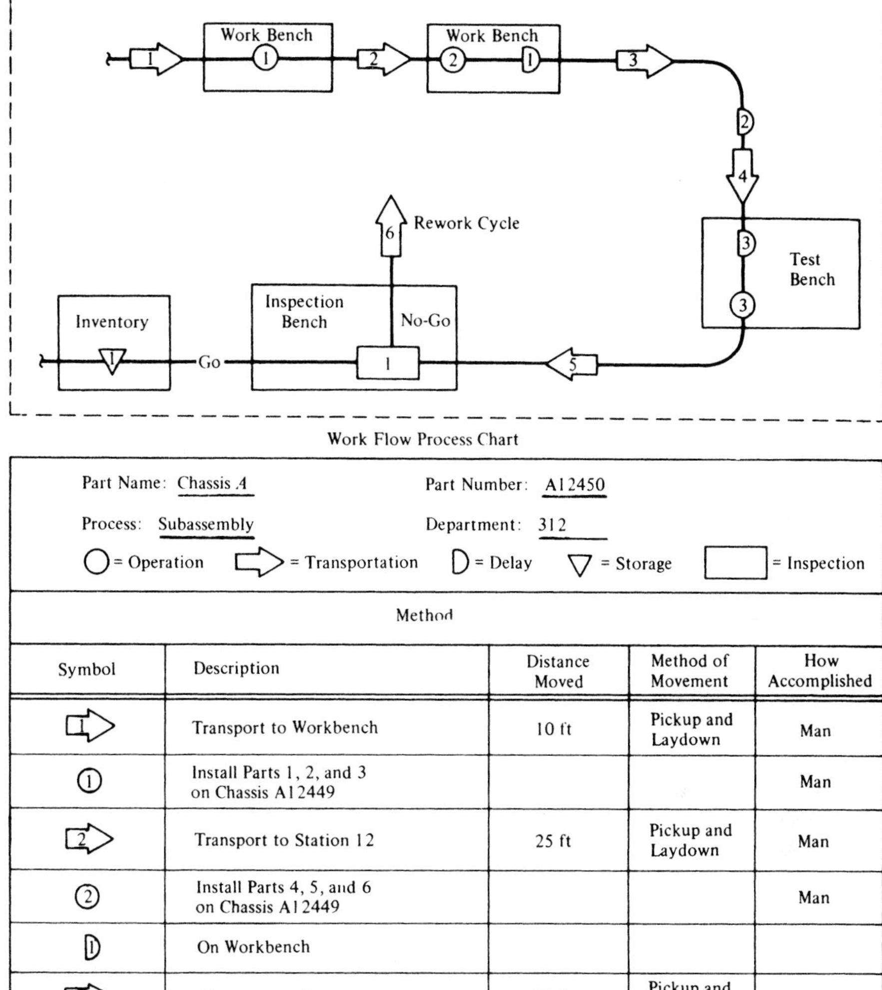

Part Name: Chassis A   Part Number: A12450

Process: Subassembly   Department: 312

○ = Operation ⇨ = Transportation D = Delay ▽ = Storage ☐ = Inspection

Method

Symbol	Description	Distance Moved	Method of Movement	How Accomplished
⇨ 1	Transport to Workbench	10 ft	Pickup and Laydown	Man
○ 1	Install Parts 1, 2, and 3 on Chassis A12449			Man
⇨ 2	Transport to Station 12	25 ft	Pickup and Laydown	Man
○ 2	Install Parts 4, 5, and 6 on Chassis A12449			Man
D 1	On Workbench			
⇨ 3	Transport to Conveyor	10 ft	Pickup and Laydown	Man
D 2	On Conveyor			Power

Figure 6.7 Work process approach.

large system by itself and attempt to evaluate input-output requirements for different internal production configurations. In any event, designing a production capability will lead to many questions and numerous decisions impacting cost must be made. From the perspective of logistics, those design activities relating to material flow, physical distribution, and transportation will have a great impact on the overall support of the system.

6.3 QUALITY CONTROL

The quality of a product is the degree to which it satisfies the desires of a specific consumer or user. It pertains to developing the right product that will satisfy the customer's needs, meet all customer expectations in a timely manner, and is economical in terms of its utilization and support. Quality is a function of system design in terms of the components selected and the characteristics of the configuration; production in terms of the manufacturing, fabrication, assembly, and test processes used; and system support in terms of the maintenance and logistics practices and processes used throughout the programmed life cycle. Quality must be addressed on a life-cycle basis, and not just as a characteristic of the production process as has been the case in the past.[11]

The realization of good quality can be acquired through the implementation of a *total quality management* (TQM) approach as described in Chapters 1 and 5. TQM, a concept initiated by the Department of Defense, is an integrated management approach directed at achieving total customer satisfaction through the continuous improvement of system characteristics by improving design, organizational activities, and so on. The concept of TQM includes the judicious application of control methods and the use of evaluation tools for the purposes of system assessment.[12]

Although the subject of quality is broad and should be addressed in the context of TQM, those aspects pertaining to the production process are of particular interest here. From the standpoint of measurement and evaluation, quality may be stated in terms of a measure of the degree to which a product conforms to specification and workmanship standards. It can be expressed in terms of a given set of attributes required to meet system operational requirements. These attributes may include size, weight, shape, durability, hardness, performance, reliability, maintainability, supportability, attractiveness, and so on, and should be defined in meaningful measurable quantitative terms. Measurement accuracies and tolerances (upper and lower limits) should be defined for each attribute. Quality level is measured by the percentage acceptable (or defective) in a given lot or population. Operating characteristic (OC) curves are developed for each case, and control charts ($\bar{X}$ charts, R charts, etc.) are used to facilitate the measurement task.

A primary objective of quality control is to assure that these attributes are maintained throughout the production cycle. As variations occur as a result of design decisions (in the initial selection of components), the selection of different suppliers, the use of different machines, the involvement of different people, the use of different manufacturing processes, and so on, it is essential that these variations be detected and

[11] The subjects of quality and quality control are rather extensive and far beyond the summary discussion presented herein. Four excellent references are (a) Duncan, A. J., *Quality Control and Industrial Statistics*, 5th Ed., Richard D. Irwin, Inc., Homewood, Ill., 1986; (b) Feigenbaum, A. V., *Total Quality Control*, 3rd Ed., McGraw-Hill Book Company, New York, N.Y., 1991; (c) Grant, E. L., and Leavenworth, R. S., *Statistical Quality Control*, 6th Ed., McGraw-Hill Book Company, New York, N.Y., 1988; and (d) Juran J. M., and Gryna, F. M., *Quality Planning and Analysis: From Product Development through Use*, 3rd Ed., McGraw-Hill Book Company, New York, N.Y., 1993. Refer to Appendix H for additional references.

[12] TQM is covered further in (a) DOD 5000.51G, *Total Quality Management: A Guide for Implementing Total Quality Management*, Department of Defense, Washington, D.C.; and (b) SOAR-7, *A Guide for Implementing Total Quality Management,* Reliability Analysis Center, Rome Air Development Center, Griffiss AFB, New York, N.Y., 1990.

eliminated as early as practicable. Reducing the number of variances possible will result in improved quality. This is accomplished through the application of statistical methods and the implementation of statistical process control (SPC) procedures. SPC procedures are used to identify problem areas (i.e., out-of-tolerance conditions), and to help isolate possible causes. Through this approach, continuous improvement can be realized through the ongoing incorporation of changes to design, manufacturing processes, organizational procedures, and so on.[13]

The conformance to good quality (the adherence of a product to specification requirements) is particularly significant for the following reasons:

1. It reduces item rejects and the recycling for repair, rework, testing, and inspection. The reworking of an item requires additional manpower and machine usage and results in a high material waste. Good quality causes a reduction in production costs.
2. It reduces the probability of requiring an extensive amount of maintenance in the field shortly after an item is delivered for operational use. An item may pass a series of inspections in the production cycle and yet fail shortly after delivery because of one or more manufacturing defects. The added maintenance burden will result in a need for additional logistic support resources or cause those resources available to be depleted earlier than expected. Good quality causes a reduction in system operation and maintenance costs.
3. It reduces the probability of product liability suits, particularly if warranty provisions prevail. Manufacturing defects may cause accidents resulting in damage to equipment and facilities, or result in the death of personnel. Good quality reduces the risk to the producer.
4. It causes a favorable impact on future reprocurements and new sales. Complete customer satisfaction is essential.

Items 1 and 2 both significantly affect logistic support in the production facility and in the field, respectively. The initial engineering design may be suitable; however, unless the equipment is producible and the production process is adequate, the product output will not necessarily reflect what was initially intended. Without adequate standards and controls, each individual equipment item may turn out to be somewhat different, operational effectiveness may fall short of initial objectives, the reliability of equipment may degrade, and so on. Good quality control is necessary to ensure that both prime equipment and logistic support objectives, defined through the design and development process, are realized.

The evaluation of a product for quality is accomplished through (1) the establishment of quality tests, setting up quality standards and acceptance criteria, and interpreting quality data; and (2) the subsequent implementation of a sustaining quality control effort constituting a combination of tests and inspections. The development of

[13] The student would benefit at this point by a review of the basic principles of statistical quality control, the application of various control charts, the use of OC curves, and so on. The acceptance of bad products, a function of the established quality level, will directly affect logistic support.

quality standards and acceptance criteria is accomplished in conjunction with the activities of plant engineering, manufacturing engineering, methods engineering, and production control. Processes are analyzed and quality levels are established. A quality level is chosen by considering the probabilities of accepting bad quality products as good and of rejecting good quality products as bad. Selecting the proper level is dependent on the costs of inspection and risk factors. Once quality levels are identified for the attributes to be measured, accuracies and tolerances must be specified and acceptance criteria are established. Tests and inspections are then planned at different times in the production cycle to ensure proper quality standards at the various stages of production (e.g., after fabrication, subassembly, or final assembly). These requirements are then included in the applicable process specifications developed as a manufacturing engineering output.

The second aspect of quality control includes the function of inspection and test. Inspection is the process of verifying the quality of an item with respect to some standard. Inspection serves several purposes. The first is to determine the quality of some portion of the items being turned out by a process. The problem is to determine whether to accept or reject a production lot on the basis of an analysis of the selected items. This is often referred to as *acceptance control*. The second purpose is to determine the condition of an operation or process in terms of its quality-producing performance. In this instance, the intent is *process control*. A given inspection may satisfy both purposes.

There are two different categories of inspection. The first is 100% *inspection or screening*, and the second is *sampling inspection*. For critical components, 100% inspection is often desirable. This includes those instances where measurement accuracies and tolerances of an item are critical or when a given production process is suspect. Each and every item in the production lot is inspected against the process specification. In sampling inspection, a random sample of items is selected from a production lot (a statistical population), and each item in the sample is inspected in accordance with the requirements of the process specification. An accept decision constitutes the acceptance of the entire lot. The frequency of sampling and the sample size are dependent on the type of equipment, the length of the production run, the number of items produced per lot, and the associated risks.

As indicated earlier, the purpose of inspection and test is to ensure that a specified level of product quality is maintained *throughout* the production process. From the standpoint of logistic support, there are many system attributes considered significant which must be closely monitored and controlled. For instance:

1. The equipment reliability (MTBF), demonstrated during Type 2 testing, should be maintained for all production models. To assure that the required MTBF is attained, a sample number of equipments may be selected from each production lot and subjected to the reliability sequential test plan described in Section 5.7. In some instances, lower MTBF will be evident which means that reliability degradation is occurring through the production process. This will necessitate an adjustment to input quality standards (e.g., improve component acceptance criteria, tighten manufacturing processes, increase testing, etc.) in such a manner as to improve product reliability. Conversely, the test results may indicate a higher MTBF which demonstrates equipment maturity and reliability growth.

2. Certain equipment performance parameters (i.e., range, accuracy, flow rate, etc.) are critical to success of the system mission in the field. Operational test and support equipment has been identified by the supportability analysis (SA) and developed to check the prime equipment in terms of these parameters. A test, accomplished as part of Type 2 testing, initially ensures the compatibility between the prime equipment and its associated test and support equipment. It is required that this compatibility be maintained for all production items; thus, it may be feasible to select certain equipments at random from different production lots and repeat the test.

3. One of the important aspects of maintainability and logistic support is *interchangeability*—that is, an item can be removed and a like item installed in its place without significantly affecting equipment performance operation. The replacement must be compatible in form, fit, and function. With the variations often occurring in production processes, it may be feasible to accomplish maintainability demonstrations on a select number of equipments from each production lot to ensure that configuration control and interchangeability features are being maintained.

There are numerous examples and considerations which apply to quality control in the production operation. The measure of quality is a function of specified product standards and the established test and inspection requirements. It is not the intent nor is it necessarily feasible to seek perfection as the associated production and inspection costs may be prohibitive. On the other hand, poor quality is costly for the reasons stated above. An optimum quality level is defined at that point where all product operational and performance characteristics are attained and the total production cost is minimized. The relationship between levels of quality and cost is illustrated in Figure 6.8. Quality is assessed by inspecting the product (to varying degrees) at one or more points in the production process. Accept–reject decisions are made and defects are

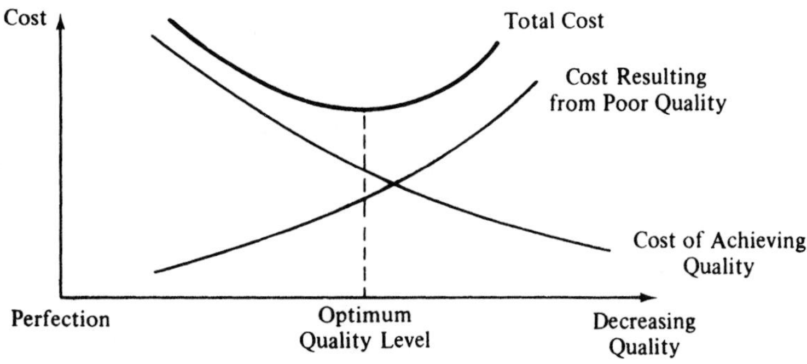

Figure 6.8 Relationship between levels of quality and cost.

recycled for corrective action. The object is to establish a production process with the necessary controls that will respond to customer requirements.[14]

6.4 PRODUCTION OPERATIONS

Implementation of the functions defined in the previous paragraphs results in the initial design of the production capability. Materials and services are acquired and actual production commences in accordance with the prescribed quantities and schedule. Production includes the following:

1. Fabrication, assembly, inspection, and test of prime equipment
2. Manufacture and test of small consumable products
3. Fabrication, assembly, inspection, and test of logistic support items designated for operational use in the field. This includes test and support equipment, transportation and handling equipment, training equipment, spare/repair parts, etc.
4. Assembly and inspection of software (e.g., computer program tapes) and operation and maintenance data for use in the field
5. Provisioning, procurement, and acquisition of commercial off-the-shelf (COTS) items, associated software, and data (that which is already available in some inventory and is acquired through procurement)

Conventionally, one views production as being primarily associated with the manufacture of prime equipment only. Although the prime equipment is generally the most significant item, the various elements of logistic support must also be considered in the production process. In some instances, the same producer and physical facility are employed in the production of both prime equipment and logistic support items. In other cases, different production facilities are involved. In any event, the preproduction planning and design effort (i.e., plant engineering, manufacturing engineering, methods engineering, production control, and quality control) must address all elements of the system.

Throughout the production operation many varied activities occur. There are fabrication and assembly functions, inventory and material handling functions, test and inspection functions, and so on. These functions in themselves require work benches, tools and test equipment, handling provisions, material inventories, personnel, and data. So in reality, one is dealing with two aspects of logistics—the items of logistic support that are produced for field use (identified by the SA in Chapter 4) and the logistic support required to accomplish production.

When considering the production operation as a system in itself, one is dealing with the product that is required by the consumer (to include all applicable categories above such as prime equipment, spare/repair parts, software, etc.) and the processes, materials, and services necessary to manufacture the product. The combination con-

[14] Fabrycky, W. J., Ghare, P. M., and Torgersen, P. E., *Applied Operations Research and Management Science*, Prentice Hall, Inc., Upper Saddle River, N.J., 1984.

stitutes the production capability discussed earlier and illustrated in Figure 6.6. At designated points in the production process, various facets of the operation are measured and evaluated to ensure that a proper level of effectiveness is being maintained. In the design of the production capability, input-output requirements are established for each major function. These requirements may constitute the assembly, inspection, or test of n gidgets per unit of calendar time; the fabrication of x parts per a given level of cost; the percentage of items reworked; the quantity of labor-hours expended per item subassembly; and so on. In addition, the items involved must exhibit a specified level of quality. The combining of these requirements results in a total production capability, and this capability must be assessed on a continuing basis to verify that consumer requirements are being met in an efficient manner.

The measurement and evaluation occur through the application of inspection and test functions and work sampling techniques. Inspection and test provisions, including a combination of acceptance control and sampling inspections, verify that product performance and operational characteristics (i.e., proper level of product quality) are being maintained. Work sampling techniques are employed to evaluate production operations.

Work sampling is a method of analyzing and assessing the time (clock time and manhours) required for the performance of work tasks, and is accomplished through a series of random observations which are extended throughout the production period. These observations may be acquired by establishing timekeeping procedures by job and analyzing the collected time data; observing jobs and monitoring task sequences and time with a stopwatch or through the use of video; taking motion pictures of a job and synchronizing camera speed and film frames with time; or by making recordings of certain data and relating the output to probability distributions that are representative of various machine and human activity functions. Each of the available techniques is advantageous in certain situations. For instance, in analyzing an organizational entity which is responsible for a number of functions within the overall production operation, the timekeeping procedure (measuring labor-hours expended for a given level of work) may be preferable. When establishing a detailed time standard or measuring the time for mounting one part on another, monitoring and recording time with a stopwatch may be appropriate. Conversely, in the assembly of an item, particularly if the item is large and the assembly procedure is relatively complex, the video or motion picture approach may be the most feasible. The use of video or motion pictures allows for the evaluation of task sequences, personnel actions and error rates, and task times. The selection of the technique(s) to be employed is obviously dependent on the type of data desired.

The evaluation of production operations is accomplished by using a combination of inspection and test results and work sampling data. The output data are compared with the initially established quality requirements and manufacturing standards. Corrective action is initialized in areas of noncompliance (i.e., when the operation is not producing the desired results), or when changes can be incorporated to improve the efficiency of production. The incorporation of changes must be accomplished in accordance with the configuration management procedures described in Chapter 5.6.

Figure 6.9 illustrates the basic production cycle and the feedback corrective-action loop. That is, requirements are initially established, a production capability is

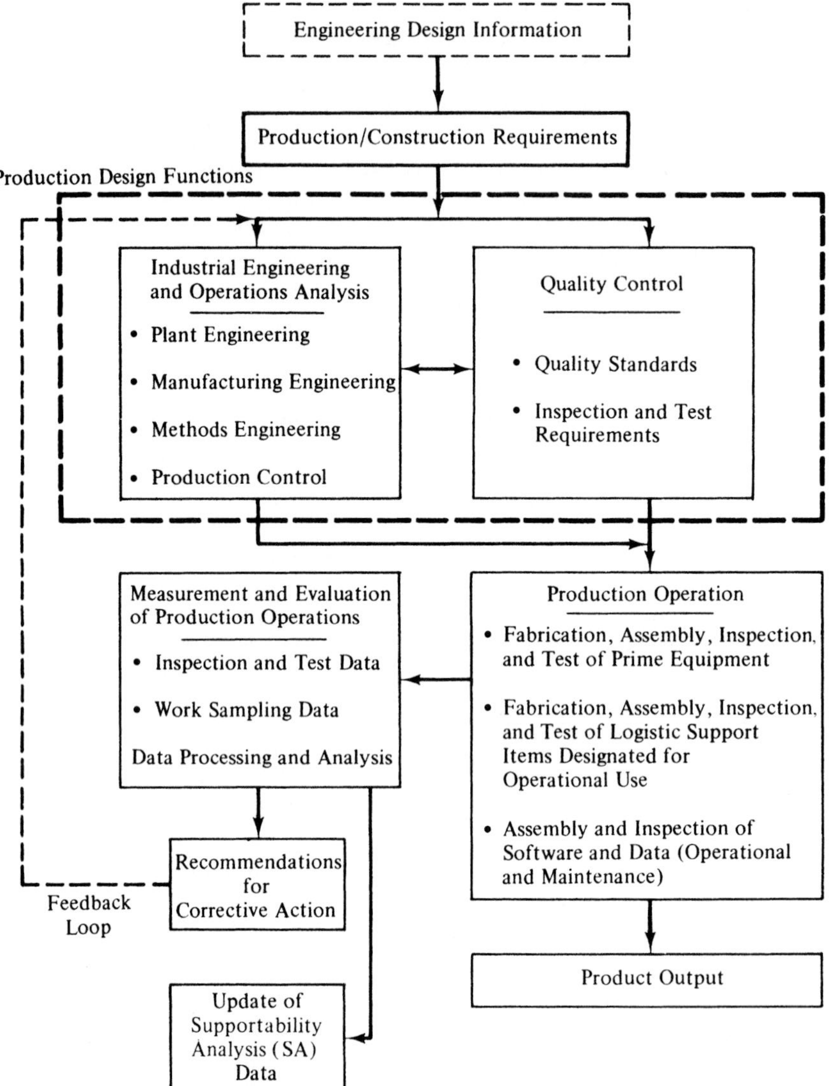

Figure 6.9 Basic production cycle.

designed, production commences, the overall process is assessed on a continuing basis to ensure adequacy, and improvements are incorporated where appropriate.

Referring to Figure 6.9, it should be noted that measurement and evaluation data covering production operations serve several purposes. The first is to assess the overall production capability discussed earlier. The second purpose is to update the SA where appropriate. Many of the assembly, inspection, and test procedures and quality standards used in production operations are identical to those proposed for operation and support of the system in the field during the operational use phase. In the performance of maintenance actions, throughout the system life cycle, there are disassembly, reassembly, inspection, and test requirements. Although the environmental conditions differ somewhat, the methods of task accomplishment are directly comparable in terms of task times, sequences, and logistic support needs. Monitoring and collecting data covering these functions provide results which are directly applicable. Thus, it is beneficial to identify and observe applicable functions and gain insight relative to future requirements. This information is then used to update the SA where appropriate. Results of the SA are evaluated (on a continuing basis) to assess the product in terms of design for supportability and producibility.

6.5 TRANSITION FROM PRODUCTION TO USER OPERATION

After the system, equipment, product, and/or associated elements of logistic support are produced, the appropriate items are then distributed for consumer use. This distribution process may involve a large geographical area and it may include a number of activities such as packaging, storing, warehousing, shipping or transporting, handling, item installation and checkout, and customer service. Further, this process may vary considerably depending on the type of item being distributed and the nature of the consumer and the marketplace. In the interests of simplicity, it seems appropriate to cover this facet of activity in two segments: the distribution of large-scale systems and equipment, and the distribution of small consumable products for the commercial markets.

Distribution of Large-Scale Systems and Equipment

The transition process from production/construction to full-scale operational use is critical since the prime equipment, software, and logistic support must be delivered on a concurrent basis and in a timely manner. The goal is to meet user requirements and to provide an operational system ready to go at the required point in time.

The designated operational sites and activation dates are specified early in the program through the definition of operational requirements (refer to Chapter 3).[15] The transition from production to user operation is covered in the formal logistic support plan prepared during the preliminary system design phase (refer to Chapter 9). This plan supplements the operational requirements data and includes the following:

1. The identification of operational sites, location, and activation date (both national and international locations).
2. The identification of items to be delivered at each site, the time and route of delivery, and the mode of transportation used for delivery.
3. The requirements and procedures used to install equipment and make the system completely operational once the appropriate items have been delivered. This may entail the establishment of a maintenance shop, installation and checkout of test and support equipment, initiation of a supply support capability, and so on.

[15] The activation date is considered as the point in time when the applicable site has initially attained full operational status in accordance with user requirements. This date may be different for each site if two or more sites are involved.

4. The type and extent (in terms of time duration) of logistic support provided by the producer after the system activation date has passed and the system is considered fully operational. This may entail the immediate transfer of support responsibility from the producer to the user, or the provision of a sustaining support capability by the producer for an interim period of time or throughout the system life cycle (i.e., contractor logistic support).

The period of transition will vary in time depending on the system type and complexity. If the system is relatively simple to install, operate, and maintain, the transition may require only a few days before the user is able to assume full responsibility for system operation and support. Conversely, if the system is large and complex, the transition may be much longer and accomplished on a gradual basis. An extensive amount of training may be required, producer field service engineers may be assigned to user site locations to assist in system operation, and the producer may provide an interim repair capability until such time that the user is fully qualified in all aspects of system support.[16] The objective is to accomplish transition at the proper time such that program risks are minimized.[17] Logistic support in this transitional stage addresses two basic areas:

1. The packing and shipping of the product from the producer's facility to the user's operational site, and the inspecting, installing, and checkout of the product at the site. The product may include prime equipment, support equipment, spare/repair parts, and so on. Logistics involves the consideration of

(a) Methods of packing and type of containers used for shipping.

(b) Mode of transportation used in shipment (e.g., cargo aircraft, truck, railroad, pipeline, and/or ship), and the procedures for the processing of items to their respective destinations.[18]

(c) Test and support equipment, personnel, facilities, and technical data required for inspection, installation, and checkout of the product at the operational site. Hopefully, the logistic items assigned to support the prime equipment for the life cycle will be adequate to accomplish these functions.

[16] For some systems the producer will retain full support of the system or equipment throughout the life cycle under a leasing agreement or a repair/rework contract. The reference here pertains to the point in time until the ultimate support capability (whichever policy is followed) is attained.

[17] In certain instances, it may be preferable to defer the transition of a support capability to the user until such time that the user is able to provide such a capability without causing a degrading effect on system operation. The proper time phasing of transitioning different aspects of the total logistic support capability (often referred to as *phased logistic support*) should be considered from a cost-effective basis and covered in the integrated logistic support plan.

[18] Logistics and the distribution of products has become an "international" business involving many nationalities. This, in turn, requires some familiarization with international law, customs requirements for different countries, cultural habits, and so on. Thus, in the planning for logistics requirements, the specific locations where the system is to be operated and the methods for routing and transportation need to be specified clearly.

2. The provision of a logistic support capability for the system at each operational site until full activation is attained (i.e., for the duration of the transition period). Although many of the elements of support are identical to those specified for the system life cycle, there are three basic considerations that should be dealt with in arriving at this interim capability:

(a) Depending on the type of equipment, established production and quality standards, and equipment maturity acquired through testing during production operations, there may be a greater number of corrective maintenance actions occurring immediately after equipment delivery than the estimated quantity derived from reliability predictions. Reliability predictions, which form the basis for the frequency of maintenance, often assume a constant failure rate. At this point in the equipment life, the failure rate may not be at the constant state. Figure 6.10 illustrates a typical failure-rate curve (introduced in Figure 2-4). The frequency of corrective maintenance is a function of primary failures (inherent reliability characteristics), dependent or secondary failures, manufacturing defects, operator and maintenance induced faults, and so on. The primary failure rate, the largest contributor to the corrective maintenance factor, is based on the value specified at point *B* in Figure 6.10. Equipment delivery may occur at point *A* before the system assumes a mature steady-state condition, and a greater quantity of initial failures will occur. Early logistic support must consider the possibility of realizing a burden beyond that which is initially planned.

(b) When considering the introduction of a new system in the inventory, there is a period of time where formal training and equipment familiarization occurs. Until such time when user operator and maintenance personnel are thoroughly familiar with the system, there will be a certain number of operator-induced faults and the system will not be fully utilized in an effective manner. In addition, the equipment may not be properly maintained, and it is quite probable that total maintenance and support requirements will exceed initial expectations.

(c) If the system is large and complex in nature, the user may be able to successfully operate the system at the time of activation, but is unable to provide full logistic support at all levels of maintenance. For instance, material procurement lead times may force the delay in establishing a necessary capability in the perfor-

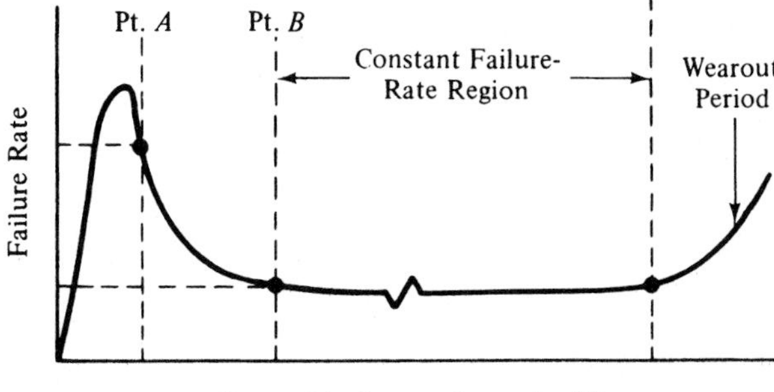

Figure 6.10 Reliability failure-rate curve.

mance of certain maintenance functions; thus, an interim support capability of some sort must be maintained to ensure successful system operation.

Logistic support requirements for the initial start-up period, as determined by the supportability analysis, must address the considerations listed above. In some cases, however, the support policy may vary somewhat from that recommended for operational use (see Chapter 3). For instance, it may be feasible for a pre-established period of time to accomplish all depot-level maintenance at the producer's manufacturing facility in lieu of at user's depot installation; it may be desirable for the producer to train and locate experienced field service engineers at the operational sites and intermediate maintenance shops to facilitate user personnel training and to assist in the performance of system operating functions; or it may be feasible to provide only higher level spares (e.g., unit spares in lieu of both unit and assembly spares) at the intermediate shop for a limited period of time and repair the assemblies at the producer's facility until user personnel are adequately trained in the maintenance of assemblies. These and other policy diversions may occur to satisfy the needs during the transition from production/construction to user operation. However, the ultimate objective is to phase into full operational status defined by the maintenance concept and the SA.

Distribution of Consumable Products for Commercial Markets

The distribution requirements pertaining to consumable items for the commercial markets are based on a completely different set of assumptions. A market analysis is usually accomplished at program inception to determine the size of the potential market, geographical location, and the market share for a given product. In general, one is dealing with a large number of individual consumers throughout the country (or world), and the marketplace may be subdivided into designated geographical territories, regions, districts, and the like. Further, there are larger numbers of end items with which to be concerned, product turnover is usually greater, and the potential for future sales becomes a driving force. The challenge is to provide the right product (in the proper quantities) at the place of need in a timely manner, and with a minimum expenditure of resources. The distribution of too many products, too few products, products to the wrong location, and so on, may be quite costly.

When dealing with the distribution of consumable products, one must be concerned with the activities conveyed by the network illustrated in Figure 6.11 (an extension of Figure 6.1) . Specifically, one must determine

1. The quantity and distribution of potential consumers or customers.
2. The number and location of potential retail outlets for the product.
3. The number, location, and size of product warehousing facilities. Should the facilities be completely automated, semiautomated, or operated through manual means?
4. The size and mix of product inventories at each location (i.e., retail outlet, warehouse, producer's manufacturing plant).
5. The requirements for product packaging, preservation, and storage.

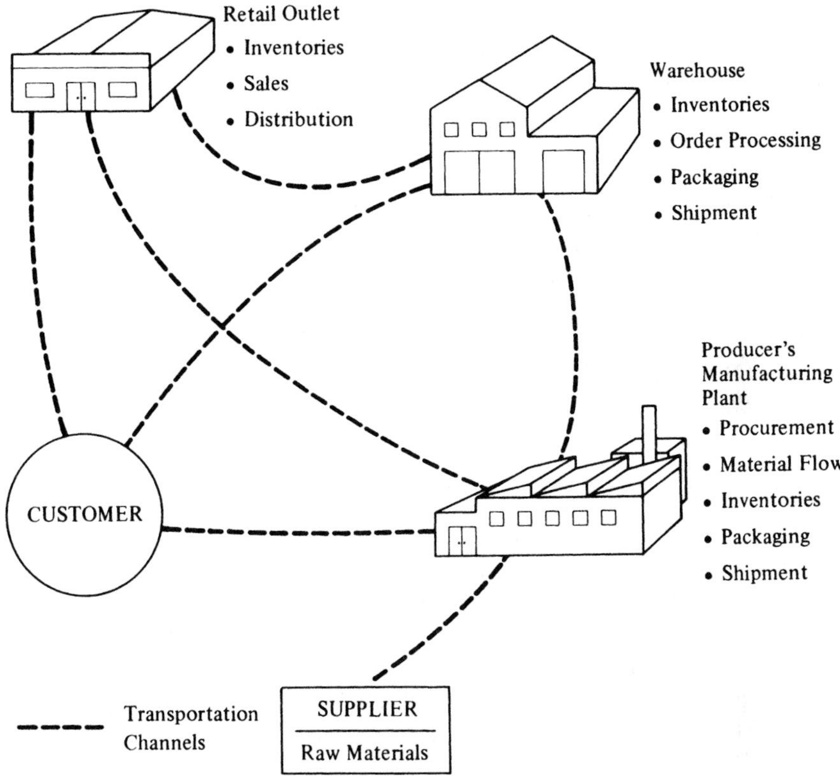

Figure 6.11 Distribution of consumable products for commercial markets.

6. The required mode, or modes, of transportation (i.e., aircraft, truck, railroad, pipeline, or ship).

The response to these questions leads to the evaluation of many possible alternatives involving different mixes of consumer concentrations, retail outlets, product warehouses, transportation and handling modes, and so on. The application of simulation techniques, optimizing techniques, location models, inventory models, transportation models, and so on, is common in arriving at a preferred solution. In essence, one is dealing with a large network which involves many variables and is dynamic in nature. Figure 6.12 illustrates the results of a trade-off involving procurement, inventory, and transportation cost considerations.[19]

[19] As you are aware, the activity discussed herein and illustrated in Figure 6.11 is included within the broad spectrum of *industrial* (or *business*) *logistics* defined in Chapter 1 and referenced in the introduction of this chapter.

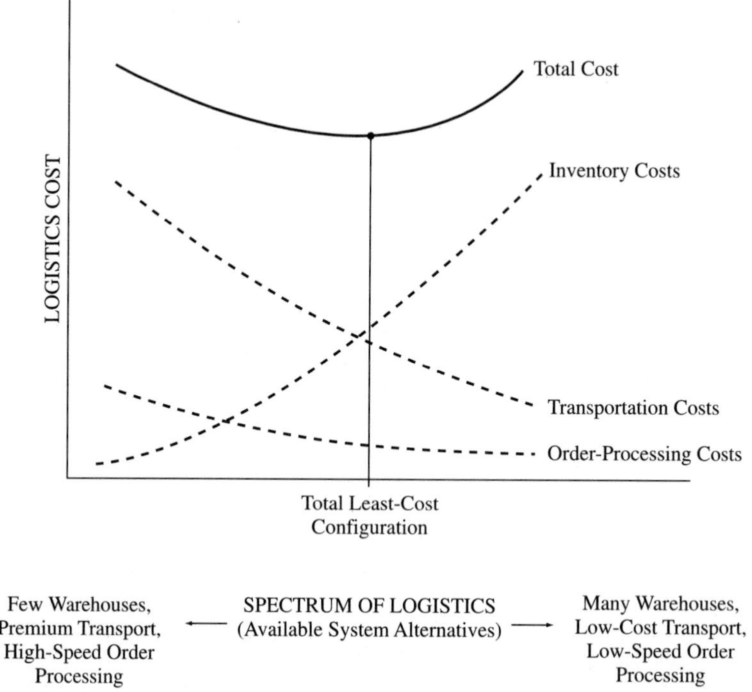

Figure 6.12 A simplified illustration showing hypothetical tradeoffs among transportation, inventory, and order-processing costs. (Source: Glaskowsky, N. A., D. R. Hudson, and R. M. Ivie, *Business Logistics*, 3rd ed., The Dryden Press, Harcourt Brace Jovanovich Publisher, Orlando, FL, 1992, Figure 1-7.)

6.6 SUMMARY

This chapter briefly deals with (1) the production and distribution of logistics elements designated for use in support of the system in the field throughout its life cycle; and (2) the logistic support involved in the production process itself. The first category covers one of several phases in the total consideration of logistics from program inception to system operational use and ultimate equipment phase-out from the inventory. This constitutes the design, development, and production of logistic support elements on a concurrent basis with the development of prime equipment. The second category basically constitutes industrial logistics that is well covered in the literature through other references. This facet of logistics is primarily oriented to the production phase only and does not deal with the overall system life cycle.

Both aspects of logistic support as discussed in this chapter are necessary in the transition from a fixed design configuration of some system to a finished product delivered to the user. In both instances, the concepts of application are comparable; that is, one is dealing with the test and support equipment, personnel and training, spare parts and inventories, data, facilities, and resources required to accomplish the goal of producing an entity. In many cases, the actual resources required are identical even though the applications may vary somewhat. In any event, the concepts employed in the identification of requirements, design, evaluation and test, and production are directly related.

QUESTIONS AND PROBLEMS

1. The initial requirements for production/construction are based on what factors?
2. Define logistic support in the context of the production/construction phase. What are the elements of logistic support?
3. What considerations are necessary in the design of a production capability? What are the basic production design functions?
4. If an equipment is to be produced in multiple quantities, what considerations are necessary in the design of a production capability?
5. In the production of multiple quantities of a given item, what are the advantages of continuous production over discontinuous production? How is logistic support affected?
6. What is the basic difference between a job-shop operation and a flow-shop operation? How is logistic support affected?
7. How are production standards established and maintained?
8. How would you measure and assess the effectiveness of a production capability?
9. Define *quality*. What is included? How would you determine the level of quality control that should be applied to a given product? Discuss the factors that should be considered.
10. Control charts used for quality and inspection purposes can be categorized in two general classes: (1) control charts for inspection by attributes (count), and (2) control charts for inspection by variables (measurement). Attribute control charts include (select the best alternative):
    (a) p charts, np charts, c charts, and u charts.
    (b) $\overline{X}$ charts, p charts, np charts, and R charts.
    (c) u charts, R charts, c charts, and p charts.
    (d) R charts, np charts, $\overline{X}$ charts, and moving-range charts.
11. Variability usually exists in the production of goods, and control charts are often used to detect any unusual variation(s). The control charts most commonly used are $\overline{X}$ and R charts. These charts are used to (select the best alternative):
    (a) Control dispersion (or spread) of the production process and the central tendency (or level) of the process, respectively.
    (b) Control the central tendency (or level) of the process and the dispersion (or spread) of the process, respectively.
    (c) Control the upper and lower limits of the process, respectively.
    (d) Control the lower and upper limits of the process, respectively.
12. How does quality control affect logistic support in the field? Give some specific examples.
13. What is the purpose of operating characteristics (OC) curves in production quality control? (Select the best alternative.)
    (a) To predict lot tolerance percent defectives of a production lot based on sample defectives.
    (b) To depict expected results of proposed sample plans.
    (c) To determine how many sample defectives should be set as the maximum allowable in accepting a lot.
    (d) To indicate the risks of accepting lots that are actually unacceptable and rejecting lots that are actually acceptable.

14. Refer to the operating characteristic curve in Figure 6.13. Producer's risk (α) and consumer's risk (β), respectively, are identified by (select the best alternative):
   (a) A, C.
   (b) C, D.
   (c) B, D.
   (d) D, C.
   Assume that AQL reflects the *acceptable quality level* and LTPD represents the *lot tolerance percent defective.*

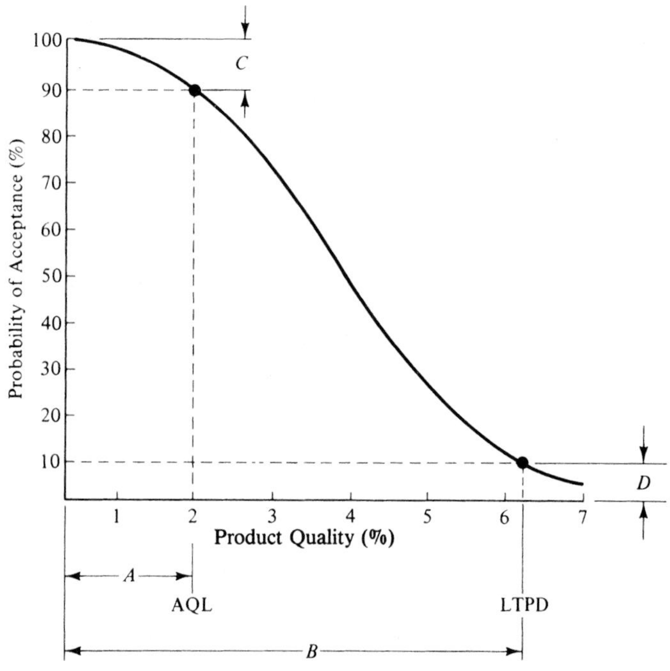

Figure 6.13 Operating characteristic curve.

15. Can the production operation affect the inherent characteristics of reliability and maintainability in the equipment? How?
16. In Figure 6.14, part (a) represents the results of a reliability sequential test at the start of a production run and part (b) represents the results of testing after 1 year of equipment production. Based on the test data, you would probably conclude that (select the best alternative):
   (a) Reliability growth has occurred through the production process and the requirements for logistic support are likely to be greater than originally planned (based on available data).
   (b) Reliability degradation has occurred through the production process and the requirements for logistic support are likely to be greater than originally planned.
   (c) Reliability growth has occurred through the production process and the requirements for logistic support are likely to be less than originally planned (based on available data).
   (d) Reliability degradation has occurred through the production process and the requirements for logistic support are likely to be less than originally planned.

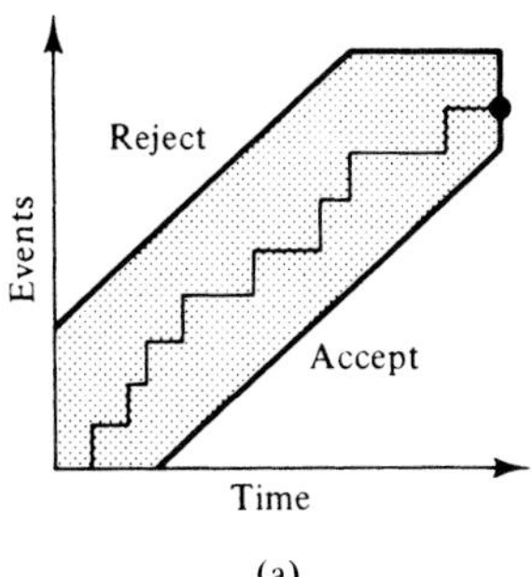

(a)

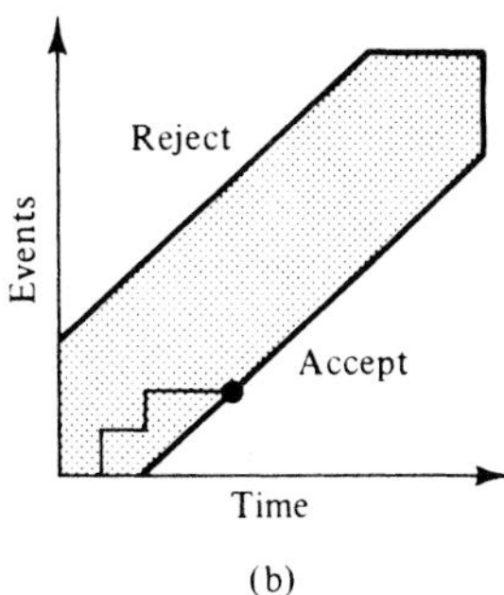

(b)

Figure 6.14 Reliability sequential test results.

17. What are the differences in producing Assembly Part 231 as installed in the prime equipment and Assembly Part 231, which is designated as a spare part? Comment on manufacturing standards, quality standards, and inspection processes.

18. Refer to footnote 4. What are some of the specific factors that need to be considered when evaluating alternative procurement policies for spare/repair parts?

19. Can test and support equipment designated for operational use in the field be employed in the production process? If so, how, and what benefits are derived?

20. In what ways are functions performed in the production process similar to operational and maintenance functions which are anticipated for the field?

21. Why is configuration control important to logistic support? What happens to logistic support when the proper level of control is not maintained?

22. You are considering a design change in the prime equipment. What factors should be evaluated in the decision-making process?

23. How would a design change in the prime equipment affect test and support equipment? Spare/repair parts? Facilities? Personnel and training? Technical data? Why is change control important?

24. Why is the transition process from production/construction to user operation so important?

25. What are the logistic support considerations in the transition from production/construction to user operation? Discuss the elements of support.

26. Is the level of logistic support provided when the system/equipment is initially deployed to the various user operational sites the same as or consistent with that recommended for subsequent periods in the life cycle? Why?

27. What is *producibility*?

28. What is *phased logistic support*? What is the major consideration involved?

29. Figure 6.10 represents a reliability characteristic curve for an equipment item. The current delivery requirements and production process results in an item whose reliability characteristics are represented at point A. Assuming that you are allowed to extend the delivery schedule, what would you do to produce an item with a reliability value as represented at point B?

30. What is the main objective of production planning? What is included?

31. Figure 6.15 represents four sequential production activities. The sequence of activities is (select the best alternative):

(a) Transportation, operation, storage, and inspection.

(b) Storage, transportation, operation, and inspection.

(c) Operation, inspection, transportation, and storage.
(d) Inspection, transportation, operation, and storage.

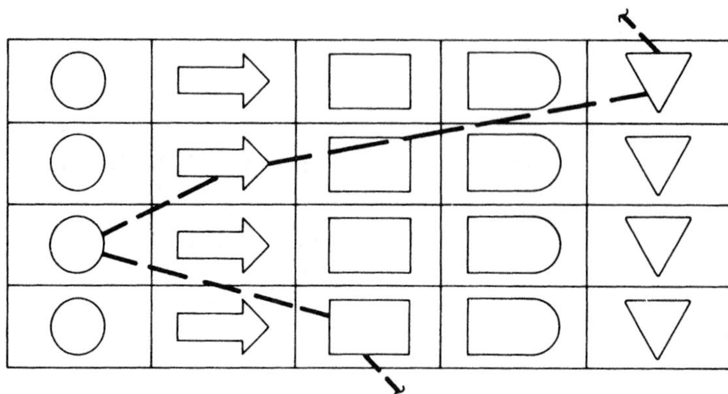

Figure 6.15 Activity flow.

32. In the accomplishment of equipment assembly/disassembly tasks as part of production operations (i.e., during the repair and rework cycle), it was noted that the equipment design lacked certain necessary inherent diagnostic and accessibility features. What would you do to correct the situation?

33. How does the evaluation of production features relate to the SA?

CHAPTER 7

LOGISTICS IN THE UTILIZATION AND SUPPORT PHASE

When a system is delivered and installed for operational use by the customer, there is an ongoing need for system support. Initially, as systems first become operational, a number of problems are likely to occur, requiring procedural adjustments and modifications to correct deficiencies. Experience has indicated that many of the problems in the past, particularly for large systems, have related to the inadequacies of product support, and fall into one of the following categories:

1. Individual elements of logistics are incompatible with the prime elements of the system; for example, the prescribed test equipment will not perform the proper functions in verifying prime equipment operation or in the performance of maintenance.
2. The depth and extent of support provided are insufficient; for example, there are not enough of the right type of spare parts available, the personnel assigned to operate and maintain the equipment are inadequately trained for the job, and so on.
3. The level of support in certain areas is greater than what is actually required; for example, facilities, personnel, and support equipment are not being fully utilized, or there are too many spare parts of a certain type, which results in a higher inventory cost than necessary.
4. The elements of logistic support are not compatible with each other; for example, the maintenance procedures do not cover the tasks being performed at a given level or the support equipment used in task accomplishment.

Although it is an objective to provide the right level of support at the right location and in a timely manner, these initial incompatibilities are not unrealistic and the necessary adjustments must be made accordingly. The type of activities here include the appropriate revisions to the SA, changes in provisioning and procurement data, equipment/software modifications, and so on.

At a later stage after the initial incompatibilities have been eliminated, the nature of system support changes. There may be an ongoing need to train customer personnel in the operation of a system, there may be a need to provide a full level of maintenance and support throughout the planned life cycle of the system, there may be a need to provide a limited level of spare parts at designated locations, or any combination of these requirements. Through evaluation and assessment of the system in operational use, the logistic support requirements established earlier are verified or updated to reflect current needs. The intent is to ensure that the system will perform as intended on a continuing basis.

This chapter deals primarily with system/product life-cycle support, along with the ongoing evaluation of the overall system support capability. The evaluation activity includes the establishment of a data collection and analysis function, and the subsequent continuous improvement of the system through the incorporation of changes as appropriate.

7.1 SYSTEM/PRODUCT SUPPORT

The requirements for the sustaining support of a system are initially identified through the maintenance concept described in Chapter 4, refined through the supportability analysis (SA) described in Chapter 4, and included in the integrated logistic support plan (ILSP), which is revised during the detail design and development phase.

The ILSP will identify the levels of maintenance, the maintenance activities that are to be accomplished at each level, the responsibilities for maintenance (customer maintenance, contractor logistic support, third-party maintenance, etc.), the detailed procedures to be followed, and the anticipated resource requirements (personnel quantities and skill levels, test and support equipment, technical data, etc.). The plan will also identify the number and location of customer operational sites, the number of customer service representatives (or field service engineers) and their assigned geographical areas of responsibility, and the number and location of intermediate-level maintenance shops and the customer activities supported by each. In essence, the ILSP will include a detailed *maintenance plan* that will serve as the basis for the planning and implementation of future system support requirements.

When addressing system/product support responsibilities, several different policies may be adopted depending on the type of system, the duration of the operational life cycle, the extent of support anticipated, and the capabilities of the user (consumer) and the producer organizations. Example approaches are noted.

1. All system maintenance activities (at each level of maintenance) will be accomplished by the producer throughout the planned life cycle. This activity is often known as *contractor logistic support* (CLS).
2. All system maintenance activities (at each level of maintenance) will be accomplished by the producer for an x period after the system first becomes operational. Subsequently, the user organization will accomplish all maintenance activities.
3. All system maintenance activities (at each level of maintenance) will be accomplished by an outside contractor, under the guidance of the user organization.

4. The user organization will accomplish all maintenance activities designated at the organizational and intermediate levels, while the producer organization will accomplish all maintenance activities designated at the manufacturer or depot levels.
5. The producer will accomplish all maintenance activities associated with items *A*, *B*, *C*, and *D*; the user will accomplish all maintenance activities covering all other items.
6. All system maintenance activities (at each level of maintenance) will be accomplished by the user organization.

In essence, the day-to-day maintenance activities for a given system may be accomplished entirely by the user organization, by the producer, by an outside contractor, or a combination thereof. The decision will be based on a number of factors such as the anticipated maintenance work load, contractual warranty provisions associated with certain elements of the system, the legal implications of product liability, and the general interest on the part of the producer (or user) to provide the services required. In addition, the responsibilities for maintenance may change as the user becomes more familiar with the system and is able to provide the support required. Maintenance and support for a system may be handled any number of ways, either through an existing function within the user's or producer's organization(s) or through a subcontracting arrangement with an outside agency.

7.2 TOTAL PRODUCTIVE MAINTENANCE (TPM)

Although the terminology may vary from one application to the next, some of the same concepts and principles that have been applied within the context of integrated logistic support (ILS) for defense systems have also been implemented in the commercial sector for manufacturing systems. Of particular significance is the concept of *total productive maintenance* (TPM), introduced by the Japanese in the early 1970s and being practiced today in many countries throughout the world (refer to Section 1.5). TPM represents an integrated life-cycle approach to the maintenance and support of a manufacturing plant.[1]

Experience has indicated that many factories in the past have been operated at less than full capacity, productivity has been low in general, and the costs of factory operations have been high. Further, a large portion of the total cost of doing business was due to "production losses." These costs, in turn, were transferred to the cost of the product being manufactured. In other words, the cost of many of the products being delivered for customer use was high because of the cost of manufacturing, which, in turn, has had a negative impact in the commercial market place.[2] This past experience resulted in an intensive effort initiated by the Japanese to increase productivity and reduce the costs of manufacturing, with the issue of *maintenance* being a "target" of opportunity.

[1] Refer to Nakajima in footnote 21, Chapter 1.

[2] According to a survey conducted by R. K. Mobley in the 1980s, between 15% and 40% of the total cost of the finished goods produced could be attributed to the maintenance costs in the factory. Where in the light process industries (e.g., food-related products) this percentage was relatively low, the percentage was high in heavy industries (e.g., iron and steel, pulp and paper). Refer to Mobley, R. K., *An Introduction to Predictive Maintenance*, Van Nostrand Reinhold, New York, N.Y., 1990.

Although the issue of maintenance was the initial "motivator," the implications relative to implementing the concepts and principles of TPM are more far reaching. In essence, TPM is an approach for improving the overall effectiveness and efficiency of a manufacturing plant. More specifically, the objectives are to

1. Maximize the overall effectiveness of manufacturing equipment and processes. This pertains to maximizing the availability of the production process through improvement of equipment reliability and maintainability, with the goal of minimizing downtime. It includes *maintenance prevention* (MP) and *maintainability improvement* (MI) which consider the incorporation of reliability, maintainability, and supportability characteristics in design (refer to Chapter 2).
2. Establish a life-cycle approach in the accomplishment of preventive maintenance. This pertains to the application of a reliability-centered maintenance (RCM) approach to justify the requirements for preventive maintenance (refer to Section 4.2.5).
3. Involve all operating departments/groups within a manufacturing plant organization in the planning for and subsequent implementation of a maintenance program (i.e., representation from engineering, opeartions, testing, marketing, and maintenance). The objective is to gain full *commitment* throughout the manufacturing organization.
4. Involve employees from the plant manager to the workers on the floor. There must be commitment from the top down in the organizational hierarchy.
5. Initiate a program based on the promotion of maintenance through "motivation management" and the development of autonomous small-group activities (i.e., the accomplishment of maintenance through a "team" approach). The operator must assume greater responsibility for the operation of his/her own equipment, and must be trained to detect equipment abnormalities, understand the cause-and-effect relationships, and accomplish minor repair. The operator is then supported by other members of the team for the accomplishment of higher-level maintenance activities. The application of the FMECA and the Ishikawa diagram can help in identifying cause-and-effect relationships (refer to Section 4.2.2).

The measure of total productive maintenance (TPM) can be expressed in terms of *overall equipment effectiveness* (OEE) which is a function of *availability, performance rate*, and *quality rate*. In Equation 7.1,

$$\text{OEE} = (\text{availability})(\text{performance rate})(\text{quality rate}) \tag{7.1}$$

where

$$\text{Availability } (A) = \frac{\text{loading time} - \text{downtime}}{\text{loading time}} \tag{7.2}$$

with *loading time* referring to the total time available for the manufacture of products (expressed in days, months, or minutes) and *downtime* referring to stoppages resulting from breakdown losses (failures), losses resulting from setups/adjustments, etc.

and

$$\text{Performance rate } (P) = \frac{(\text{output})(\text{actual cycle time})}{(\text{loading time} - \text{downtime})} \times \frac{(\text{ideal cycle time})}{(\text{actual cycle time})} \tag{7.3}$$

with *output* referring to the number of products produced; *ideal cycle time* referring to the ideal or designed cycle time per product (e.g., minutes per product); and *actual cycle time* referring to the actual time experienced (minutes per product). This metric addresses the process and rate of performance.

$$\text{Quality rate } (Q) = \frac{\text{input} - (\text{quality defects} + \text{startup defects} + \text{rework})}{\text{input}} \tag{7.4}$$

with *input* referring to the total number of products processed and *defects* referring to the total number of failures for one reason or another.

In essence, one is dealing with the manufacturing plant as a *system*, with the OEE factor being a measure of system effectiveness, as discussed in Section 2.12. As such, the manufacturing plant can be defined in functional terms, metrics can be allocated for each functional element, high-cost contributors can be identified, cause-and-effect relationships can be established, and recommendations for design improvement can be initiated as appropriate. Thus, although this subject is being addressed in this chapter (because of the fact that *maintenance* was the initial "driver"), many of the concepts and principles discussed earlier are applicable to the design, development, construction, operation, and support of a manufacturing plant.

The TPM approach first became popular in the United States in the late 1980s, particularly with those companies which are producing products in a highly competitive international environment and desire to reduce costs. In the early 1990s, the American Institute of Total Productive Maintenance (AITPM) was established, annual conferences have been scheduled, and monthly newsletters have been published. Although the terminology is different, the goals and objectives of TPM are consistent with the concepts and principles of integrated logistics support (ILS) and the design for system supportability.[3]

7.3 DATA COLLECTION, ANALYSIS, AND SYSTEM EVALUATION

The assessment of the performance and effectiveness of a system requires the availability of operational and maintenance histories of the various system elements. Performance and effectiveness parameters are established early in the life cycle with the development of operational requirements and the maintenance concept. These parameters describe the characteristics of the system that are considered paramount in fulfilling the need objectives. Now with the system deployed and in full operational status, the following questions arise:

[3] The Japanese established an initial OEE benchmark of 85%. As this text went to press, a few companies have attained this goal; however, the measured OEE for most of those companies who have applied TPM principles has been between 55% and 65%.

1. What is the *true* performance and effectiveness of the system?
2. What is the *true* effectiveness of the logistic support capability?
3. Are the initially specified requirements being met?
4. Is the system cost-effective?
5. Are all customer expectations being fulfilled?

Providing answers to these questions requires a formalized data information feedback subsystem with the proper output. A data subsystem must be designed, developed, and implemented to achieve a specific set of objectives, and these objectives must relate directly to the foregoing questions or whatever other questions the system manager needs to answer. The establishment of a data subsystem capability is basically a two-step function to include (1) the identification of requirements and the applications for such, and (2) the design, development, and implementation of a capability that will satisfy the identified requirements.

Requirements

The purpose of the data information feedback subsystem is twofold.

1. It provides ongoing data that are analyzed to evaluate and assess the performance, effectiveness, operations, maintenance, logistic support capability, and so on, for the system in the field. The system manager needs to know exactly how the system is doing and needs the answer relatively soon. Thus, certain types of information must be provided at designated times throughout the system life cycle.
2. It provides historical data (covering existing systems in the field) that are applicable in the design and development of new systems/equipments having a similar function and nature. Our engineering growth and potential in the future certainly depends on our ability to capture experiences of the past and subsequently be able to apply the results in terms of "what to do" and "what not to do" in the new design.

Supporting such a subsystem requires a capability that is both responsive to a repetitive need in an expeditious and timely manner (i.e., the manager's need for assessment information), and that incorporates the provisions for data storage and retrieval. It is necessary to determine the specific elements of data required, the frequency of need, and the associated format for data reporting. These factors are combined to identify total volume requirements for the subsystem and the type, quantity, and frequency of data reports.[4]

The elements of data identified are related to the operational and support requirements for the system. An analysis of these elements will provide certain evalu-

[4] Data collection, processing, storage, and retrieval functions must be compatible with continuous acquisition and life-cycle support (CALS) requirements (refer to section 4.4).

ative and verification type functions covering the characteristics of the system that are to be assessed. A listing of sample applications appropriate in the assessment of system support characteristics is presented in Table 7.1.

Referring to the table, when determining the reliability of an item (i.e., the probability that an item will operate satisfactorily in a given environment for a specified period of time), the data required will include the system operating time-to-failure and the time-to-failure distribution which can be generated from a history of that particular item or a set of identical and independent items. When verifying spare/repair part demand rates, the data required should include a history of all item replacements, system/equipment operating time at replacement, and disposition of the replaced items (whether the item is condemned or repaired and returned to stock). An evaluation of organizational effectiveness will require the identification of assigned personnel by quantity and skill level, the tasks accomplished by the organization, and the labor hours and elapsed time expended in task accomplishment. Assessment objectives such as these (comparable to those listed in the table) serve as the basis for the identification of the specific data factors needed.

Design, Development, and Implementation of a Data Subsystem Capability

With the overall subsystem objectives defined, the next step is to identify the specific data factors that must be acquired and the method for acquisition. A format for data collection must be developed and should include both *success* data and *maintenance* data. Success data constitute information covering system operation and utilization on a day-to-day basis, and the information should be comparable to the factors listed in Table 7.2.

Maintenance data cover each event involving scheduled and unscheduled maintenance. The events are recorded and referenced in system operational information reports, and the factors recorded in each instance are illustrated in Table 7.3.

The format for data collection may vary considerably, as there is no set method for the accomplishment of such, and the information desired may be different for each system. However, most of the factors in Tables 7.2 and 7.3 are common for all systems and must be addressed in the design of a new data subsystem. In any event, the following provisions should apply:

1. The data collection forms should be simple to understand and complete (preferably on single sheets) as the task of recording the data may be accomplished under adverse environmental conditions by a variety of personnel skill levels. If the forms are difficult to understand, they will not be completed properly (if at all) and the needed data will not be available.
2. The factors specified on each form must be clear and concise, and not require a lot of interpretation and manipulation to obtain. The right type of data must be collected.
3. The factors specified must have a meaning in terms of application. The usefulness of each factor must be verified.

TABLE 7.1 Data Information System Applications (Requirements)

1. *General Operational and Support Factors*
   (a) Evaluation of mission requirements and performance measures.
   (b) Verification of system utilization (modes of operation and operating hours).
   (c) Verification of cost/system effectiveness, operational availability, dependability, reliability, maintainability.
   (d) Evaluation of levels and location of maintenance.
   (e) Evaluation of function/tasks by maintenance level and location.
   (f) Verification of repair level policies.
   (g) Verification of frequency distributions for unscheduled maintenance actions and repair times.
2. *Test and Support Equipment*
   (a) Verification of support equipment type and quantity by maintenance level and location.
   (b) Verification of support equipment availability.
   (c) Verification of support equipment utilization (frequency of use, location, percent of time utilized, flexibility of use).
   (d) Evaluation of maintenance requirements for support equipment (scheduled and unscheduled maintenance, downtime, logistic resource requirements).
3. *Supply Support (Spare/Repair Parts)*
   (a) Verification of spare/repair part types and quantities by maintenance level and location.
   (b) Evaluation of supply responsiveness (is a spare available when needed?).
   (c) Verification of item replacement rates, condemnation rates, attrition rates.
   (d) Verification of inventory turnaround and supply pipeline times.
   (e) Evaluation of maintenance requirements for shelf items.
   (f) Evaluation of spare/repair part replacement and inventory policies.
   (g) Identification of shortage risks.
4. *Personnel and Training*
   (a) Verification of personnel quantities and skills by maintenance level and location.
   (b) Verification of elapsed times and manhour expenditures by personnel skill level.
   (c) Evaluation of personnel skill mixes.
   (d) Evaluation of personnel training policies.
   (e) Verification of training equipment and data requirements.
5. *Transportation and Handling*
   (a) Verification of transportation and handling equipment type and quantity by maintenance level and location.
   (b) Verification of availability and utilization of transportation and handling equipment.
   (c) Evaluation of delivery response times.
   (d) Evaluation of transportation and handling procedures.
6. *Facilities*
   (a) Verification of facility adequacy and utilization (operation, maintenance, and training facilities).
   (b) Evaluation of logistics resource requirements for support of operation, maintenance, and training facilities.
7. *Technical Data*
   (a) Verification of adequacy of data coverage (level, accuracy, and method of information presentation) in operating and maintenance manuals.
   (b) Verification of adequacy of field data, collection, analysis, and corrective action subsystem.
8. *Software*
   (a) Verification of adequacy of software (level, accuracy, detail, comprehensiveness).
   (b) Verification of compatibility of software with other elements of the system.

TABLE 7.2 System Success Data

System Operational Information Report
1. Report number, report date, and individual preparing report.
2. System nomenclature, part number, manufacturer, serial number.
3. Description of system operation by date (mission type, profiles and duration).
4. Equipment utilization by date (operating time, cycles of operation, etc.).
5. Description of personnel, transportation and handling equipment, and facilities required for system operation.
6. Recording of maintenance events by date and time (reference maintenance event reports).

TABLE 7.3 System Maintenance Data

Maintenance Event Report
1. *Administrative data*
(a) Event report number, report date, and individual preparing report.
(b) Work order number.
(c) Work area and time of work (month, day, hour).
(d) Activity (organization) identification.
2. *System factors*
(a) Equipment part number and manufacturer.
(b) Equipment serial number.
(c) System operating time when event occurred (when discovered).
(d) Segment of mission when event occurred.
(e) Description of event (describe symptom of failure for unscheduled actions).
3. *Maintenance factors*
(a) Maintenance requirement (repair, calibration, servicing, etc.).
(b) Description of maintenance tasks.
(c) Maintenance downtime (MDT).
(d) Active maintenance times (Mct_i and Mpt_i)
(e) Maintenance delays (time awaiting spare part, delay for test equipment, work stoppage, awaiting personnel assistance, delay for weather, etc.).
4. *Logistics factors*
(a) Start and stop times for each maintenance technician by skill level.
(b) Technical manual or maintenance procedure used (procedure number, paragraph, date, comments on procedure adequacy).
(c) Test and support equipment used (item nomenclature, part number, manufacturer, serial number, time of item usage, operating time on test equipment when used).
(d) Description of facilities used.
(e) Description of replacement parts (type and quantity).
(i) Nomenclature, part number, manufacturer, serial number, and operating time on replaced item. Describe disposition.
(ii) Nomenclature, part number, manufacturer, serial number, and operating time on installed item.
5. *Other information*
Include any additional data considered appropriate and related to the maintenance event.

These considerations are extremely important and cannot be overemphasized. All of the analytical methods, prediction techniques, models, and so on, discussed earlier have little meaning without the proper input data. Our ability to evaluate alternatives and predict in the future depends on the availability of good historical data, and the source of such stems from the type of data information feedback subsystem developed at this stage. The subsystem must not only incorporate the forms for recording the right type of data, but must consider the personnel factors (skill levels, motivation, etc.) involved in the data recording process. The person who must complete the appropriate form(s) must understand the system and the purposes for which the data are being collected. If this person is not properly motivated to do a good thorough job in recording events, the resulting data will of course be highly suspect.

Once the appropriate data forms are distributed and completed by the responsible line organizations, a means must be provided for the retrieval, formatting, sorting, and processing of the data for reporting purposes. Field data are collected and sent to a designated centralized facility for analysis and processing. The results are disseminated to management for decision making and entered into a data bank for retention and possible future use.

System Evaluation and Corrective Action

Figure 7.1 illustrates the system evaluation and corrective-action loop. The evaluation aspect responds to the type of subjects listed in Table 7.1, and can address both the system as an entity or individual segments of the system on an independent basis. Figure 7.2 presents some typical examples of evaluation factors. The evaluation approach and the analytical techniques (i.e., tools) used are basically the same as described in Chapter 4, with the only difference being the data input. The evaluation effort can be applied on a continuing basis to provide certain system measures at designated points throughout the life cycle (see *A* and *C* of Figure 7.2), or it may constitute a *one-time* investigation.

Problem areas are identified at various stages in the evaluation, and are reviewed in terms of the necessity for corrective action. Referring to Figure 7.2(b), Subsystem *E* is a likely candidate for investigation since the MLH/OH for that item constitutes 31.7% of the total system value. In Figure 7.2(c), one may wish to investigate Program Time Period 3 to determine why the mission reliability was so poor at that time. Corrective action may be accomplished in response to a system/equipment deficiency (i.e., the equipment fails to meet the specified requirements), or may be accomplished to improve system performance, effectiveness, or logistic support. If corrective action is to be accomplished, the necessary planning and implementation steps are a prerequisite to ensure the complete compatibility of all elements of the system throughout the change process.

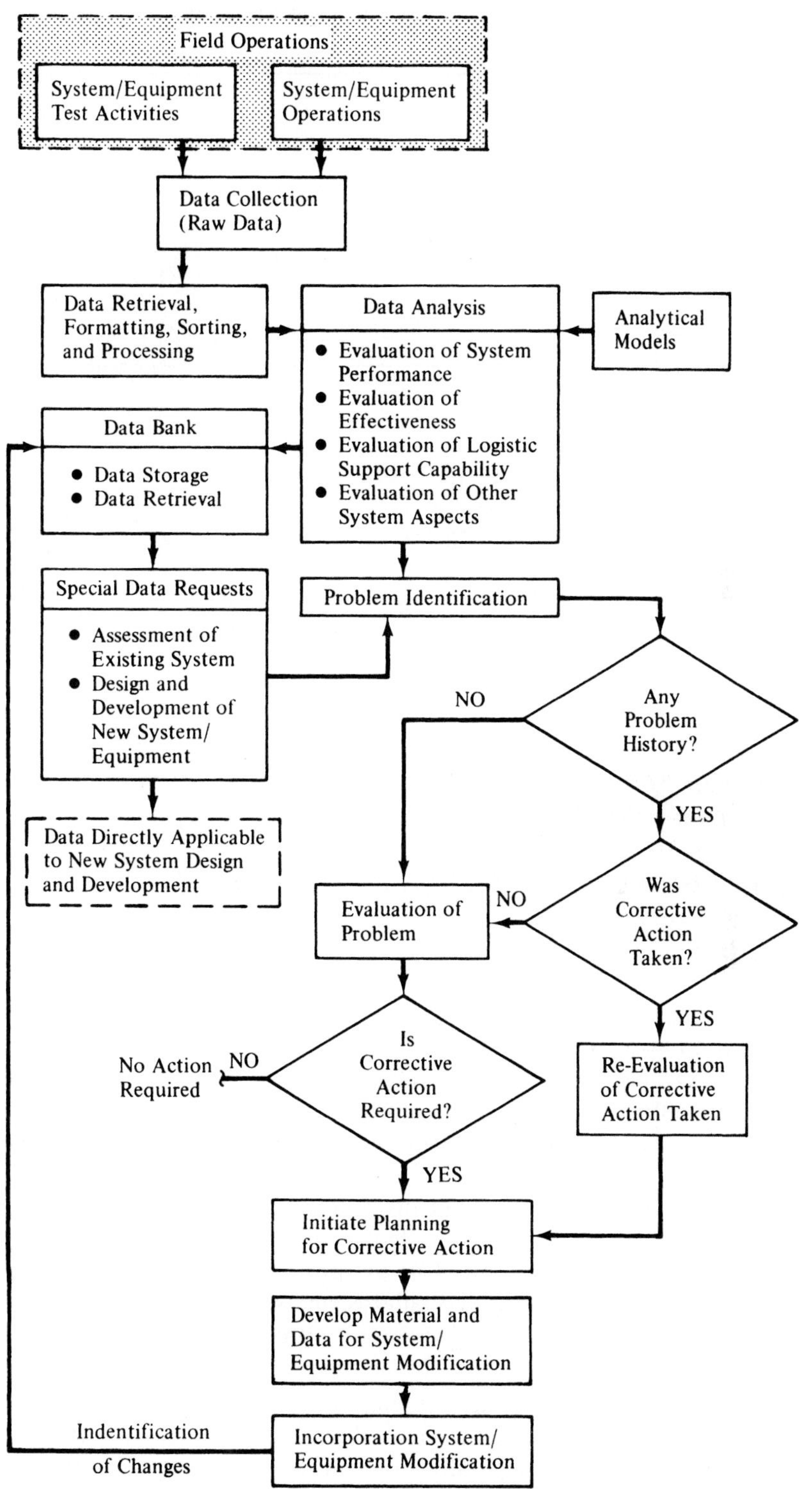

Figure 7.1 System evaluation and corrective-action loop.

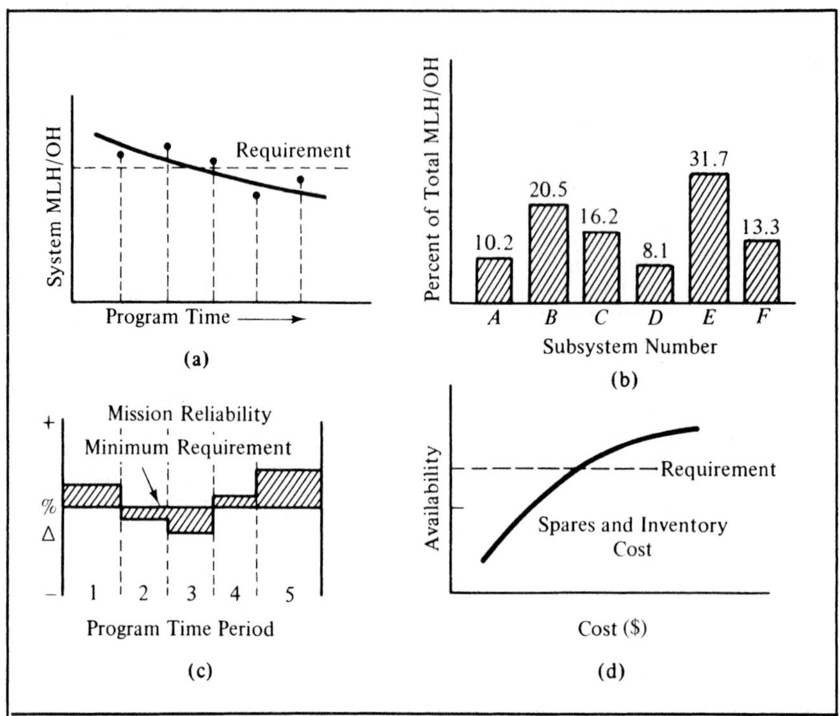

Figure 7.2 Typical evaluation factors.

7.4 EVALUATION OF LOGISTIC SUPPORT ELEMENTS

Test and Support Equipment

The evaluation of test and support equipment involves a comparison of what is procured and available at each level of maintenance (i.e., type and quantity of each item derived through analysis) and what is actually required. The initial determination of requirements in this area is accomplished in the Supportability Analysis discussed in Chapter 4. These requirements are based on the following (as applicable):

1. Equipment arrival rates or the predicted quantity of items returned for maintenance. This factor is generally determined from an exponential distribution for major items (e.g., Unit *A* of System *XYZ*) considering primary failures, secondary failures, suspected failures, and so on. A Poisson probability distribution can be used as a density function to predict the quantity of returns within each major item.
2. Equipment item active maintenance times. The time required to test the arriving items is also predicted through the use of a probability density function. Automatic testing is essentially constant, but the other aspects of the maintenance cycle may follow a log-normal distribution.

3. Support equipment design characteristics. This includes support equipment availability, equipment warm-up time, test setup time, and the ease in performing the actual maintenance operations.
4. The external work and environmental factors to include the length of the workday, personnel effectiveness, noise, temperature, and related human factors.

Through the use of a Monte Carlo sampling process, a model may be generated to simulate time demands on an item of support equipment. Probabilities are developed to cover queue length (i.e., waiting line) and the turnaround time of a typical item scheduled for maintenance. Loading studies are accomplished and an optimum quantity of support equipment is recommended and provisioned for operational use.

When the system becomes operational, it is imperative that an assessment of the above factors be made. Historical data are collected using a recording format similar to Table 7.4. After a designated period of time, arrival rates, the queue length or waiting time, and support equipment utilization are determined from actual field experience. In addition, support equipment availability and the logistics resources required for support equipment maintenance can be evaluated.

TABLE 7.4 Support Equipment Usage Record

Item Returned for Maintenance or Requiring Support	Time in Work Area		Time on Support Equipment		Time Off Support Equipment	
	Date	Hour	Date	Hour	Date	Hour
Unit *A*	12/10/82	10:30 a.m.	12/11/82	11:30 a.m.	12/11/82	1:45 p.m.
Assembly 2345	1/3/83	2:30 p.m.	1/3/83	2:30 p.m.	1/3/83	3:52 p.m.

If an existing item of support equipment at a given level of maintenance is not adequately utilized, it may be appropriate to recommend a change in the detailed maintenance plan. With a lower number of arrivals than initially anticipated, it might be feasible to shift the repair of the applicable items to a higher level of maintenance (e.g., from intermediate to depot). This will reduce the quantity of support equipment in the field and thus result in a probable savings in life-cycle cost. The final determination will result from a level of repair analysis (see Chapter 4) using the actual field data collected.

Conversely, if the queue is long and the support equipment availability is inadequate, then it may be feasible to produce an additional quantity of like items for the same geographical installation. The intent is to evaluate each facet of system operation and maintenance and verify that the proper type and quantity of support equipment are available at the right location. The results of this evaluation must be consistent with the overall system performance and effectiveness requirements, and yet support an approach that reflects the lowest life-cycle cost.

Supply Support

Supply support constitutes all materials, data, personnel, and related activities associated with the requirements, provisioning, and acquisition of spare/repair parts and the sustaining maintenance of inventories for support of the system throughout its life cycle. Specifically, this includes:

1. Initial and sustaining requirements for spares, repair parts, and consumables for the prime equipment. Spares are major replacement items and are repairable while repair parts are nonrepairable smaller components. Consumables refer to fuel, oil, lubricants, liquid oxygen, nitrogen, and so on.
2. Initial and sustaining requirements for spares, repair parts, material, and consumables for the various elements of logistic support (i.e., test and support equipment, transportation and handling equipment, training equipment, and facilities).
3. Facilities required for the storage of spares, repair parts, and consumables. This involves consideration of space requirements, location, material handling provisions, and environment.
4. Personnel requirements for the accomplishment of supply support activities such as provisioning, cataloging, receipt and issue, inventory management and control, shipment, and disposal of material.
5. Technical data and software requirements for supply support to include initial and sustaining provisioning data, catalogs, material stock lists, receipt and issue reports, material disposition reports, and so on.

Supply support is applicable to the early test and evaluation activities (described in Chapter 5), sustaining system/equipment operations, and during equipment phasedown (or phaseout) toward the end of the life cycle. Supply support is *dynamic* and must be responsive to changes in system operation, deployment, utilization, effectiveness requirements, repair policies, and environment. It also must be responsive to prime equipment modifications.

1. *Determination of requirements.* Initial requirements for spares, repair parts, and consumables stem from the supportability analysis (described in Chapter 4). The SA supports the maintenance concept in identifying functions and tasks by level, repair policies (i.e., repair or discard at failure), individual spare/repair part types, and item replacement frequencies. These factors are combined to indicate supply support requirements for each geographical location and for the system as an entity. Major considerations include (a) spares and repair parts covering actual item replacements occurring as a result of corrective and preventive maintenance actions; (b) an additional stock level of spares to compensate for repairable items in the process of undergoing maintenance; (c) an additional stock level of spares and repair parts to compensate for the pipeline and procurement lead times required for item acquisition; and (d) an additional stock level of spares to compensate for the condemnation or scrapage of repairable items.

Addressing spare/repair part requirements from an optimum standpoint consists of solving three basic problems: (1) determine the range or variety of spares; (2) determine the optimum quantity for each line item; and (3) evaluate the impact of item selection and quantities on the effectiveness of the system. Items must be justified by establishing a demand prediction and identifying the consequences of not having the spare/repair part in stock. Demand predictions are based on the Poisson distribution approach discussed in Chapter 2. Some items are considered more critical than others in terms of impact on mission success. The criticality of an item is generally based on its function in the equipment and not necessarily its acquisition cost. However, the justification for these critical items may vary somewhat depending on the type of system and the nature of its mission.

The objectives in determining spare/repair part requirements are to identify item replacements; determine item replacement and repair frequencies, repair and resupply cycle times, condemnation factors, unit cost; and to develop a supply support capability that will:

(a) Not impair the effectiveness of the system by having equipment on a nonoperationally ready status due to supply.

(b) Reflect a least-cost inventory profile by not having unnecessary items in the inventory, and minimize outstanding back orders. Large inventories may be costly.

In the early design and development stages of the system life cycle, the objectives listed above are evaluated through analyses and the use of models (see Chapter 4). The supply support capability is simulated with the intent of arriving at an optimum balance between a stock-out situation and the level of inventory. Stock-out conditions promote the *cannibalization* of parts from other equipment (which may have further detrimental effects on the system), or the necessity of initiating high priority orders from the supplier (requiring expeditious and special handling). Both options are costly.

The purpose of the analysis and simulation effort is to evaluate the best among a number of alternative approaches. When arriving at a recommended solution, the results are used in the initial provisioning of spares, repair parts, and consumables. In other words, the appropriate material items are acquired for support of the system for a specified period of time when the system first becomes operational. Reprovisioning is then accomplished as appropriate, using experience data from the field, to cover successive periods throughout the life cycle.

2. *Provisioning and acquisition of material.* Given the basic requirements for spare/repair parts and consumables, a plan is developed as part of the integrated logistic support plan for the provisioning of the appropriate material. Provisioning constitutes the source coding of items, the preparation of stock lists and procurement documentation, and the acquisition and delivery of material.[5] In general, it is not fea-

[5] Source coding basically applies to the determination of whether an item is repairable or nonrepairable and, if repairable, where it is to be repaired. This information is determined by the SA (see Chapter 4), and the results are coded and included in the provisioning documentation. Procurement documentation includes identification of suppliers, pricing information, scheduling data, lead times, and so on.

sible to provision enough support for the entire life cycle of the system as too much capital would be tied up in inventory. The cost of inventory maintenance is high and much waste could occur, particularly if equipment changes are implemented and certain components become obsolete. In addition, initial provisioning is generally based on the estimated maintenance factors provided in the SA (i.e., replacement rates, repair and recycle times, pipeline times, etc.). These are only estimates derived from predictions and as such may be in error.[6] Estimation errors will of course have a significant impact on the quantity of items in inventory; thus, provisioning should be accomplished at shorter intervals to allow for the necessary adjustments based on actual field experience.

Figure 7.3 illustrates the possible differences between an early estimate of the maintenance replacement factor, actual operational experience, and a normalized set of values based on the operational experience. The figure shows the variation of estimates over time. An original estimate and its anticipated variation with time are made and used in the SA and associated simulation efforts involving supply support. Initial provisioning is accomplished using this estimate as the basis for determining the quantity of different items in the inventory.[7] The system is then deployed and operational field data may indicate a different level of item usage. Actual usage may be erratic at first, but ultimately may be represented by a normalized curve as shown in the figure. Assuming that initial provisioning is based on the period of time indicated, there will be an excess of items in the inventory. Maintenance factors can be adjusted for the next provisioning period and the excess stockage can be used during that time. Conversely,

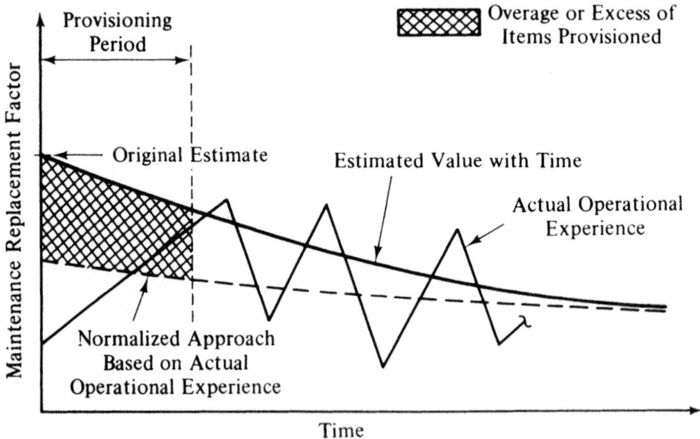

Figure 7.3 Maintenance replacement factors.

[6] Experience on many systems has indicated that the correlation between predicted maintenance factors and actual field results has been rather low. As we acquire more data and experience relative to the input requirements and their relationships, it is hoped that our predictions will continue to improve in the future as they have in the past decade.

[7] Each item type may be represented by a different curve (as illustrated in Figure 7.3).

the provisioning of too little support results in the probability of causing the system to be inoperative due to stock-out, which can also be costly.

Generally, a realistic period of time is specified (e.g., 6 months, 1 year, 2 years, etc.) and enough spare/repair parts are procured to cover the demand requirements anticipated. Sometimes the quantities of different items procured will be adjusted for economic reasons. For instance, it may be feasible to acquire a set quantity of an item in order to realize a price break or to avoid excessive supplier production startup and shutdown costs. If the quantity is excessive and the applicable items are not consumed during the selected support period, the residue may be carried into the next period of support.

The provisioning concept will vary somewhat between major high-value items and smaller repair parts.[8] In addition, the procurement policies may differ for items with different usage rates. Fast-moving items may be procured locally near the point of usage such as the intermediate maintenance shop, while slower moving items stocked at the depot may be acquired from a remotely located supplier as the pipeline and procurement lead times are not as critical.

The planning for major high-value items is quite detailed and generally receives management scrutiny at every step in the process. This is required for both strategic and economic reasons. First, the supply status of these items influences to a great extent the supply levels of a large portion of the smaller repair parts in the inventory. Second, the dollar value of the larger items is significant and may even exceed the total value of the hundreds of repair parts and accessories needed for their continued support. In other words, a relatively small number of items may represent a large percentage of the total inventory value. Thus, greater emphasis is placed in the computation of replacement factors and consumption rates to arrive at a *true demand*. In addition, the provisioning time periods may vary from item to item to assure an economic order quantity that is compatible with the actual usage rate. In some instances, a given quantity of items is maintained in stock to compensate for repair and recycle times, pipeline and procurement lead times, and so on, and new quantities are ordered on a one-for-one basis as existing items are withdrawn from the inventory.

For other items such as common and standard spares and repair parts, particularly where large quantities are involved over the life cycle, the economic order quantity (EOQ) principle can be applied. The ultimate goal is to have the correct amount and type of supplies available for the lowest total cost. Procurement costs vary with the quantity of orders placed. The principle involves the optimization between the placing of many orders resulting in high material acquisition costs and placing orders less frequently while maintaining a higher level of inventory causing increasing inventory maintenance and carrying costs. In other words, ordering creates procurement cost while inventory creates carrying cost. The economic order principle equates the *cost to order* to the *cost to hold*, and the point at which the combined costs are at a minimum

[8] High-value items are those components with a relatively high unit acquisition cost, and should be provisioned on an individual basis. In addition, special packing and handling may be required. The classification of high value items will vary with the program, and may be established at a certain dollar value (i.e., all components whose unit cost exceeds x dollars are considered as high-value items).

indicates the desired size of order. The economic order principle and inventory profiles are discussed in Section 2.3.

The illustration in Figure 2.21 is a theoretical representation of an inventory cycle for a given item. Actually, demands are not always constant and quite often the reorder cycle changes with time. Figure 2.22 presents a situation that is more realistic.

In the accomplishment of provisioning, the most feasible approach is to employ the theoretical EOQ model illustrated in Figure 2.21 for each applicable class of items until such time that enough data are collected to enable a definition of an actual inventory profile. When this is available, the factors in Figure 2.21 can be adjusted as applicable to fit the actual situation.

The demand for spare/repair parts and the provisioning factors illustrated in Figures 2.21 and 2.22 lead to the preparation of stock item lists, procurement data, inventory records, and so on. These data items are oriented to the line organization responsible for system support. The factors used will be different for the intermediate shop and the depot since the demands, procurement lead times, pipeline, and so on, vary.[9]

3. *Inventory maintenance and control.* As the system progresses through its life cycle, data are collected and analyzed to assess the effectiveness of the supply support capability. The object is to establish a historical base for individual item demand rates and to adjust the provisioning factors as required to improve supply support.

The operational demand based on experience may assume a jagged plot as illustrated in Figure 7.3. Trends and averages may be identified using statistical techniques (e.g., exponential smoothing, method of least squares, etc.), and the results are included in an updating of the provisioning data for future spare/repair part reorders. At the same time, the inventory is evaluated in terms of current assets, average months of supply on hand, costs to procure material and maintain the inventory, quantity of orders, and so on. A sample data sheet format is presented in Table 7.5. Inventory status for each item at each supply point is maintained on a continuing basis.

In the event that historical experience indicates that the level of inventory is insufficient to support operational needs, the EOQ and quantity of orders are recom-

TABLE 7.5 Inventory Status Record

Item	Annual Demands (D)	Average Months of Supply on Hand (Month)	Quantity of Orders per Year (N)	Average Active Inventory ($)	Cost to Procure ($)	Cost of Inventory Maintenance ($)
Part 234	12	0.3	2	721	35	21
Assembly *A*	18	6.7	3	681	72	43

[9] The provisioning approach discussed in this section basically assumes the ongoing availability of procurement sources. If it appears that future suppliers will be limited, it may be necessary or desirable to assume the risk and procure enough spares for the entire system life cycle.

puted and the required additional stock is procured. On the other hand, if it appears that the stock level is excessive, either one of two actions may be required. First, the provisioning factors are adjusted and future orders are curtailed until the inventory is depleted to a desired level. Second, a destockage plan may be implemented to remove a designated quantity of the serviceable assets on hand. In any event, criteria should be established covering both the desired upper and lower levels of inventory, and the evaluative and control aspects should employ these criteria in the measurement of supply support effectiveness.

4. *Disposition of material.* Repairable spares that are condemned (for one reason or another) and nonrepairable parts, when removed from the system and replaced, are generally shipped to the depot or the supplier facility for disposition. These items are inspected, disassembled where possible, and the items that may have further use will be salvaged, reclaimed, and recycled. The residue will be disposed of in an expedient and economical manner consistent with environmental and ecological requirements (refer to Chapter 8).

Personnel and Training

Personnel and training requirements are initially derived through (1) an operator and maintenance functional analysis, (2) a detailed task analysis identifying the quantity and skills required, and (3) a comparison of the personnel requirements for the system with the personnel quantities and skills available in the user organization. Operator personnel requirements are usually derived from a human factors task analysis, and maintenance personnel requirements are identified in the SA. The difference between system requirements and the user personnel skill levels scheduled to operate and maintain the system in the field is covered through a combination of formal and on-the-job training (OJT). Training requirements include both the training of personnel initially assigned to the system and the training of replacement personnel throughout the life cycle. Training in both operator and maintenance functions (at all levels) is accomplished.

When the system becomes operational, data are collected at all levels for each organization responsible for the accomplishment of system operation and maintenance functions. The object is to assess organizational effectiveness on a continuing basis. Specifically, one is interested in the quantity of personnel required to do the job, the maintenance labor hours expended, the attrition rate and whether it increases or decreases with time, error rates in the performance of job functions, and general morale. An analysis of such data will indicate whether the initial personnel selection criteria are adequate and whether the training program as designed is satisfactory. If error rates are high, either additional training is required or a change in the basic system/equipment design is necessary. High attrition rates may necessitate a continuous formal training effort. On the other hand, if the system is not complex and attrition and error rates are low, it may be feasible to eliminate the program for formal training and maintain only the requirement for a specified amount of OJT. Personnel operations in the field are evaluated using the same approach as the industrial engineering and quality control organizations use in monitoring production or construction operations.

Technical Data

Technical data include all operating and maintenance procedures, special test procedures, installation instructions, checklists, change notices and change procedures, and so on, for the prime equipment, support equipment, training equipment, transportation and handling equipment, and facilities. The requirements for data coverage stem from engineering design data and the SA. Operator and maintenance functions by organization, repair policies, and logistic support resource requirements are identified. Technical data are prepared to cover all system operations and logistic support requirements in the field.[10]

Subsequent to data preparation, the appropriate documentation is used in support of test and evaluation activities, production/construction functions, and field operations. Problems are noted relative to actual documentation errors, and difficulties concerning prime equipment and logistic support activities are recorded. For instance, if a particular function is difficult to accomplish, it may be due to the fact that the documented procedure is inadequate relative to the scope of coverage; thus, additional information should be prepared and incorporated through a change notice. In addition, technical data should be updated as required to cover prime equipment modifications, revised maintenance policies, and changes in the elements of logistic support.

Summary

The preceding paragraphs discuss some of the major factors involved in the evaluation of the elements of logistic support. Although not all elements are covered, the information presented does cover facets that require particular attention. The objective is to design and implement a data information feedback subsystem that will enable a true assessment of the logistic support capability. All elements of logistics must be evaluated both individually and on an integrated basis. Each individual element directly impacts the others; thus, when changes occur (to correct deficiencies or for system improvement), the impact of the change must be reflected throughout.

7.5 SYSTEM MODIFICATIONS

When a change occurs in a procedure, the prime equipment, or an element of logistic support, the change in most instances will affect many different elements of the system. Procedural changes will impact personnel and training requirements and necessitate a change in the technical data (equipment operating and/or maintenance instructions). Hardware changes will affect spare/repair parts, test and support equipment, technical data, and training requirements. Each change must be thoroughly evaluated in terms of its impact on other elements of the system prior to a decision on

[10] This pertains to the information package that defines the system and its elements, provides system operating and maintenance instructions, and so on. In some instances, much of this information may be presented in the form of an electronic database.

whether or not to incorporate the change. The feasibility of incorporating a change will depend on the extensiveness of the change, its impact on the system's ability to accomplish its mission, the time in the life cycle when the change can be incorporated, and the cost of change implementation. A minimum amount of evaluation and planning are required in order to make a rational decision on whether the change is feasible.

If a change is to be incorporated, the necessary change control procedures must be implemented. The various components of the system (including all of the elements of logistic support) must track. Otherwise, the probability of inadequate logistic support and unnecessary waste is high. The configuration change control and the change cycle are discussed in Section 5.6. The concepts and concerns in the production/construction stage are equally appropriate in the operational phase. In fact, a change involving the operational system is more critical since we are dealing with a late stage in the life cycle. Changes will usually be more costly as the life cycle progresses.

QUESTIONS AND PROBLEMS

1. Assuming that the objectives of logistic support have been met in a satisfactory manner in the earlier program phases, what are the logistic support functions in the system utilization and support phase?

2. Briefly describe total productive maintenance (TPM) and its objectives, and identify the metrics pertaining to TPM.

3. Compare the concepts of ILS and TPM. Identify similiarities and differences between the two.

4. How would you verify the adequacy of the following?
 (a) Spare and repair parts.
 (b) Test and support equipment.
 (c) Frequency distributions for unscheduled maintenance actions.
 (d) Personnel quantities and skill levels.
 (e) Operating and maintenance procedures.

5. What data are required to measure the following?
 (a) Cost effectiveness.
 (b) System effectiveness.
 (c) Operational availability.
 (d) Life-cycle cost.
 (e) Overall equipment effectiveness

6. A good field data subsystem serves what purposes?

7. Why is system success data important? What type of information is required?

8. Why is change control so important? How do system changes affect logistic support?

9. How does the design of test and support equipment affect the quantity of same?

10. How does the type and quantity of support equipment available affect spare and repair part requirements?

11. What is a *queue*? How does it influence spare and repair part requirements?

12. How do the maintenance requirements for support equipment affect prime equipment availability?

13. What considerations must be addressed in determining the proper quantity of spares and repair parts for a given stock location? How are spare part requirements justified?
14. Some spare/repair parts are considered to have a higher priority (in importance) than others. What factor(s) determines this priority?
15. If the cannibalization of items from other equipment is practiced on a regular basis, what will be the likely effects on the system?
16. How are maintenance replacement factors determined? How are they adjusted to reflect a realistic support posture? How do replacement factors relate with MTBM and MTBF?
17. Referring to Figure 2.21, what happens to the EOQ when the demand increases? What happens when there are outstanding backorders? What factors are included in production lead time?
18. What factors determine the level of safety stock?
19. What are *high-value items*? How are they classified?
20. How would you measure organizational effectiveness?
21. How would you measure human reliability?
22. Select a system of your choice and develop a data collection and feedback capability. Include an example of the data collection format proposed. Describe the data analysis, evaluation, and corrective action loop. How would you make it work?
23. How would you consider equipment disposability in the design process? How would you determine the cost of disposal?
24. Refer to Figure 2.21 and assume that production of an item occurs incrementally and that

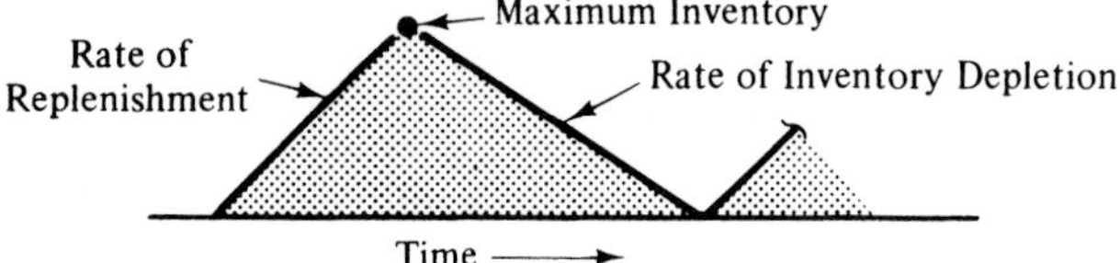

Figure 7.4 Inventory cycle.

inventory replenishment is accomplished as each item is produced. The inventory level for the economic lot size model follows the illustration in Figure 7.4.
Assume that d is the annual rate of demand for the item and is constant, p the annual rate of production for the item and is constant, Q the size of the production lot (quantity of units), and L lead time and is constant. Write the equation (in terms of the factors above) for:
    (a) The rate of inventory replenishment.
    (b) The rate of inventory depletion.
    (c) Maximum inventory value.
    (d) Average inventory value.
    (e) Total annual cost.
25. What is meant by *spare/repair parts provisioning*?
26. Identify and describe possible alternative maintenance support policies for a system of your choice (refer to Section 7.1). What are the advantages/disadvantages of each?

CHAPTER 8

LOGISTICS IN THE SYSTEM RETIREMENT, MATERIAL RECYCLING, AND DISPOSAL PHASE

System retirement, phaseout, and the recycling or disposal of material no longer required in the operational inventory are subjects not too well covered in the literature. It is common to address the design and development of a system and the operation of that system; however, quite often the phaseout and subsequent disposition of the system (and its components) are not adequately considered until the time arrives to do something about it!

With the increasing concern for the environment, and the possibility of environmental degradation, it is no longer feasible to ignore this area of activity. Referring to Figure 8.1, the activity represented by the shaded area constitutes a life cycle in itself and must be addressed along with the design and development of the prime mission-related components of the system, the design for producibility, and the design for supportability. In other words, an appropriate level of emphasis should be placed initially on the design for *disposability* or *recycling/reuse*, and later on developing the appropriate logistics infrastructure for the subsequent processing of items to be discarded. The "design" issue is addressed in Section 5.2.7, while the second part is briefly covered in the chapter.

8.1 SYSTEM REQUIREMENTS

The requirements associated with this area of activity may be classified in three categories:

1. Those elements or components of the system that are replaced at designated times for the purposes of upgrading or modifying the system for sustainability or improvement. This includes the insertion of new technologies because of the

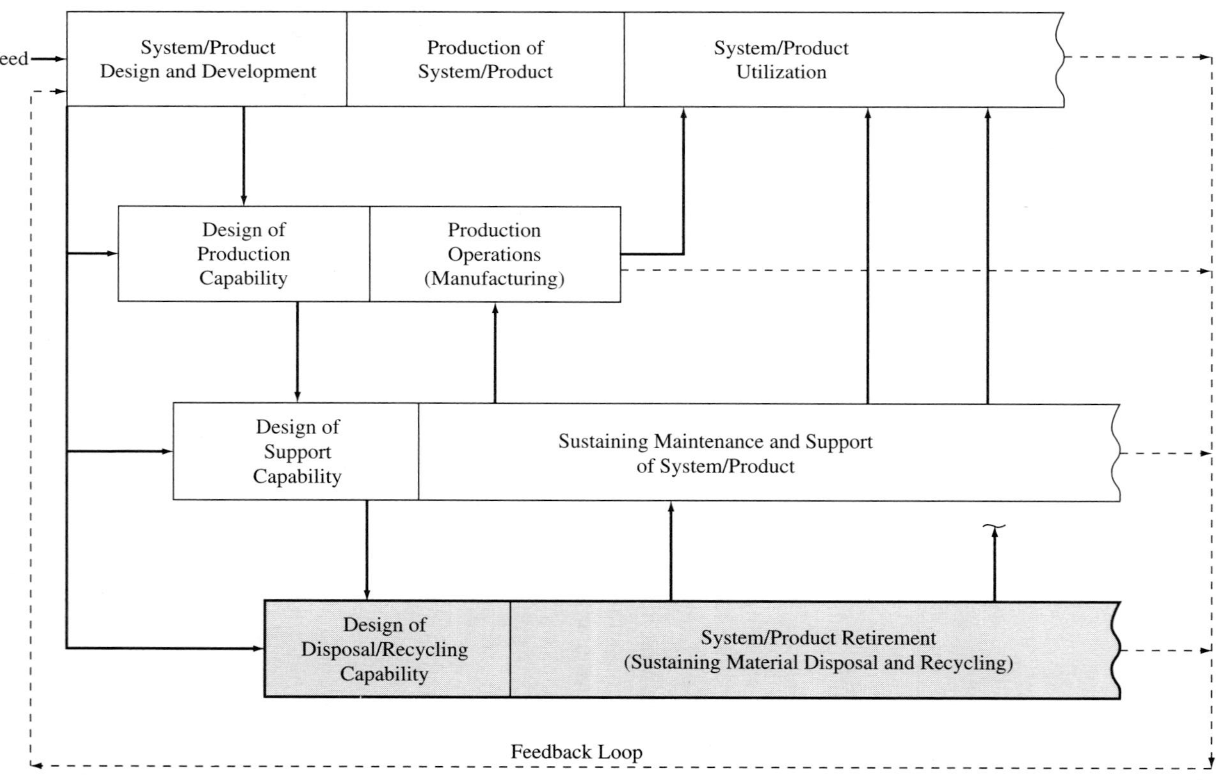

Figure 8.1 The life cycles and their interrelationships. (reference: Figure 1.5).

short life cycles of those components that were designed in from the beginning. Referring to Figure 1.8, there may be numerous points in the overall system life cycle when items are removed and require further disposition.

2. Those nonrepairable items that are removed and replaced in the performance of corrective maintenance (i.e., when system failures occur) or preventive maintenance (when critical items are scheduled for replacement).
3. Those elements of the system that are retired and removed from the operational inventory at the end of their respective life cycles when there is no longer a functional need for the system.

The planning for system retirement, phaseout, and material recyling and disposal should be responsive to the following questions:

1. What items of equipment, software, materials, data, elements of support, and so on, are likely to be phased out of the inventory, and when is this expected to occur?
2. What should be done to these items (i.e., disposition)?
3. Where should this be accomplished and by whom?
4. To what extent can the items being removed from the inventory be decomposed and recycled for reuse?
5. Are the methods used for decomposition, recycling, and material disposal consistent with ecological and environmental requirements? What are the impacts on the environment?
6. What logistic support requirements are necessary to accomplish the retirement, phaseout, material recycling, and disposal functions?
7. What metrics should be applied to this area of activity (i.e., turnaround times, recycle rates, process times, economic, and effectiveness factors, etc.)?

8.2 MATERIAL RECYCLING AND DISPOSAL

The specific logistic support requirements for this phase of the life cycle are based on the results of the supportability analysis (SA), and are derived in a manner similar to what is accomplished for the system utilization and support phase. Referring to Section 3.6, the functional analysis must include those activities dealing with system upgrades and modifications for improvement, the ongoing maintenance activities that are accomplished throughout the life cycle, and those activities dealing with system retirement and the ultimate disposition of its elements/components. From the identification of major functions, the specific resource requirements associated with each function are to be determined from the process illustrated in Figure 3.25. The development of maintenance functional flow diagrams (see Figure 3.23) and the subsequent accomplishment of a maintenance task analysis (refer to Section 4.2.4) can be accomplished to facilitate the identification of the resources required for the purposes of maintenance.

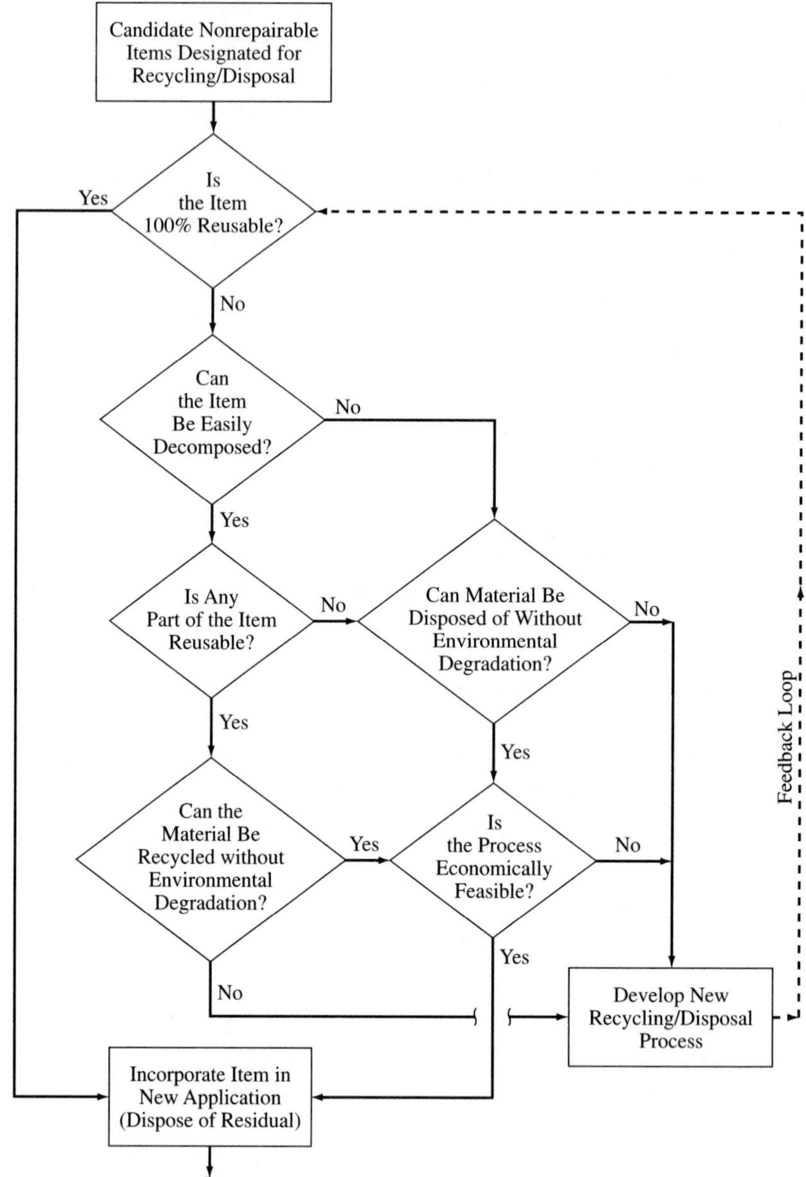

Figure 8.2 The material recycling/disposal process.

Given that the requirements for item removals/replacements have been identified, one should evaluate each of these items following the process illustrated in Figure 8.2. Referring to the figure, can the item be reused as an entity for any other application? If not, has it been designed in such a way that it can be easily decomposed, and can any of its parts be used? If an item cannot be reused as is, can it be recycled and modified for use? Further, if recycling and reuse are not feasible, can the material in question be disposed of without causing a negative impact on the environment (refer to Figure 5.22)? Finally, is the process economically feasible?

A major objective, as part of the system engineering process described in Chapters 3 through 5, is to design the system and its elements for recycling, reuse, and disposability. From a supportability perspective, the accomplishment of such an objective should result in a minimum requirement for follow-on logistics resources (i.e., spares/repair parts, test equipment, software, personnel, facilities, data, etc.). When such resources are necessary, the use of commercial and standard items is preferred. In essence, the same principles applicable in the design for reliability, maintainability, and supportability should be implemented here. Just as in the "introduction and build-up stage" of the system life cycle, attention must be given to the "phaseout and recycling" stage.[1]

QUESTIONS AND PROBLEMS

1. What is meant by *design for the environment? Design for disposability? Design for decomposition? Design for recycleability?* How is logistics impacted? Provide some examples.
2. Identify some of the issues that need to be addressed when designing for the environment?
3. What steps would you pursue in determining the logistic support requirements for this phase of the life cycle? What methods/tools would you employ in the process?
4. Select a system (or major element) of your choice and complete a functional analysis covering the activities in this phase of the life cycle.
5. Select a major function from the results of Problem 4, accomplish a detailed task analysis, and identify the logistic support resources required.
6. Accomplish a life cycle cost analysis for the system selected in Problem 4. Show the breakout of costs for the various activities in this phase of the life cycle.

[1] There have been numerous examples in which the lack of such has resulted in the requirements for a large new facility necessary for material decomposition, special test equipment, special transportation and handling equipment, high-skilled personnel, and so on.

CHAPTER 9

LOGISTICS MANAGEMENT

Logistics management involves the planning, organizing, directing, and controlling of all activities necessary to fulfill the requirements discussed throughout the earlier chapters of this text. Logistics terms and definitions are introduced in Chapters 1 and 2, requirements throughout the system life cycle are described in Chapters 3 through 8, and the implementation of these requirements from a managerial perspective is discussed in this chapter.

The fulfillment of logistics objectives, presented in today's environment, provides some interesting challenges. More specifically:

1. Logistics is life-cycle oriented. There are logistics functions in all phases of the system life cycle, and an overall understanding of the broad spectrum of program activities across the board is required.
2. A wide variety of subdisciplines is inherent within the overall logistics spectrum. There are planning functions, design activities, analysis and data processing tasks, provisioning and procurement actions, production and distribution requirements, product support and customer service activities, and so on. Human resource requirements will vary significantly in terms of personnel backgrounds, skills, and individual training requirements. Logistics, as a discipline, is not at all homogeneous in context. The management of such an activity requires the appropriate leadership characteristics and good internal communications.
3. The interdisciplinary nature of logistics requires the establishment of an effective communications link with many external organizational entities. The interfaces between the customer, the producer, and suppliers are numerous, and the successful completion of logistics tasks requires good planning and effective managerial controls.

4. The configuration makeup of many systems being developed, or currently in operational use, includes components from a wide mix of suppliers located throughout the world. The fulfillment of logistics objectives requires the establishment of good contracting arrangements with many different categories of suppliers. Further, a large proportion of this supplier activity is outside the United States. Logistics has assumed international dimensions, particularly in recent years as the trends toward globalization increase. Knowledge of customs, transportation regulations, monetary exchange rates, and the cultural peculiarities of many different nationalities is required.

The characteristics of the logistics field are such that it offers many opportunities as well as challenges for the future. A managerial approach to logistics, with these characteristics in mind, is presented in the sections to follow.

9.1 LOGISTICS PLANNING

9.1.1 Logistics in the System Life Cycle

The life cycle for a typical system will evolve from the definition of need and progress through a series of phases as illustrated in Figures 1.6 and 1.7. These phases reflect distinct categories of activity, leading from the initial determination of system requirements to the description of a specific design configuration, the production and/or construction of the system, system utilization by the customer, sustaining maintenance and support, and ultimate system retirement.

Although the acquisition process for all systems, whether large or small, will involve transition through these basic phases, the specific format and depth of activity may vary from program to program, depending on the type of system and its mission, the extent to which new design is required and the complexity of such, schedule and budget constraints, and many other related factors. Thus, the detailed requirements for a given program must be appropriately *tailored* to the particular need. There will always be planning requirements, design requirements (either at the system level or component level), production/construction requirements, and so on. However, the specific nature of such requirements will vary.

In addressing the subject of logistics, reference is made to the system acquisition process in Figure 9.1. The figure presents the basic phases of activity discussed throughout this book, identifies a few typical program milestones, and illustrates the life-cycle involvement of logistics. In conceptual design, logistics planning is initiated at program inception, and the specific requirements for logistics are developed and included in the System Type *A* Specification. These requirements are derived through the performance of a feasibility analysis, the definition of operational and maintenance concepts, and the accomplishment of early supportability analysis (SA) functions. The results of this effort form the baseline for all subsequent logistics activity for a given program. The specific tasks completed during the conceptual design phase are identified in

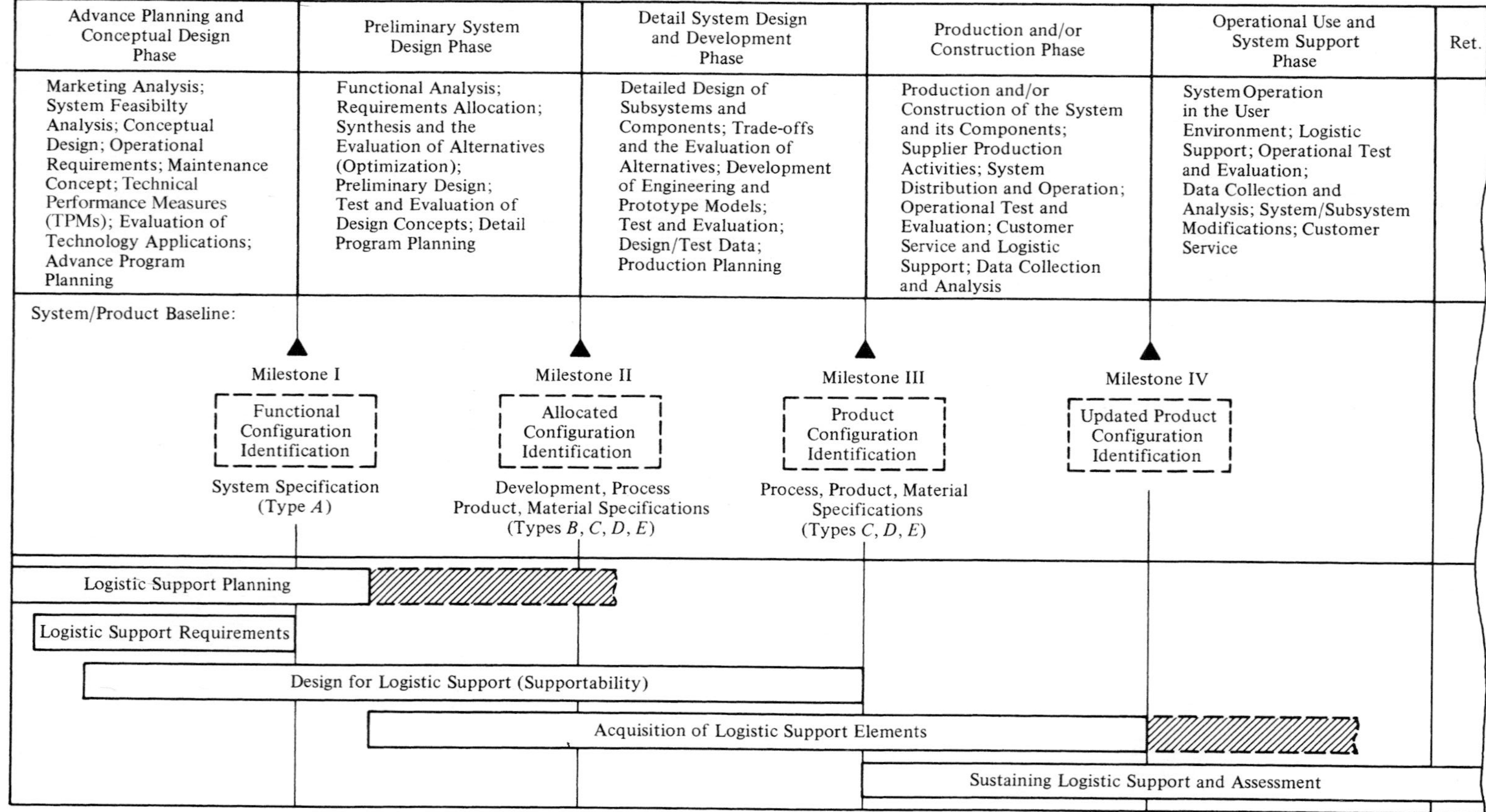

Figure 9.1 System acquisition process.

Figure 9.2 and are covered through the material included in Chapters 3 to 5. These tasks must, of course, be tailored to the particular program need.

As the program evolves through the preliminary system design phase, logistics planning continues, leading to the preparation of the formal integrated logistic support plan (ILSP) describing all follow-on logistic support activities as applicable and identified in Figure 9.2. Design-related activities include functional analysis and the allocation of requirements, the accomplishment of trade-off studies and supportability analysis (SA), design liaison and support, and conducting formal design reviews. These activities are discussed in Chapters 3 to 5.

In the detail design and development phase, significant activities include the continuation of design liaison and support functions, the supportability analysis (SA), conducting formal design reviews, and system test and evaluation. The SA includes two major thrusts. The first is the accomplishment of design trade-off studies, level of repair analyses, life-cycle cost analyses, and related activities directed toward the objective of *designing for supportability*. The second involves the evaluation of the system design configuration, as it exists at the time, with the objective of defining logistic support resource requirements (i.e., spare/repair parts, test and support equipment, maintenance personnel quantities and skill levels, etc.). With the specific requirements for logistics identified, the process of provisioning, procurement, and acquisition commences.

Logistics in the production/construction phase includes (1) the acquisition of the required elements of system support; (2) the flow of system components and supporting materials from the supplier, through the production process, and to the consumer in the field; and (3) customer service activities in support of user operations at various field sites. Given an approved design configuration, it is first necessary to acquire and then provide the necessary system support on a sustaining basis. Acquisition constitutes the provisioning, procurement, and/or manufacture of spare/repair parts, special test and support equipment, technical data, computer resources, and so on. Material flow requirements include the packaging, handling, transportation, and distribution of system components designated for operational use, along with those items required in support of maintenance activities in the field. Customer service refers to the initial installation and checkout of the system at the user site, sustaining on-site maintenance support and training (i.e., field services often provided by the producer), system evaluation and supportability assessment, and the incorporation of system modifications as required.

During the operational use and system support phase, there is an ongoing level of effort associated with system maintenance in the field, along with the continuing evaluation and assessment of the overall support capability. This assessment is provided through a data collection, analysis, and feedback capability as described in Chapter 7. Operational and maintenance data are collected, analyzed, and the results are used to update the supportability analysis (SA) data developed during the earlier phases of the life cycle. The revised SA data may be utilized for the purposes of reprovisioning and the procurement of additional spare/repair parts. Relative to evaluation, major problem areas are noted, and recommendations for corrective action and/or

Conceptual Design	Preliminary Design	Detail Design and Development	Production and Construction	System Operation, Maintenance and Support	System Retirement
• Initial logistics planning and management • Requirements analysis (customer needs study) • Feasibility analysis (tech. opportunities) • System operational requirements • System maintenance concept • Functional analysis (system level) • Supportability analysis (design criteria) • System specification (Type *A*) • Conceptual design review (functional baseline)	• Logistics planing and management — integrated logistic support plan • Functional analysis • Requirements allocation • Design participation and support activities • Supportability analysis (design tradeoffs, analyses, predictions, definition of system support requirements) • Logistics management information (design/ logistics data) • Developmental, product, process specifications (Types *B*, *C* and *D*) • Developmental test and evaluation of system components • System design reviews (allocated baseline)	• Logistics planning and management (manufacturing/customer requirements) • Design participation and support activities • Supportability analysis (design trade-offs, analyses, predictions, definition of system support requirements) • Logistics management information (design/ logistics data) • Development, product, process, material specifications (Types *B*, *C*, *D*, and *E*) • Provisioning, procurement, and acquisition of system support elements • System test and evaluation • Equipment/ software and critical design reviews	• Logistics planning and management (customer requirements) • Manufacture, procure, and/or construct prime elements of the system (materials flow, inventory control, transportation, distribution) • Manufacture, procure, and/or construct elements of system support (materials flow, inventory control, transportation, distribution) • System modifications as required • Logistics management information (revision of logistics data-supportability assessment) • System test and evaluation • Customer service (site installation and checkout, field services, and support)	• Logistics planning and Management (customer requirements) • Maintain and support the system at user sites in the field (consumer activities) • Customer service (services and materials provided by manufacturer) • Data collection, analysis, evaluation, and feedback capability (update support-ability analysis data as required) • System modification a required	• Logistics planning and management • Material phaseout, recycling, and/or disposal • Logistic support elements (as required) • Data collection, analysis, and feedback capability

Figure 9.2 Logistics activities in the life cycle.

improvement are initiated as appropriate. Finally, as the system (or a component thereof) is retired, there is a logistics function associated with the recycling or disposal of material (refer to Chapter 8).

As reflected in Figure 9.1, logistics activities extend throughout the entire system life cycle. The specific nature of these activities, as they apply to a given program, will be based on the technical requirements included in the System Type *A* Specification and the program requirements described in the integrated logistic support plan (ILSP). The system specification provides overall top-level guidance for the design and development of a new system, and is the basis for the preparation of subordinate specifications and standards applicable to that system. A simplified hierarchy of specifications is illustrated in Figure 9.3. The ILSP is discussed further in Section 9.1.3.

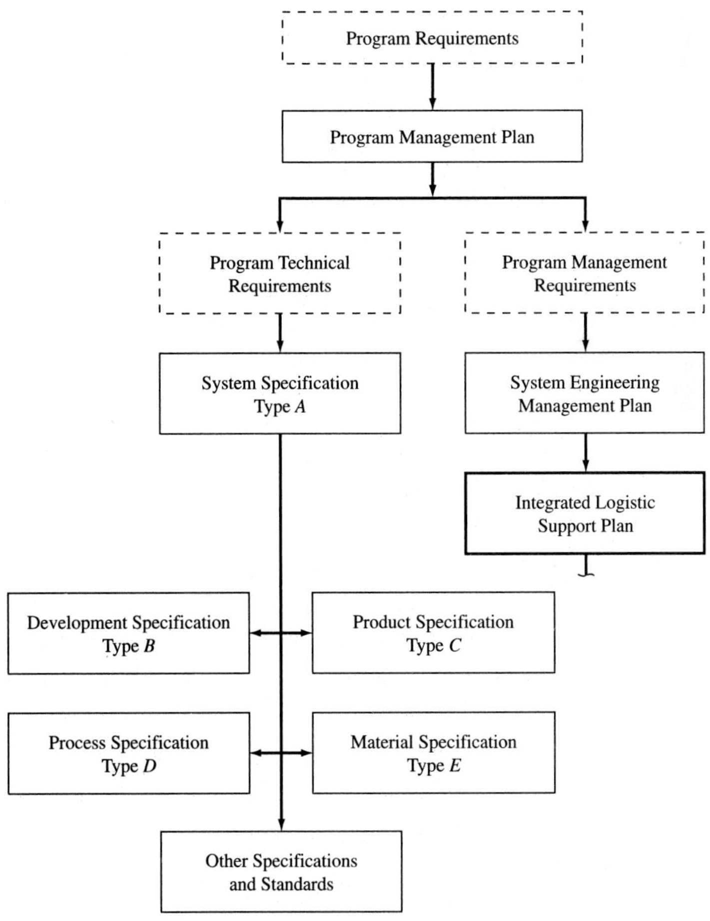

Figure 9.3 Hierarchy of technical specifications.

9.1.2 Development of Specifications

Referring to Figure 3.1, there are different classes of specifications as one proceeds through the life cycle. These are shown in a hierarchical manner as illustrated in Figure 9.3. A brief description of each class is presented.

1. *System specification (Type A)*. Includes the technical, performance, operational, and support characteristics for the system as an entity. Operational requirements, the maintenance concept, the results from the feasibility analysis, and a description of system requirements in functional terms are presented. Additionally, the logistics requirements to which the system is to be designed are described. Figure 9.4 presents a sample outline of what may be contained within the system specification.
2. *Development specification (Type B)*. Includes the technical requirements for any element or component below the system level where research, design, and development are accomplished. This may cover an equipment item, assembly, computer program, facility, critical item of support, and so on. Each development specification must include the performance, effectiveness, and the supportability characteristics that are required in the evolving of design from the system level and down.
3. *Product specification (Type C)*. Includes the technical requirements for any item below the top system level that is currently in the inventory and can be procured *off-the-shelf.* This may cover standard system components (equipment, assemblies, units, cables), a specific element of software, a spare part, a tool, and so on.
4. *Process specification (Type D)*. Includes the technical requirements that cover a process associated with any element or component of the system (e.g., machining, bending, welding, plating, heat treating, sanding, marking, packing, shipping, and so on).
5. *Material specification (Type E)*. Includes the technical requirements that pertain to raw materials, mixtures (e.g., paint, chemical compounds), and/or semifabricated materials (electrical cable, piping) that are used in the manufacture and assembly of a product.

In the preparation of specifications, which is a key design engineering task, it is important to ensure that the technical requirements covering supportability and logistics are not only included in the system specification, but are appropriately reflected in all applicable subordinate specifications. At the same time, the detailed requirements in the development, product, process, and material specifications must directly support the criteria included in the System Type *A* Specification. It is through these requirements that a life-cycle approach to the *design for supportability* is considered.

9.1.3 Development of Planning Documentation

Referring to Figure 9.1, logistics planning is initiated early in the conceptual design phase when the technical requirements for the system are being defined. Although these technical requirements are presented through the Type *A* System Specification

System Specification

1.0 Scope

2.0 Applicable Documents

3.0 Requirements

3.1 System Definition
- 3.1.1 General Description
- 3.1.2 Operational Requirements (Need, Mission, Utilization Profile, Distribution, Life Cycle)
- 3.1.3 Maintenance Concept (Levels of Repair)
- 3.1.4 Functional Analysis and System Definition
- 3.1.5 Allocation of Requirements
- 3.1.6 Functional Interfaces and Criteria
- 3.1.7 Environmental Conditions

3.2 System Characteristics
- 3.2.1 Performance Characteristics
- 3.2.2 Physical Characteristics
- 3.2.3 Effectiveness Requirements
- 3.2.4 Reliability
- 3.2.5 Maintainability
- 3.2.6 Usability (Human Factors)
- 3.2.7 Supportability
- 3.2.8 Transportability/Mobility
- 3.2.9 Flexibility
- 3.210 Other

3.3 Design and Construction
- 3.3.1 CAD/CAM/CALS Requirements
- 3.3.2 Materials, Processes, and Parts
- 3.3.3 Mounting and Labeling
- 3.3.4 Electromagnetic Radiation
- 3.3.5 Safety
- 3.3.6 Interchangeability
- 3.3.7 Workmanship
- 3.3.8 Testability
- 3.3.9 Economic Feasibility

3.4 Documentation/Data

3.5 Logistics
- 3.5.1 Maintenance Requirements
- 3.5.2 Supply Support
- 3.5.3 Test and Support Equipment
- 3.5.5 Facilities and Equipment
- 3.5.6 Packaging, Handling, Storage, and Transportation
- 3.5.7 Computer Resources (Software)
- 3.5.8 Technical Data and Information Systems
- 3.5.9 Customer Services

3.6 Producibility

3.7 Disposability

3.8 Affordability

4.0 Test and Evaluation

5.0 Quality Assurance Provisions

6.0 Distribution and Customer Service

Figure 9.4 Example Type *A* System Specification format.

(refer to Figure 9.3), the program planning requirements are identified through the development of a preliminary *integrated logistic support plan* (ILSP). This document is a "living" plan, updated on a progressive basis, and leads to the preparation of a formal ILSP during the preliminary system design phase.[1]

Logistics planning begins with the identification of program objectives and projected tasks by phase. Typical tasks for a large-scale program are presented in Figure 9.2. In essence, the objective is to (1) define the requirements and plan for logistics; (2) design the system for supportability; (3) identify, acquire, and distribute the required elements of logistics for the sustaining life-cycle support of the system during operational use; and (4) measure and evaluate, or assess, the effectiveness of the system support capability throughout the consumer use period.

These objectives and program requirements are fulfilled through the implementation of the ILSP. Figure 9.5 depicts the basic areas that should be covered in a for-

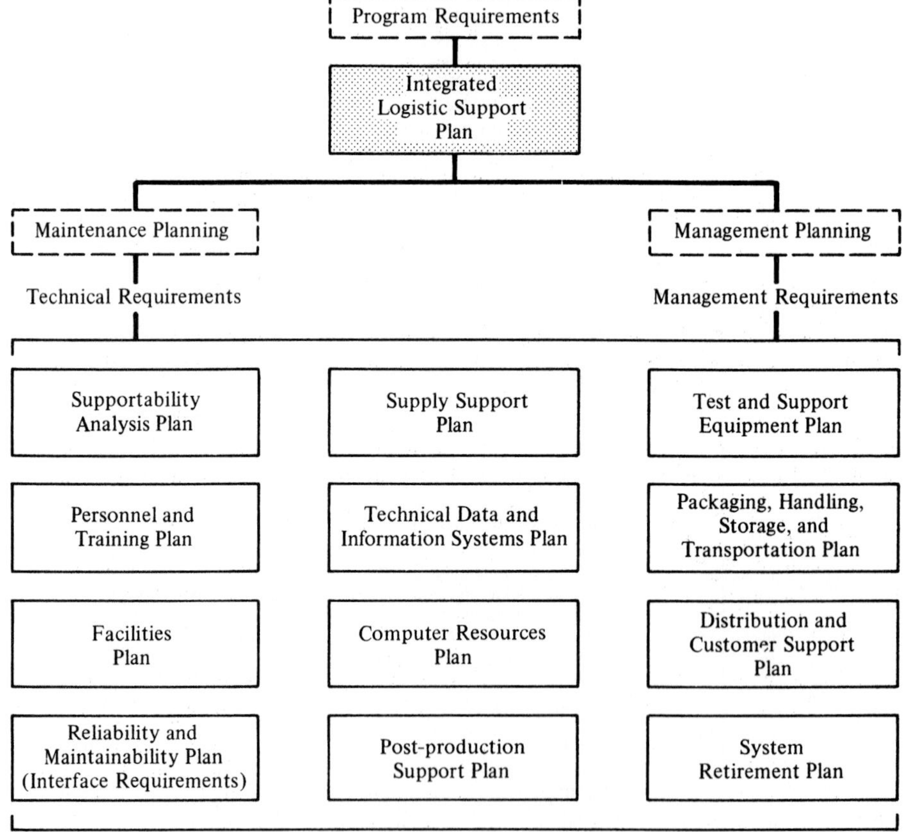

Figure 9.5 Integrated logistic support plan development.

[1]The ILSP, as described herein, is considered to be the top planning and management document for logistics requirements as they are applied to a large-scale program. It is recognized that, for some programs, logistics planning may be covered by an *integrated support plan* (ISP) or a *supportability analysis plan* (SAP), depending on the program size, status, and nature of the system being developed. Whatever the case, a planning document tailored to the program need must be developed.

Integrated Logistic Support Plan

1.0 Introduction
- 1.1 Purpose and Scope
- 1.2 System Description and Background
- 1.3 Logistics Acquisition Strategy
- 1.4 References and Specifications

2.0 System Characteristics
- 2.1 System Operational Requirements
- 2.2 System Maintenance Concept
- 2.3 Functional Analysis and Allocation
- 2.4 Logistics Research Results
- 2.5 Feasibility Studies
- 2.6 Other Requirements

3.0 ILS Planning and Management
- 3.1 Program Requirements
- 3.2 ILS Organizational Structure
- 3.3 Supplier/Subcontractor Organization(s)
- 3.4 Organizational Interfaces
- 3.5 Work Breakdown Structure
- 3.6 Task Schedules and Major Milestones
- 3.7 Cost Projections
- 3.8 Program Reviews, Evaluation, and Control
- 3.9 Technical Communications (Reports and Documentation)
- 3.10 Risk Management

4.0 ILS Element Plans
- 4.1 Supportability Analysis Plan
- 4.2 Reliability and Maintainability Plan
- 4.3 Supply Support Plan
- 4.4 Test and Support Equipment Plan
- 4.5 Personnel and Training Plan
- 4.6 Technical Data and Information Systems Plan
- 4.7 Packaging, Handling, Storage, and Transportation Plan
- 4.8 Facilities Plan
- 4.9 Computer Resources Plan
- 4.10 Distribution and Customer Support Plan
- 4.11 Post-production Support Plan
- 4.12 System Retirement Plan

Figure 9.6 Integrated logistic support plan format (example).

mal logistics plan, and Figure 9.6 presents a sample ILSP outline. For the purposes of further clarification, a brief description of selected sections of proposed document content follows.

Detailed maintenance plan. The *maintenance plan*, as defined here, refers to the overall plan for system/product support, primarily from the technical requirements point of view. It constitutes the technical baseline used in the development of the detailed plans identified in Figure 9.5. Specifically, it includes a description of the proposed levels of maintenance, the responsibilities for maintenance (producer versus customer functions), system-level criteria for the development of the various elements of logistics (e.g., spare/repair parts, test and support equipment, personnel quantities and skills), effectiveness factors pertaining to the support capability, the overall distribution and flow of materials, the maintenance environment, and so on. Where the maintenance concept (described in Chapter 3) is primarily a before-the-fact input to the design process, the detailed maintenance plan is based on the results of design (i.e., a given system configuration with supporting analysis data).

The maintenance plan, while evolving from the maintenance concept generated during conceptual design, stems from the results of the supportability analysis (SA) and is updated on a progressive basis throughout the detail design and development phase. The plan serves as a basis for system life-cycle support, it covers the transition from contractor to customer support, and it includes procedures pertaining to maintenance operations in the field.

Referring to the ILSP outline in Figure 9.6, Section 2.0 is intended to provide the necessary technical information required for maintenance planning. Initially in the development of the ILSP, it is appropriate to include (in a summary manner) the system operational requirements, the maintenance concept, a top-level functional analysis, the results from logistics research and feasibility studies, and the like. As system development progresses and the design configuration becomes better defined, this section of updated versions of the ILSP may evolve into a detailed maintenance plan.

Supportability analysis plan (SAP). The SAP serves as a basic planning and management document for those logistics activities that are accomplished primarily in the system acquisition process (as compared to the total spectrum of logistics functions throughout the life cycle). These activities address the logistics requirements definition process, major design interface functions, trade-offs and analyses, logistics modeling, data collection and processing, design review and evaluation, and so on. The SAP is generally applicable in the acquisition of large-scale defense systems, prepared during the conceptual design phase and updated during the preliminary system design phase, and is implemented by the prime contractor. It may be included as part of the ILSP, or utilized as a separate planning document in the acquisition of major developmental items. In any event, extreme care must be exercised to ensure that the SAP directly supports the ILSP (where applicable), and that redundancies between the two are eliminated.

Reliability and maintainability plan. Reliability and maintainability requirements are inherent to meeting the objectives of logistics. As noted throughout this book, reliability and maintainability characteristics in design are closely interrelated with the goals in *design for supportability*. Reliability and maintainability analyses and predictions are an integral part of the supportability analysis (SA) effort. These disciplines, along with human factors and others, must be integrated with those design-related logistics functions accomplished throughout the system acquisition process.

In many organizations, reliability and maintainability program tasks are accomplished by departments or groups outside the logistics organization, and individual plans are developed in response to these requirements. These plans cover the activities described in Chapters 3 to 5: reliability and maintainability allocation, analyses and predictions, design review, test and demonstration, and so on. As such, it is particularly important that they be closely integrated with the other element plans within the ILSP. In other words, the ILSP should either include a reliability and maintainability plan, or a section covering the interface requirements and referencing these plans.

Supply support plan. The requirements for supply support are determined through the supportability analysis (SA) process. This plan identifies these requirements and describes the methods/procedures for the provisioning (item identification, cataloging, source coding), procurement, and acquiring of spare/repair parts, inventories, and consumable materials for both the early initial interim support and for the long-term support of the system throughout its planned life cycle. This plan, covering both situations as applicable, includes the following:

1. A summary listing of significant spare/repair parts and the consumable materials required for each level of maintenance (i.e., organizational, intermediate, depot, or supplier).
2. A plan for the procurement and acquisition of new (nonstocklisted) spares and consumable materials for supporting the prime mission-oriented equipment, test and support equipment, training equipment, facilities, and software. Special supplier requirements dealing with manufacturing and test, packaging and handling, transportation, and related issues should be addressed.
3. A plan for the procurement and acquisition of common and standard off-the-shelf (COTS) spares and consumable materials for the prime mission-oriented equipment, test and support equipment, training equipment, facilities, and software. Special provisions associated with *government furnished equipment* (GFE), for defense systems, should be covered.
4. Warehousing and accountability functions associated with the ongoing maintenance and support of the system. This includes the initial cataloging and stocking, inventory maintenance and control, procurement of replacement items, associated facility requirements, and the disposition of condemned items and residual assets.
5. A plan for data collection, analysis, and the updating of spare/repair part (and consumable materials) demand factors necessary for improving the procurement cycles and reducing waste. This constitutes the necessary feedback process required for the true assessment of the system support capability and the updating of the SA. Compatibility with the CALS requirements (as applicable) is essential.[2]

[2] Recently, experiences have pointed to instances where, in the defense sector, there are conditions of overstocking of spares, the procurement of spare/repair parts too early and/or too late, the stocking of

Test and support equipment plan. The requirements for test equipment, support equipment, ground handling equipment, calibration equipment, fixtures, and so on, are determined through the supportability analysis (SA) process. This plan identifies these requirements and describes the methods and procedures for acquiring the necessary equipment, testing it, and distributing it to the consumer in the field. Specifically, this includes the following:

1. A summary listing of all recommended test and support equipment, ground handling and calibration equipment, and significant tools for each level of maintenance (i.e., organizational, intermediate, depot, or supplier).
2. A plan for the procurement and acquisition of new items to be designed and developed, tested, produced, and delivered for consumer use. This should reference test requirements specifications (TRSs), the results of *make-or-buy* decisions, supplier contractual requirements, quality assurance provisions, warranty requirements, and maintenance requirements for each item being acquired (e.g., calibration requirements).
3. A plan for the procurement and acquisition of common and standard off-the-shelf test and support equipment and associated accessories. This should reference test requirements specifications (TRSs), supplier contractual requirements, quality assurance provisions, warranty requirements, and the maintenance requirements for each item (e.g., calibration requirements). Special provisions associated with government furnished equipment (GFE) for defense systems should be covered.
4. A plan for the acquisition of computer resources (software) as required to support the newly developed and common/standard test and support equipment identified under items 2 and 3.
5. A plan for integration, test, and evaluation to ensure compatibility between the prime elements of the system and the elements of support (equipment, software, personnel, and procedures).
6. A plan for the delivery of test and support equipment, associated software, data, and so on, to the customer, and the follow-on installation and checkout at each geographical location.

Test and support equipment requirements can range from a large intermediate-level shop full of rather sophisticated test stations to a few common and standard items of checkout equipment. In the event of the first case, this plan may be rather extensive (e.g., a complete ILSP just for the intermediate-level shop alone). However, in most instances, the coverage herein should be adequate.

Personnel and training plan. The specific requirements for maintenance and support personnel in terms of quantities, skill levels, and job classifications by location are determined through the supportability analysis (SA) process. These requirements are

spares that are no longer needed, and so on. This, of course, results in unnecessary waste. Thus, it is important that the feedback process, conveying information on actual experiences in the field, be effective.

compared with the quantities, skills, and job classifications currently within the user's organization, and the results lead to the development of a personnel training plan (i.e., the formal training necessary to bring the user personnel skills to the level specified for the system). Although maintenance training is emphasized here, system operator training is sometimes included. This plan should cover:

1. The training of system operators—type of training, length, basic entry requirements, brief program/course outline, and output expectations. System operator requirements are often determined through system engineering and/or human factors program requirements.
2. The training of maintenance personnel for all levels—type of training, length, basic entry requirements, brief program/course outline, and output expectations.
3. Training equipment, devices, aids, simulators, computer resources, facilities, and data required to support operator and maintenance personnel training.
4. Proposed schedule for initial operator and maintenance personnel training, and for the accomplishment of replenishment training throughout the system life cycle (for replacement personnel).

Technical data plan. Technical data constitute system operating instructions, maintenance and servicing procedures, installation and checkout procedures, calibration procedures, overhaul instructions, change notices and data covering system modifications, and so on. The final data format may vary from the traditional text material to a digital-formatted database developed for compatibility with CALS requirements. This plan should include the following:

1. A description of the technical data requirements for each level of activity (operator plus levels of maintenance) by system element. This may include the operating instructions for the system, maintenance procedures for Unit *B*, overhaul instructions for Assembly 1, and so on. It may be appropriate to present these data items in the form of a documentation tree, or to relate them in some form of a hierarchial structure, particularly if precedence exists.
2. A schedule for the development of each significant data item.
3. A plan for the verification and validation of operating and maintenance procedures.
4. A plan for the preparation of change notices and for the incorporation of changes/ revisions to the technical manuals.

Packaging, handling, storage, and transportation plan. This plan is developed to cover the basic distribution and transportation methods/ procedures for the shipment of system components from the producer to the consumer, for the shipment of elements of logistic support to operational sites, and for the shipment of items requiring maintenance support. Specific transportation and handling requirements are initially derived from the maintenance concept and supported through the supportability analysis (SA) process. This plan includes the following:

1. A summary listing of special categories of items requiring transportation.

2. Proposed mode(s) of transportation based on anticipated demand rates, available routing, weight and size of items, cost-effectiveness criteria, and so on. This shall consider both first destination requirements as dictated by material delivery schedules and recurring transportation requirements based on maintenance support needs.
3. Proposed methods of packaging items for shipment (types of containers—reusable versus nonreusable, security measures, environmental protection provisions).
4. Safety criteria, precautions, and provisions for handling, storage, and preservation of material.

In today's environment, logistics is an international business, and transportation requirements often involve the worldwide distribution of materials. Systems and equipment are often shipped across national boundaries, over land and sea, and so on. In the planning for such, one needs to be somewhat conversant with international law, local transportation and the customs requirements in different countries, and the economic factors that relate to the transportation of goods and services across nationalistic boundaries.

Facilities plan. This plan is developed to identify all real property, plant, warehouse and/or maintenance facilities, and capital equipment required to support system testing (as related to supportability assessment), personnel training, operation, and logistics functions. The plan must contain sufficient qualitative and quantitative information to allow facility planners to

1. Identify requirements for facilities (location, environmental concerns, type of construction, space, and layout), and define utility requirements such as power and electricity, environmental controls, water, telephone, and so on.
2. Identify capital equipment needs (material handling provisions, assembly and test processing equipment, warehousing shelves, etc.).
3. Evaluate existing facilities and associated capital items, and assess adequacies/inadequacies in terms of meeting the need.
4. Estimate the cost of facility acquisition (cost of constructing a new facility or modifying an existing facility), and the cost of capital equipment to meet the needs of the system.

The facilities plan should include appropriate criteria to ensure that facility design is completely compatible with the prime equipment and its support elements. The plan must include the necessary scheduling information to permit the proper and timely implementation of any required civil engineering activity.

Computer resources plan. With the advent of the ever-increasing amount of software utilized in today's programs, particularly for large systems, it is appropriate to include a computer resources plan within the ILSP. This plan includes the following:

1. Identification of all computer programs and software required for system support. This covers automated condition monitoring programs, maintenance diagnostic routines, information processing systems involving logistics data, and so on.
2. Definition of computer language requirements, specifications, and compatibility requirements with existing programs (to include CALS).
3. Acquisition procedures for the development of or procurement of new software.
4. Software configuration management procedures and quality assurance provisions.
5. Software change procedures and change management.
6. Hardware (i.e., computers and associated accessories) required to interface with the software requirements.

For some systems, the software element represents a large segment of total system support cost and the necessary planning and management control of such is essential.

Distribution and customer support plan. For large systems, in particular, a significant transition often occurs when evolving from production/construction operations to a full-scale user capability. A rapid buildup of equipment in the field often occurs while, at the same time, there is a lack of adequately trained personnel and not enough spare/repair parts, support equipment, technical data, and so on, available for adequate system support. There is usually a designated period of time after the system initially becomes "operational" when an *interim contractor support capability* is required. The purpose of this plan is to define the requirements for distributing system components, to describe the extent to which initial customer support is needed (schedule, duration, quantity of personnel and spares, etc.), and to describe the process for transitioning to full-scale user operations.

Postproduction support plan. This plan is directed toward covering the logistic support for a system (or components thereof) after the basic production capability for that system has been discontinued and phased out. For instance, where will the consumer be able to acquire additional spare parts 8 to 10 years from now when the initial producer is no longer in the business? This plan must address contingencies of this nature.

System retirement plan. This plan covers the retirement and phaseout of equipment, the disassembly and reclamation of material items as appropriate, and the ultimate disposal of residual material. In some cases, the process of phaseout could be rather extensive, including the requirements for logistic support. The plan should identify the requirements for transportation and handling, the support equipment necessary for material processing, personnel, facilities, and the data necessary for the processing of items out of the inventory.

9.2 DEVELOPMENT OF A WORK BREAKDOWN STRUCTURE (WBS)[3]

Given the identification of logistics program tasks in Figure 9.2, it is appropriate to integrate these tasks into a work breakdown structure (WBS). The WBS is a product-oriented family tree that leads to the identification of the activities, functions, tasks, work packages, and so on, that must be performed for the completion of a given program. Tasks are evaluated (in terms of type, commonality, and complexity), and are combined into work packages. Individual work packages are then identified with specific blocks in the WBS.

The WBS is a logical separation of work-related units, linking objectives and activities with resources. During the early stages of program planning, a *summary work breakdown structure* (SWBS) is generally prepared by the customer and is included in a request for proposal (RFP) or in an invitation for bid (IFB). This structure is developed from the top down for budgetary and reporting purposes, covers all program functions, and generally includes three levels of activity. As program planning progresses and individual contract negotiations are consummated between the customer and the contractor, the SWBS is developed further and adapted to a particular contract or procurement action, resulting in a *contract work breakdown structure* (CWBS). The CWBS is tailored to specific contract requirements and will be different for each phase of program activity. Figure 9.7 illustrates the process of WBS development, and Figure 9.8 shows a partial CWBS.

Referring to Figure 9.8, work packages are identified against each block (see block 3A1200). These packages and blocks are then related to organizational groups, departments, branches, suppliers, and so on. Cost estimating is accomplished for each work package and identified by WBS block. The WBS is structured and coded in such a manner that program costs may be initially targeted and then collected against each block. Costs may be accumulated both vertically and horizontally to provide summary figures for various categories of work. The cost data, combined with the appropriate schedules, provide management with the necessary tools for project evaluation and control.[4]

9.3 SCHEDULING OF LOGISTICS TASKS

The scheduling of logistics tasks may be accomplished using one or a combination of techniques such as the simple bar chart, milestone chart, program network, Gantt chart, and line of balance. Several of these techniques are discussed in the sections to follow.

[3] The subjects of *work breakdown structure* (WBS) and *work packaging* are covered in most textbooks dealing with project management. A good reference is Kerzner, H., *Project Management: A Systems Approach to Planning, Scheduling, and Controlling*, 5th Ed., Van Nostrand Reinhold Company, Inc., New York, N.Y., 1995.

[4] In terms of cost collection, the WBS and cost breakdown structure used in life-cycle cost analyses are quite similar. The WBS serves as a program management tool and is often used for contracting purposes. The cost breakdown structure covers all costs and is used in performing system/equipment analysis. Although their purposes are quite different, many of the same cost input factors are the same. The two efforts should be coordinated to the maximum extent possible to ensure compatibility where applicable.

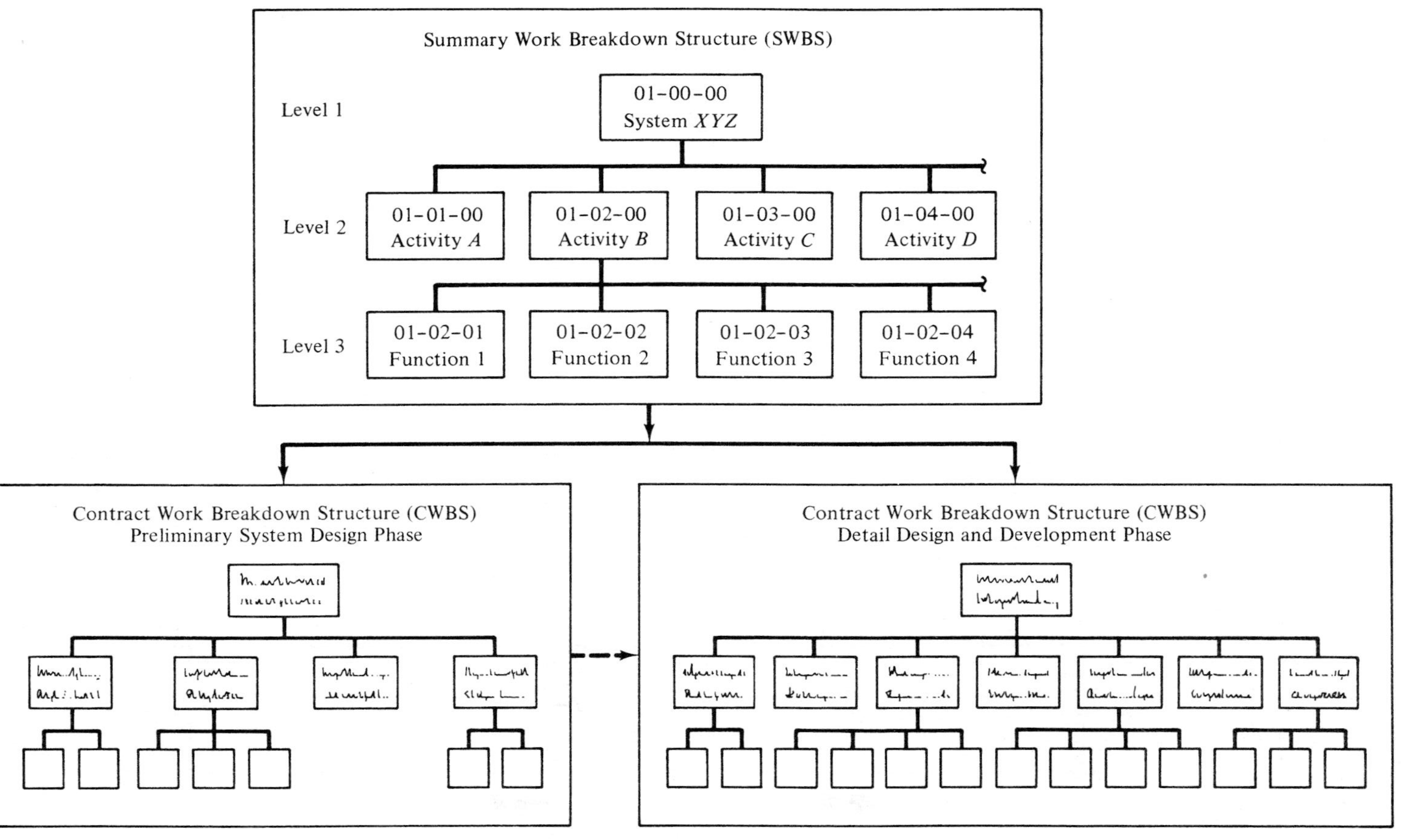

Figure 9.7 Work breakdown structure development (partial).

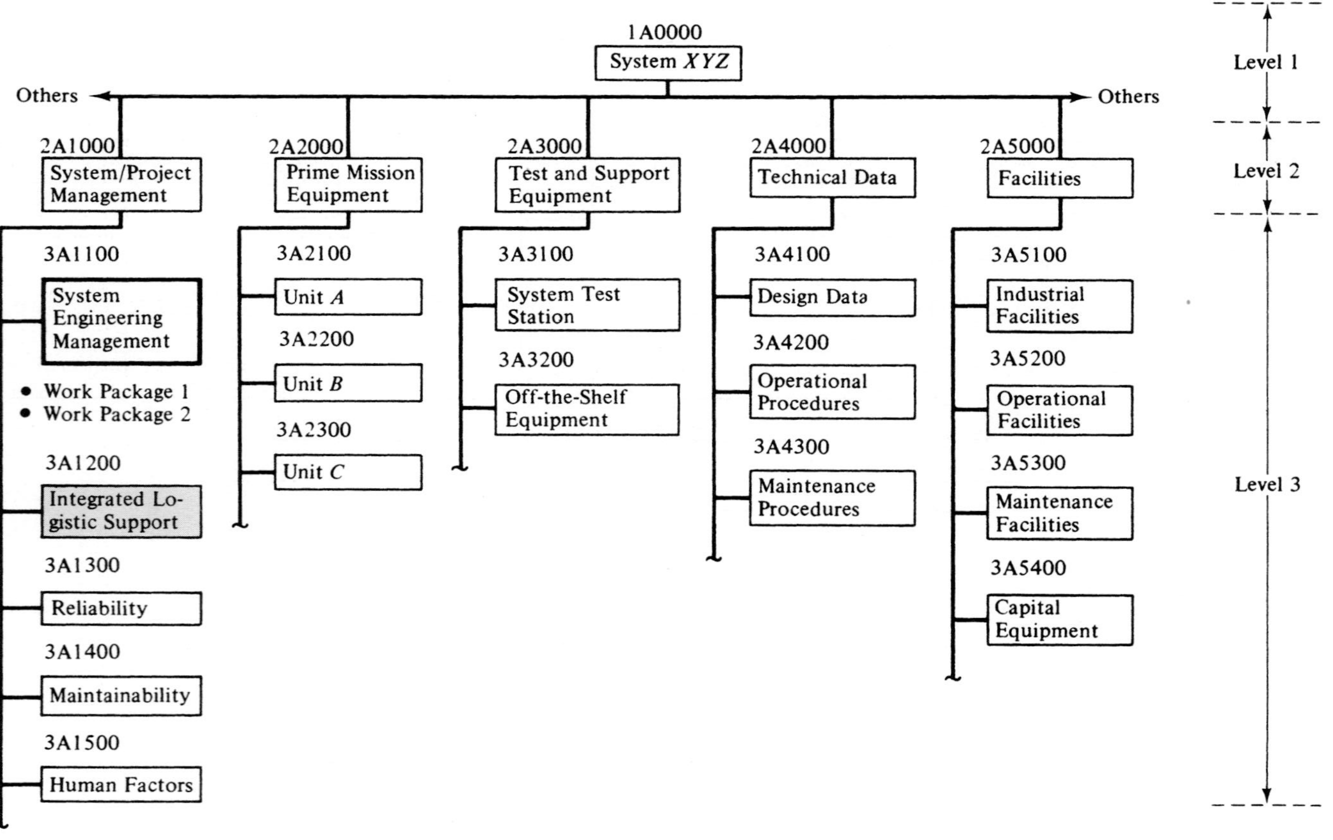

Figure 9.8 Partial work breakdown structure (sample).

Bar Chart/Milestone Chart

A simple bar chart includes the presentation of tasks in a sequential format showing the time span of effort for each activity. The milestone chart includes a presentation of program events in terms of required start and completion times. Specific identifiable outputs are shown. An example of each is illustrated in Figure 9.9.

One of the most popular scheduling approaches is to combine the two formats as shown in Figure 9.10. The information presented in the figure represents an extension of the logistics tasks in Figure 9.2, laid out in a timeline through several phases of the life cycle.

Although the bar chart/milestone chart format is one of the most popular scheduling techniques for engineering design and early logistic support activities, this technique does not always force the integration (or show the interrelationships) of the many and varied tasks required in fulfilling the requirements for most logistics programs. Logistics includes a wide variety of functions, involving many different organizational entities, and the interrelationships are numerous. Thus, it may be appropriate to utilize a different technique such as the program network approach discussed subsequently.

Program Network Scheduling

The *program evaluation and review technique* (PERT) is a scheduling approach that combines events and activities into a program network. This technique is ideally suited to early planning where precise time data are not readily available and is effective in showing the interrelationships of combined activities. PERT introduces the aspects of probability by which better management decisions can be made. With these characteristics, the technique is well suited to defining a logistics program, particularly regarding the numerous and varied program activities that must be properly integrated. A partial example of PERT scheduling is illustrated in Figure 9.11.[5]

In applying PERT to a project, one must identify all interdependent activities and events for each applicable phase of the project. Events (indicated by the circles in Figure 9.11) are related to program milestone dates that are based on management objectives. Managers and programmers work with project organizational elements to define these objectives and identify tasks and subtasks. When this is accomplished to the necessary level of detail, networks are developed, starting with a summary network and working down to detailed networks covering specific segments of a program. The development of PERT networks is a team effort.[6]

[5] Three good references covering scheduling methods are (1) Kerzner, H., *Project Management: A Systems Approach to Planning, Scheduling, and Controlling*, 5th Ed., Van Nostrand Reinhold, New York, N.Y., 1995; (2) Ullmann, J. E., (Ed.), *Handbook of Engineering Management*, John Wiley, New York, N.Y., 1986; (3) and Cleland, D. I., and King, W. R., *Project Management Handbook*, 2nd Ed., Van Nostrand Reinhold, New York, N.Y., 1989.

[6] The level of detail and depth of PERT network development (i.e., the number of activities and events included) are based on the criticality of tasks and the extent to which program monitoring and control is desired. Milestones which are critical in meeting the objectives of the program should be included along with activities that require extensive interaction for successful completion. The number of events will of course vary with each project.

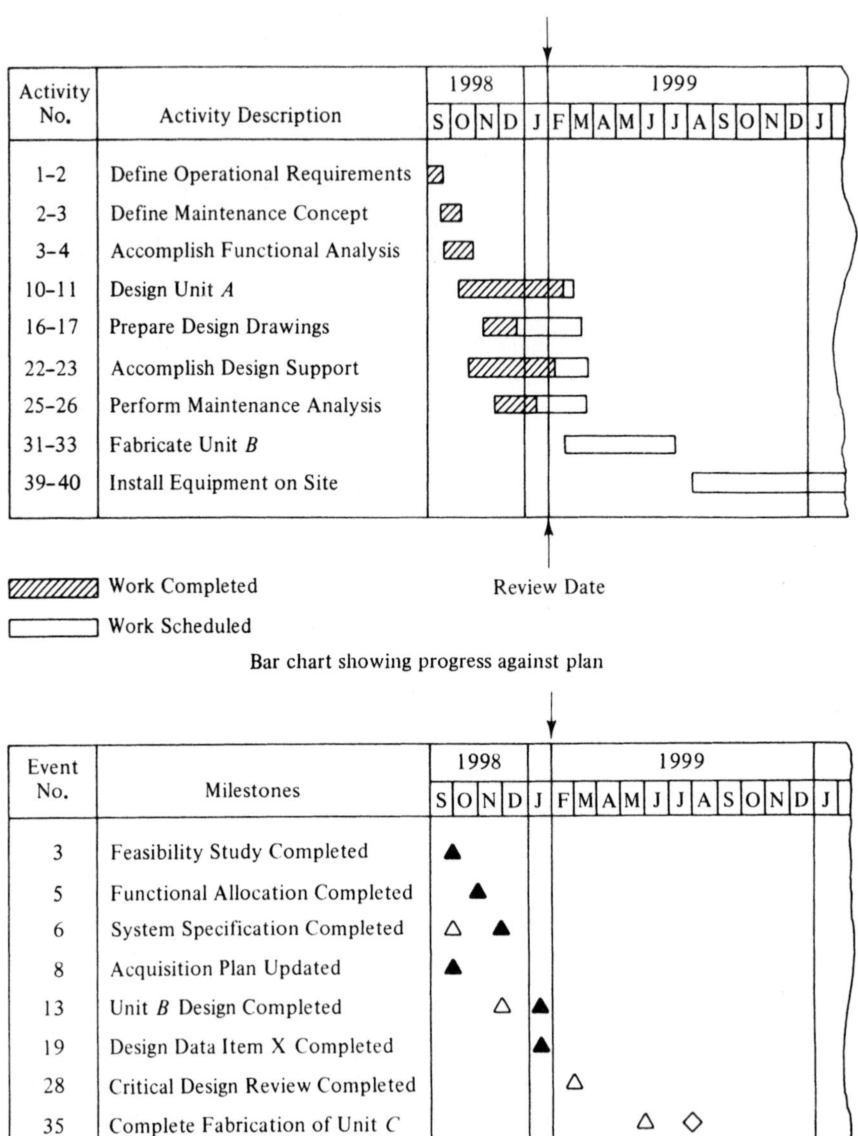

Figure 9.9 Partial bar and milestone charts.

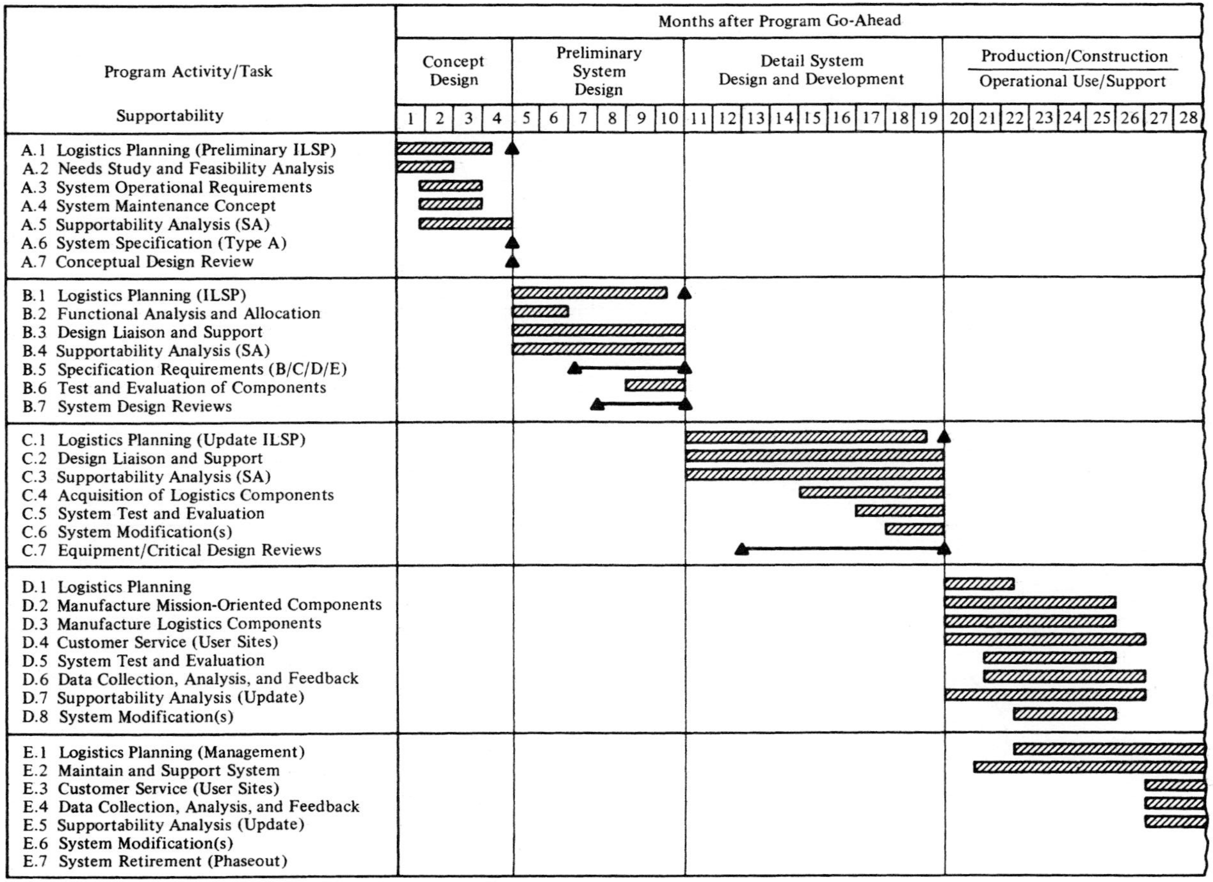

Program Activity/Task	Months after Program Go-Ahead			
	Concept Design	Preliminary System Design	Detail System Design and Development	Production/Construction — Operational Use/Support
Supportability	1 2 3 4	5 6 7 8 9 10	11 12 13 14 15 16 17 18 19	20 21 22 23 24 25 26 27 28
A.1 Logistics Planning (Preliminary ILSP)				
A.2 Needs Study and Feasibility Analysis				
A.3 System Operational Requirements				
A.4 System Maintenance Concept				
A.5 Supportability Analysis (SA)				
A.6 System Specification (Type A)				
A.7 Conceptual Design Review				
B.1 Logistics Planning (ILSP)				
B.2 Functional Analysis and Allocation				
B.3 Design Liaison and Support				
B.4 Supportability Analysis (SA)				
B.5 Specification Requirements (B/C/D/E)				
B.6 Test and Evaluation of Components				
B.7 System Design Reviews				
C.1 Logistics Planning (Update ILSP)				
C.2 Design Liaison and Support				
C.3 Supportability Analysis (SA)				
C.4 Acquisition of Logistics Components				
C.5 System Test and Evaluation				
C.6 System Modification(s)				
C.7 Equipment/Critical Design Reviews				
D.1 Logistics Planning				
D.2 Manufacture Mission-Oriented Components				
D.3 Manufacture Logistics Components				
D.4 Customer Service (User Sites)				
D.5 System Test and Evaluation				
D.6 Data Collection, Analysis, and Feedback				
D.7 Supportability Analysis (Update)				
D.8 System Modification(s)				
E.1 Logistics Planning (Management)				
E.2 Maintain and Support System				
E.3 Customer Service (User Sites)				
E.4 Data Collection, Analysis, and Feedback				
E.5 Supportability Analysis (Update)				
E.6 System Modification(s)				
E.7 System Retirement (Phaseout)				

Figure 9.10 Combined bar/milestone chart (example).

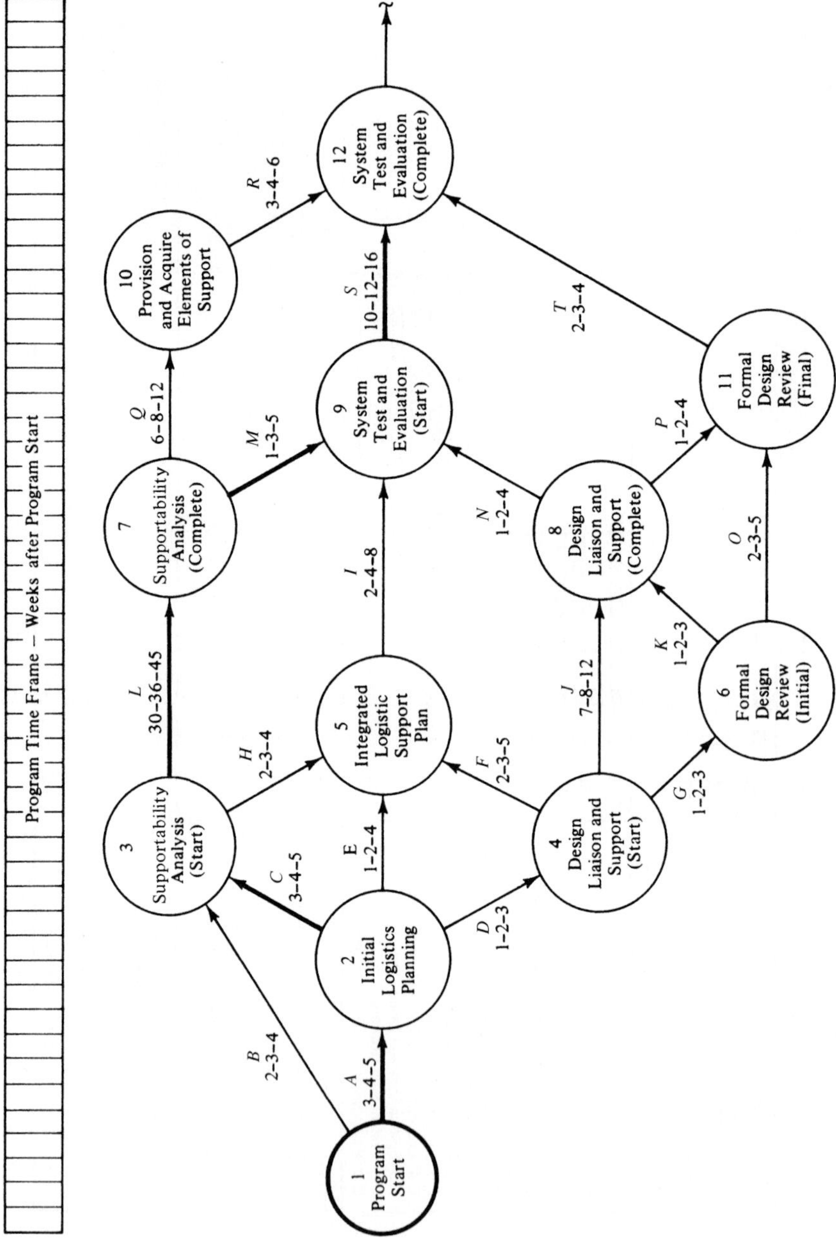

Figure 9.11 Partial summary program network.

When constructing PERT networks, one starts with an end objective (e.g., Event 12 in Figure 9.11) and works backward in developing the network until Event 1 is identified. Each event is labeled, coded, and checked in terms of program time frame. Activities (represented by the lines between the event circles in the figure) are then identified and checked to ensure that they are properly sequenced. Some activities can be performed on a concurrent basis, while others must be accomplished in series. For each completed network, there is *one beginning event* and *one ending event*, and all activities must lead to the ending event. Figure 9.12 is presented to identify the sample activities illustrated in Figure 9.11.

The next step in developing a program network is to estimate activity times, and to relate these times in terms of probability of occurrence. An example of the calculations that support a typical PERT network is presented in Figure 9.13 and described subsequently. The following steps are appropriate:

1. *Column 1.* List each event, starting from the last event and working backward to the beginning (i.e., from Event 12 to Event 1 in Figure 9.11).
2. *Column 2.* List all previous events that lead into, or are shown as being prior to, the event listed in Column 1 (e.g., Events 9, 10, and 11 lead to Event 12).
3. *Columns 3 to 5.* Determine the optimistic time (t_a), the most likely time (t_b), and the pessimistic time (t_c) in weeks or months for each activity. Optimistic time means that there is very little chance that the activity can be completed before this time, while pessimistic time means that there is little likelihood that the activity will take longer. The most likely time (t_b) is located at the highest probability point or the peak of the distribution curve. These times may be predicted by someone who is experienced in estimating. The time estimates may follow different distribution curves, where P represents the probability factor.

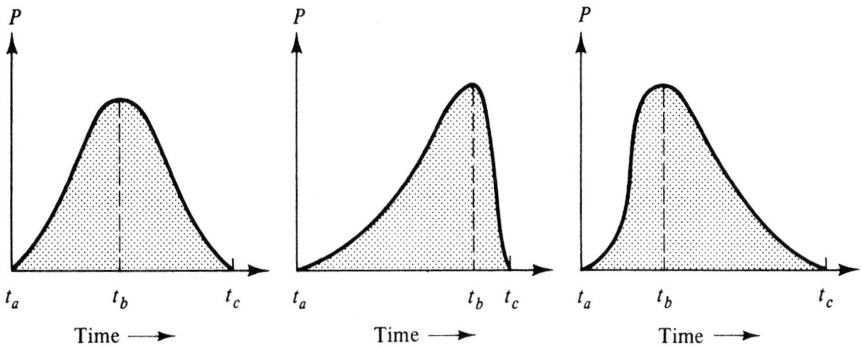

4. *Column 6.* Calculate the expected or mean time, t_e, from

$$t_e = \frac{t_a + 4t_b + t_c}{6} \tag{9.1}$$

5. *Column 7.* In any statistical distribution, one may wish to determine the various probability factors for different activity times. Thus, it is necessary to compute

Activity	Description of Program Activity
A	Complete initial logistics planning activity.
B	Perform needs analysis and feasibility study; define system operational requirements and maintenance concepts; and commence with the supportability analysis (SA).
C	Prepare the supportability analysis plan (SAP), based on early logistics program planning information.
D	Identify design liaison and support activities, and commence with design support task.
E	Prepare the formal integrated logistic support plan (ILSP), based on early logistics program planning information.
F	Prepare system design data input for the formal integrated logistic support plan (ILSP).
G	Prepare design material for formal design review.
H	Prepare supportability analysis (SA) data for formal integrated logistic support plan (ILSP).
I	Prepare integrated system test and evaluation plan, based on integrated logistic support planning information.
J	Complete design liaison and support activity.
K	Prepare design review and evaluation recommendations for design consideration.
L	Complete the supportability analysis (SA) task, as it relates to design evaluation and the determination of specific logistic support requirements.
M	Prepare supportability analysis (SA) data required for the system test and evaluation plan.
N	Prepare design material for the system test and evaluation plan.
O	Prepare for the final formal design review, based on the results of the initial formal design review.
P	Prepare design material for the final formal design review (requiring data bases, drawings/layouts, part lists, prediction and analysis results, etc.).
Q	Provision and acquire (through "make or buy") the necessary elements of logistic support required for system test and evaluation, based on supportability analysis (SA) data.
R	Provide the required elements of logistic support for system test and evaluation (for the assessment of system supportability).
S	Complete the system test and evaluation activity, and prepare a final test report.
T	Prepare design documentation, material, recommendation, etc., resulting from the formal design review for consideration in the overall system test and evaluation effort.

Figure 9.12 List of activities in program network.

the variance (σ^2) associated with each mean value. The square root of the variance, or the standard deviation, is a measure of the dispersion of values within a distribution, and is useful in determining the percentage of the total population

1	2	3	4	5	6	7	8	9	10	11	12
Event Number	Previous Number	t_a	t_b	t_c	t_e	σ^2	TE	TL	TS	TC	Probability (%)
12	11	2	3	4	3.0	0.111	–	–	–	55	4.40%
	10	3	4	6	4.2	0.250	–	–	–	60	52.80%
	9	10	12	16	12.3	1.000	59.8	59.8	0	65	96.77%
11	8	1	2	4	2.2	0.250	16.7	56.8	40.1		
	6	2	3	5	3.2	0.250	–	–	–		
10	7	6	8	12	8.3	1.000	52.8	55.6	2.8		
9	8	1	2	4	2.2	0.250	–	–	–		
	7	1	3	5	3.0	0.444	47.5	47.5	0		
	5	2	4	8	4.3	1.000	–	–	–		
8	6	1	2	3	2.0	0.111	–	–	–		
	4	7	8	12	8.5	0.694	14.5	45.3	30.8		
7	3	30	36	45	36.5	6.250	44.5	44.5	0		
6	4	1	2	3	2.0	0.111	8.0	43.3	35.3		
5	4	2	3	5	3.2	0.250	–	–	–		
	3	2	3	4	3.0	0.111	11.0	43.2	32.2		
	2	1	2	4	2.2	0.250	–	–	–		
4	2	1	2	3	2.0	0.111	6.0	36.8	30.8		
3	2	3	4	5	4.0	0.111	8.0	8.0	0		
	1	2	3	4	3.0	0.111	–	–	–		
2	1	3	4	5	4.0	0.111	4.0	4.0	0		

Note: Assume that time values for t_a, t_b, etc., are in weeks.

Figure 9.13 Example of network calculations.

sample that falls within a specified band of values. The variance is calculated from Equation (9.2):

$$\sigma^2 = \left(\frac{t_c - t_a}{6}\right)^2 \tag{9.2}$$

6. *Column 8.* The earliest expected time for the project, TE, is the sum of all times, t_e, for each activity, along a given network path, or the cumulative total of the expected times through the preceding event remaining on the same path throughout the network. When several activities lead to an event, the highest time value (t_e) will be used. For instance, in Figure 9.11, Path 1, 3, 7, 9, 12 totals 54.8; Path 1, 2, 3, 7, 9, 12 totals 59.8; and Path 1, 2, 4, 8, 11, 12 totals 19.7. The highest value for TE (if one were to check all network paths) is 59.8 weeks, and this is the value selected for Event 12. The TE values of Events 11, 10, and so on, are calculated in a similar manner, working backward to Event 1.
7. *Column 9.* The latest allowable time for an event, TL, is the latest time for completion of the activities that immediately precede the event. TL is calculated by starting with the latest time for the last event (i.e., where TE equals 59.8 in Figure 9.13 and working backward subtracting the expected time (t_e) for each activity, remaining on the same path. The TL values for Events 11, 10, and so on, are calculated in a similar manner.

8. *Column 10.* The slack time, TS, is the difference between the latest allowable time (TL) and the earliest expected time (TE):

$$\text{TS} = \text{TL} - \text{TE} \tag{9.3}$$

9. *Columns 11 and 12.* TC refers to the required scheduled time for the network based on the actual need. Assume that management specifies that the project reflected in Figure 9.11 must be completed in 55 weeks. It is now necessary to determine the likelihood, or probability (P), that this will occur. This probability factor is determined as follows:

$$Z = \frac{\text{TC} - \text{TE}}{\sqrt{\Sigma \text{ path variance}}} \tag{9.4}$$

where Z is related to the area under the normal distribution curve, which equates to the probability factor. The *path variance* is the sum of the individual variances along the longest path, or the critical path, in Figure 9.11 (i.e., Path 1, 2, 3, 7, 9, 12).

$$Z = \frac{55 - 59.8}{\sqrt{7.916}} = -1.706$$

Referring to the normal distribution tables in Appendix G, the calculated value of -1.706 represents an area of approximately 0.044; that is, the probability of meeting the scheduled time of 55 weeks is 4.4%. If the management requirement is 60 weeks, the probability of success would be approximately 52.8%; or if 65 weeks were specified, the probability of success would be around 96.77%.

When evaluating the resultant probability value (Column 12 of Figure 9.13), management must decide on the range of factors allowable in terms of risk. If the probability factor is too low, additional resources may be applied to the project in order to reduce the activity schedule times and improve the probability of success. On the other hand, if the probability factor is too high (i.e., there is practically no risk involved), this may indicate that excess resources are being applied, some of which could be diverted elsewhere. Management must assess the situation and establish a goal.

The application of PERT scheduling is appropriate for both small- and large-scale projects, and is of particular value in one-of-a-kind projects or for those programs where repetitive tasks are not predominant. PERT can be easily adapted to engineering research and development projects where unknowns exist and there is a certain degree of associated risk. Additionally, the network approach is well suited to the logistic field, where the activity interrelationships and organizational interfaces are many.

The *critical path method* (CPM) is a network scheduling technique with activities and events analogous to PERT, except that emphasis is placed on *critical* activities, or those requiring the greatest amount of time for completion. A network similar to the one presented in Figure 9.11 is developed and slack times are estimated as presented in Figure 9.13 (Column 10). The critical path is the longest path through the network where slack times are zero, or the path represented by the heavy line in Figure 9.11 (i.e., the path including Events 1, 2, 3, 7, 9, and 12). The critical path identifies those activities which may pose problems of both a technical and an administrative nature if schedule slippage occurs. In other words, these are the activities that must be closely monitored and controlled throughout the program.

The network paths representing other program activities shown in Figure 9.11 include slack time which constitutes a measure of program scheduling flexibility. The slack time is the interval of time where an activity might actually be delayed beyond its earliest scheduled start without necessarily delaying the overall program completion time. Total slack time is available with the objective of possibly effecting a trade-off and the reallocation of resources. Program scheduling improvements may be possible by shifting resources from activities with slack time to activities along the critical path.

Given the PERT and CPM networks, these can now be extended to include cost. A cost network can be superimposed upon a PERT/CPM network by estimating the total cost and the cost slope for each activity line as illustrated in Figure 9.14. The figure reflects Activity L (Figure 9.11), and the estimated cost is $25,675. The curve in the figure indicates an estimated rate of expenditure.

When implementing this technique, there is always the time-cost option, which enables management to evaluate alternatives relative to the allocation of resources for activity accomplishment. In many instances, time can be saved by applying more resources or, conversely, cost may be reduced by extending the time to complete an activity. The time-cost option can be attained by applying the following general approach:

1. Determine alternative time and cost estimates for all activities, and select the lowest-cost alternative in each instance.
2. Calculate the critical path time for the network. If the calculated value is equal to or less than the total time permitted, select the lowest-cost option for each activity, and check to ensure that the total of the incremental activity times does not exceed the allowable overall program completion time. If the calculated value exceeds the program time permitted, review the activities along the critical path and select the alternative with the lowest cost slope, and reduce the time value to be compatible with the program requirement.

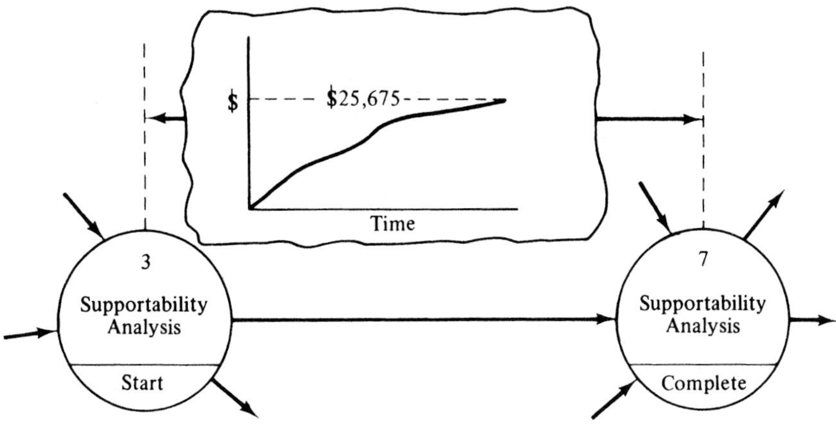

Figure 9.14 Event-cost projection.

3. After the critical path has been established in terms of the lowest cost option, review all the network paths with slack time, and shift activities to extend the times and reduce costs wherever possible. Activities with the steepest cost-time slopes shall be addressed first.

9.4 COST ESTIMATING AND CONTROL

One of the major challenges in logistics management constitutes the development of cost information for the logistics tasks in Figure 9.2 as they are applied to a given program. In the preparation of cost estimates extreme care must be exercised to ensure that the following areas are properly addressed:

1. A logistics program includes a number of tasks that are initiated during the conceptual design phase, and are progressively updated as the system development effort evolves through the preliminary system design and detail design and development phases. For example, there are some tasks within the Supportability analysis (SA) that may be rather extensive and ultimately require a significant amount of design data for backup. Given that the required design data are usually not available until late in the detail design and development phase, there is a tendency to delay the accomplishment of the SA until the last minute. Although this may appear to be an economical approach from one perspective, the SA information may not be available in a timely manner to support design decisions. On the other hand, complying with every SA requirement to great depth in the conceptual design phase is not always practical. There is an evolutionary process in the design and development of new systems, and various baselines are established along the way (e.g., functional baseline, allocated baseline, etc.). Logistics program tasks must be tailored to these baselines, and to the depth of system definition at the time. Many of these tasks are accomplished progressively, and care must be taken to ensure the proper level of effort—not too much or too little!

2. Within the overall logistics domain, there are many tasks (and data item requirements) that, although they are oriented to different end items, share some commonalty in their completion. For instance, there is a specified level of analysis required in the definition of spare/repair parts, a specified level of analysis required in the definition of test and support equipment, a specified level of analysis required in the determination of personnel quantities and skills, and so on. In response to individual organizational demands, data formatting procedures, and so on, these various analyses are often accomplished on an individual-by-individual basis. Yet, experience has indicated that there is a great deal of commonalty across the board in the analysis area. Although individually formatted data output reports are ultimately channeled in different directions, much of the basic analysis activity can be combined and integrated. When preparing cost estimates covering the different facets of logistics, recognition of these areas of commonality is necessary in arriving at a cost-effective approach.

3. The full implementation of logistics tasks may involve a significant amount of supplier activity, particularly for large-scale systems, and a major portion of this activity may be of an international nature. Further, the establishment of formal contractual relationships with suppliers can occur at many different points in a given program. Because a large proportion of the logistics activity is heavily dependent on supplier requirements, it is essential that these requirements be thoroughly defined as early in the program as practical. In developing cost information, it is relatively easy to add contingencies because of the unknowns, and redundancies in cost are introduced. In this era of increased competition in a highly dynamic international environment, it is imperative that cost duplication be avoided.

4. The implementation of a logistics program is often accomplished through a matrix type of organization structure. If care is not taken, this approach can result in additional costs. Both the project and functional organizations may require parallel administrative structures, and there may be a double counting in the estimation of costs for a given program. Thus, it is essential that logistics tasks and responsibilities be completely defined.

Good cost control is important to any organization regardless of size. This is particularly true in our current environment where resources are limited and competition is high. Cost control starts with the initial development of cost estimates for a given program and continues with the functions of cost monitoring and the collection of data, the analysis of such data, and the initiation of corrective action in a timely manner. Cost control implies good overall cost management, which includes cost estimating, cost accounting, cost monitoring, cost analysis, reporting, and the necessary control functions. More specifically, the following activities are applicable:[7]

1. *Define the elements of work.* Develop a statement of work (SOW), describe the tasks that are to be completed, and schedule the selected tasks as discussed in Section 9.3.
2. *Integrate tasks into the work breakdown structure (WBS).* Combine the appropriate project tasks into work packages, and integrate these elements into the work breakdown structure (WBS) as described in Section 9.2.
3. *Develop cost estimates for each project task.* Prepare a cost projection for each project task, identify the necessary cost accounts, and relate this to the appropriate element(s) of the WBS.
4. *Develop a cost data collection and reporting capability.* Develop a method for cost accounting (i.e., the collection and presentation of project costs), data analysis, and the reporting of cost data for management information purposes. Major

[7] Two references covering the subject of cost estimating are Stewart, R. D., and Wyskida, R. M., *Cost Estimator's Reference Manual*, 2nd ed., John Wiley & Sons, Inc., New York, 1995; and Ostwald, P. F., *Cost Estimating*, 3rd ed., Prentice Hall, Inc., Upper Saddle River, N.J., 1992. The emphasis here is primarily oriented to the costing of all project tasks identified in the WBS versus the development of cost estimating relationships (CERs) for life-cycle cost analysis purposes. LCC coverage is included in Appendix E. Cost control is covered in most textbooks on program/project management. One good reference is Kerzner, H., *Project Management: A Systems Approach to Planning, Scheduling, and Controlling*, 5th ed., Van Nostrand Reinhold Company, Inc., New York, N.Y., 1995.

areas of concern are highlighted (i.e., current or potential cost overruns, high-cost "drivers," etc.).

5. *Develop a procedure for evaluation and corrective action.* Inherent within the overall requirement for cost control is the provision for feedback and corrective action. As deficiencies are noted, or potential areas of risk are identified, project management must initiate the necessary corrective action in an expeditious manner.

The extent to which a manager will be able to control the program depends on the visibility that he or she has in assessing the various program elements, particularly those involving high risk. This visibility is based on the WBS makeup, the depth of cost estimating, the accounting or cost collection system, and the design and effectiveness of the management information and reporting system. Assurance that the proper tools are available for adequate program control stems from the initial planning effort conducted during the early part of the conceptual design phase. The planner needs to respond to two basic questions:

1. What are the program requirements, and what does the manager need to know to ensure that these requirements are met?
2. How will the manager obtain the information needed at the proper time?

Answering these questions should lead to the development of a management information system which will provide the right data at the time needed. Included in these requirements are the needs of the logistics manager, who must be aware of the status of each logistic support task at all times. Such awareness is necessary if one is to assure that program requirements will be met.

Regarding information relevant to the logistics manager, he or she needs to have access to the standard cost/schedule data usually available through the project status reporting system. Figure 9.15 represents a typical output from a PERT/CPM/COST reporting system. Additionally, the logistics manager requires some visibility of the status pertaining to significant system technical performance measures (TPMs). Figure 9.16 constitutes a cumulative summary of life-cycle cost projections, and compares the results with initial target objectives. Figure 9.17 illustrates the results of an evaluation of three specific system parameters. In each instance, problem areas can easily be identified.

Although the aspects of cost reporting and control are highly significant, these factors must be related to other parameters of the system. From the standpoint of logistics, all of these relevant system parameters must be viewed on an integrated basis.

9.5 MAJOR INTERFACES WITH OTHER PROGRAM ACTIVITIES

The number of interfaces that exist between logistics and other program activities are numerous. Logistics, with its highly interdisciplinary nature, is dependent on many other organizations. Because of this, good logistics planning should not only encompass those activities described in Section 9.1, but must tie in with other planning activ-

Network/Cost Status Report

Project: System *XYZ*				Contract Number: 6BSB-1002					Report Date: 4/10/98			
Item/Identification				Time Status					Cost Status			
WBS No.	Cost Account	Beginning Event	Ending Event	Exp. Elap. Time (t_e) (Weeks)	Earliest Completion Date (D_E)	Latest Completion Date (D_L)	Slack $D_L - D_E$ (Weeks)	Actual Date Completed	Cost Est. ($)	Actual Cost to Date ($)	Latest Revised Est. ($)	Overrun (Underrun) ($)
4A1210	3310	8	9	4.2	3/4/91	4/11/98	11.6	4/4/98	2500	2250	2250	(250)
4A1230	3762	R100	R102	3.0	5/15/91	4/28/98	-3.3		4500	4650	5000	500
5A1224	3521	7	9	20.0	6/20/91	8/3/98	0		6750	5150	6750	0

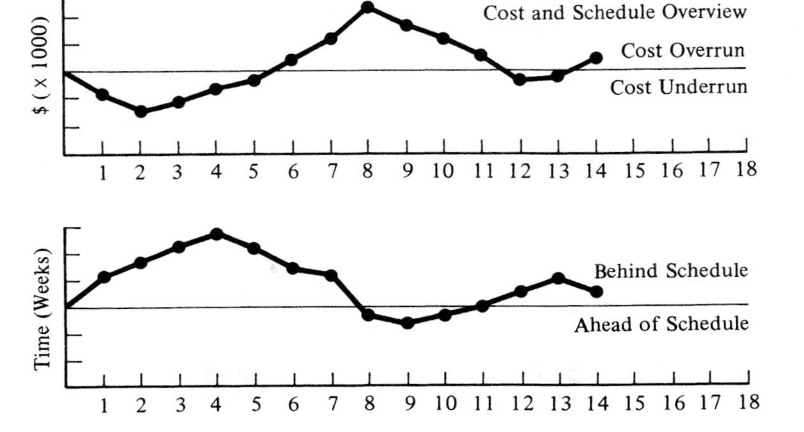

Figure 9.15 Program cost/schedule reporting.

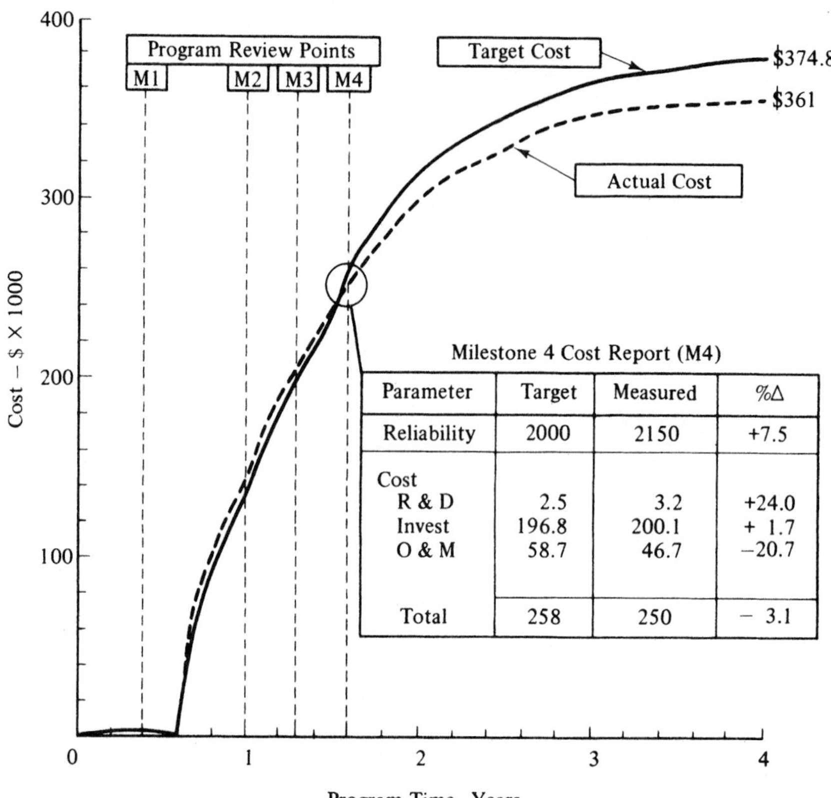

Parameter	Target	Measured	%Δ
Reliability	2000	2150	+7.5
Cost			
R & D	2.5	3.2	+24.0
Invest	196.8	200.1	+ 1.7
O & M	58.7	46.7	–20.7
Total	258	250	– 3.1

Figure 9.16 Program cost projection.

ities. An effective communication link between the ILSP and other key program plans must be provided. Of particular interest are the following:

1. *System engineering management plan* (SEMP). System engineering constitutes the process of bringing a system into being (definition of system requirements, functional analysis and allocation, synthesis, trade-offs and design optimization), and includes the integration of design disciplines into the overall engineering effort. The logistics engineering activity (i.e., the design interface) should be covered in the SEMP, along with reliability engineering, maintainability engineering, human factors, safety engineering, electrical and mechanical engineering, and so on. The SEMP is the top design integration plan and must support the *design for supportability* objectives in the ILSP.[8]

[8] System engineering program requirements (and the SEMP) are covered further in Blanchard, B. S., and Fabrycky, W. J., *Systems Engineering and Analysis*, 3rd Ed., Prentice Hall, Inc., Upper Saddle River, N.J., 1998; Defense Systems Management College, *Systems Engineering Management Guide*, DSMC, Fort Belvoir, Va.; and Blanchard, B. S., *System Engineering Management*, 2nd ed., John Wiley & Sons, Inc., N.Y., 1998.

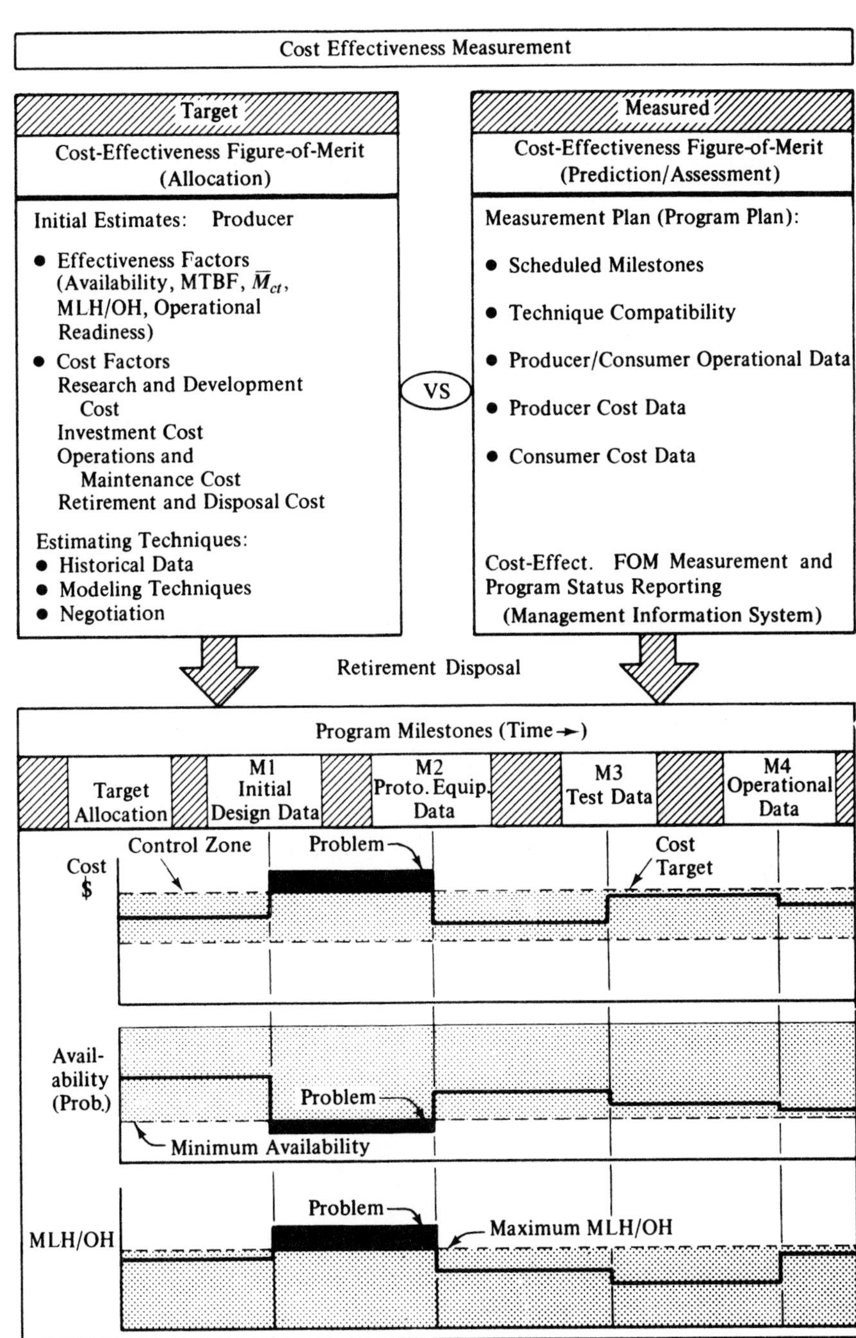

Figure 9.17 Parameter measurement and evaluation.

2. *Configuration management plan.* Configuration management constitutes the process that identifies the functional and physical characteristics of an item during its life cycle, controls changes to these characteristics, and records and reports the processing of changes and their implementation status. It is a process of baseline management (considering the functional, allocated, and product baselines), defining the system configuration at any point during the design and development process. This definition, and the monitoring of changes, is critical to the logistics area.[9]
3. *Test and evaluation master plan (TEMP).* This plan, usually developed for most large-scale programs during the conceptual design phase, identifies the need for system test and evaluation, describes test and evaluation requirements, identifies testing procedures, defines resource requirements, and describes data recording, analysis, and reporting requirements. The requirements for Type 1 to 4 testing, as described in Section 5.7, are covered. This plan covers many of the activities that are associated with the actual assessment (i.e., measurement) of system supportability characteristics.[10]
4. *Production/manufacturing plan.* A production/manufacturing plan is developed as the system design configuration is defined and make-or-buy decisions lead to the identification of production requirements. Production requirements not only include prime mission-oriented system components, but selected elements of support (e.g., spare/repair parts, special test equipment, etc.). This plan also includes criteria pertaining to design for producibility, requirements dealing with materials handling, and procedures covering the distribution of products to the consumer. These activities relate directly to logistics requirements.[11]
5. *Total quality management plan (TQMP).* This plan covers the requirements for total quality control (TQC), quality assurance, statistical quality control (SQC), statistical process control (SPC), and all of those activities that support the overall quality of the system and its components. In addition to the prime-mission–oriented components of the system, this plan should cover "quality" requirements as they pertain to all elements of logistics.[12]

9.6 ORGANIZATION FOR LOGISTICS

Organization is the combining of resources in such a manner as to fulfill a need. Organizations constitute groups of individuals of varying levels of expertise combined into a social structure of some type to accomplish one or more functions. Organizational structures will vary with the functions to be performed, and the results will depend on

[9] Refer to EIA-IS-649, "Configuration Management," Electronic Industries Association, Arlington, Va., 1997; and MIL-HDBK-61, "Configuration Management," Department of Defense, Washington, D.C. (latest edition).

[10] For large-scale defense systems a good reference is Defense Systems Management College, *Test and Evaluation Management Guide*, DSMC, Fort Belvoir, Va.

[11] A good reference is Defense Systems Management College, *Manufacturing Management: Guide for Program Managers*, DSMC, Fort Belvoir, Va.

[12] Appendix H includes some excellent references covering the subject of quality.

the established goals and objectives, the resources available, the communications and working relationships of the individual participants, motivation, and many other factors. The ultimate objective, of course, is to achieve the most cost-effective utilization of human, material, and monetary resources through the establishment of decision-making and communications processes designed to accomplish specific objectives.

In logistics, acquiring and maintaining the proper balance of resources to meet the stated objectives is a challenge. The nonhomogeneity of functions and the diversity of personnel backgrounds and skills necessitates a well-integrated, highly interdisciplinary, controlled team approach. Further, the requirements in implementing a given logistics program will vary from one phase to the next, creating a dynamic condition. For a large project, the logistics organization may take the form of a pure functional structure. Conversely, a matrix approach may be more appropriate for addressing logistics requirements on smaller projects. Again, a tailored application is necessary.[13]

Consumer, Producer, and Supplier Relationships

To address the subject of *organization for logistics* properly, one needs to understand the environment in which logistics functions are performed. Although this may vary somewhat depending on the size of the project and the stage of design and development, the discussion herein is primarily directed to a large project operation characteristic for the acquisition of many large-scale systems. By assuming a large project approach, it is hoped that a better understanding of logistics in a somewhat complex environment will be provided. The reader must, of course, adapt to his or her program requirements.

For a relatively large project, the logistics activity may appear at several levels as shown in Figure 9.18. The customer (or consumer) may establish a logistics organization to accomplish the tasks identified in Figure 9.2, or these tasks may be relegated to the producer through some form of contractual structure. The question is: Who is responsible and has the authority to perform the logistics functions as defined?

In some instances, the consumer may assume full responsibility for the overall design and integration of the system. Top system-level design, preparation of the system specification, preparation of the ILSP, and completion of the tasks in Figure 9.2 are accomplished by the customer's organization. Individual producers, subcontractors, and component suppliers will accomplish supporting tasks as necessary.

In other cases, although the customer provides overall guidance in terms of issuing a general statement of work (SOW) or an equivalent contractual document, the producer (or prime contractor) is held responsible for system design and integration. In this situation, the definition and expansion of system requirements is accomplished by the producer (to include the definition of overall logistics requirements and the preparation of the ILSP), with supporting tasks being conducted by individual suppliers as required. In other words, although both the customer and the producer have established logistics organizations, the basic responsibility (and authority) for fulfilling

[13] The level of discussion on organizational concepts in this chapter is rather cursory in nature, and it is intended to provide the reader with an overview of some of the key points with respect to logistics (as compared to an in-depth treatise on the subject). A more comprehensive coverage of organization and management theory is provided in a few of the references in Appendix H.

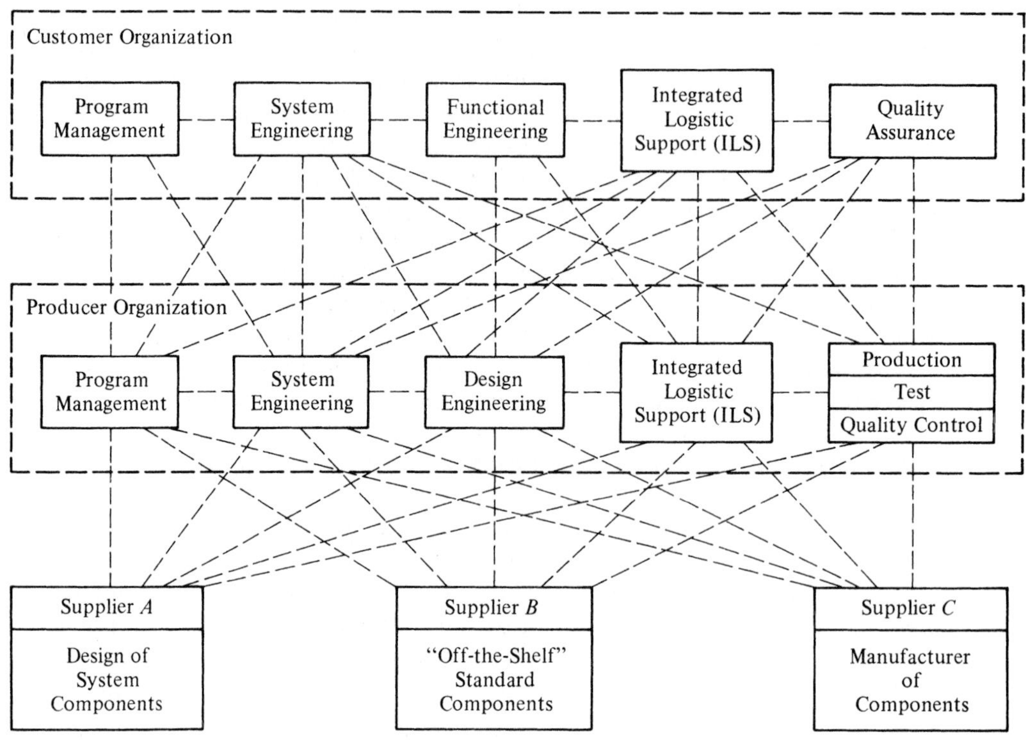

Figure 9.18 Consumer/producer/supplier interfaces.

the objectives described throughout this book lies within the producer's organization. This is the model that will serve as the basis for much of the discussion from here on.

Referring to Figure 9.18, the main point of emphasis relates to the extensive amount of communications that must exist between the logistics organization and other organizations within the project. The interfaces are many, and the successful accomplishment of logistics tasks requires an in-depth understanding of the responsibilities and tasks performed by these many and varied organizational entities. Not only are the communication requirements extensive within the producer's organization per se, but the necessary communications must be established both upward from producer to customer and downward from producer to supplier.

Customer Organization and Functions (Consumer)

The customer/consumer organization may range from one small group of individuals to an industrial firm, a commercial business, an academic institution, a government laboratory, or a military service such as the U.S. Air Force (or the U.S. Army or U.S. Navy). The customer may be the ultimate "user" of the system or may be the procuring agency for a user.

There are a variety of approaches, and associated organizational relationships, involved in the design and development of new systems. The objective is to identify the overall *program manager* and to pinpoint the responsibility and authority for *logistics management*. In the past, there have been many instances where the procuring agency has initiated a contract with an industrial firm (i.e., the producer) for the design and development of a large system, but has not delegated complete responsibility (or authority) for logistics management. The company has been held responsible for the design, development, and delivery of a system in response to certain specified requirements. However, the customer has not always provided the producer with the necessary data and/or controls to allow for the development effort to proceed in accordance with good design practices—to include the design for supportability. At the same time, the customer has not accomplished the necessary functions of logistics management. The net result has been the development of a system without incorporation of the characteristics discussed throughout the earlier chapters: that is, a system that is unreliable, not maintainable, not supportable, not cost-effective, and is not responsive to the needs of the customer.

The fulfillment of logistics objectives is highly dependent on a *commitment* from the top down. These objectives must be recognized from the beginning by the customer, and an organization needs to be established to ensure that these objectives are met. The customer must create the appropriate environment and take the lead by initiating either one of the following courses of action:

1. Accomplish the necessary logistics functions within the customer's organizational structure (refer to Figure 9.18). Preparation of the system specification, the ILSP, the supportability analysis (SA), the conductance of design reviews, and the completion of the major tasks identified in Figure 9.2 are fulfilled by the customer. In other words, the complete job of system-level design, integration, and support is accomplished by the customer or procuring agency.
2. Accomplish the necessary logistics functions within an industrial firm (i.e., the producer's organization depicted in Figure 9.18). Preparation of the system specification, the ILSP, the supportability analysis (SA), and the completion of the tasks identified in Figure 9.2 have been delegated to the producer.

As the split in authority, responsibilities, and subsequent duties may not be as clean-cut as implied, it is important that the responsibility for logistics be established at the beginning. The customer must clarify system objectives and program functions, and the requirements for logistics must be well defined.

In the event that the logistics responsibility is delegated to the producer (i.e., the second option earlier), the customer must support this decision by providing the necessary top-down guidance and managerial backing. Responsibilities must be properly delineated, system-level design information generated through early customer activities must be made available to the producer (e.g., the results of feasibility studies, the documentation of operational requirements), and the producer must be given the necessary leeway relative to making design decisions at the system level. The challenge for the customer is to prepare a good comprehensive, well-written, and clear statement of work to be implemented by the producer. This, in turn, forms the basis for the negotiation of a contractual arrangement between the customer and the producer (or contractor).

Producer Organization and Functions (Contractor)

For most large-scale projects, the producer (or the *contractor*) will undertake the bulk of activities associated with the design and development of a new system. The customer will specify system-level and program requirements through the preparation of a request for proposal (RFP) or an invitation for bid (IFB), and various industrial firms will respond in terms of a formal proposal. As there may be a number of responding proposals, a formal competition is initiated, individual proposals are reviewed and evaluated, negotiations are consummated, and a selection is made. The successful contractor will then proceed with the proposed level of effort.

In addressing program requirements, it is essential that the successful contractor have access to all information and data leading up to the requirements specified in the technical portions of the RFP/IFB. In several instances, the RFP will include a system specification covering the technical aspects of system design and development, along with a statement of work (SOW) directed toward program tasks and the management aspects of a program. As part of the overall package for some programs, the customer may include an ILSP. In any event, the transition process from customer to contractor is one of the most critical steps in the implementation of a program. It is here that the necessary communications must be established.

Given that system-level requirements are properly defined and that a prime contractor has been selected to design and develop a new system, the next step is to discuss the subject of logistics in the context of the contractor's organization. Organizational structures will vary from the pure *functional* approach to *product line,* combined *product line/functional, matrix,* and so on. These organizational patterns are discussed in the sections to follow.

1. *Functional organization structure.* The primary building block for most organizational patterns is the *functional* structure reflected in Figure 9.19. This approach, sometimes referred to as the *classical* or *traditional* approach, involves the grouping of specialties or disciplines into separately identifiable entities. The intent is to perform similar activities within one organizational component. For example, all engineering work would be the responsibility of one executive, all logistics work would be the responsibility of another executive, and so on.

As for any organizational structure, there are advantages and disadvantages. Figure 9.20 identifies some of the pros and cons associated with the pure functional approach illustrated in Figure 9.19. As shown, the president (or general manager) has under his or her control all of the functional entities necessary to design and develop, produce, deliver, and support a system. Each department maintains a strong concentration of technical expertise and, as such, a project can benefit from the most advanced technology in the field. Additionally, levels of authority and responsibility

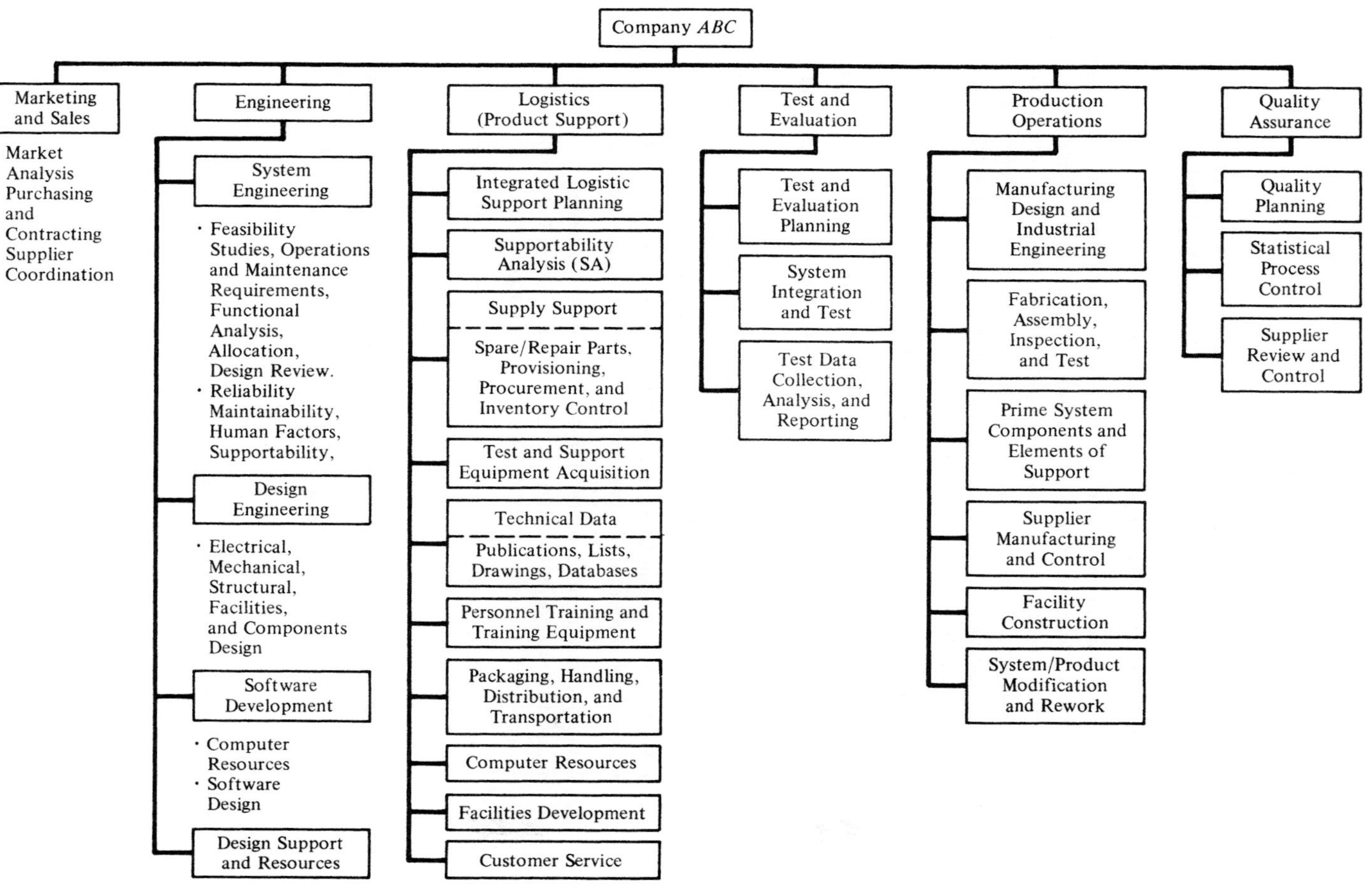

Figure 9.19 Producer's functional organization.

Advantages

1. Enables the development of a better technical capability for the organization. Specialists can be grouped to share knowledge. Experiences from one project can be transferred to other projects through personnel exchange. Cross-training is relatively easy.
2. The organization can respond quicker to a specific requirement through the careful assignment (or reassignment) of personnel. There are a larger number of personnel in the organization with the required skills in a given area. The manager has a greater degree of flexibility in the use of personnel and a broader labor base with which to work. Greater technical control can be maintained.
3. Budgeting and cost control is easier due to the centralization of areas of expertise. Common tasks for different projects are integrated, and it is easier not only to estimate costs but to monitor and control costs.
4. The channels of communication are well established. The reporting structure is vertical, and there is no question as to who is the boss.

Disadvantages

1. It is difficult to maintain an identity with a specific project. No single individual is responsible for the total project or the integration of its activities. It is hard to pinpoint specific project responsibilities.
2. Concepts and techniques tend to be functionally oriented with little regard toward project requirements. The tailoring of technical requirements to a particular project is discouraged.
3. There is little customer orientation or focal point. Response to specific customer needs is slow. Decisions are made on the basis of the strongest functional area of activity.
4. Because of the group orientation relative to specific areas of expertise, there is less personal motivation to excel and innovation concerning the generation of new ideas is lacking.

Figure 9.20 Advantages and disadvantages of functional organization.

are clearly defined, communication channels are well structured, and the necessary controls over budgets and costs can easily be established. In general, the organization structure is well suited for a single-project operation.

Conversely, the pure functional structure may not be as appropriate for large multiproduct firms or agencies. Where there are many different projects, each competing for special attention and the appropriate resources, there are some disadvantages. The main problem is that there is no strong central authority or individual responsible for the total project. As a result, the integration of activities, crossing functional lines, often becomes difficult. Conflicts occur as each functional activity struggles for power and resources, and decisions are often made on the basis of what is best for the functional group rather than what is best for the project. Further, the decision-making processes are sometimes slow and tedious because all communications must be channeled through upper-level management. Projects may fall behind and suffer in the classical functional organization structure.

2. *Product line/project organization structure.* As industrial firms grow and there are more products being developed, it is often convenient to classify these products into common groups and to develop a product-line organization structure as shown in Figure 9.21. A company may become involved in the development of communication systems, transportation systems, and electronic test and support equipment. Where there is functional commonality, it may be appropriate to organize the company into three divisions, one for each product line. Further, these divisions may be geographically separated, and each may serve as a functional entity with operations similar to those illustrated in Figure 9.19.

In divisions where there are large systems being developed, the product-line responsibilities may be subdivided into projects as illustrated in Figure 9.21 (Company *GHI*). In such cases, the project will be the lowest entity. A project organization is one that is solely responsive to the planning, design and development, production, and support of a single system or a large product. It is time limited, directly oriented to the life cycle of the system, and the commitment of personnel and material is purely for the purposes of accomplishing tasks peculiar to that system. Each project will contain its own engineering function, logistics capability, production operation, and so on. The project manager has the authority and the responsibility for all aspects of the project, whether it is a success or a failure.

In the case for both the product-line and the project structures, the lines of authority and responsibility for a given project are clearly defined, and there is little question as to priorities. On the other hand, there is a potential for the duplication of activities within a firm that can be quite costly. Emphasis is on individual projects as compared to the overall functional approach illustrated in Figure 9.19. Some of the advantages and disadvantages of project/product-line structures are presented in Figure 9.22.

3. *Matrix organization structure.* The matrix organization structure is an attempt to combine the advantages of the pure functional organization and the pure project organization. In the functional organization, technology is emphasized while project-oriented tasks, schedules, and time constraints are sacrificed. For the pure project, technology tends to suffer since there is no single group for the planning and development of such. Matrix management is an attempt to acquire the greatest amount of technology, consistent with project schedules, time and cost constraints, and related customer requirements. Figure 9.23 presents a typical matrix organization structure (Company *JKL*). Project responsibilities extend horizontally, and functional responsibilities are vertical.

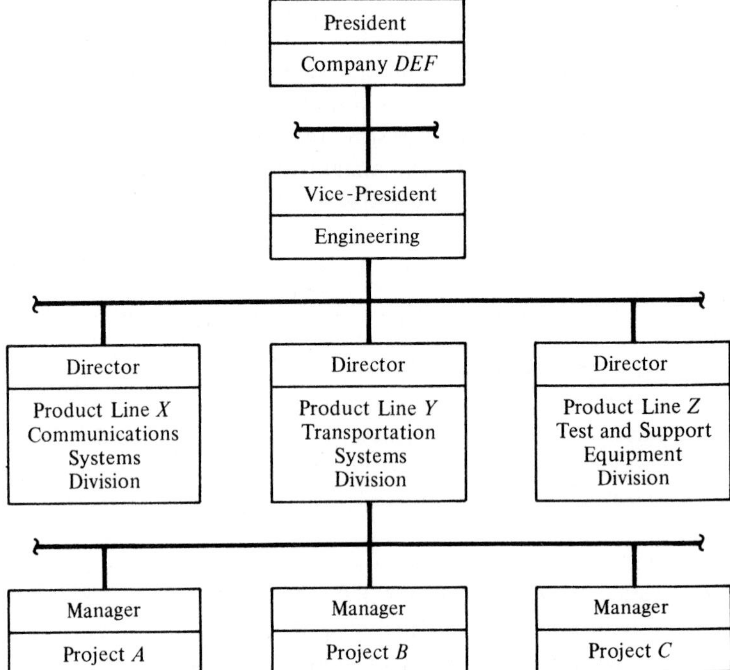

Product line organization with project subunits.

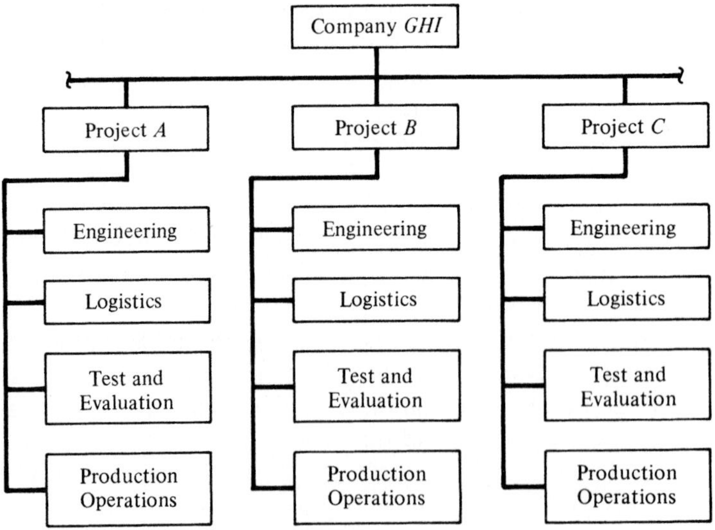

Project organizational structure

Figure 9.21 Product line/project organizational structures.

Advantages

1. The lines of authority and responsibility for a given project are clearly defined. Project participants work directly for the project manager, communication channels within the project are strong, and there is no question as to priorities. A good *project* orientation is provided.
2. There is a strong customer orientation, a company focal point is readily identified, and the communication processes between the customer and the contractor are relatively easy to maintain. A rapid response to customer needs is realized.
3. Personnel assigned to the project generally exhibit a high degree of loyalty to the project, there is strong motivation, and personal morale is usually better with product identification and affiliation.
4. The required personnel expertise can be assigned and retained exclusively on the project without the time sharing that is often required under the functional approach.
5. There is greater visibility relative to all project activities. Cost, schedule, and performance progress can easily be monitored, and potential problem areas (with the appropriate follow-on corrective action) can be identified earlier.

Disadvantages

1. The application of new technologies tends to suffer without strong functional groups and the opportunities for technical interchange between projects. As projects go on and on, those technologies that are applicable at project inception continue to be applied on a repetitive basis. There is no perpetuation of technology, and the introduction of new methods and procedures is discouraged.
2. In contractor organizations where there are many different projects, there is usually a duplication of effort, personnel, and the use of facilities and equipment. The overall operation is inefficient and the results can be quite costly. There are times when a completely decentralized approach is not as efficient as centralization.
3. From a managerial perspective, it is difficult to utilize personnel effectively in the transfer from one project to another. Good qualified workers assigned to projects are retained by project managers for as long as possible (whether they are being effectively utilized or not), and the reassignment of such personnel usually requires approval from a higher level of authority, which can be time-consuming. The shifting of personnel in response to short-term needs is essentially impossible.
4. The continuity of an individual's career, his or her growth potential, and the opportunities for promotion are often not as good when assigned to a project for an extended period of time. Project personnel are limited in terms of opportunities to be innovative relative to the application of new technologies. The repetitiousness of tasks sometimes results in stagnation.

Figure 9.22 Advantages and disadvantages of project/product line organization.

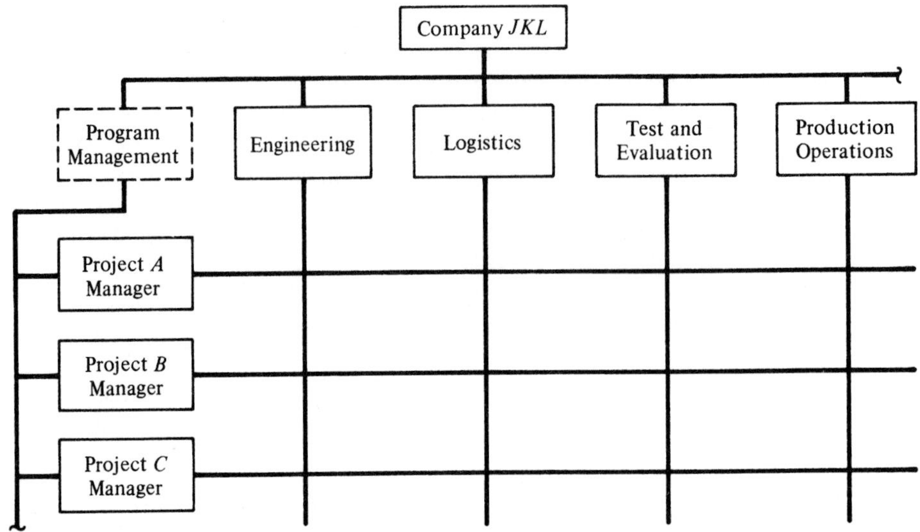

Figure 9.23 Matrix organizational structure.

Referring to the figure, each project manager reports to a vice-president (or equivalent), and has the overall responsibility and is accountable for project success. At the same time, the functional departments are responsible for maintaining technical excellence and for ensuring that all available technical information is exchanged between projects. The functional managers, who also report to a vice-president, are responsible to ensure that their personnel are knowledgeable in their respective fields of endeavor.

The matrix organization, in its simplest form, can be considered as being a two-dimensional entity with the projects representing potential profit centers and the functional departments identified as cost centers. For small industrial firms, the two-dimensional structure may reflect the preferred organizational approach because of the flexibility allowed. The sharing of personnel and the shifting back and forth are often inherent characteristics. Conversely, for large corporations with many product divisions, the matrix becomes a somewhat complex multidimensional structure.

To ensure success in the implementation of matrix management, a highly cooperative and mutually supportive environment must be created within the company. Managers and workers alike must be committed to the objectives of matrix management. A few key points are noted subsequently.

(a) Good communication channels (vertical and horizontal) must be established to allow for a free and continuing flow of information between projects and the functional departments. Good communications must also be established from project to project.

(b) Both project and functional department managers should participate in the initial establishment of company-wide and program-oriented objectives. Further, each must have an input and become involved in the planning process. The purpose is to help ensure the necessary commitment on both sides. Additionally, both project and functional managers must be willing to negotiate for resources.

(c) In the event of conflict, a quick and effective method for resolution must be established. A procedure must be developed with the participation and commitment from both project and functional managers.

(d) For personnel representing the technical functions and assigned to a project, both the project manager and the functional department manager should agree to the duration of assignment, the tasks to be accomplished, and the basis upon which the individual(s) will be evaluated. The individual worker must know what is to be expected of him or her, the criteria for evaluation, and which manager will be conducting the performance review (or how the performance review will be conducted). Otherwise, a two-boss situation (each with different objectives) may develop, and the employee will be caught in the middle.

The matrix structure provides the best of several worlds; that is, it is a composite of the pure project approach and the traditional functional approach. The main advantage pertains to the capability of providing the proper mix of technology and project-related activities. At the same time, a major disadvantage relates to the conflicts that arise on a continuing basis as a result of a power struggle between project and functional managers, changes in priorities, and so on. A few advantages and disadvantages are noted in Figure 9.24.

4. *Logistics organization.* The preceding sections provide an overview of the major characteristics of the functional, project, and matrix organization structures. The advantages and disadvantages of each need to be assessed in arriving at a preferred organization for logistics.

Regarding the broad field of logistics, there are many activities where a centralized integrated functional organization approach is preferred, and there are other areas where a strong project orientation appears to be desirable. For instance, in the processing of supportability analysis (SA) data and in the production of technical manuals, it is often more efficient to integrate these activities across the board for all projects. This allows for the utilization of data from one project to help fulfill the requirements for another. In the publications area, the combining of equipment and associated resources provides for a better overall production capability.

Conversely, for activities that are directly related to the design and development of a particular system, it is more appropriate to assume a project organizational approach. The accomplishment of reliability, maintainability, human factors, supportability, and those SA tasks that are related to a particular design configuration can be more effective through a pure project structure. In fact, it is often appropriate to physically locate those individuals who are assigned to design-oriented tasks in the same area with the responsible system/product designers. The objective is to promote the team approach to design and to enhance the communications among its members.

Given the wide variety of functions within the overall logistics domain, it is often feasible to assume an organizational configuration similar to the structure illustrated in Figure 9.25. In this instance, a large proportion of the logistics activities is accomplished through a functionally oriented logistics department. However, there are

Advantages

1. The project manager can provide the necessary strong controls for the project while having ready access to the resources from many different functionally oriented departments.
2. The functional organizations exist primarily as support for the projects. A strong technical capability can be developed and made available in response to project requirements in an expeditious manner.
3. Technical expertise can be exchanged between projects with a minimum of conflict. Knowledge is available for all projects on an equal basis.
4. Authority and responsibility for project task accomplishment are shared between the project manager and the functional manager. There is mutual commitment in fulfilling project requirements.
5. Key personnel can be shared and assigned to work on a variety of problems. From the company top-management perspective, a more effective utilization of technical personnel can be realized and program costs can be minimized as a result.

Disadvantages

1. Each project organization operates independently. In an attempt to maintain an identity, separate operating procedures are developed, separate personnel requirements are identified, and so on. Extreme care must be taken to guard against possible duplication of efforts.
2. From a company viewpoint, the matrix structure may be more costly in terms of administrative requirements. Both the project and the functional areas of activity require similar administrative controls.
3. The balance of power between the project and the functional organizations must be clearly defined initially and closely monitored thereafter. Depending on the strengths (and weaknesses) of the individual managers, the power and influence can shift to the detriment of the overall company organization.
4. From the perspective of the individual worker, there is often a split in the chain of command for reporting purposes. The individual is sometimes ''pulled'' between the project boss and the functional boss.

Figure 9.24 Advantages and disadvantages of matrix organization.

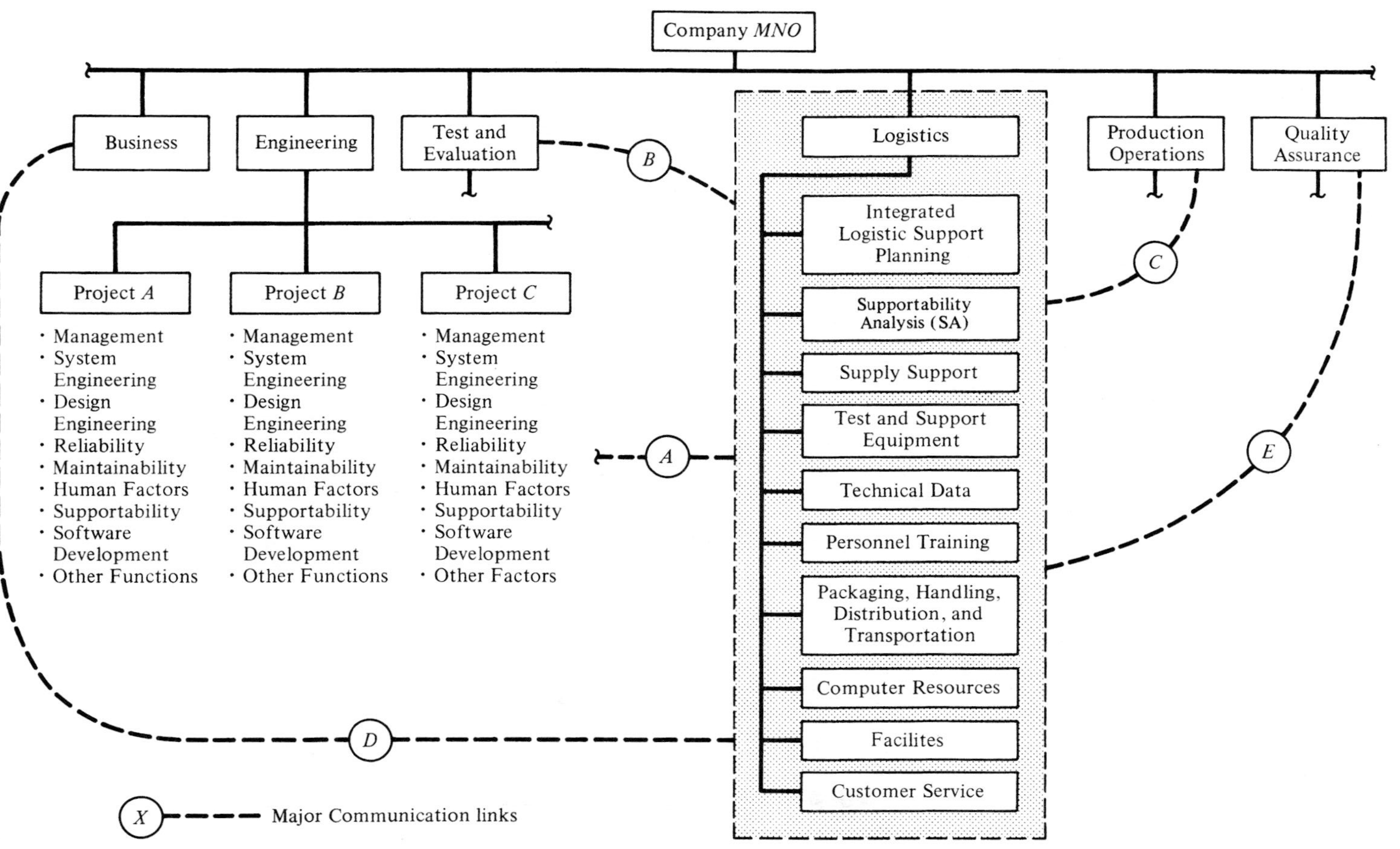

Figure 9.25 Combined project/staff organization (partial).

design-related tasks accomplished for each project, and there are many additional tasks that require support from marketing, purchasing and contracts, test and evaluation, production operations, and quality assurance. Some of the major activity interfaces are identified in Figure 9.26.

Referring to Figure 9.25, there may be a variation in approaches within the same firm. One or two large projects may exist along with numerous smaller projects. The large projects will tend to support an organizational structure similar to that presented in Figure 9.21, whereas the smaller projects will probably follow a matrix orientation. Where the larger projects can afford to support significant numbers of personnel on a full-time basis, the smaller projects may only be able to support a select number of individuals on a part-time basis. The specific requirements are dictated through the generation of tasks by the project organization; that is, a request for assistance is initiated by the project manager, with the task(s) being completed within the functional department.

Project size will not only vary with the type and nature of the system being developed, but with the specific stage of development. A large-scale system in the early stages of conceptual design may be represented by a small project structure, selectively utilizing the resources from various functional organizations as required (refer to Figure 9.25). As system development progresses into the preliminary system design and detail design and development phases, the organization structure may shift somewhat, replicating the configuration in Figure 9.21. In other words, the characteristics and structure of organizations are often *dynamic* by nature. The organization structure must be adapted to the needs of the project at the time, and these needs may shift from one phase to the next.

Although there are differences from one program to the next, the logistics tasks identified in Figure 9.2 are assumed as being somewhat typical. These tasks cover the major areas of activity described throughout the earlier chapters of this text, and their application must be *tailored* to the particular program need. It is not intended to imply that the logistics organization does everything. Further, the successful completion of these tasks does not mean to convey a tremendous level of effort. Many of these tasks are iterative in nature, starting with a minimal amount of activity during conceptual design and progressively expanding to include more in-depth coverage of the entire system and its components in the detail design and development phase. The emphasis is on the process and on maintaining continuity throughout.

Referring to Figures 9.25 and 9.26, there are numerous interfaces that exist within a traditional company organization. The resources required to fulfill logistics objectives are located throughout many departments (e.g., marketing, purchasing, contract management, engineering, production operations, distribution and transportation, test and evaluation, quality assurance, etc.). The ultimate success in meeting logistics objectives is, of course, highly dependent on managerial support from the top down! The president (or general manager), vice president of engineering, vice president of production, Project *B* manager, and other high-level managers must each *understand* and *believe in* the concepts and objectives of logistics. If the logistics manager is to be successful, these higher-level managers must be directly supportive 100 percent of the time. There will be many occasions where individual design engineers or middle-level managers

Communication Channel (Figure 9.25)	*Supporting Organization (Interface Requirements)*
A	*Project Activities:* 1. *Project management:* to determine specific logistics requirements for a given project (i.e., tasks, schedules, cost, etc.), and to establish continuing liaison and close communications to ensure the successful completion of project tasks. 2. *System engineering:* to provide initial design input and to acquire documentation from feasibility analyses, system operational requirements, the maintenance concept, functional analyses, and the allocation of requirements; to provide assistance in the preparation of the system specification and subordinate specifications as applicable; and to participate in formal design reviews. 3. *Design engineering:* to provide initial design criteria for supportability; to provide day-to-day assistance in design; to acquire design information/data for logistic support analyses, review, and evaluation; and to initiate recommendations for improvement/corrective action as feasible. 4. *Reliability engineering:* to establish and maintain close liaison in the initial determination of reliability requirements for a given program and in the subsequent completion of reliability tasks; and to acquire reliability information/data for the logistic support analysis. 5. *Maintainability engineering:* to establish and maintain close liaison in the initial determination of maintainability requirements for a given program and in the subsequent completion of maintainability tasks; and to acquire maintainability information/data for the logistic support analysis. 6. *Human factors engineering:* to establish and maintain close liaison in the initial determination of human factors requirements for a given program and in the subsequent completion of human factors tasks; and to acquire human factors information/data for the logistic support analysis. 7. *Supportability engineering:* to determine supportability requirements for a given program and to accomplish logistics engineering functions in design. 8. *Software development:* to provide initial design criteria for software requirements as they pertain to system maintenance and support, and ensure ultimate compatibility between system software and computer resources for logistics.

Figure 9.26 Major logistics activity interfaces.

Communication Channel (Figure 9.25)	*Supporting Organization (Interface Requirements)*
B	*Test and Evaluation:* to establish initial requirements for supportability test and evaluation; to directly support system test and evaluation activities; and to review and evaluate test and evaluation results for supportability assessment.
C	*Production Operations:* (1) to assist in establishing the initial requirements for and to maintain ongoing support relative to the flow of materials in the manufacture, assembly and test of prime mission-oriented equipment; (2) to establish the initial requirements for and to provide ongoing support in the manufacture, assembly, and test of spare/repair parts, special test equipment, etc.; and (3) to provide assistance relative to the packaging, handling , distribution, warehousing, and transporting of system components to the consumer.
D	*Business Operations:* 1. *Marketing and sales:* to acquire and sustain the necessary communications with the customer. Supplemental information pertaining to customer requirements, system operational and maintenance support requirements, changes in requirements, outside competition, etc., is needed. This is above and beyond the formal "contractual" channel of communications. 2. *Accounting:* to acquire both budgetary and actual cost data in support of economic analysis efforts (e.g., life -cycle cost analyses). 3. *Purchasing:* to assist in the identification, evaluation, and selection of component suppliers (with regard to supportability implications) and in the provisioning and procurement of components. 4. *Human resources (personnel):* to solicit assistance in the initial recruiting and hiring of qualified logistics personnel, and in the subsequent training and maintenance of personnel skills. To conduct training programs across the board relative to logistics concepts, objectives, and the implementation of program requirements. 5. *Contract management:* to keep abreast of contract requirements (of a technical nature) between the customer and the contractor. To ensure that the appropriate contract structure and relationships are established and maintained with suppliers in meeting the logistics' needs for a given program.
E	*Quality Assurance:* to establish initial quality requirements for the elements of logistic support, and to maintain quality standards in the acquisition of these elements through supplier activity. To provide support for the company-wide total quality management (TQM) program.

Figure 9.26 *(continued)* Major logistics activity interfaces.

will go off on their own, making decisions that will conflict with logistics objectives. When this occurs, the logistics manager must have the necessary support to ensure that actions are taken to get things back on track.

5. *Integrated product and process development* (IPPD). The ultimate objective in the design and development of any system is to establish a *team* approach, with the appropriate communications, enabling the application of concurrent engineering methods throughout. However, there are often problems when dealing with large programs involving many different functional organizational units. Barriers are developed which tend to inhibit the necessary day-to-day close working relationships, the timely transfer of essential information, and the communications discussed earlier. As there are many companies organized strictly along functional lines (versus the project-staff configuration), there is a need to ensure that the proper level of communications is maintained between all applicable organizational units, regardless of where they may be permanently located within the overall company structure.

With this objective in mind, the Department of Defense (DOD) initiated the concept of *integrated product and process development* (IPPD) in the mid-1990s. IPPD can be defined as a "management technique that simultaneously integrates all essential acquisition activities through the use of multidisciplinary teams to optimize the design, manufacturing and supportability processes."[14] This concept promotes the communications and integration of the key functional areas, as they apply to the various phases of program activity. While the specific nature of the activities involved and the degree of emphasis exerted will change somewhat as the system design and development effort evolves, the structure conveyed in Figure 9.27 is maintained throughout in order to foster the necessary communications across the more traditional functional lines of authority. In this regard, the concept of IPPD is directly in line with systems engineering and ILS objectives.

Inherent within the IPPD concept is the establishment of *Integrated Product Teams* (IPTs), with the objective of addressing certain designated and well-defined issues.[15] An IPT, constituting a selected team of individuals from the appropriate disciplines, may be established to investigate a specific segment of design, a solution for some outstanding problem, design activities which have a large impact on a high-priority TPM, and so on. The objective is to create a *team* of qualified individuals that can effectively work together to solve some problem in response to a given requirement. Further, there may be a number of different teams established to address issues at different levels in the overall system hierarchical structure; i.e., issues at the system-level, subsystem-level, and/or component-level. Refining to Figure 9.27, an IPT may be established to concentrate on those activities that significantly impact selected *performance* factors, *cost-of-ownership, integrated data,* and *configuration management*. There may be another IPT assigned strictly to "track"

[14] DOD Regulation 5000.2, "Manditory Procedures For Major Defense Acquisition Programs (MDAPs) and Major Automated Information System (MAIS) Acquisition Programs," Department of Defense, Washington, DC, March 1996.

[15] The term IPT is also used as a designator for "Integrated Process Team."

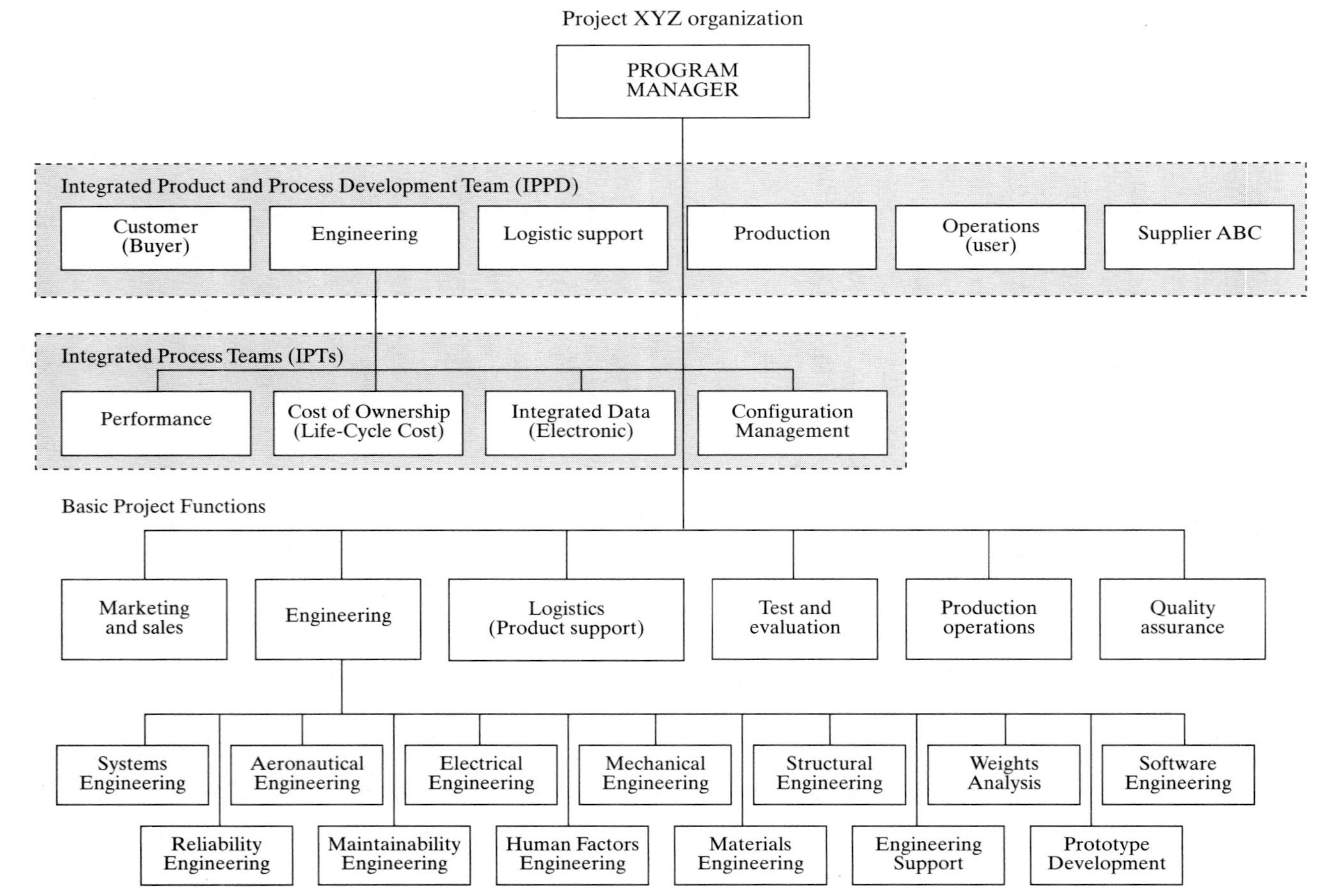

Figure 9.27 Functional organization with IPPD/IPT activities.

the *integrated data environment* issue. The objective is to provide the necessary emphasis in critical areas, and to reap the benefits of a *team* approach in arriving the best solution possible.

IPTs are often established by the Program Manager, or by some designated high-level authority in the organization. The representative team members must be well-qualified in their respective areas of expertise, must be empowered to make on-the-spot decisions when necessary, must be proactive relative to team participation, success-oriented, and be resolved to addressing the problem assigned. The Program Manager must clearly define the objectives for the team, expectations in terms of results, and the team members must maintain a continuous "up-the-line" communications channel. The longevity of the IPT will depend on the nature of the problem and the effectiveness of the team in progressing toward meeting its objective. Care must be taken to avoid the establishment of too many teams, as the communication processes and interfaces become too complex when there are many teams in place. Additionally, there often are conflicts when it comes to issues of importance and which issue is "traded-off" as a result. Further, as the team ceases to be effective in accomplishing its objectives, it should be disbanded accordingly. An established team that has "outlived its usefulness" can be counter-productive. The IPPD/IPT concept may be superimposed on any of the organizational structures described earlier on an "as-required" basis.

Supplier Organization and Functions

The term *supplier*, as defined herein, refers to a broad category of organizations that provide system components to the producer (i.e., the contractor). These components may range from a large element of the system (e.g., a facility, an intermediate maintenance shop full of test equipment, Unit *B* of System *XYZ*) to a small nonrepairable item (e.g., a resistor, fastener, bracket, cable). In some instances, the component may be newly developed and require detailed design activity. The supplier will design and produce the component in the quantities desired. In other cases, the supplier will serve as a manufacturing source for the component. The supplier will produce the desired quantities from a given data set. In other words, the component design has been completed (whether by the same firm or some other), and the basic service being provided is "production." A third scenario may involve the supplier as being an inventory source for one or more common and standard commercial off-the-shelf (COTS) components. There are no design or production activities, but just distribution and materials handling functions. As one can see, individual supplier roles can vary significantly.

In the process of identifying and selecting suppliers to provide system components, there are several issues that must be addressed. In addition to having the right component and demonstrating a desired level of technical competency, there are economic considerations, political considerations, environmental considerations, and so on. Suppliers are often selected based on geographic location with economic need in mind. It may be desirable to establish a new manufacturing capability to help stimulate the economy in a depressed area. Suppliers may be selected because of their location within a politician's jurisdiction. The selection of a supplier may be based on an environmental issue, particularly if there is a production process that impacts the environment in a detrimental manner. More recently, there has been a significant trend toward "globalization," and suppliers have been selected based on nationality. This globalization and international exchange is likely to expand further in view of the growth in activities along the Pacific Rim, with the advent of the European Union, and when considering the technology advances in communications, data processing methods, and transportation systems.

The quantity of suppliers and the nature of their activities are a function of the type and complexity of the system being developed. For large, highly complex systems, there may be many different suppliers located throughout the world, as reflected in Figure 9.28. Some of these suppliers may be heavily involved in the design and development of major elements of the system, whereas other suppliers serve as sources for manufacturing and for selected inventories. For a system of this type, there is usually a large variety and mix of activities.

With regard to logistics, the requirements specified by the customer and imposed on the system prime contractor must be passed on to the various suppliers as applicable. The contractor will prepare an appropriate specification (Type *B*, *C*, *D*, or *E*) for each supplier requirement, along with a supporting Statement of Work (SOW) with specific supplier program tasks identified. Supplier responses, in the context of formal proposals, are submitted to the contractor and evaluated, selections are made, contracts are negotiated, and programs are implemented as a result.

In dealing with the suppliers of standard off-the-shelf components, it is important to prepare a good specification for the initial procurement of these items. Input-output parameters, size and weight, shape, density, and so on, must be covered in detail

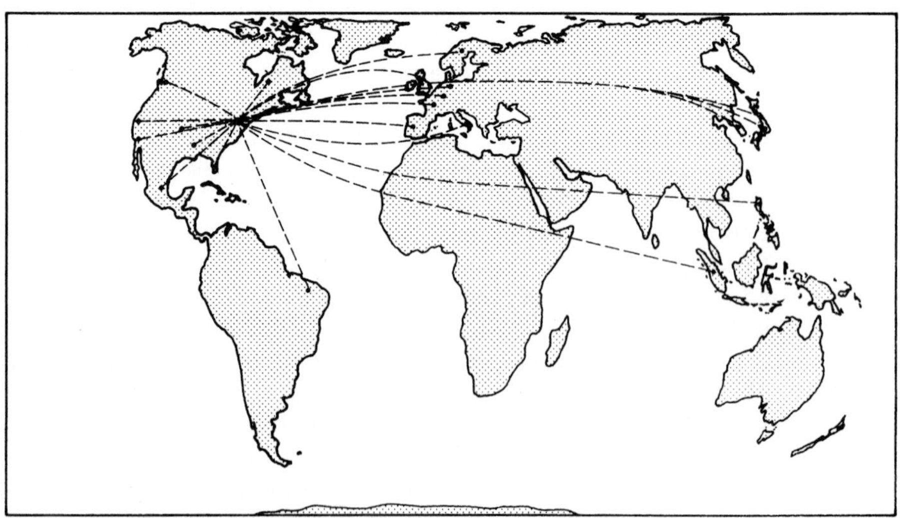

Figure 9.28 Potential suppliers for System *XYZ*.

along with allowable tolerances. Uncontrolled variances in component characteristics can have a significant impact on total system effectiveness and quality. Complete electrical, mechanical, physical, and functional interchangeability must be maintained where applicable. Although there are many different components that are currently in inventory and fall under the category of "commercial off-the-shelf (COTS)" components, extreme care must be exercised to ensure that like components (i.e., those with the same part number) are actually manufactured to the *same* standards. Also, the allowable variances around key parameters must be minimized.

In addressing the category of "supplier," one often finds a layering effect as illustrated in Figure 9.29. There is a wide variety of suppliers with varying objectives and organizational patterns. Although many are functionally oriented, the practices and procedures of each will be different. With regard to logistics per se, it is unlikely that the supplier organization will include a department, group, or section identified as such. However, accomplishment of the functions described previously (as applicable) is necessary. Although the overall responsibilities for logistics are assigned to the prime contractor, there must be some identifiable organizational element responsible for the *technical integration* of those tasks assigned to the supplier. It is important that logistics concepts and objectives be established and understood from the beginning, and the contribution of each supplier is essential to the realization of these objectives.

From the perspective of the prime contractor, the organization and management of supplier activity present major challenges. Not only are there many different types of suppliers with varying levels of responsibility, but these suppliers may be located across the United States, Canada, Mexico, Africa, Asia, Australia, Europe, South America, and so on. This is particularly true for large projects where the supplier is responsible for major elements of the system.

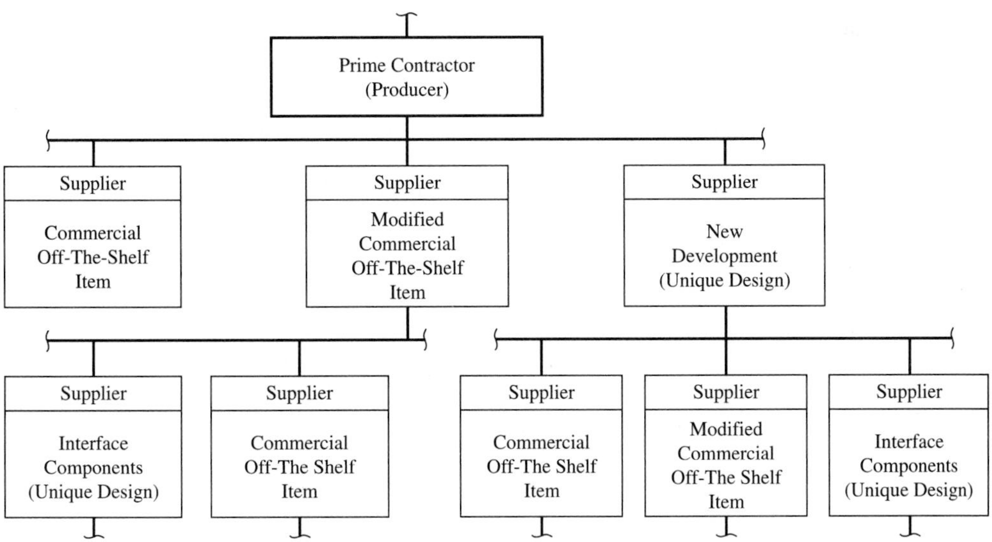

Figure 9.29 Supplier layering (refer Figure 5.24)

In such instances, it is necessary to ensure that the appropriate level of design integration is maintained throughout the system development process. This commences with the initial definition of supplier requirements and the preparation of specifications. Subsequently, design integration is realized through good communications and the periodic review of design documentation, and the conductance of design reviews. The establishment of successful communications is usually dependent on a complete understanding of the environment in which the supplier operates. For instance, in the international arena, the logistics engineer should have a basic understanding of the culture, customs and practices, export-import requirements, geographic layout, communication and transportation links, and available resources within each of the nationalities providing system components. Some knowledge of the language, and interpretations thereof, as it is applied in specifications and design data is important. In any event, this area of activity is increasing in importance as the emphasis on globalization becomes greater.[16]

Staffing the Organization

The interdisciplinary nature of logistics is such that a broad spectrum of personnel backgrounds and skills is required for the staffing of a logistics organization. There are planning and management functions, detail design engineering functions, analysis functions, technical writing functions, provisioning and procurement functions, transportation and distribution functions, maintenance and customer service functions, and so on. These various functions are dispersed throughout the system life cycle, may differ from one program phase to another, and the logistics manager must be particularly sensitive to these nonhomogeneous characteristics in the recruiting, hiring, and placement of personnel. For the purposes of further discussion, logistics personnel requirements have been combined into four basic categories:

1. *Logistics requirements and planning.* A person who has an in-depth understanding of logistics functions in the system life cycle, can analyze technical requirements and define design criteria for system supportability, and who is aware of the interrelationships among the elements of logistics and with other facets of a program. Specifically, this person must understand system requirements, the system design and development process, customer/user operations and organization (the customer environment), and must be knowledgeable in the basic principles of management (planning, organizing, scheduling, cost estimating, task implementation, and control). This person should possess the following general attributes: (a) approach work in a logical and organized manner; (b) have a good understanding of the numerous functions and activities required to make a program successful; (c) be able to evaluate technical requirements, specifications, work statements, and develop a program that will be responsive; (d) have vision, be creative, and be self-motivated; (e) be communicative, diplomatic, and apply good judgment in planning and performing logistics functions; and (f) have some prior experience in program planning.

[16] Refer to Appendix B for checklist questions pertaining to supplier selection and evaluation.

The qualifications of an individual in this category appear to be quite constraining since the desired personal attributes are many. However, it should be recognized that logistic representatives in the past, particularly in the early program phases, have in essence been considered in the class of "second-rate citizens." When system acquisition costs are anticipated as being high, logistic planning and the associated early program tasks are the first to be eliminated. Many program managers are not concerned with logistics at an early stage since they do not have to worry about it until later, and they feel that someone else will inherit any resulting problems that may occur; thus, a person who has experience, is technically astute, has visualization, possesses good judgment, and is persistent (not easily discouraged) is desired.

2. *System/product design support.* A design-oriented person, preferably with some design experience, who is knowledgeable of current design methods (e.g., CAD, CAM, CALS) and has an in-depth knowledge of consumer operations, organization, facilities, personnel skills, and the user environment in general. This person must have an engineering background, be able to converse fluently with the design engineer (on the latter's terms) and must be able to translate system requirements and field experience into design criteria. He or she is inclined to be a specialist in a given technical area and must be up to date relative to state-of-the-art advances in that field. The person in this category works with the design engineer on a day-to-day basis and should possess a high degree of ingenuity, creativeness, intelligence, precision, and general technical ability.

3. *Supportability analysis.* A person with a combination of skills in maintenance, maintenance analysis, statistical analysis, technical documentation, provisioning and procurement, logistics modeling, and data processing. This person must be capable of reviewing and evaluating design data, identifying maintenance tasks and the specific requirements for logistic support (e.g., spare/repair parts, test and support equipment, technical data, facilities, personnel and training), developing logistics management information, and in the processing of supportability analysis (SA) data (e.g., CALS application). He or she must be thoroughly familiar with the customer environment, levels of maintenance, user operating and maintenance procedures, organization structure and personnel skills, and should have actual hands-on maintenance experience in the field. Those involved in maintenance analyses must be able to visually project (into the future) a system, or a component thereof, and its support, and to capture that projection in writing and through the generation of a database. The specialists covering each of the logistics elements are included within this category, and the individuals selected must be analytically inclined and possess the ability to write. A good background in mathematics, statistics, modeling techniques, and computer applications is desirable.

4. *Distribution and customer support.* A person knowledgeable in the different facets of material flow and product support. Material flow includes the transportation and distribution of products, evolving from the supplier and moving through the production stage, warehousing, and delivery to the user. Personnel in this area should be conversant in business practices, operations research methods, principles of the inventory control, mathematics, and statistics. The other major thrust within this category is product support, or the maintenance and support of the system/product throughout the period of customer use. Such support may be in the form of a technician dispatched to the field to perform maintenance on the equipment (i.e., troubleshooting, repair, testing, etc.). Personnel background should include the successful completion of a technical school, along with some hands-on field experience on comparable equipment.

The requirements for staffing an organization stem from the results of early planning and, subsequently, from the labor projections included in the ILSP. Tasks are identified, combined into work packages and the work breakdown structure (WBS), and the work packages are grouped and related to specific position requirements. The positions are, in turn, arranged within the organizational structure considered to be most appropriate for the need (refer to Figures 9.19, 9.21, 9.23, 9.25 and 9.27). Although the personnel loading requirements will be oriented to the specific program, a generic approach, utilizing the four categories described above, is presented in Figure 9.30.

Referring to the figure, the personnel assigned to the first category, *logistics requirements and planning*, will initially be involved in the activities described throughout Chapter 3. This includes the determination of system support requirements through the feasibility analysis, the definition of operational requirements and the maintenance concept, and the preparation of the system specification. At the same time, there are some early planning activities and the development of the ILSP. The

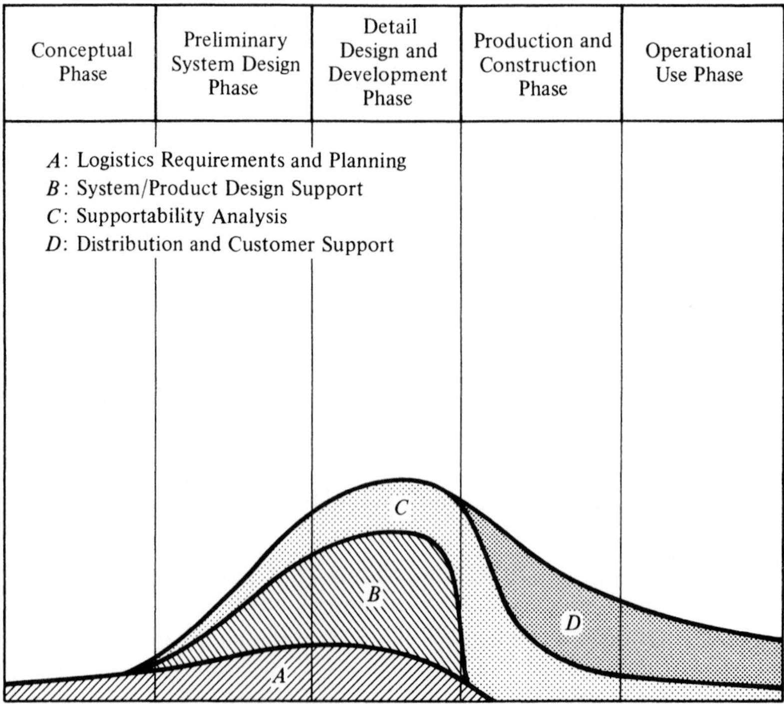

Figure 9.30 Typical personnel-loading curve.

system/product design support category addresses those design participation activities described in Chapter 5. Many of these activities are closely related to the reliability, maintainability, and human factors requirements discussed throughout. The third category, *supportability analysis*, covers those design evaluation and analytical functions described in Chapter 4 and Appendices C to E. The last category, *distribution and customer support*, can be broken down into two distinct areas. The first involves material flow and the distribution of products, as emphasized in the business logistics approach described in Chapter 1. The second deals with the maintenance and support of systems/products at customer locations in the field (i.e., field engineering, field service, customer service, or equivalent).

In staffing an organization, the logistics manager needs to develop the appropriate mix of personnel backgrounds and skills. This, of course, will require a close working relationship with the Human Resources Department in establishing the initial requirements for personnel, in developing position descriptions and advertising material, in recruiting and the conducting of interviews, in the selection of qualified candidates, and in the final hiring of individuals for employment within the logistics organization.[17]

9.7 MANAGEMENT AND CONTROL

Successful program implementation includes the day-the-day managerial functions associated with the influencing, guiding, and supervising of personnel in the accomplishment of the organizational activities identified in the ILSP. This requires a continuing awareness, on the part of management, of the "specifics" pertaining to the tasks being performed and whether the results are fulfilling the program objectives in an effective manner. The results desired include those factors identified through the definition of system operational requirements, the maintenance concept, the prioritization of the technical performance measures, and the functional analysis and allocation process (refer to Sections 3.3 through 3.7).

The "controlling" aspect pertains to the (1) ongoing monitoring of program activities in order to assess the status of the tasks being performed at designated times in the program and the results of those tasks in terms of the desired characteristics and metrics pertaining to the system being developed or reengineered; and (2) the initiation of corrective action as required to overcome any noted deficiencies in the event that program objectives are not being met.

Periodic program management reviews are usually conducted throughout the system design and development process. The nature and frequency of these will vary with each program and will be a function of design complexity, the number and location of organizations participating in the program, the number of suppliers, and the number

[17] In most companies the human resources department is responsible for establishing job classifications and salary structures for the recruiting and hiring of personnel, for initiating employee benefit coverage, for providing employee opportunities for education and training, and so on. It is incumbent on the logistics department manager to ensure that his or her organizational requirements are initially understood and subsequently met through recruiting, employment, and training activities.

of problems being encountered. Systems being developed, incorporating new technologies and where the risks are greater, will require more frequent reviews than for other programs.

The objective of these management-oriented reviews is to determine the status of selected program tasks, assess the degree of progress with respect to fulfilling contractual requirements, review program schedule and cost data, review supplier activities, and coordinate the results of the technical design reviews (refer to Section 5.5). These reviews are directed toward the dealing with issues of a "program" nature, and are not intended to duplicate the formal design review process which deals with design-related problems of a "technical" nature. However, of mutual interest is the "tracking" of the critical TPMs, as illustrated in Figure 9.31 (also refer to Figure 9.17).

The data system that includes the information needed for these program reviews must be designed in such a way as to provide the right type of information, to selected locations and in the proper format, in a timely manner and at the desired frequency, with the right degree of reliability, and at the right cost. The nature and the content of the data will be based on several factors, both of a technical nature (e.g., TPMs) and of a program nature (e.g., high-volume suppliers). It is not uncommon for one to go ahead and design and develop, or procure, a large computer-based management information system (MIS) capability with good intent, but that turns out to be too complex, nonresponsive by providing the wrong type of information, and very costly to operate. Such a capability must be tailored to the situation and sensitive to the needs of the particular project in question.

Regarding the more conventional type of reporting requirements, Figure 9.9 shows the current status of program activities (the sample bar graph at the top) and the status regarding major milestones (at the bottom). Figure 9.10 combines these into a master schedule, and Figure 9.11 presents milestones and activities in the form of a CPM network. This leads to the reporting of costs in Figure 9.15, where the activities are tied to both the CWBS and a PERT/CPM program network. Figure 9.16 shows a projected cost stream based on life-cycle considerations.

The data from a management information system should readily point out existing problem areas. Also, potential areas where problems are likely to occur, if program operations continue as originally planned, should become visible. To deal with these situations, planning should establish a corrective-action procedure that will include the following:

1. Problems are identified and ranked in order of importance.
2. Each problem is evaluated on the basis of the ranking, addressing the most critical ones first. Alternative possibilities for corrective action are considered in terms of (a) effects on program schedule and cost, (b) effects on system performance and effectiveness, and (c) the risks associated with the decision as to whether to take corrective action.
3. When the decision to take corrective action is reached, planning is initiated to take steps to resolve the problem. This may be in the form of a change in management policy, a contractual change, an organizational change, or a system/equipment configuration change.

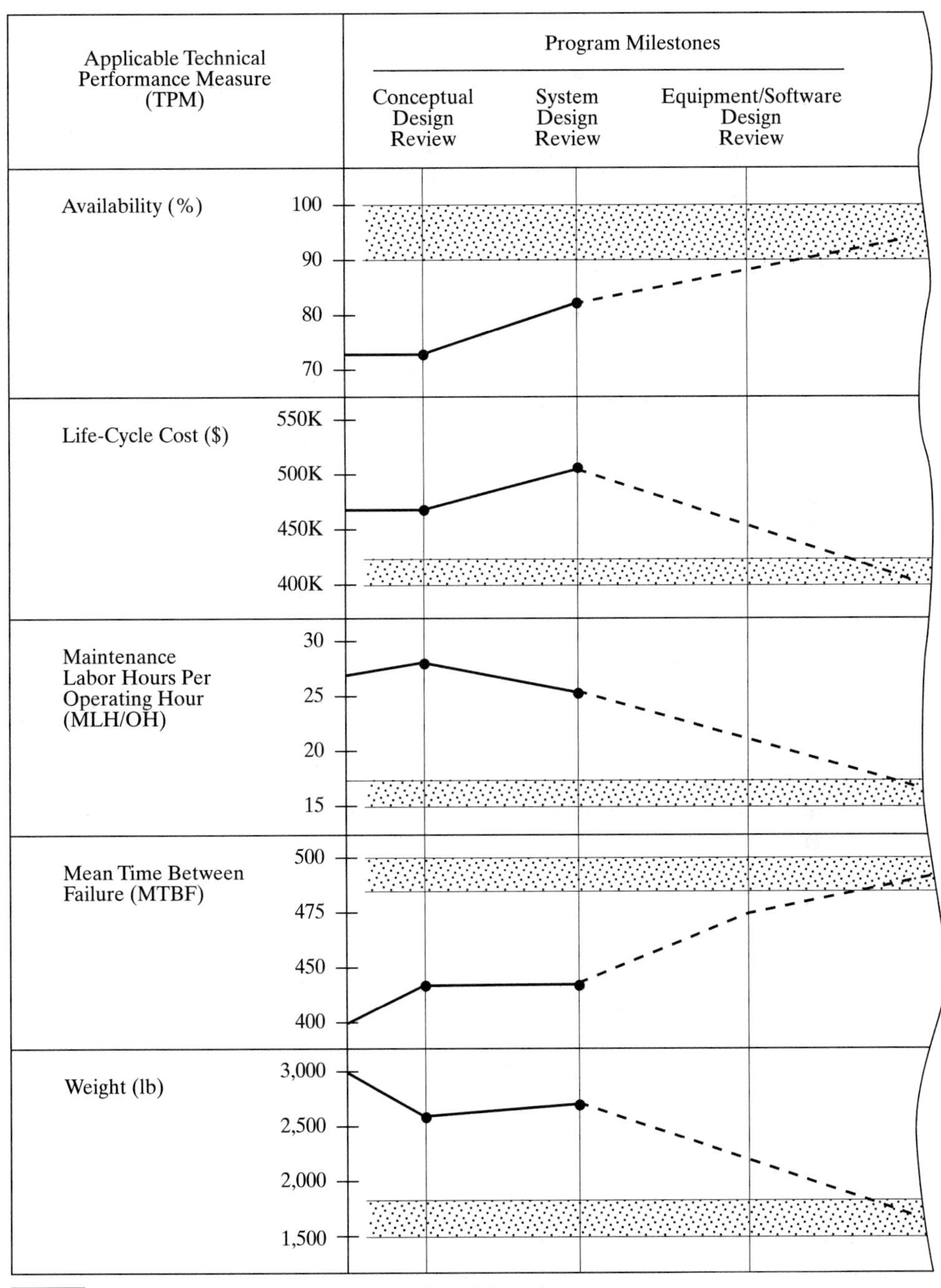

Figure 9.31 System parameter measurement and evaluation at design review.

4. After corrective action has been implemented. some follow-up activity is required to (a) ensure that the incorporated change(s) actually has resolved the problem, and (b) assess other aspects of the program to ensure that additional problems have not been created as a result of the change.

The activities associated with *management* and *control* are iterative by nature and continue throughout the system life cycle.

9.8 SUMMARY

The successful accomplishment of the objectives defined throughout this book is dependent on

1. Proper planning for logistic support in all program phases.
2. Incorporation of appropriate supportability characteristics in system design and development.
3. Identification, provisioning, and timely acquisition of support elements required for prime equipment operation and maintenance.
4. Continued assessment of the overall effectiveness of the system throughout its life cycle, and the initiation of modifications for corrective action or product improvement (as required).

Logistics must be considered on an equivalent basis along with performance, schedule, and cost implications. The appropriate trade-offs must be accomplished early in the life cycle and the proper balance attained if the product output is to be effective.

Logistics must be recognized by all levels of program management as being a major contributor in each phase of system development. The elements of support must be properly identified, and item acquisition should be accomplished in a timely manner. The contracting for logistics should consider the application of incentive and penalty clauses in areas where special emphasis is required. Logistics covers a broad spectrum of interests and should be properly integrated with all other facets of program activity.

QUESTIONS AND PROBLEMS

1. What are the basic requirements of program management? List components and criteria.
2. What is meant by *organization*? What are its characteristics, objectives, and so on?
3. There are various types of organizational structures including the *pure functional*, the *product line*, the *project*, and the *matrix*. Briefly describe the structure and identify some of the advantages and disadvantages of each.
4. Refer to Question 3. Which type of organizational structure is preferred from a logistics perspective? Why?
5. Assume that you have been assigned the responsibility for the design and development of a new system. Develop a project organizational structure (using any combination of approaches desired), identify the major elements contained therein and describe some of the

key interfaces, and describe how you plan to organize for logistics. Construct an organizational chart.

6. Refer to Question 5. Describe the logistics functions (or tasks) that should be accomplished.

7. From an organizational standpoint, identify and describe some of the conditions that must exist in order to accomplish logistics objectives in an effective manner.

8. Describe some of the major task interfaces between the logistics organization and each of the following: (a) system engineering; (b) design engineering; (c) reliability and maintainability engineering; (d) production (manufacturing); (e) configuration management; (f) purchasing; (g) contract management; (h) total quality management (quality control).

9. Refer to Question 5. Develop an ILSP in outline form, identify the applicable program phases and major functions/tasks by phase, and develop a bar/milestone chart (or program network).

10. Using the information identified in Problem 9, prepare a summary program network. The following data are available:

Event	Previous Event	t_a	t_b	t_c
8	7	20	30	40
	6	15	20	35
	5	8	12	15
7	4	30	35	50
	3	3	7	12
6	3	40	45	65
	2	25	35	50
5	2	55	70	95
4	1	10	20	35
3	1	5	15	25
2	1	10	15	30

(a) Construct a PERT/CPM chart from the preceding data.
(b) Determine the values for standard deviation, TE, TL, TS, TC, and *P*.
(c) What is the critical path? What does this value mean?

11. When employing PERT/COST, the time-cost operation applies. What is meant by the time-cost option? How can it affect the critical path?

12. Select a system of your choice and develop a system specification.

13. What is the basic difference between a development specification and a procurement specification? A process specification and a material specification? What is a *specification tree*? Why is it important?

14. What is a *work package*? What is the purpose of a WBS? How do work packages relate to the WBS? Construct a WBS for a program of your choice.

15. What is the difference between a SWBS and a CWBS?

16. Describe the relationships between the WBS, program functions/tasks, cost estimates, cost control numbers, and organizational responsibilities.

17. How does the WBS relate to the cost breakdown structure utilized in life-cycle cost and cost-effectiveness analyses?

18. Define *management by objectives* and discuss how the principles can be applied to logistics.

19. What type of personnel are needed to perform logistics functions (backgrounds and skills)?

20. As a manager of a logistics organization, one of your objectives is to *motivate* your employees to produce a high-quality output in an effective and efficient manner. Describe your approach. What factors would you consider? (*Hint:* Undertake some research relative to Maslow and Herzberg.)
21. As manager of the logistics department, what steps would you take to ensure that your organization maintains a leading position relative to technical competency?
22. When managing logistics activities throughout program implementation, what measures would you employ to ensure that logistics is being properly addressed?
23. In Problem 22, if an unsatisfactory condition is detected, how would you initiate corrective action? List and briefly discuss the steps involved.
24. How would you emphasize logistics in contracting for an item?
25. What factors are important in the application of incentive and penalty factors for logistics?
26. Why is the establishment of a good management information system so important?
27. Indentify some of the problems that can develop if logistic support is not properly planned and implemented.

APPENDICES

APPENDIX A

SYSTEM DESIGN REVIEW CHECKLIST

On a periodic basis throughout the design and development process, it is beneficial to accomplish an informal review to assess (1) whether the necessary tasks have been accomplished, and (2) the extent that supportability characteristics have been considered and incorporated in the system design. Questions, presented in a checklist format, have been developed to reflect certain features. Reliability, maintainability, human factors, and supportability considerations are included. Not all questions are applicable in all reviews; however, the answer to those questions that are applicable should be YES to reflect desirable results. For many questions, a more indepth study of applicable reliability, maintainability, human factors, and supportability criteria documentation will be required prior to arriving at a decision. These questions directly support the abbreviated checklist presented in Figure 5.8 (Chapter 5, page 239). Appendix B includes a checklist pertaining to supplier selection and evaluation.

General Requirements

1.0 System Operational Requirements

1.1. Has the mission been defined? Mission scenarios?

1.2. Have all basic system performance parameters been defined?

1.3. Has the planned operational deployment been defined (quantity of systems per location)?

1.4. Has the system life cycle been defined?

1.5. Have system utilization requirements been defined? This includes hours of system/equipment operation or quantity of operational cycles per a given time period. Operational scenarios?

1.6. Has the operational environment been defined in terms of temperature extremes, humidity, shock and vibration, storage, transportation, and handling?

2.0 Effectiveness Factors

2.1. Have system availability, dependability, readiness, or equivalent operational effectiveness factors been identified?

2.2. Have the appropriate quantitative reliability, maintainability, and supportability factors been specified? This includes MTBF, MTBM, λ MDT, $\overline{M}$, $\overline{M}ct$, $\overline{M}pt$, M_{max}, MLH/OH, Cost/OH, Cost/MA, and so on.

2.3. Have effectiveness factors for the system support capability been defined?

3.0 System Maintenance Concept

3.1. Have the echelons or levels of maintenance been specified and defined?

3.2. Have basic maintenance functions been identified for each level?

3.3. Have quantitative parameters been established for turnaround time (TAT) at each level and logistics pipeline time between levels?

3.4. Has the logistics pipeline time between levels been minimized to the extent feasible considering cost? The lack of adequate supply responsiveness has a major detrimental effect on total logistic support.

3.5. Have level of repair policies been established? Repair versus discard? Repair at intermediate/depot level?

3.6. Have the criteria for level of repair decisions been adequately defined?

3.7. Has the level of maintenance (organizational, intermediate, depot, or supplier) been defined for each repairable item?

3.8. Have criteria been established for test and support equipment at each level of maintenance? Software?

3.9. Have criteria been established for personnel quantities and/or skills at each level of maintenance?

3.10. Have the responsibilities for maintenance been established?

4.0 Functional Analysis and Allocation

4.1. Have system operational and maintenance functions been defined? Have the appropriate functional relationships and interfaces been established?

4.2. Have reliability, maintainability, and supportability factors been allocated to the appropriate system elements (e.g., unit, assembly, subassembly, etc.)?

4.3. Have cost factors been allocated to the appropriate system elements?

4.4. Have the appropriate factors been allocated to the logistic support infrastructure where applicable?

5.0 Supportability Analysis (SA)

5.1. Have trade-off evaluations and analyses been accomplished to support all logistic support requirements?

5.2. Have the applicable analyses described in Figure 4.9 been accomplished and appropriately integrated into the SA?

5.3. Does the supportability analysis data package (logistics management information) justify system design for supportability? Is the data package compatible with CALS?

5.4. Have trade-off evaluations and analyses been adequately documented?

6.0 Logistic Support Operational Plan

6.1. Has a plan been developed for the design, production, acquisition, deployment, interim contractor support, postproduction support, and integration of the prime equipment and logistic support elements in the field? This includes a preliminary logistic support plan and an integrated logistic support plan (ILSP).

6.2. Has a plan been developed for the handling of system modifications in the field?

6.3. Has a plan been developed covering system/equipment retirement and material phaseout?

Logistic Support Elements

1.0 Test and Support Equipment

1.1. Have the test and support equipment requirements been defined for each level of maintenance?

1.2. Have standard test and support equipment items been selected? Newly designed equipment should not be necessary unless standard equipment is unavailable.

1.3. Are the selected test and support equipment items compatible with the prime equipment? Does the test equipment do the job?

1.4. Are the test and support equipment requirements compatible with the supportability analysis?

1.5. Have test and support equipment requirements (both in terms of variety and quantity) been minimized to the greatest extent possible?

1.6. Are the reliability, maintainability, and supportability features in the test and support equipment compatible with those equivalent features in the prime equipment? It is not practical to select an item of support equipment which is not as reliable as the item it supports.

1.7. Have logistic support requirements for the selected test and support equipment been defined? This includes maintenance tasks, calibration and test equipment, spare/repair parts, personnel and training, data, software, and facilities.

1.8. Is the test and support equipment selection process based on cost-effectiveness considerations (i.e., life-cycle cost)?

1.9. Have test and maintenance software requirements been adequately defined?

2.0 Supply Support (Spare/Repair Parts)

2.1. Are the types and quantity of spare/repair parts compatible with the level of repair analysis?

2.2. Are the types and quantity of spare/repair parts designated for a given location appropriate for the estimated demand at that location? Too many or too few spares can be costly.

2.3. Are spare/repair part provisioning factors consistent with supportability analysis?

2.4. Are spare/repair part provisioning factors directly traceable to reliability and maintainability predictions?

2.5. Are the specified logistics pipeline times compatible with effective supply support? Long pipeline times place a tremendous burden on logistic support.

2.6. Have spare/repair parts been identified and provisioned for preoperational support activities (e.g., interim contractor support, test programs, etc.)?

2.7. Have spare/repair part requirements been minimized to the maximum extent possible?

2.8. Have test and acceptance procedures been developed for spare/repair parts? Spare/repair parts should be processed, produced, and accepted on a similar basis with their equivalent components in the prime equipment.

2.9. Have the consequences (risks) of stock-out been defined in terms of effect on mission requirements and cost?

2.10. Has an inventory safety stock level been defined?

2.11. Has a provisioning or procurement cycle been defined (procurement or order frequency)?

2.12. Has a supply availability requirement been established (the probability of having a spare available when required)?

2.13. Have post-production support requirements been defined?

3.0 Personnel and Training

3.1. Have operational and maintenance personnel requirements (quantity and skill levels) been defined?

3.2. Are operational and maintenance personnel requirements minimized to the greatest extent possible?

3.3. Are operational and maintenance personnel requirements compatible with the supportability analysis and with human factors data? Personnel quantities and skill levels should "track" both sources.

3.4. Are the planned personnel skill levels at each location compatible with the complexity of the operational and maintenance tasks specified?

3.5. Has maximum consideration been given to the use of existing personnel skills for new equipment?

3.6. Have personnel attrition rates been established?

3.7. Have personnel effectiveness factors been determined (actual time that work is accomplished per the total time allowed for work accomplishment)?

3.8. Have operational and maintenance training requirements been specified? This includes consideration of both initial training and replenishment training throughout the life cycle.

3.9. Have specific training programs been planned? The type of training, frequency of training, duration of training, and student entry requirements should be identified.

3.10. Are the planned training programs compatible with the personnel skill level requirements specified for the performance of operational and maintenance tasks?

3.11. Have training equipment requirements been defined? Procured?

3.12. Have maintenance provisions for training equipment been planned?

3.13. Have training data requirements been defined?

3.14. Are the planned operating and maintenance procedures (designated for support of the system throughout its life cycle) utilized to the maximum extent possible in the training program(s)?

4.0 Technical Data (Operating and Maintenance Procedures)

4.1. Have operating and maintenance procedure requirements been defined? Have the necessary procedures been prepared?

4.2. Are operating and maintenance procedures compatible with the supportability analysis data? This pertains particularly to the logic troubleshooting flow diagrams, task sequences, and support requirements defined in the maintenance analysis described in Appendix C.

4.3. Are operating and maintenance procedures as brief as possible without sacrificing necessary information?

4.4. Are operating and maintenance procedures adequate from the standpoint of presenting simple step-by-step instructions; including appropriate use of illustrations; and including tables for presenting data?

4.5. Are operating and maintenance procedures compatible with the level of activity performed at the location where the procedures are used? Depot maintenance instructions should not be included in manuals which are used at the intermediate level of maintenance. The maintenance procedures should be compatible with the level of repair analysis and the maintenance-concept.

4.6. Are operating and maintenance procedures written to the skill level of the individual accomplishing the functions covered by the procedures? Procedures should be written in a simple, clear, and concise manner for low-skilled personnel.

4.7. Do the operating and maintenance procedures specify the correct test and support equipment, spare/repair parts, transportation and handling equipment, and facilities?

4.8. Do the procedures include special warning notices in areas where safety is a concern?

4.9. Are the designated operating and maintenance procedures used in system/equipment test programs?

5.0 Facilities and Storage

5.1. Have facility requirements (space, volume, capital equipment, utilities, etc.) necessary for system operation been defined?

5.2. Have facility requirements (space, volume, capital equipment, utilities, etc.) necessary for system maintenance at each level been defined?

5.3. Have operational and maintenance facility requirements been minimized to the greatest extent possible?

5.4. Have environmental system requirements (e.g., temperature, humidity, and dust control) associated with operational and maintenance facilities been identified?

5.5. Have storage or shelf-space requirements for spare/repair parts been defined?

5.6. Have storage environments been defined?

5.7. Are the designated facility and storage requirements compatible with the supportability analysis and human factors data?

5.8. Have the warehousing requirements for material flow and distribution been defined?

6.0 Transportation and Handling

6.1. Are transportation and handling requirements for both operational and maintenance functions defined? This includes transportation of prime equipment, test and support equipment, spares, personnel, and data. National and international requirements should be identified.

6.2. Are transportation and handling environments (temperature, shock and vibration, exposure to dust and salt spray, storage, etc.) defined?

6.3. Are the modes (air, ground vehicle, rail, sea, pipeline, or a combination) of transportation known? A profile or scenario, similar to that accomplished for mission definition, should be developed showing the various transportation and handling requirements.

6.4. Are the requirements for reusable containers known? Design information should be developed on reusable containers.

6.5. Are the requirements for packing known? This includes labor, material, preservation, storage limitations, and the processing of an item for shipment.

Design Features

1.0 Selection of Parts

1.1. Have appropriate standards been consulted for the selection of components?

1.2. Have all component parts and materials selected for the design been adequately evaluated prior to their procurement and application? Evaluation should consider performance parameters, reliability, maintainability, supportability, and human factors.

1.3. Have supplier sources for component part procurement been established?

1.4. Are the established supplier sources reliable in terms of quality level, ability to deliver on time, and willingness to accept part warranty provisions?

1.5. Have the reliability, maintainability, human factors, and logistics engineers been consulted in the selection and application of parts? Reliability is concerned with part failure rates, stresses, tolerances, allowable temperature extremes, signal ratings, and so on. Maintainability and human factors are concerned with the part effects on maintenance times, mounting provisions, human interfaces, and so on.

2.0 Standardization

2.1. Are standard equipment items and parts incorporated in the design to the maximum extent possible (except for items not compatible with effectiveness factors)? Maximum standardization is desirable.

2.2. Are the same items or parts used in similar applications?

2.3. Are the number of different part types used throughout the design minimized? In the interest of developing an efficient supply support capability, the number of different item spares should be held to a minimum.

2.4. Are identifying equipment labels and nomenclature assignments standardized to the maximum extent possible?

2.5. Are equipment control panel positions and layouts (from panel to panel) the same or similar when a number of panels are incorporated and provide comparable functions?

3.0 Test Provisions (Testability)

3.1. Have self-test provisions been incorporated where appropriate?

3.2. Is the extent or depth of self-testing compatible with the level of repair analysis?

3.3. Are self-test provisions automatic?

3.4. Have direct fault indicators been provided (either a fault light, an audio signal, or a means of determining that a malfunction positively exists)? Are continuous performance monitoring provisions incorporated where appropriate?

3.5. Are test points provided to enable check-out and fault isolation beyond the level of self-test? Test points for fault isolation within an assembly should not be incorporated if the assembly is to be discarded at failure. Test point provisions must be compatible with the level of repair analysis.

3.6. Are test points accessible? Accessibility should be compatible with the extent of maintenance performed. Test points on the operator's front panel are not required for a depot maintenance action.

3.7. Are test points functionally and conveniently grouped to allow for sequential testing (following a signal flow), testing of similar functions, or frequency of use when access is limited?

3.8. Are test points provided for a direct test of all replaceable items?

3.9. Are test points adequately labeled? Each test point should be identified with a unique number, and the proper signal or expected measured output should be specified on a label located adjacent to the test point.

3.10. Are test points adequately illuminated to allow the technician to see the test point number and labeled signal value?

3.11. Can every equipment malfunction (degradation beyond specification tolerance limits) which could possibly occur in the equipment be detected through a no-go indication at the system level? This is a measure of test thoroughness.

3.12. Will the prescribed maintenance software provide adequate diagnostic information?

4.0 Packaging and Mounting

4.1. Is functional packaging incorporated to the maximum extent possible? Interaction effects between modular packages should be minimized. It should be possible to limit maintenance to the removal of one module (the one containing the failed part) when a failure occurs and not require the removal of two, three, or four modules.

4.2. Is the packaging design compatible with level of repair analysis decisions? Repairable items are designed to include maintenance provisions such as test points, accessibility, plug-in components, and so on. Items classified as discard at failure should be encapsulated and relatively low in cost. Maintenance provisions within the disposable module are not required.

4.3. Are disposable modules incorporated to the maximum extent practical? It is highly desirable to reduce overall support through a no-maintenance design concept as long as the items being discarded are relatively high in reliability and low in cost.

4.4. Are plug-in modules and components utilized to the maximum extent possible (unless the use of plug-in components significantly degrades the equipment reliability)?

4.5. Are accesses between modules adequate to allow for hand grasping?

4.6. Are modules and components mounted such that the removal of any single item for maintenance will not require the removal of other items? Component stacking should be avoided where possible.

4.7. In areas where module stacking is necessary because of limited space, are the modules mounted in such a way that access priority has been assigned in accordance with the predicted removal and replacement frequency? Items requiring frequent maintenance should be more accessible.

4.8. Are modules and components, not of the plug-in variety, mounted with four fasteners or less? Modules should be securely mounted, but the number of fasteners should be held to a minimum.

4.9. Are shock-mounting provisions incorporated where shock and vibration requirements are excessive?

4.10. Are provisions incorporated to preclude installation of the wrong module?

4.11. Are plug-in modules and components removable without the use of tools? If tools are required, they should be of the standard variety.

4.12. Are guides (slides or pins) provided to facilitate module installation?

4.13. Are modules and components labeled?

4.14. Are module and component labels located on top or immediately adjacent to the item and in plain sight?

4.15. Are the labels permanently affixed and unlikely to come off during a maintenance action or as a result of environment? Is the information on the label adequate? Disposable modules should be so labeled.

4.16. In equipment racks, are the heavier items mounted at the bottom of the rack? Unit weight should decrease with the increase in installation height.

4.17. Are operator panels optimally positioned? For personnel in the standing position, panels should be located between 40 and 70 inches above the floor. Critical or precise controls should be between 48 and 64 inches above the floor. For personnel in the sitting position, panels should be located 30 inches above the floor. Refer to your latest reference on anthropometric data.

5.0 Interchangeability

5.1. Are modules and components having similar functions electrically, functionally, and physically interchangeable?

5.2. Are components with the same part number but provided by different suppliers completely interchangeable?

6.0 Accessibility

6.1. Are access doors provided where appropriate? Are hinged doors utilized?

6.2. Are access openings adequate in size and optimally located for the access required?

6.3. Are access doors and openings labeled in terms of items that are accessible from within?

6.4. Can access doors that are hinged be supported in the open position?

6.5. Are access door fasteners minimized?

6.6. Are access door fasteners of the quick-release variety?

6.7. Can access be attained without the use of tools?

6.8. If tools are required to gain access, are the number of tools held to a minimum? Are the tools of the standard variety?

6.9. Are accesses between modules and components adequate?

6.10. Are access requirements compatible with the frequency of maintenance? Accessibility for items requiring frequent maintenance should be greater than that for items requiring infrequent maintenance.

7.0 Handling

7.1. For heavy items, are hoist lugs (lifting eyes) or base-lifting provisions for fork-lift-truck application incorporated? Hoist lugs should be provided on all items weighing more than 150 pounds.

7.2. Are hoist and base lifting points identified relative to lifting capacity? Are weight labels provided?

7.3. Are packages, units, components, or other items weighing over 10 pounds provided with handles? Are the proper-size handles used, and are they located in the right position? Are the handles optimally located from the weight-distribution standpoint? (Handles should be located over the center of gravity.)

7.4. Are packages, units, or other items weighing more than 40 pounds provided with two handles (for two-person carrying capability)?

7.5. Are containers, cases, or covers provided to protect equipment vulnerable areas from damage during handling?

8.0 Fasteners

8.1. Are quick-release fasteners used on doors and access panels?

8.2. Are the total number of fasteners minimized?

8.3. Are the number of different types of fasteners held to a minimum? This relates to standardization.

8.4. Have fasteners been selected based on the requirement for standard tools in lieu of special tools?

9.0 Panel Displays and Controls

9.1. Are controls standardized?

9.2. Are controls sequentially positioned?

9.3. Is control spacing adequate?

9.4. Is control labeling adequate?

9.5. Have the proper control/display relationships been incorporated?

9.6. Are the proper type of panel switches used?

9.7. Is the control panel lighting adequate?

9.8. Are the controls placed according to frequency of use?

9.9. Has a human factors engineer been consulted relative to controls and panel design?

10.0 Adjustments and Alignments

10.1. Are adjustment requirements and frequencies known?

10.2. Have adjustment requirements been minimized?

10.3. Are adjustment points accessible?

10.4. Are adjustment-point locations compatible with the maintenance level at which the adjustment is made?

10.5. Are adjustment interaction effects eliminated?

10.6. Are factory adjustments specified?

10.7. Are adjustment points adequately labeled?

11.0 Cables and Connectors

11.1. Are cables fabricated in removable sections?

11.2. Are cables routed to avoid sharp bends?

11.3. Are cables routed to avoid pinching?

11.4. Is cable labeling adequate?

11.5. Is cable clamping adequate?

11.6. Are connectors of the quick-disconnect variety?

11.7. Are connectors that are mounted on surfaces far enough apart so that they can be firmly grasped for connecting and disconnecting?

11.8. Are connectors and receptacles labeled?

11.9. Are connectors and receptacles keyed?

11.10. Are connectors standardized?

11.11. Do the connectors incorporate provisions for moisture prevention?

12.0 Servicing and Lubrication

12.1. Have servicing requirements been held to a minimum?

12.2. When servicing is indicated, are the specific requirements identified? This includes frequency of servicing and the materials needed.

12.3. Are procurement sources for servicing materials known?

12.4. Are servicing points accessible?

12.5. Have personnel and equipment requirements for servicing been identified? This includes handling equipment, vehicles, carts, and so on.

12.6. Does the design include servicing indicators?

13.0 Calibration

13.1. Have calibration requirements been held to a minimum?

13.2. Are calibration requirements known?

13.3. Are calibration frequencies known?

13.4. Are calibration tolerances known?

13.5. Are standards available for calibration?

13.6. Are calibration procedures prepared?

13.7. Is traceability to the National Institute of Standards and Technology (NIST) possible?

13.8. Have the facilities for calibration been identified?

13.9. Are the calibration requirements compatible with the supportability analysis and the maintenance concept?

14.0 Environment

14.1. Has the equipment design considered the following: temperature, shock, vibration, humidity, pressure, wind, salt spray, sand and dust, rain, fungus, and radiation? Have the ranges and extreme conditions been specified and properly addressed in design?

14.2. Have provisions been made to specify and control noise, illumination, humidity, and temperature in areas where personnel are required to perform operating and maintenance functions?

15.0 Storage

15.1. Can the equipment and spare parts be stored for extended periods of time without excessive degradation (beyond specification limits)?

15.2. Have scheduled maintenance requirements for stored equipment been defined?

15.3. Have scheduled maintenance requirements for stored equipment been eliminated or minimized?

15.4. Have the required maintenance resources necessary to service stored equipment been identified?

15.5. Have storage environments been defined?

15.6. Has the need for specialized environmentally controlled facilities been eliminated where possible?

16.0 Transportability

16.1. Have transportation and handling requirements been defined?

16.2. Have transportation requirements been considered in the equipment design? This includes consideration of temperature ranges, vibration and shock, humidity, and so on. Has the possibility of equipment degradation been minimized if transported by air, ground vehicle, ship, or rail?

16.3. Can the equipment be easily disassembled, packed, transported from one location to another, reassembled, and operated with a minimum of performance and reliability degradation?

16.4. Have container requirements been defined?

16.5. Have the requirements for ground handling equipment been defined?

16.6. Was the selection of handling equipment based on cost-effectiveness considerations?

17.0 Producibility

17.1. Does the design lend itself to economic production? Can simplified fabrication and assembly techniques be employed?

17.2. Has the design stabilized (minimum change)? If not, are changes properly controlled through good configuration management methods?

17.3. Is the design such that rework requirements are minimized? Are spoilage factors held to a minimum?

17.4. Has the design been verified through prototype testing, environmental qualification, reliability qualification, maintainability demonstration, and the like?

17.5. Is the design such that many models of the same item can be produced with identical results? Are fabrication steps, manufacturing processes, and assembly methods adequately controlled through good quality assurance procedures?

17.6. Has adequate consideration been given to the application of just-in-time (JIT), Taguchi, material requirements planning (MRP), and related methods in the production process?

17.7. Are production drawings, CAD/CAM/CALS data, material lists, and so on, adequate for production needs?

17.8. Can currently available facilities, standard tools, and existing personnel be used for fabrication, assembly, manufacturing and test operations?

17.9. Is the design such that automated manufacturing processes (e.g., CAM, numerical control techniques) can be applied for high-volume repetitive functions?

17.10. Is the design definition such that two or more suppliers can produce the system/product from a given set of data with identical results?

18.0 Safety

18.1. Has an integrated safety engineering plan been prepared and implemented?

18.2. Has a hazard analysis been accomplished to identify potential hazardous conditions? Is the hazard analysis compatible with the reliability FMECA/FMEA (where applicable)?

18.3. Have system/product hazards from heat, cold, thermal change, barometric change, humidity change, shock, vibration, light, mold, bacteria, corrosion, rodents, fungi, odors, chemicals, oils, greases, handling and transportation, and so on, been eliminated?

18.4. Have fail-safe provisions been incorporated in the design?

18.5. Have protruding devices been eliminated or are they suitably protected?

18.6. Have provisions been incorporated for protection against high voltages?

18.7. Are all external metal parts adequately grounded?

18.8. Are sharp metal edges, access openings, and corners protected with rubber, fillets, fiber, or plastic coating?

18.9. Are electrical circuit interlocks employed?

18.10. Are standoffs or handles provided to protect system components from damage during the performance of shop maintenance?

18.11. Are tools that are used near high-voltage areas adequately insulated at the handle or at other parts of the tool that the maintenance person is likely to touch?

18.12. Are the environments such that personnel safety is ensured? Are noise levels with a safe range? Is illumination adequate? Is the air clean? Are the temperatures at a proper level? Are OSHA requirements being maintained?

18.13. Has the proper protective clothing been identified for areas where the environment could be detrimental to human safety? Radiation, intense cold or heat, gas, loud noise, and so on, are examples.

18.14. Are safety equipment requirements identified in areas where ordinance devices (and the like) are activated?

19.0 Reliability

19.1. Is the design simple? Have the number of component pasts been kept to a minimum?

19.2. Are standard high-reliability parts being utilized?

19.3. Are item failure rates known? Has the mean life been determined?

19.4. Have parts been selected to meet reliability requirements?

19.5. Have parts with excessive failure rates been identified (unreliable parts)?

19.6. Have adequate derating factors been established and adhered to where appropriate?

19.7. Have the shelf life and wearout characteristics of parts been determined?

19.8. Have all critical-useful-life items been eliminated from the design? If not, have they been identified with inspection/replacement requirements specified? Has a critical-useful-life analysis been accomplished?

19.9. Have critical parts that require special procurement methods, testing, and handling provisions been identified?

19.10. Has the need for the selection of "matching" parts been eliminated?

19.11. Have fail-safe provisions been incorporated where possible (protection against secondary failures resulting from primary failures)?

19.12. Has the use of "adjustable" components been minimized?

19.13. Have safety factors and safety margins been used in the application of parts?

19.14. Have component failure modes and effects been identified? Has a FMEA, a FMECA, or a fault-tree analysis (FTA) been accomplished?

19.15. Has a stress-strength analysis been accomplished?

19.16. Have cooling provisions been incorporated in design "hot spot" areas? Is cooling directed toward the most critical items?

19.17. Has redundancy been incorporated in the design where needed to meet specified reliability requirements?

19.18. Are the best available methods for reducing the adverse effects of operational and maintenance environments on critical components being incorporated?

19.19. Have the risks associated with critical-item failures been identified and accepted? Is corrective action in design being taken?

19.20. Have reliability requirements for spares and repair parts been considered?

19.21. Have reliability predictions been accomplished? Have reliability testing requirements been defined? Test requirements in design? Test requirements in production/construction? Have they been covered in the Test and Evaluation Master Plan (TEMP)?

19.22. Has a reliability failure analysis and corrective action capability been installed?

20.0 Software

20.1. Have all system software requirements for operating and maintenance functions been identified? Have these requirements been developed through the system-level functional analysis (i.e., is there traceability indicated)?

20.2. Is the software complete in terms of scope and depth of coverage?

20.3. Is the software compatible relative to the equipment with which it interfaces? Is operating software compatible with maintenance software? With other elements of the system?

20.4. Are the language requirements for operating software and maintenance software compatible?

20.5. Has the software been packaged in functional modules for rapid removal and replacement?

20.6. Is all software adequately covered through good documentation (i.e., logic functional flows and coded programs)?

20.7. Has the software been adequately tested and verified for accuracy (performance), reliability, and maintainability?

21.0 Disposability

21.1. Has the equipment been designed for disposability (e.g., selection of materials, packaging)?

21.2. Have procedures been prepared to cover system/equipment/component disposal?

21.3. Can the components or materials used in system/equipment design be recycled for use in other products?

21.4. If component/material recycling is not feasible, can decomposition be accomplished?

21.5. Can recycling or decomposition be accomplishing using existing logistic support resources?

21.6. Are recycling or decomposition methods and results consistent with environmental, ecological, safety, political, and social requirements?

21.7. Is the method(s) used for recycling or decomposition economically feasible?

APPENDIX B

SUPPLIER EVALUATION CHECKLIST

This checklist, to be applied in the evaluation of suppliers, is "tailored" and is a supplemental version of the in-depth design review checklist presented in Appendix A. Not all of the questions are applicable in all situations; however, the answer to those questions that are applicable should be YES to reflect the desired results.

B.1 General Criteria

B.1.1 Has a technical performance specification been prepared covering the product being acquired? Is this specification "supportive" of and "traceable" from the System Specification?

B.1.2 Is the product a commercial off-the-shelf (COTS) item requiring no adaptation, modification, or rework for installation?

B.1.3 Has the COTS item been assessed in terms of effectiveness and life-cycle cost?

B.1.4 If the product is a COTS item and requires some modification for installation,

1. Has the degree of modification been clearly defined and minimized to the extent possible?
2. Has the impact of the modification been assessed in terms of effectiveness and life-cycle cost? Has the life-cycle cost been minimized to the extent possible?
3. Can the modification be accomplished easily and with a minimum of interaction effects?
4. Have common and standard parts, reusable software, recycleable material, and so on, been incorporated in the modification/interface package or kit?

B.1.5 Have alternative sources of supply for the same product been identified?

B.1.6. If new design is required, has it been justified to the extent that the COTS, modified COTS, and comparable options are not feasible?

B.2 Product Design Characteristics

B.2.1 Technical Performance Parameters

1. Does the product fully comply with the functional performance specification (i.e., development, product, process, or material specification as applicable)?
2. Has the applicable mission scenario (or operational/utilization profile) been defined for the product?
3. Are the product's design characteristics responsive to the prioritized technical performance measures (TPMs)? Does the design reflect the most important features?
4. Were the design characteristics derived through the use of a QFD (or equivalent) approach?
5. Are the performance requirements easily traceable from those specified for the system level?
6. Are the performance requirements measurable? Can they be verified or validated?

B.2.2 Technology Applications

1. Does the design utilize state-of-the-art and commercially available technologies?
2. Do the technologies utilized have a life cycle that is at least equivalent to the product life cycle?
3. Have "short-life" technologies been eliminated? If not, have such applications been minimized?
4. Has an "open-architecture" approach been utilized in the design such that new technologies can be inserted without causing a redesign of other elements of the product?
5. Have alternative sources for each of the technologies being utilized been identified?
6. Have the technologies being utilized reached a point of maturity/stability relative to their applications?

B.2.3 Physical Characteristics

1. Is the product both functionally and physically interchangeable?
2. Can the product be physically removed and replaced with a like item without requiring any subsequent adjustments or alignments? If not, have such interaction effects been minimized?
3. Does the product design comply with the physical requirements in the technical specification (i.e., size, shape, and weight)?

B.2.4 Effectiveness Factors

1. Have the appropriate effectiveness factors been defined and included in the technical specification (i.e., TPMs applicable to the product being acquired)?

2. Can the effectiveness requirements be traced back to comparable requirements specified at the system level?
3. Has the supplier provided a measure of reliability for the product (e.g., R, failure rate, or MTBF)? Is this figure of merit based on actual field experience?
4. Have the applicable reliability requirements been considered in the product design?
5. Has the supplier provided a measure of maintainability for the product (e.g., MTBM, MLH/OH, $\overline{\mathrm{M}}\mathrm{ct}$, $\overline{\mathrm{M}}\mathrm{pt}$ MDT, or equivalent)? Is this figure or merit based on actual field experience?
6. Have the applicable maintainability requirements been considered in the product design?
7. Have the applicable human-factors requirements been considered in the product design?
8. Have the applicable safety requirements been considered in the product design?
9. Have the applicable supportability/serviceability requirements been considered in the product design.
10. Have the applicable quality requirements been considered in the product design?

B.2.5 Producibility Factors

1. Has the product been designed for producibility?
2. Is the design data/documentation such that any other supplier with comparable facilities/equipment, capabilities, and experience can manufacture the product in accordance with the specification?

B.2.6 Disposability Factors

1. Has the product been designed for disposability?
2. Has the supplier developed the appropriate planning documentation and procedures covering the disposal or recycling of the product?

B.2.7 Environmental Factors

1. Has the product been designed with ecological and environmental requirements in mind?
2. Has the supplier prepared an environmental impact statement for the introduction of the product?

B.2.8 Economic Factors

1. Has the product been designed with economic considerations in mind.
2. Has the supplier conducted a life-cycle cost analysis for the product? Are the results realistic? Refer to Appendix E.

B.3 Product Maintenance and Support Infrastructure

B.3.1 Maintenance and Support Requirements

1. Does the supplier have an established maintenance and support infrastructure in place?
2. Has the supplier defined the maintenance concept/plan for the product.

3. Have the appropriate supportability "metrics" been established for the product and included in the maintenance concept/plan (i.e., response time, turnaround time, maintenance process time, test equipment reliability and maintainability factors, facility utilization, spare parts demand rates and inventory levels, transportation rates and times, etc.)?
4. Does the maintenance concept/plan facilitate or allow for the required degree of *responsiveness* on the part of the supplier?
5. Have the preventive maintenance requirements been established for the product (if any)? Have these requirements been justified through a reliability-centered maintenance (RCM) approach?
6. Have the product maintenance and support resource requirements been defined (i.e., spares, repair parts, and associated inventories; personnel quantities, skill levels, and training; test and support equipment; facilities; packaging, transportation and handling; technical data; and computer resources)? Have these requirements been adequately justified through a maintenance engineering analysis (MEA), a supporability analysis (SA), or equivalent?

B.3.2 Data/Documentation

1. Does the supplier have a computerized maintenance management data capability in place? Is this capability being affectively utilized for the purposes of *continuous product/process improvement?* Does it provide the visibility relative to how well the product is performing in the field?
2. Does the supplier have in place a reliability data collection, analysis, feedback, and corrective-action process? Are product failures properly recorded and are they traceable to the "cause"?
3. Is the supplier monitoring and measuring the effectiveness of its preventive maintenance program? Where applicable, have the preventive maintenance requirements been revised to reflect a more cost-effective approach?

B.3.3 Warranty/Guarantee Provisions

1. Have product warranties/guarantees been established?
2. Have the established warranty provisions been adequately defined through some form of a contractual mechanism?
3. Are the warranty provisions consistent with the defined maintenance concept?

B.3.4 Customer Service

1. Does the supplier have an established customer service capability in place?
2. Will the supplier provide assistance in the installation and checkout of the product at the producer's site or the user's site (if required)?
3. Will the supplier provide on-site field service support if required?
4. Does the supplier provide operator and maintenance training at the producer's site and/or the user's site when necessary? Is this training available "on call?" Will it be available throughout the product life cycle?
5. In support of training activities, will the supplier provide the necessary data, training manuals, software, aids, equipment, simulators, and so on? Will the supplier provide updates/revisions to the training material as applicable?
6. Does the supplier have a program for measuring training effectiveness?

B.3.5 Economic Factors

1. Is the product support infrastructure cost-effective?
2. Have the requirements been based on life-cycle cost objectives?

B.4 Supplier Qualifications

B.4.1 Planning/Procedures

1. Does the supplier have a standard policies and procedures manual/guide?
2. Are the appropriate management procedures properly documented and followed on a day-to-day basis?
3. Are the procedures/processes periodically reviewed, evaluated, and revised as necessary for the purposes of *continuous process improvement?*
4. Has the supplier identified the activities and tasks that are essential in the successful accomplishment of system and supportability requirements?

B.4.2 Organizational Factors

1. Has the supplier's organization been adequately defined in terms of activities, responsibilities, interface requirements, and so on?
2. Does the organizational structure support the overall program objectives for the system? Is it compatible with the producer's organizational structure?
3. Has the supplier identified the organizational element responsible for the accomplishment of logistics engineering tasks (as applicable)?

B.4.3 Available Personnel and Resources

1. Does the supplier have the available personnel and associated resources to assign to the task(s) being contracted? Will these personnel/resources be available for the duration of the program?
2. Do the personnel assigned have the proper background, experience, and training to do the job effectively?

B.4.4 Design Approach

1. Has the supplier implemented the system engineering process in the design of its products?
2. Has an effective design database been established, and is it compatible with the system-level database established by the producer (prime contractor)?
3. Does the supplier have in place a configuration management program, along with a disciplined change-control process? Has a configuration "baseline" approach been implemented in the development and growth of the product?
4. Has the supplier's design process been enhanced through the use of such tools as CAD, simulation, rapid prototyping, and so on?

B.4.5 Manufacturing Capability

1. Does the supplier have a well-defined manufacturing process in place?
2. Does the process incorporate the latest technologies and computer-aided methods (i.e., robotics, the use of CAD or CIM technology, etc.)?
3. Is the process flexible and does it support an "agile" and/or "lean" manufacturing approach?

4. Does the supplier utilize materials requirements planning (MRP), capacity planning (CP), shop floor control (SFC), just-in-time (JIT), master production scheduling (MPS), statistical process control (SPC), and other such methods in the manufacturing process?
5. Has the supplier implemented a formal quality program in accordance with ISO-9000 (or equivalent)? Is the supplier ISO-9000 certified? Does the supplier have a formal procedure in place for correcting deficiencies?
6. Has the supplier implemented a total productive maintenance (TPM) program within its manufacturing plant? Has a TPM measure of effectiveness been established (i.e., OEE, or overall equipment effectiveness)?

B.4.6 Test and Evaluation Approach

1. Has the supplier developed an integrated test and evaluation plan for the product?
2. Have the requirements for testing been derived in a logical manner, and are they compatible with the identified technical performance measures (TPMs) for the system, and as allocated for the product?
3. Does the supplier have the proper facilities and resources to support all product testing requirements (i.e., people, facilities, equipment, data)?
4. Does the supplier have in place a data collection, analysis, and reporting capability covering all testing activities?
5. Does the supplier have a plan for "retesting" if required?

B.4.7 Management Controls

1. Has the supplier incorporated the necessary controls for monitoring, reporting, providing feedback, and intiating corrective action with regard to technical performance measurement, cost measurement, and scheduling?
2. Has the supplier implemented a configuration management capability?
3. Has the supplier implemented an integrated data management capability?
4. Has the supplier developed a risk management plan?

B.4.8 Experience Factors

1. Has the supplier had experience in designing, testing, manufacturing, handling, delivering, and supporting this product before?
2. Has the supplier utilized experiences from other projects to help respond to the requirements for this program—that is, the transfer of "lessons learned"?

B.4.9 Past Performance

1. Has the supplier successfully completed similar projects in the past?
2. Has the supplier been responsive to all of the requirements for past projects?
3. Has the supplier been successful in delivering products in a timely manner and within cost?
4. Has the supplier delivered reliable and high-quality products?
5. Has the supplier been responsive in initiating any corrective action that has been required to correct deficiencies?
6. Has the supplier stood behind all product warranties/guarantees?
7. Does the supplier's organization reflect stability, growth, and high quality?

8. Is the supplier's business posture good?
9. Does the supplier enjoy an excellent reputation?

B.4.10 Maturity

1. Has the supplier established a process for benchmarking?
2. Has the supplier implemented an organizational self-assessment program

B.4.11 Economic Factors

1. Has the supplier implemented a life-cycle cost-analysis approach for all of its functions, products, processes, and so on?
2. Has the supplier implemented an "activity-based costing (ABC)" approach with the objective of acquiring full visibility relative to the high-cost contributors, cause-and-effect relationships, and leading to the implementation of improvements for cost-reduction purposes?

MAINTENANCE TASK ANALYSIS (MTA)

The maintenance analysis is an inherent part of the supportability analysis, and the material presented in this appendix complements the information described in Chapter 4. The analysis provides an engineering data package covering all scheduled and unscheduled maintenance functions, tasks, and associated logistic support requirements for the system or equipment being analyzed. The type of information output provided by the analysis is presented in Figures 4.3 and 4.7.

Maintenance analysis data cover:

1. All significant repairable items (i.e., system, subsystem, assembly, and subassembly). Items, identified through the level of repair analysis, which are relatively complex and require an analysis to determine the type and extent of support needed should be addressed.
2. All maintenance requirements (i.e., troubleshooting, remove and replace, repair, servicing, alignment and adjustment, functional test and checkout, inspection, calibration, overhaul, etc.).

The maintenance analysis is developed on an iterative basis throughout system definition and design. Basically, the analysis stems from the maintenance concept and is dependent on engineering design data, reliability and maintainability analyses and predictions, human factors data, and so on. Maintenance analysis data are used to support design decisions and to serve as the basis for the subsequent provisioning and acquisition of specific logistic support items needed for the operation and maintenance of the system when deployed in the field.

The maintenance analysis data format is by no means fixed. It can vary from program to program, and is tailored to the specific information desired and the time in the life cycle when needed. When developing analysis data in support of early design decisions (involving the evaluation of alternatives—see Chapter 4), the amount of information needed is less precise and not as extensive as the data required for a fixed configuration going into a final design review. In addition, the early analysis efforts must provide the right information expeditiously as the designer is required to make many decisions in a relatively short period of time. If this information is not available in a timely manner, the decisions will be made anyway and the results may not prove to be as beneficial. Thus, the maintenance analysis may assume several different postures as system/equipment development progresses.

The material presented herein is an extension of the MTA example described in Section 4.2.4. Figure 4.27 illustrates a specific situation where the application of the MTA is particularly appropriate. Figures 4.28, 4.29, and 4.30 illustrate the data formats used in the case study presented. The material in this appendix describes the process used in completing the information needed.

Logic Troubleshooting Diagram (Figures 4.28, C.1, C.2, and C.3)

In the performance of corrective maintenance, the analyst must visualize what the maintenance technician will experience in the field. At random points in time when the system is operating, failures are likely to occur and will be detected by the operator through visual, audio, and/or physical means. The operator proceeds to notify the appropriate maintenance organization that a problem exists.

The maintenance technician assigned to deal with the problem must analyze the situation and verify that the system is indeed faulty. In some instances, the fault will be obvious, particularly in dealing with mechanical or hydraulic systems when a structural failure has occurred or a fluid leak takes place. On other occasions, the technician must operate the system and attempt to repeat the condition leading to failure occurrence. This is often the case for electronic equipment when the failure is not always obvious.

In any event, corrective maintenance generally commences with the identification of a failure symptom such as the system does not work, the hydraulic system leaks, the engine does not respond in terms of power output, no voltage indication on the front panel meter, and so on. Based on a symptom of this nature, the maintenance technician proceeds to troubleshoot and accomplish the necessary repair actions (refer to Figure 2.11).

Troubleshooting may be extremely simple or quite complex. If a hydraulic leak is detected, the source of the leak is often quite easily traced. On the other hand, the failure of a small component in a radar or computer equipment is not readily identified. In this instance, the technician must accomplish a series of steps in a logical manner which will lead him or her directly to the faulty item. At times, these steps are not adequately defined and the technician is forced into a trial-and-error approach to maintenance. A good example is when the technician starts replacing parts on a mass basis (without analyzing cause-and-effect relationships) hoping that the problem will disappear in the process. This, of course, affects maintenance downtime and spare/repair part needs, as the technician may replace many parts when only one of them is actually faulty.

To preclude the possibility of wasting time and resources when the system is deployed in the field, the system design must provide the necessary characteristics to enable the maintenance technician to proceed in an accurate and timely manner in identifying the cause of failure. Such characteristics may constitute a combination of go/no-go lights, test points, meters, and other readout devices providing the necessary information which allows the technician to go from step to step with a high degree of confidence that he or she is progressing in the right direction. This objective is one of the goals of the maintainability design effort. Given a design approach, the maintenance analysis is accomplished to verify that the design is supportable. This facet of the analysis (i.e., the diagnostic aspect) is best accomplished through the development of logic troubleshooting flow diagrams, including go/no-go solutions on a step-by-step basis, and supported by diagnostic software where applicable. Three examples are presented in Figures C.1, C.2, and C.3 (also refer to Figure 4.28).

The analyst should review reliability failure mode and effect analysis data to determine cause and effect relationships, and then proceed to list all of the major

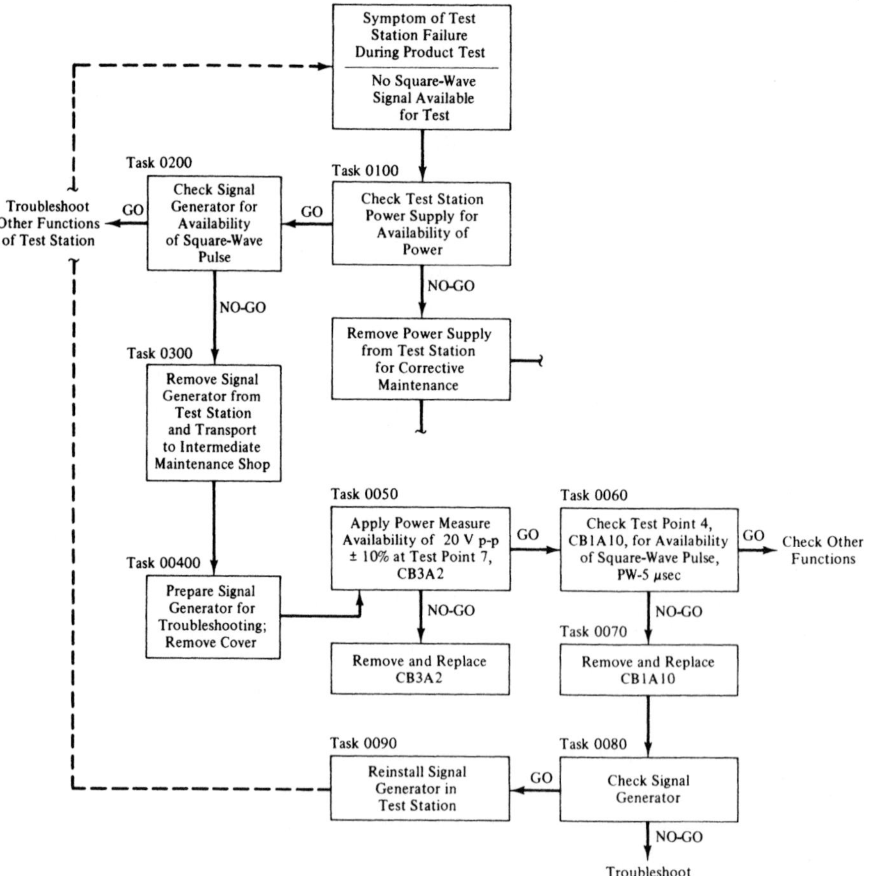

Figure C.1 Abbreviated logic troubleshooting flow diagram (example 1).

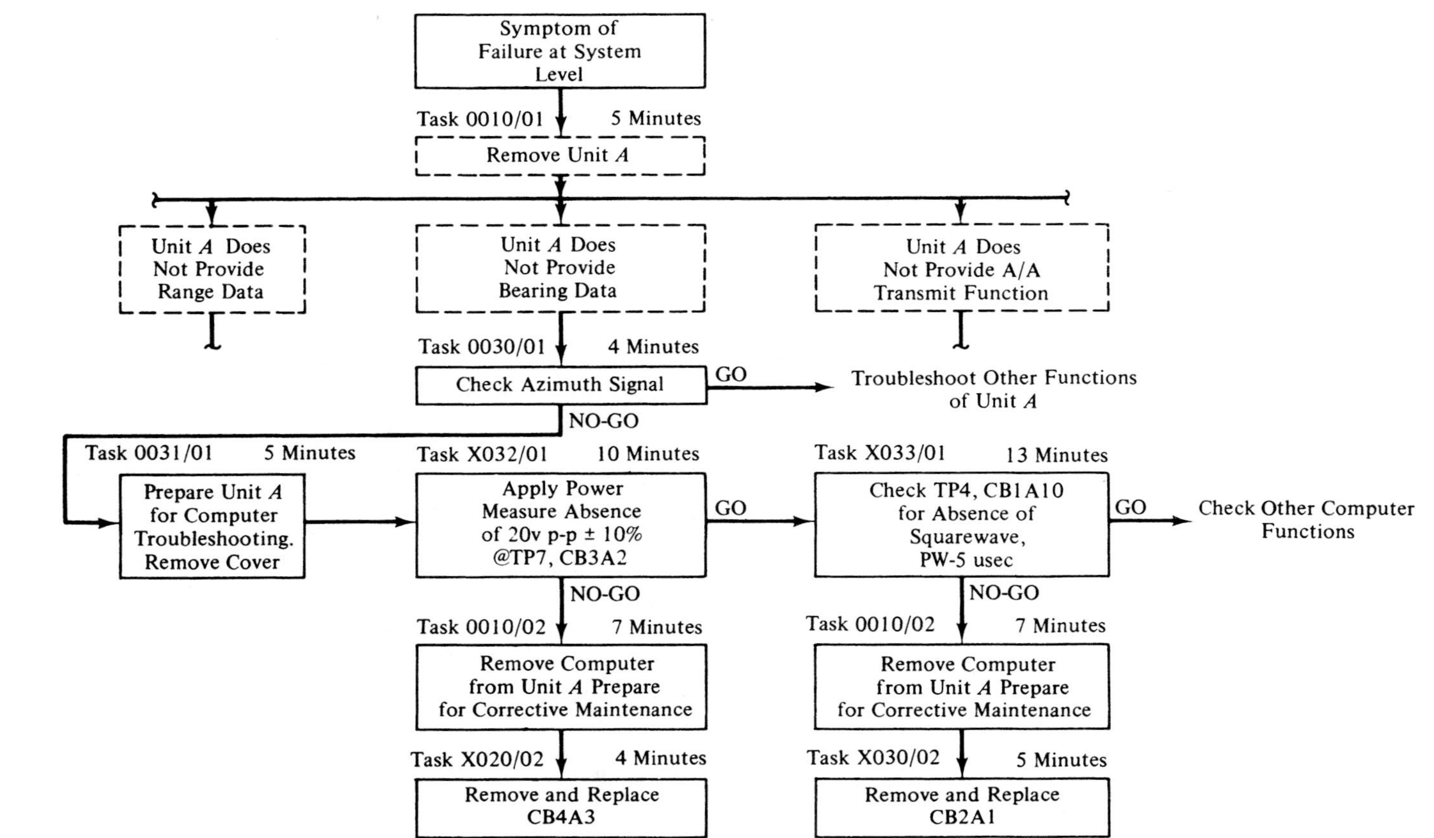

Figure C.2 Logic troubleshooting flow diagram (example 2).

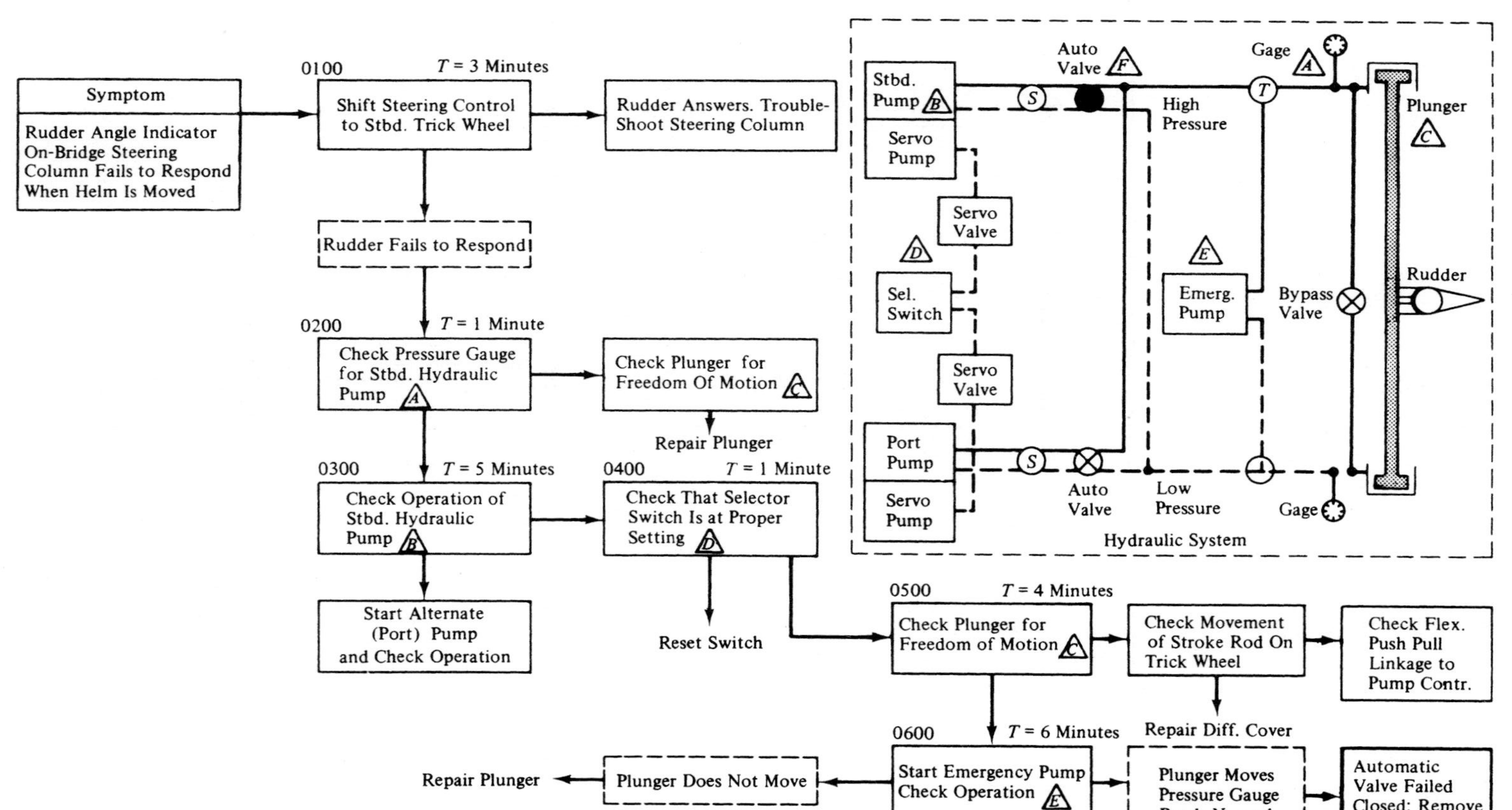

Figure C.3 Logic troubleshooting flow for ship steering system (example 3).

symptoms which the system is likely to experience.[1] For each symptom, various troubleshooting approaches are analyzed in terms of maintenance time and logistics resources, and the best approach is selected.[2] The analysis process is accomplished through the generation of logic troubleshooting flow diagrams in conjunction with the completion of maintenance task analysis sheets 1 and 2 for the troubleshooting requirement.

Maintenance Task Analysis—Sheet 1 (Figures 4.29, C.4, and C.5)

The following instructions should be followed when completing sheet 1 of the maintenance task analysis form:

1. *Block 1: System.* Enter the proper nomenclature of the system or end item covered by the overall analysis.

2. *Block 2: Item name/part number.* Enter the name and manufacturer's part number for the repairable item being covered by this task analysis sheet. This may constitute a subsystem, assembly, subassembly, and so on.

3. *Block 3: Next-higher assembly.* Include the name and part number of the next higher assembly. This constitutes one higher indenture level of hardware above the item listed in block 2. If a subassembly is being analyzed, the assembly should be identified in block 3.

4. *Block 4: Description of requirement.* A technical description and justification for the maintenance requirement identified in block 6 should be included. The need for performing maintenance must be clearly established. Include references to related requirements.

5. *Block 5: Requirement number.* A number is assigned for each requirement applicable to the item being analyzed. Requirements are identified in block 6, and may be sequentially numbered from 01 to 99 as necessary.

6. *Block 6: Requirement.* The requirement nomenclature should be entered in this block. Generally, maintenance requirements fall into one of the following areas:

(a) *Adjustment/alignment.* To line up, balance, or alter as necessary to make an item compatible with system requirements. This is a maintenance requirement when the primary cause of the maintenance action is to adjust or align, or to verify adjustment/alignment of an item. Adjustment/alignment accomplished subsequent to repair of a given item is not considered as a separate requirement, but is included as a task in the repair requirement.

[1] It may impossible to cover all symptoms of failure; however, the analyst should be able to identify tha major ones or those occurring most frequently.

[2] The preferred approach is the one which is consistent with performance and effectiveness requirements and which reflects the lowest life-cycle cost.

Maintenance Task Analysis (Sheet 1) Page 1 of 2

1. System	2. Item Name/Part No.	3. Next Higher Assy.	4. Description of Requirement
System Test	Test Station/A12345	System Text/A12300	During final manufacturing test of product "X" (serial no. 25610), the system test station failed to produce a square-wave signal neccessary for performance checkout. The requirement for troubleshooting and test station repair exists.

5. Req. No.	6. Requirem't	7. Req. Freq.	8. Maint. Level	9. Ma Cont. No.
01/02	Troubleshoot and Repair	0.0105	Organizational Intermediate	A20000

10. Task Number	11. Task Description	12. Elapsed Time (min) 2 4 6 8 10 12 14 16 18 20 22 24 26 28 30 32 34 36 38	13. Total Elap. Time	14. Task Freq.	Personnel (man-min) 15. B	16. I	17. S	18. Total
0100	Check Test Station Power Supply for	(1)	4	0.0105	4			4
	Availability of Power							
0200	Check Signal Generator for Availability	(1)	10	0.0105	10			10
	of Square-wave Pulse	(2)			10			10
0300	Remove Signal Generator from Test Station	(1)	22	0.0105	22			22
	and Transport to Intermediate-Level							
	Maintenance Shop							
0400	Prepare Signal Generator for			0.0105				
	Troubleshooting	(3) (Cycle 2)	6		6			6
0500	Apply Power, Measure Availability of	(3)	12	0.0105	12			12
	20 V P-P ± 10% at T.P. 7, CB3A2	(4)					12	12
0600	Check T.P. 4, CB1A10, for Availability of	(3)	14	0.0105	14			14
	Square-wave Pulse, PW – 5 μsec.	(4)					14	14
0700	Remove and Replace CB1A10			0.0105				
		(3) (Cycle 3)	16		16			16
0800	Checkout Signal Generator	(3)	8	0.0105	8			8
		(4)						
0900	Reinstall Signal Generator in Test Station	(2)	20	0.0105		20		20
		Σ	19. 112	Σ	20. 102	21. 20	22. 26	23. 148

Prepared by: Blanchard Date 9/30/77 Maintenance Task Analysis (Sheet 1)

Figure C.4 Maintenance task analysis—sheet 1 (example 1) (reference: Figure C.1).

1. System	2. Item Name/Part No.	3. Next Higher Assy.	4. Description of Requirement:
Ship Steering	Hydraulic System/A12345	Ship Steering /A45400	A faulty starboard automatic valve in the hydraulic system must be repaired to restore ship's steering to full operational capability. Repair is accomplished on shipboard

5. Req. No.	6. Requirem't	7. Req. Freq.	8. Maint. Level	9. Ma Cont. No.
02	Repair	0.000486	Intermediate	A10000

10. Task Number	11. Task Description	12. Elapsed Time (min) 2 4 6 8 10 12 14 16 18 20 22 24 26 28 30 32 34 36 38	13. Total Elap. Time	14. Task Freq.	Personnel (Man-min) 15. B	16. I	17. S	18. Total
0010	Actuate Valve 2. Shut off Pressure from Stbd. Pump		2	0.000486		2		2
0020	Operate Emergancy Handpump		23		23			23
0030	Remove 1½" External Tubing		10		10	10		20
0040	Remove ¾" External Tubing		4		4	4		8
0050	Remove Automatic Valve Assy. from System		2		2	2		4
0060	Install Flanges on Tubing. Stop Handpump.		7			7		7
	Actuate Valve 3. Start Port Pump (Pressure On Syst.)							
0070	Transport Valve Assy. to Intermedate Shop		13		13	13		26
0080	Disassemble Valve Assy. & Remove Valve Rod &	(Cycle 2)	18			18		18
	Piston. Remove Valve P/N 16742-1.							
0090	Transport Valve to Machine Shop		8		8			8
0100	Machine & Clean Valve	(Cycle 3)	33				33	33
0110	Clean Valve Rod, Piston, Spring Assy.		19			19		19
0120	Install Valve, Valve Rod, Piston, Spring,	(Cycle 4)	17			17		17
	Gaskets & O-Rings into Valve Assy.							
0130	Checkout Automatic Valve Assy.		12		12	12		24
0140	Transport Valve Assy. to System		13		13	13		26
0150	Stop Port Pump. Actuate Valve 3 & Start	(Cycle 5)	23			5		23
	Handpump							
0160	Remove Flanges from Tubing. Install Valve Assy.		6		6	6		12
0170	Connect ¾" External Tubing		5		5	5		10
0180	Connect 1½" External Tubing		7		7	7		14
0190	Stop Handpump. Actuate Valve 2. Start		4	0.000486		4		4
	Stbd. Pump							
		Σ	19. 184	Σ	20. 121	21. 144	22. 33	23. 298

Prepared by: Blanchard Date: 9/30/77

Figure C.5 Maintenance task analysis—sheet 1 (example 2) (reference: Figure C.3).

(b) *Calibration.* A maintenance requirement whenever an item is checked against a working standard, a secondary standard, or a primary standard. Calibration generally applies to precision measurement equipment (PME) and can be accomplished either on a scheduled basis or on an unscheduled basis subsequent to the accomplishment of a repair action on a PME item. Calibration provides the necessary test accuracy and traceability to the National Institute of Standards and Technology (e.g., a working standard is checked against a secondary standard that in turn is checked against a primary standard).

(c) *Functional test.* A system or subsystem operational checkout either as a condition verification after the accomplishment of item repair or as a periodic scheduled requirement.

(d) *Inspection.* A maintenance requirement when the basic objective is to ensure that a requisite condition or quality exists. To inspect for a desired condition, it may be necessary to remove the item, to gain access by removing other items, or to partially disassemble the item for inspection purposes. In such instances, these associated actions which are necessary to accomplish the required inspection would be specific tasks.

(e) *Overhaul.* A maintenance requirement whenever an item is completely disassembled, refurbished, tested, and returned to a serviceable condition meeting all requirements set forth in applicable specifications. Overhaul may result from either a scheduled or unscheduled requirement and is generally accomplished at the depot maintenance facility or the producer's facility.

(f) *Remove.* A maintenance requirement when the basic objective is to remove an item from the next-higher assembly.

(g) *Remove and reinstall.* A maintenance requirement when an item is removed for any reason, and the same item is later reinstalled.

(h) *Remove and replace.* Constitutes the removal of one item and replacement with another like item (i.e., spare part). Such action can result from a failure or from a scheduled replacement of a critical-useful-life item.

(i) *Repair.* Constitutes a series of corrective maintenance tasks required to return an item to a serviceable condition. This involves the replacement of parts, the alteration of material, fixing, sealing, filling-in, and so on.

(j) *Servicing.* Includes the maintenance operations associated with the application of lubricants, gas, fuel, oil, and so on. Servicing may require removal, disassembly, reassembly, adjustment, installation, or a lesser number of these tasks. Servicing may also be included as a task under a requirement to repair, calibrate, or test an item; however, it will not be classified under servicing in this instance.

(k) *Troubleshooting.* Involves the logical process (series of tasks) which leads to the positive identification of the cause of a malfunction. It includes localization and fault isolation. Troubleshooting is best illustrated by the logic troubleshooting flow diagram illustrated in Figures 4.28, C.1, C.2, and C.3.

Whenever one of the foregoing requirements is the basic underlying cause, purpose, or objective to be satisfied, the operation becomes a maintenance requirement.

7. *Block 7: Requirement frequency.* The frequency at which the requirement is expected to occur is entered in this block. If the requirement is "Repair" and the item is the "Hydraulic system," the analyst must determine or predict how often repair of the hydraulic system is to be accomplished. This value is dependent on the task frequencies in block 14, and is generally expressed either in number of actions per hour of equipment operation or in terms of a given calendar time period.

The need for establishing a frequency factor for a requirement is to provide a basis for determining associated logistics resource demands. In other words:

(a) How often do we expect to need a spare/repair part?

(b) How often do we need test and support equipment for maintenance?

(c) How often will maintenance personnel be required?

(d) How often will an equipment item be down for maintenance?

(e) How often will servicing be required?

(f) How often will overhaul be accomplished?

Once these questions are answered, it becomes possible to project downtime and determine logistics resource requirements.

When determining the intervals or frequencies representing unscheduled maintenance actions, the inherent reliability characteristics of an item may be significant, but they are not necessarily dominant considerations. For some equipment, field data have indicated that less than one-half of all unscheduled maintenance actions are attributed to random catastrophic (primary) failures; therefore, it is obvious that consideration must be given to other contributing factors as well. On the other hand, there are many instances where reliability failure rates dominate all considerations. In this event, a proportional amount of attention must be given to the significance of random failures. In either case, the maintenance frequency factor must consider the following (refer to Chapter 2, Table 2.1):

(a) *Inherent reliability characteristics.* This category covers primary failures based on the physical makeup of the item and the stresses to which the item is subjected. Part catastrophic failures and out-of-tolerance conditions are included. Primary failures are derived through reliability prediction data and are usually based on a constant failure rate.

(b) *Dependent failures (chain effect).* This category includes those secondary failures which occur as a result of a primary catastrophic failure. In other words, the failure of one item may cause other items to fail. The frequency of dependent failures is based on the extent of fail-safe characteristics inherent in the design. In electronic equipment, circuit protection provisions may be incorporated to reduce the probability of dependent failures. The reliability failure mode, effect, and criticality analysis (FMECA) is the best data source for reflecting dependent failure characteristics.

(c) *Manufacturing defects or burn-in characteristics.* Quite often when an equipment item first comes off the production line, there are a rash of failures until the system is operated for a short period of time and a constant failure rate is realized. This is particularly true for electronic equipment when certain corrective actions are necessary to attain system stabilization. The extent of difficulty at this early point in the operational use phase is dependent on the amount of equipment operation and the type of testing accomplished in the production/construction phase. If enough hours of equipment operation are attained through testing, the quantity of failures after delivery will probably be less. In any event, when new equipment first enters the inventory, there may be more initial failures than indicated by the predicted reliability failure rate, and this factor must be considered in determining the requirements for initial system support.

(d) *Wear-out characteristics.* After equipment has been in operational use for a period of time, various individual components begin to wear out. Some components (i.e., mechanical linkages, gears) wear out sooner than others. When this occurs, the resultant frequency of failure and corrective maintenance will increase.

(e) *Operator-induced failures.* Overstress of the system because of operator human error is a possibility and very significant in determining frequency factors. Unplanned stresses can be placed on the system as a result of different operating methods (by different operators or by the same operator), and these stresses can accumulate to the extent that the system will fail more frequently. Hopefully, conditions such as this can be minimized or eliminated through the proper emphasis on human factors in the design process.

(f) *Maintenance-induced failures.* Damage to equipment caused by human error during maintenance actions may result from not following the proper maintenance procedures, improper application and use of tools and test items, losing components or leaving parts out when reassembling items after repair, forcing replacement items to fit when physical interferences are present, causing physical damage to components lying adjacent to or near the item being repaired, and many others. Induced faults of this type may occur due to improper maintenance environment (inadequate illumination, uncomfortable temperatures, high noise level), personnel fatigue, inadequately trained personnel performing the maintenance tasks, equipment sensitivity or fragility, poor internal accessibility to components, and not having the right elements of logistic support available when needed. Maintenance-induced faults are possible both during the accomplishment of corrective maintenance and preventive maintenance. For instance, during a scheduled calibration when performing a fine adjustment, a screwdriver slip may cause damage in another area resulting in corrective maintenance.

(g) *Equipment damage due to handling.* Another important aspect is the probability of equipment damage due to bumping, dropping, shoving, tossing, and so on. These factors are particularly relevant during transportation modes and often contribute to subsequent equipment failures. The extent to which consideration is given toward transportability in the equipment design will influence the effects of handling on the equipment failure rate. The analyst should evaluate anticipated transportation and handling modes, review the design in terms of effects, and assign an appropriate frequency factor.

System failures (of one type or another) may be detected through built-in self-test go/no-go indicators, performance monitoring devices, and/or through actual visual

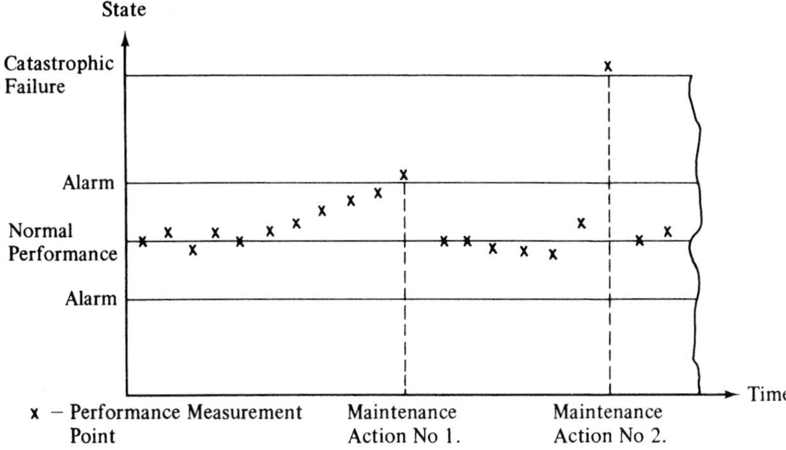

Figure C.6 Performance monitoring.

indication of a faulty condition. Performance monitoring devices may allow for the measurement of certain designated system parameters on a periodic basis, thus permitting the observation of trends. The concept of performance monitoring is illustrated in Figure C.6, where measurement points are indicated in terms of system operating time. When certain failure trends are noted, maintenance actions may be initiated even though an actual catastrophic failure has not occurred. In other words, the anticipation of future maintenance causes a maintenance action to be accomplished at an earlier time. These instances must also be considered in determining requirement frequency.

The foregoing considerations are reviewed on an individual basis and may be combined to provide an overall factor for a given corrective maintenance action, such as illustrated in Table 2.1 (Chapter 2). The frequency factor in block 7 represents the overall number of instances per equipment operating hour (or calendar period) for the maintenance requirement defined in block 6. The factor is determined from the frequency of the individual tasks identified in blocks 10 and 11. The maintenance requirement may constitute a series of tasks, each one of which is accomplished whenever the requirement exists. In this instance, the frequency factor for each task (block 14) will be the same as the factor entered in block 7. This is illustrated in Figure C.4. On the other hand, it may be feasible to include a series of *either–or* tasks on a single maintenance task analysis sheet. For instance, when a troubleshooting requirement exists, tasks may be identified as follows:

[10]Task Number	[11]Task Description	[14]Task Frequency
0010	Localize fault to Unit *A*	0.000486
0020	Remove Unit *A* access cover	0.000486
X030	Isolate to Assembly *A*	0.000133
X040	Isolate to Assembly *B*	0.000212
X050	Isolate to Assembly *C*	0.000141
0060	Install Unit *A* access cover	0.000486
0070	Check out Unit *A*	0.000486

The troubleshooting requirement will apply to Assembly *A* in 0.000133 instance per equipment operating hour, Assembly *B* in 0.000212 instance, and so on. The sum of the various alternatives (*x* tasks) will constitute the overall frequency, or the value in block 7 is 0.000486. Tasks 0010, 0020, 0060, and 0070 are applicable for each occurrence of Tasks X030, X040, and X050. This method of data presentation provides a simplified approach for covering the numerous possible alternatives often applicable to the diagnostic aspects of corrective maintenance, particularly for electronic equipment. A clearer picture of the alternative courses of action is illustrated in the logic troubleshooting flow diagram in Figure C.1.

8. *Block 8: Maintenance level.* The level of maintenance at which the requirement will be accomplished is identified. This may be organizational, intermediate, depot, or supplier. Refer to Chapter 3 for a description of the levels of maintenance.

9. *Block 9: MA control number.* Enter the maintenance analysis (MA) control number for the item identified in block 2. The control numbering sequence is a numerical breakdown of the system by hardware indenture level and is designed to provide traceability from the highest maintenance significant item to the lowest repairable unit. An example of MA control number assignment is as follows:

Equipment Item	MA Control Number
System *XYZ*	A0000
Unit *A* (in System *XYZ*)	A1000
Assembly 2 (in Unit *A*)	A1200
Subassembly *C* (in Assy. 2)	A1230

An MA control number is assigned to each maintenance significant item. Maintenance tasks and logistic support resource requirements are identified against each control number and may be readily tabulated from the lowest item to the system level. MA control numbers are assigned by function and not necessarily by part number. If a modification of Assembly 2 is accomplished or an alternative configuration of Assembly 2 is employed, the manufacturer's part number changes, but the MA control number remains the same as long as the basic function of the applicable item is unchanged.

10. *Block 10: Task number.* A four-digit alphanumeric designation is assigned to each task. Sequential tasks (those applicable each time that the maintenance requirement is accomplished) are identified with a "0" as the first digit. Tasks which are optional (either–or) depending on their applicability are designated with an *X*. The last three digits indicate the sequence of task accomplishment (e.g., 0010, 0020, 0030, etc.). Subtasks may be identified further, if desired, through the use of the fourth digit (e.g., 0021, 0022, etc.). The alphanumeric designation simplifies task identification when maintenance analysis data are computerized.

11. *Block 11: Task description.* Tasks should be stated in a concise technical manner, and should be defined to the extent necessary to enable development of a complete maintenance checklist. Significant tasks include those whose accomplishment requires a spare, an item of support equipment, or some related element of logistic support. Including an extensive amount of detail is not necessary; however, enough detail should be provided to enable the identification of logistic support requirements.

12. *Block 12: Elapsed time (minutes).* For each task listed in blocks 10 and 11, an elapsed time line should be projected. If two or more people are involved in a task, it may be feasible to construct a separate time line for each individual (e.g., ▨ ① for worker 1, ▨ ② for worker 2, etc.). If the task extends beyond the 40-minute period indicated on the form, the time line is continued on the next line commencing at zero and a new cycle is indicated. In instances where times are extensive (hours instead of minutes), the basic scale can readily be converted to hours. The benefit of the layout of time lines included in Figure C.4 is to allow for the evaluation of personnel utilization and task sequencing. There may be a number of tasks which can be accomplished in parallel, or in other instances a series relationship may exist. If several equipments require maintenance at the same time, the time-line analysis in block 12 will aid the analyst in determining schedule and labor requirements. By pictorially presenting time-line data, it is possible to evaluate a number of sequences, arriving at an optimum arrangement. The objective is to minimize both elapsed time and labor time, and to make greater use of lower-skilled personnel where possible.

13. *Blocks 13 and 19: Total elapsed time.* The numerical values of elapsed time (represented by the time lines in block 12) are entered in block 13 and totaled in block 19. This constitutes the time between task start and task completion. Do not total the times of the various task segments.

14. *Block 14: Task frequency.* The frequency of each task, in terms of instances per equipment hour of operation, is recorded. This represents the number of times that a certain maintenance action is expected to occur or the anticipated demand rate for a given spare/repair part in a remove and replace action.

15. *Blocks 15–18 and 20–23 (man-minutes).* Enter the total man-minutes required per task for each of three basic skill levels. These skill levels may be defined as follows:[3]

(a) *Basic skill level (blocks 15 and 20).* A basic skill level is assumed to be a technician with the following characteristics.

Age—18 to 21 years.
Experience—no regular work experience prior to training.
Educational level—high school graduate.
General reading/writing level—ninth grade.

After a limited amount of specialized training, these personnel can perform routine checks, accomplish physical functions, use basic handtools, and follow clearly presented instructions where interpretation and decision making is not necessary. Workers in this class usually assist more highly skilled personnel, and require

[3] There may be other classifications of personnel skills depending on the user's organization structure. The intent is to define specific classifications that will be applicable in system operational use, and to evaluate maintenance tasks in terms of these classifications. Some examples of personnel classification may be found in civil service job descriptions and in military documentation.

constant supervision. For military applications, service pay grades E3 and E4 generally fall in this category.

(b) *Intermediate skill level (blocks 16 and 21).* These personnel have had a more formalized education, consisting of approximately 2 years of college or equivalent course work in a technical institute. In addition, they have had some specialized training, and 2 to 5 years of experience in the field related to the type of equipment in question. Personnel in this class can perform relatively complex tasks using a variety of test instruments, and are able to make certain decisions pertaining to maintenance and the disposition of equipment items. Military personnel grade E5, or civilian equivalent, falls in this category.

(c) *Supervisory or high skill level (blocks 17 and 22).* These personnel have had 2 to 4 years of formal college or equivalent course work in a technical institute, and possess 10 years or more of related on-the-job experience. They are assigned to supervise and train intermediate and basic-skill-level personnel, and are in the position to interpret procedures, accomplish complex tasks, and make major decisions affecting maintenance policy and the disposition of equipment. They are knowledgeable in the operation and use of highly complex precision (calibration) equipment. Military personnel grades of E6 and above, or equivalent civilian classifications, are included.

Each task in block 11 is analyzed from the standpoint of complexity and requirement for task accomplishment relative to human sensory and perceptual capacities, motor skills, mobility coordination, human physical dimensions and muscular strength, and so on. In some instances, the conductance of a human factors detailed task analysis will be necessary to provide the detail required for a skill-level assignment. An appropriate skill level is assigned to each task and the man-minutes by skill level are entered in blocks 15 to 18. A total of personnel labor time for the requirement (identified in block 6) is included in blocks 20 to 23.

Once the personnel skill levels have been assigned by task, it is necessary to compare these with the skills of user personnel scheduled (in the future) to maintain the equipment in the field. The results will dictate formal training requirements. For example, user personnel scheduled for assignment to maintain System *XYZ* in the field are currently classified as basic, and the task analysis indicates a requirement for intermediate skills. In this instance, it is obvious that some formal/on-the-job training is required to upgrade the personnel from basic to the intermediate level. Through a review of all maintenance requirements at each level, the analyst can determine the number of personnel required, the skill levels, and the areas that must be covered through formal training.

Maintenance Task Analysis—Sheet 2 (Figures 4.30, C.7, and C.8)

The instructions below should apply when completing sheet 2 of the maintenance task analysis form. Sheet 2 supplements sheet 1 (Figure C.4) through the assignment of common task numbers in block 7.

1. Item Name/Part No.	2. Req. No.	3. Requirement	4. Req. Freq.	5. Maint. Level	6. Ma Cont. No.
Test Station /A12345	01/02	Troubleshoot and Repair	0.0105	Organizational and Intermediate	A20000

7.	8.	Replacement Parts		Test and Support/Handling Equipment			16. Description of Facility	17. Special Technical Data
Task Number	Qty. Per Assy.	9. Part Nomenclature / 11. Part Number	10. Rep. Freq.	12. Qty.	13. Item Nomenclature / 15. Item Part Number	14. Use Time (min)	Requirements	Instructions
0100				1	AC-DC Meter SK 932101	4	Organizational (On-Site) Maintenance	Check Power at Front Panel T.P. #, 115 Vac ± 10%
0200				1	Signal Generator FM1291006-2	10		Check Square-wave Pulse at Front Panel T.P. 6, PW-5 μsec
0300				1	Screwdriver/732102 Dolly/24A102	22 22		Transport to Intermediate -Level Maintenance Shop
0400				1 1	Screwdriver/732100 Special Harness Assy./GM1023	6 56	Intermediate-Level Maintenance Shop	Bypass Power Disconnect Switch S-12
0500					Power Supply F102116-1 AC-DC Meter SK 932101	12 12		Measure Availability of 20V P-P ± 10% at T.P. 7, CB3A2
0600				1	Signal Generator TM10034-10	14		Check Availability of Square-wave Pulse, PW-5μsec, T.P. 4, CB1A10.
0700	1	CB1A10/GM10113-6	0.0105	1	Soldering Tool/A1047 Screwdriver/710000	16 16		Circuit Board will not be Repaired – Discard
0800				1	Signal Generator FM1291006-2	8		Refer to Task 0200
0900					Screwdriver/732102 Dolly/24A102	20 20		Accomplish Operational Check of Test Station. Refer to Procedure, TM-30

Figure C.7 Maintenance task analysis—sheet 2 (example 1) (reference: Figure C.4).

1. Item Name/Part No. Hydraulic System/A12345		2. Req. No. 02		3. Requirement Repair		4. Req. Freq. 0.000484	5. Maint. Level Intermediate	6. Ma Cont. No. A10000
7. Task Number	8. Qty. Per Assy.	Replacement Parts: 9. Part Nomenclature / 11. Part Number	10. Rep. Freq.	Test and Support/Handling Equipment: 12. Qty.	13. Item Nomenclature / 15. Item Part Number	14. Use Time (min)	16. Description of Facility Requirements	17. Special Technical Data Instructions
0010								
0020								Handpump Operating Instructions
0030				1	Wrench - 1½"/600120-2	10		Hydraulic System Maintenance
0040				1	Wrench - ¾"/645809-1	4		Procedures (MP3201)
0050				1	Handling Dolly/S101-4	2		↓
0060				1	1½" Flange/AA123	150		
				1	¾" Flange/AB145	150		
0070				1	Handling Dolly/S101-4	13	Intermediate Maint. Shop	
0080				1	Wrench - ¼"/632111/1	18	↓	
				1	Screwdriver/732102	18		
0090								
0100				1	Grinder/BN101(S&G)	33	Machine Shop (With Cleaning	Grinder/Sander Operating
					Sander/C32101(PMN)	33	Facilities)	Instructions (OP3104 & OP3107)
0110	1 Gall.	Solvent/SA123	0.000484				Intermediate Maint. Shop	Hydraulic System Maintenance
0120	1	Gasket/AN 118-1	0.000484				↓	Procedures (MP3201)
	2	"O" Ring/AN9001-2	0.000484					↓
0130					Pneu. Test Set/HPI-162	12		Pneu. Test Set Oper. Instructions
0140					Handling Dolly/S101-4	13		Hydraulic System Maintenance
0150								Procedures (MP3201)
0160					Wrench - 1½"/600120-2	6		↓
					Wrench - ¾"/645809-1	6		
0170	1	¾" Gasket/AN912	0.000484		Wrench ¾"/645809-1	5		
0180	1	1½" Gasket/AN877-1	0.000484		Wrench - 1½"/600120-2	7		
0190								

Prepared by: Blanchard Date: 10/1/xx

Figure C.8 Maintenance task analysis—sheet 2 (example 2) (reference: Figure C.5).

1. *Blocks 1 to 6.* Enter the applicable data from sheet 1.

2. *Block 7: Task number.* Identify (by use of the appropriate four-digit alphanumeric designation) each task from sheet 1 where replacement parts, test and support equipment, facilities, and/or special data instructions are required. If a task requires the replacement of an item, a spare part need should be identified along with the anticipated frequency of replacement. Similarly, if a task requires the use of an item of support equipment, the task should be entered in block 7. Whenever a logistic support requirement is anticipated, the applicable task(s) must be included.

		Replacement Parts		
[7] Task Number	[11] Task Description	[8] Qty. per Assy.	[9/10] Part Nomenclature Part Number	[10] Rep. Freq.
0010	Localize fault to Unit *A*	....	....	....
0020	Remove Unit *A* access cover	....	....	....
X030	Isolate to Assy. *A*	....	....	....
X031	Remove and Replace Assy. *A*	1	Ampl.-Modular Assy. A160189-1	0.000133
X040	Isolate to Assy. *B*	....	....	....
X041	Remove and Replace	1	Power Converter Assy. A180221-2	0.000212
X050	Isolate to Assy. *C*	....	....	....
X051	Remove and Replace Assy. *C*	1	Power Supply Assy. A21234-10	0.000141
0060	Install Unit *A* access cover	....	....	....
0070	Check out Unit *A*	....	....	

3. *Blocks 8 to 11: Replacement parts.* This category identifies all anticipated replacement parts and consumables associated with scheduled and unscheduled maintenance. Blocks 9 and 1 l define the item by nomenclature and manufacturer's part number. Block 8 indicates the quantity of items used (actually replaced or consumed) in accomplishing the task. Block 10 specifies the predicted replacement rate based on a combination of primary and secondary failures, induced faults, wear-out characteristics, condemned items scrapped, or scheduled maintenance actions. If an item is replaced for any reason, it should be covered in this category. This forms the basis for determination of MTBR.

In the event that an either-or situation exists (identified by an X task number in block 7, sheet 1), the listing of replacement parts should follow the task listing. The replacement parts are identified for all maintenance requirements at each level, and the replacement frequencies (replacements per hour of equipment operation) indicate the demand for each part application. When these factors are integrated at the system level, it is possible to determine spare/repair part requirements.

4. *Blocks 12 to 15: Test and support/handling equipment.* Enter all tools, test and support equipment, and handling equipment required to accomplish the tasks listed in block 7. Blocks 13 and 15 define the item by nomenclature and manufacturer's part number. Block 12 indicates the quantity of items required per task, and block 14 specifies item utilization.

When determining the type of test and support equipment, the analyst should address the following in the order presented.

(a) Determine that there is a definite need for the test and support/handling equipment. Can such a need be economically avoided through a design change?

(b) Given a need, determine the environmental requirements. Is the equipment to be used in a sheltered area or outside? How often will the equipment be deployed to other locations? If handling equipment is involved, what item(s) is to be transported, and where?

(c) For test equipment, determine the parameters to be measured. What accuracies and tolerances are required? What are the requirements for traceability? Can another test equipment item at the same location be used for maintenance support or does the selected item have to be sent to a higher level of maintenance for calibration? It is desirable to keep test requirements as simple as possible and to a minimum. For instance, if 10 different testers are assigned at the intermediate level, it would be preferable to be able to check them out with one other tester (with higher accuracy) in the intermediate shop rather than send all 10 testers to the depot for calibration. The analyst should define the entire test cycle prior to arriving at a final recommendation.

(d) Determine whether an existing off-the-shelf equipment item will do the job (through review of catalogs and data sheets) rather than contract for a new design, which is usually more costly. If an existing item is not available, solicit sources for design and development.

(e) Given a decision on the item to be acquired, determine the proper quantity. Utilization or use time is particularly important when arriving at the number of test and support equipments assigned to a particular maintenance facility. When compiling the requirements at the system level, there may be 24 instances when a given test equipment item is required; however, the utilization of that item may be such that a quantity of two will fulfill all maintenance needs at a designated facility. Without looking at actual utilization and associated maintenance schedules, there is a danger of specifying more test and support equipment than what is really required. This results in unnecessary cost. The ultimate objective is to determine overall use time, allow enough additional time for support equipment maintenance, and procure just enough items to do the job.

Test and support equipment requirements are listed in blocks 12-15. However, the justification and backup data covering each item are provided in a specification defining need, functions, and recommended technical approach. This may be presented in the form of a design specification if a new development effort is required, or in the form of procurement specification if an off-the-shelf item is identified. In any event, once the maintenance analysis identifies a need for a support equipment item action is taken to acquire the item in time for system evaluation and test (preferably for Type 2 testing).

5. *Block 16: Description of facility requirements.* When evaluating each task, the analyst must determine where the task is to be accomplished and the facilities required. This includes the determination of space requirements, capital equipment needs, equipment layout, storage space, power and light, telephone, water, gas, environmental controls, and so on. If a "clean room" is required for the accomplishment of precise maintenance and calibration, it should be specified. The analyst should solicit assistance from system design and human factors personnel and generate a facilities plan showing a complete layout of all essential items. The type of facility, a brief summary description, and reference to the facilities plan are included in block 16.

6. *Block 17: Special technical data instructions.* The reference procedure describing the task(s) identified in block 7 is listed in block 17. In addition, enter any special instructions (where specific detailed coverage is required), safety or caution notices, or other information which should be conveyed to the maintenance technician.

Maintenance Analysis Summary (Figure C.9)

The maintenance analysis data summary (Figure C.9) presents a composite coverage of all maintenance requirements for each significant repairable item. The data summary includes a brief description of the item function, definition of the maintenance concept, and a listing of predicted quantitative factors covering the applicable scheduled and unscheduled maintenance requirements. The summary presents an overview of the information included in the maintenance task analysis sheets (Figures C.4 and C.7).

1. *Block 1: System.* Enter the proper nomenclature of the system or end item covered by the overall analysis.

2. *Block 2: Item name/part number.* Enter the name and manufacturer's part number for the repairable item being covered by this analysis.

3. *Block 3: Next higher assembly.* Include the name and part number of the next higher assembly.

4. *Block 4: Design specification.* Enter the applicable design or performance specification covering the description and design requirements for the item listed block 2. The document listed serves to tie maintenance support requirements with basic design criteria.

5. *Block 5: MA control number.* Enter the maintenance analysis (MA) control number for the item identified in block 2.

1. System	2. Item Name/Part No.	3. Next Higher Assy.	4. Design Specification	5. Ma Control No.
CXXX Aircraft	Unit A Synchronizer / PN 1345	System XYZ	ZA88446 (10/2/72)	B12000

6. Functional Description: System XYZ is an airborne navigation subsystem installed in the CXXX Aircraft. Unit A of the navigation subsystem provides finite range and bearing information for the overall aircraft. Unit A includes Assembly A, a power supply, and Circuit Boards 2A1 and 3A2.

7. Maintenance Concept Description: Unscheduled Maintenance - On detection of a fault at the system level, the applicable fault is isolated to Unit A, B, or C (as applicable) through built-in self-test. The unit is removed and replaced at the aircraft, and the faulty item is returned to the Intermediate Shop for corrective maintenance. Fault isolation is accomplished to CB2A1, CB3A2, Assy. A, or Pwr. Supply. Assy. A Pwr. Supply are repairable while CB's are nonrepairable. Scheduled Maintenance - Periodic calibration.

8. Req. No.	9. Requirement	10. Maint. Level	11. Req. Freq.	12. Elap. Time	13. Per. Skills	14. Man-Min	15. Repl. Parts ⚠1	16. Test and Support Equip. ⚠2
01	Troubleshooting	Intermediate	0.00184	27	Intermediate	27		Oscilloscope/HPI-34
					Basic	27		Voltmeter/BN-33
								Square-wave Gen./CPP33
02	Repair	Intermediate	0.00184	18	Basic	18	Assy. A/A12345	Screwdrive, Soldering
							CB3A2/BN1456	Gun, Wire clippers
							Power Supply/PP320	
							CB2A1/BN1576	
03	Functional Test	Intermediate	0.00184	6	Intermediate	6		Oscilloscope/HPI-34
04	Calibration	Depot	0.00139	480	Supervisory	590		Precision VTVM/ASI-13
								WS Bridge/SU-123
								Oscilloscope/Tek 462
								Common Hand tools
Total			0.00321	5.31		668		

17. Notes: ⚠1 Refer to Addendum A for specific type and quantity of spare/repair parts. ⚠2 Refer to Addendum B for a detailed listing of test and support equipment.

Prepared by: Blanchard Date: 10/1/XX

Figure C.9 Maintenance analysis summary.

6. *Block 6: Functional description.* For the item identified in block 2, briefly describe the function performed (its purpose). The description should be complete enough to ascertain what the item is and what function it serves, and should tie in with the functional analysis described in Section 3.6.

7. *Block 7: Maintenance concept description.* Briefly describe the maintenance concept or policy for the item covered in block 2. This should include reference to the levels or echelons of maintenance, the depth of maintenance, the maintenance environment, and any additional information considered appropriate. Both scheduled and unscheduled requirements should be summarized in a narrative form. The description should support the information defined in Section 3.4.

8. *Block 8: Requirement number.* Enter the appropriate two-digit number for each maintenance requirement applicable to the item covered by the analysis. This may include troubleshooting, repair, servicing, inspection, and so on.

9. *Block 9: Requirement.* Include the nomenclature for each requirement number listed in block 8.

10. *Block 10: Maintenance level.* The level or echelon of maintenance at which the requirement will be accomplished is identified. This may be organizational, intermediate, depot, or supplier.

11. *Block 11: Requirement frequency.* The frequency (in terms of instances per equipment operating hour or calendar time period) at which the requirement is expected to occur is entered. This information is derived from block 7 of the maintenance task analysis, sheet 1 (Figure C.4).

12. *Block 12: Elapsed time.* The total elapsed clock time for each requirement, included in block 13 of the maintenance task analysis, sheet 1, is entered.

13. *Blocks 13 and 14: Personnel skills and man-minutes.* The information in these blocks provides an overview of the personnel skill levels (basic, intermediate, and supervisory) and maintenance labor time anticipated for each requirement and for the end item itself. This information is derived from blocks 20 to 23, maintenance task analysis, sheet 1.

14. *Blocks 15 and 16: Replacement parts/test and support equipment.* To provide an overview of spare/repair part and support equipment requirements, a summary listing is presented in blocks 15 and 16, respectively. The information presented is certainly not all inclusive in the space allowed, but serves as a checklist of major items of logistic support. More in-depth data are obtained from the maintenance task analysis, sheet 2 (Figure C.7), and from related supporting reports and program planning documentation.

15. *Block 17: Notes.* Reference any additional information as appropriate to amplify and support the entries in any of the other blocks. Use an addendum sheet if the space allowed is inadequate.

The maintenance analysis summary is developed to cover each repairable item from the lowest applicable indenture hardware component to the system level. At the system level, the requirements for logistic support are summarized for evaluative purposes.

Data Analysis and Evaluation

Figure 4.7 presents a summary listing of output factors desired from the supportability analysis, of which the maintenance analysis is an integral part. Many of these factors (primarily those which do not represent cost values), are derived from the data sheets described above. The data items presented as related to the elements of logistic support are summarized in Figure C.10.

Relative to system operational and maintenance effectiveness factors, many of the system figures of merit described in this book may be calculated from the data presented in the analysis. For instance (referring to Section 2.10),

$$A_o = \frac{\text{MTBM}}{\text{MTBM} + \text{MDT}} \tag{C.1}$$

MTBM is the reciprocal of the frequency of maintenance, which is derived from the factors in block 11 of the maintenance analysis summary (Figure C.9). MDT is the mean elapsed downtime, which can be determined from block 12 (of the same data sheet).

$$A_a = \frac{\text{MTBM}}{\text{MTBM} + \overline{\text{M}}} \tag{C.2}$$

$$A_i = \frac{\text{MTBF}}{\text{MTBF} + \overline{\text{M}}\text{ct}} \tag{C.3}$$

MTBF is the reciprocal of the failure rate, which is based on equipment inherent reliability characteristics, and can be determined from a breakout of the data in blocks 7 and 14 of the maintenance task analysis, sheet 1. $\overline{\text{M}}$ and $\overline{\text{M}}\text{ct}$ constitute active maintenance elapsed times and can be determined from block 13 of the maintenance task analysis, sheet 1, and/or block 12 of the maintenance analysis summary.

$$\text{MLH/OH} = \frac{\text{total maintenance labor hours}}{\text{equipment operating hour}} \tag{C.4}$$

MLH/OH can be derived from the factors in block 23 of the maintenance task analysis, sheet 1, or block 14 of the maintenance analysis summary, depending on the end item being evaluated.

These and other system quantitative measures can be derived from maintenance analysis data. However, the student must have a good understanding of how the various data elements are to be applied to provide meaningful results. If there is a question concerning the meaning of some of these factors, the student is advised to review Chapter 2. Review of maintenance analysis data is accomplished to determine (analytically) whether system requirements have been met. Areas where problems exist are readily identified and should be brought to the attention of the design engineer for possible corrective action. In other words, maintenance analysis is a tool that serves as a check on the design process to ensure the development of a system configuration which can be effectively and economically supported. Also, the analysis serves as the basis for the subsequent provisioning and acquisition of all logistic support elements.

Areas of Logistic Support	Data Source		
	Maintenance Task Analysis-Sheet 1	Maintenance Task Analysis-Sheet 2	Maintenance Analysis Summary
1. Test and Support/Handling Equipment	Blocks 4, 8, 11	Blocks 12, 13, 14, 15	Blocks 6, 7, 16
2. Spare/Repair Parts	Blocks 8, 11, 14	Blocks 5, 8, 9, 10, 11	Blocks 7, 15
3. Personnel and Training	Blocks 8, 11, 12 15, 16, 17, 18, 19, 20, 21, 22, 23	Block 17	Blocks 7, 13, 14
4. Facilities	Blocks 8, 11	Block 16	Blocks 6, 7
5. Technical Data	Blocks 4, 8, 11, 12, 15, 16, 17	Blocks 3, 5, 8, 9, 11, 12, 13, 15, 16, 17	Blocks 6, 7, 9, 10, 13, 15, 16
6. Operational and Maintenance Effectiveness Factors	Blocks 7, 8, 13, 14, 18, 23	Blocks 4, 5, 10, 14	Blocks 7, 10, 11 12, 14
7. Maintenance Concept and Policy Definition	Blocks 4, 6, 8, 11, 15, 16, 17	Blocks 3, 5, 9, 11, 13, 15, 16, 17	Blocks 7, 9, 10, 13, 15, 16

Figure C.10 Maintenance analysis data application.

ANALYTICAL MODELS/TOOLS

Sections 3.8 and 4.1 introduce the subject of models and discuss typical model applications. A model constitutes a simplified representation of a real-world situation, and is generally employed to aid the decision maker in arriving at a preferred solution to problems through the evaluation of alternatives. Models can vary from a basic mathematical expression to a complex computer program. The type of model required is a function of the quantity of variables, the number of alternatives, and the overall complexity of the operation. The decision maker must analyze the problem at hand, identify techniques that can be used to resolve individual segments of the defined problem, and select or develop a model that will properly employ these techniques in problem resolution. A model is a tool utilized as an aid in decision making; it is not an entity in itself.

Typical Problem Applications and Model Development

To provide an understanding of model requirements as they pertain to the field of logistics, one should identify typical problem applications that are likely to occur throughout system design and development. For the purposes of discussion, it is assumed that such applications will fall into one (or more) of the following four categories:

1. *Conceptual design and advance system planning.* During the early stages of system design, there is a requirement to evaluate and compare alternative operational concepts, mission scenarios, utilization profiles, performance factors, maintenance and logistic support policies, and related factors. Top system-level requirements are being defined, specific technology applications are being evaluated, technical design approaches are being established, functional analyses and allocations are being accomplished, and so on. At this stage in the system life cycle (and depending on the degree of design definition), the designer has to rely on such analytical techniques as simulation, linear and dynamic programming, forecasting, networking, queuing theory, etc. Those activities discussed throughout Chapters 3 to 5 are accomplished to varying degrees, and the model requirements are oriented to system-level problems. More specifically, there may be requirements for the use of life-cycle cost models, functional analysis models, level of repair analysis models, reliability and maintainability analysis and allocation models, transportation models, network models for advance planning, and the like. However, the level of complexity of the model structure will not be as great as what may be needed in subsequent phases of the program.

2. *Detail system/product design and development.* As design definition progresses, there is a requirement to evaluate alternative human-machine interface configurations, equipment packaging schemes, diagnostic and testability routines, built-in versus external test provisions, component part applications, software versus hardware approaches, fabrication and assembly processes, maintenance and repair options, and so on. Refer to Chapters 4 and 5 for additional applications.

The detailed design of units, assemblies, modules, software programs, and lower-level system components is being accomplished. Design activities not only relate to the prime-mission-oriented elements of the system, but to the production processes and the product support capability as well. Specific modeling requirements include the use of life-cycle cost analysis models, level of repair analysis models, spare/repair parts and inventory control models, reliability and maintainability prediction models, FTA/FMECA models, transportation models, maintenance shop models, and the like. While many of the analysis functions are similar to those accomplished during the conceptual and preliminary system design phases, the degree of complexity of the model design must be greater in order to handle properly the depth of design detail and the sensitivities required. For example, life-cycle cost analyses are accomplished throughout all phases of system design and development; however, the depth of analysis and supporting data will obviously be greater as one progresses from conceptual design to the detail design and development phase. Thus, model design requirements and the ultimate structure may be somewhat more complex in this instance.

3. *Evaluation of the system configuration and the determination of specific logistics resource requirements.* Given an assumed design configuration of the system, or elements thereof, a comprehensive analysis is accomplished with the objective of determining specific logistics resource requirements, such as spare/repair parts, test and support equipment, personnel quantities and skill levels, facilities, technical data, and computer resources. Refer to the supportability analysis (SA) in Chapter 4 and in Appendices C and E.

Design of the prime mission-oriented elements of the system is *mature* (i.e., approved through formal design review), and it is necessary to evaluate various alternative methods for supporting the system in the field. Modeling requirements may include the utilization of a combination of life-cycle cost models, level of repair analysis models, reliability and maintainability prediction models, spare/repair parts and inventory control models, transportation models, maintenance shop models, and so on.

In addition, models are often developed for the purposes of recording, processing, storing, and the reporting of supportability analysis data. Many organizations have developed rather large and complex computerized routines in response to SA requirements for a given program. In these instances, the modeling requirements may become extremely complex in both design and operation.

4. *System assessment and the determination of logistic support effectiveness in the field.* To ensure complete customer satisfaction, it is essential that the system be evaluated in the field in terms of its overall effectiveness and efficiency of operation. In support of this objective, it is appropriate to determine the effectiveness of the logistic support capability, as discussed in Chapter 7. Modeling requirements here will likely include the utilization of large computerized programs for the purposes of data collection, analysis, processing, storage, and reporting. These requirements should be compatible with both the SA data processing requirements and the continuous acquisition and life-cycle support (CALS) capability.

If, during the ongoing assessment process, areas of deficiency are noted and recommendations for product improvement are initiated, any proposed changes should be evaluated in terms of impact on life-cycle cost, reliability and maintainability characteristics, and so on. In such instances, it may be appropriate to utilize the models discussed earlier for the purposes of evaluation.

Having identified the different categories of problems, it is essential that the proper tools be selected. The following considerations should be addressed in anticipation of model selection:

1. There are numerous interrelated elements of a system that must be treated on an integrated basis. The interactions between reliability, maintainability, performance, effectiveness, supportability, and so on, are many, and a variation in any one will affect the others to varying degrees. The model selected must represent these conditions in a realistic manner, while allowing for a productive analysis effort overall.

2. There are many different alternatives available for consideration and requiring evaluation. With the objective of reducing risk, the decision maker will desire to investigate as many of these options as possible. The model selected must be capable of providing an assessment of each alternative on an individual basis, or a group of alternatives viewed collectively, in a timely and economical manner.

3. In reviewing the nature of the problem applications identified earlier, there are instances where the same problem category repeats as one progresses from conceptual design to the detail design and development phase (e.g., life-cycle cost analysis, level of repair analysis); however, the depth of analysis will increase as the level of design definition increases. The model selected must not only be comprehensive in structure, but its design must be flexible such that varying depths of analysis can be accomplished. Additionally, the model structure must allow for the easy incorporation of changes without destroying its overall characteristics.

4. In accomplishing an analysis effort, it is necessary that the analyst be able to (a) evaluate the system as an entity, (b) evaluate various components of the system on an individual basis and optimize component design, and (c) project the results of the component analysis in the context of the system as a whole. In other words, one must be able to view the system in entirety, break the system down into component parts, and then put the system back together again. The model selected must be designed in such a way as to enable evaluation at the system level, while allowing for the evaluation of individual elements of the system.

Modeling, in general, offers tremendous educational benefits as well as being an effective tool in decision making. The analyst can readily identify relationships between system parameters, and is better prepared to respond to the "What if?" questions. For instance, what is the impact on spare/repair parts if the equipment reliability is degraded? What is the impact on support equipment if the prime equipment packaging design changes? What is the impact on overall system maintenance if the equipment is utilized to a greater extent than initially planned? There are many questions of this nature to which an analyst with experience can readily respond, and in many instances these intuitive rapid responses are necessary to avoid possible future problems.

Models may be developed by applying a variety of analytical techniques.[1] For instance, one may employ simulation to analyze the overall behavior of a system over time under a specified set of constraints. Functional relationships exist between the solution parameters and the control variables in the model, and in many instances the solutions are stated in terms of a range of values (not point estimates) or one of a set of intervals that contain the correct answer. Conversely, straight analytical methods may be used to provide a single answer or a unique set of answers for any given set of input variables.

Models are often categorized by the analytical method(s) used to solve the problem.[2] There are reliability models based on the probability of occurrence of various events. Given the configuration, certain basic probabilities, and a distribution, such models may be used in the prediction of failure rates and the quantity of maintenance actions. In a similar manner, maintainability models may be employed to determine anticipated maintenance times.

Network models are useful in evaluating the characteristics of systems that involve the movement of material, and may be employed in evaluating the movement of prime equipment, spare/repair parts, and repairables shipped to a higher level maintenance facility (e.g., intermediate shop to depot) for corrective maintenance support. Several network models use probabilistic considerations in determining which path to take when choosing one from a number of alternatives.

Optimization techniques are used when an optimal system design or support policy is desired and there are a number of constraints. Linear programming can be employed in optimizing with linear constraints while nonlinear programming is used when the applicable functions and constraints are nonlinear. Dynamic programming is a technique utilized if the solution involves a discrete set of possible alternatives with sequentially related constraints. This method is particularly suitable when the problem at hand involves a large number of variables.

[1] For a better understanding of available analytical techniques and their specific applications, the student should review additional material covering the use of operations research methods.

[2] Paulson, R. M., Waina, R. B., and Zacks, L. H., *Using Logistics Models in System Design and Early Support Planning*, Report R-550-PR, The RAND Corporation, Santa Monica, Calif., February 1971.

Accounting models are used when the functions involved are basically additive. A good example is the determination of life-cycle cost, which is structured for the purpose of adding up numerous individual component costs.

There are many quantitative methods available for application in the development of models. Often, a model will incorporate a number of different analytical methods. For instance, optimization will be used along with linear and nonlinear programming, queueing techniques, probabilistic functions, accounting methods, and so on, in a given model. The selection of techniques is tailored to the specific problem areas being addressed and becomes dominant in the development of a model.

When developing models, certain cautions must be exercised. First, a single, complex, all-inclusive model cannot offer some of the advantages which are available through the use of multiple models designed to function as an integrated set. Computer storage may be quickly exceeded; it may be difficult to acquire the required time on the computer; the model may not incorporate the desired flexibility or growth potential; it becomes difficult and costly to incorporate changes in the model; and the model may not provide timely results. On the other hand, developing a set of models (each model in the set having the capability of evaluating a specific class of problems) does provide the necessary flexibility required to solve the large variety of problems listed above. Individual models can be tailored and are more responsive to evaluating the impact of a design change on spare/repair parts, the impact of a waiting line or repairable item demands on support equipment loading and utilization, and other similar lower-level problems. If changes to the model are required, the necessary changes can be incorporated without affecting other models in the set. Thus, when developing models, the modular approach is preferable. Both individual evaluations and the total system evaluation effort must be considered such that different models in the set may be employed singly or on a combined basis.

As a second point, in developing the model, the usual dangers inherent in abstraction are present. A mathematically feasible model or series of expressions may require gross oversimplifications. There is no guarantee that the time and effort invested in model development will provide the results desired. The analyst must be careful to ensure that the model does indeed reflect a realistic situation.

Finally, there is the danger that the analyst, after playing with the model for a long period of time, will become too attached to the model. Some analysts will become so attached that they will insist that the model *is* the real world or is directly applicable to *all* problems at hand. Again, care must be exercised to ensure that the model selected or developed is compatible with the problem(s) being solved. Models are only tools used in decision making and cannot be considered as a substitute for experience and judgment. As such, the use of models in a variety of situations has produced successful results.

Selected Models for Logistic Support Applications

Experience through the years has justified the development and utilization of a wide variety of models in support of logistics-related activities. The accomplishment of life-cycle cost analyses, level of repair analyses, reliability and maintainability analyses and predictions, spare/repair parts and inventory policy analyses, maintenance shop repair and materials flow analyses, and so on, has in the past and will continue to be a requirement for most system developmental efforts. As a result, various contractors have developed the tools necessary to respond to customer requirements.

In most instances, the models that have been developed for application are relatively complex in nature, require the use of a design workstation or a mainframe computer of some configuration, and are proprietary in terms of availability to the general public. Most organizations involved in system design have their own reliability prediction model, life-cycle cost model, and so on. The models were developed for specific system applications or in response to a particular contractual requirement.

Of a more generic nature, there are many models that have been developed for the purposes of responding to logistics-related analysis requirements and are adaptable to the personal computer (PC). These models have been developed to incorporate some of the same characteristics that are included in the larger models, but can be utilized through a remote design workstation. Further, these models can be effectively utilized in education and training programs.

Your author has identified a few models that are representative of the tools that may be utilized in responding to logistics-related analysis requirements. Of particular significance are the models utilized in accomplishing reliability and maintainability predictions, the FMECA, level of repair analysis (LORA), and life-cycle cost (LCC) analysis. A brief description of each, along with reference to the source, is included:[3, 4]

1. *Reliability prediction program (RPP).* The RPP constitutes a series of computer software routines designed to aid in the accomplishment of reliability predictions based on component part stress factors. A system may contain any number of assemblies, subassemblies, and component parts (limited only by available disk space); component failure rate information can be introduced through a series of menu prompts; part application data and stress factors can be addressed (temperature, power, etc.); and the prediction of MTBF, or λ, values can be determined for the system in question. Implementation of RPP can be accomplished using a personal computer and is available for all current versions of MIL-HDBK-217, *Reliability Prediction of Electronic Equipment*, Department of Defense, Washington, D.C. *Reference*: Powertronic Systems, Inc., 13700 Chef Menteur Highway, P.O. Box 29109, New Orleans, LA 70189.

2. *PC-availability*. A model that utilizes Markov analysis to study the influence of failure rates, repair rates, and logistic support on system availability. The objective is to provide assistance in the development of optimum system configuration design and repair policies. *Reference:* Management Sciences, Inc., 6022 Constitution Ave., N.E., Albuquerque, NM 87110.

[3] A listing of over 350 computer-based tools (models) is included in Blanchard, B. S., Fabrycky, W. J., and Verma, D. (EDS), *Application of the System Engineering Process to Define Requirements for Computer-Based Design Tools*, Monograph, International Society of Logistics (SOLE), Hyattsville, Md., 1994.

[4] It should be noted that the process of model development and utilization is highly *dynamic*. New tools are being introduced daily and old tools are being replaced. Thus, the accuracy of this list can be questioned. The purpose here is to let the reader know what has been available in the past (i.e., an example of the nature of the models that have been used in support of logistics activities).

3. *PC-predictor.* A reliability model that automatically applies the part stress analysis or parts count methods of MIL-HDBK-217E to produce equipment failure rate estimates and to accomplish reliability predictions. *Reference:* Management Sciences, Inc., 6022 Constitution Ave., N.E., Albuquerque, NM 87110.

4. *Tiger computer program.* A family of computer programs that can be used to evaluate, by Monte Carlo simulation, an equipment or a large-scale complex system in order to establish various reliability, readiness, and availability measures. Key features include the ranking of equipment by degree of unreliability and unavailability, evaluating a mission with multiphase types, and performing sensitivity analyses on a complex system by downgrading or upgrading the characteristics of each equipment. *Reference:* M. L. Buckberg, Room 5W64, Bldg. NC-2, Reliability Engineering, Naval Sea Systems Command, Department of the Navy, Washington, DC 20362.

5. *Core.* A systems engineering program that includes a top-down definition of the process that starts with a statement of the problem. It enables the user to extract the originating requirements from the source, analyze them for completeness, consistency and testability, and trace each requirement to a behavioral model which describes the interactions and process sequences. A functional analysis is included, with the user being able to allocate requirements to a physical system architecture. *Reference:* VITECH Corp., 2070 Chain Bridge Road, Suite 320, Vienna, VA 22182.

6. *Mechanical reliability prediction program (MECHREL).* This program is utilized in the performance of reliability prediction of mechanical systems. Component parts (e.g., gearboxes, values, pumps, bearings, filters) are defined in terms of stresses, failure rates, failure modes, and so on, and combined to predict system MTBF. *Reference:* Eagle Technology, Inc., 2300 S. Ninth St., Arlington, VA 22204.

7. *Maintainability effectiveness analysis program (MEAP).* This model is used to compute maintenance times for electronic and electromechanical components, and to accomplish maintainability predictions for systems in accordance with MIL-HDBK-472, *Maintainability Prediction*. Predicted values of MTTR, M_{max}, and MDT for various indenture levels of a system can be determined. *Reference:* Systems Effectiveness Associates, Inc., 20 Vernon St., Norwood, MA 02062.

8. *Maintainability prediction program (MPP).* The MPP can be used to predict the maintainability of electrical, electronic, electromechanical, and mechanical systems. Predictions of MTTR ($\overline{M}ct$), M_{max} (60th to 90th percentile), $\overline{MMH/OH}$, and $\overline{MMH/MA}$ factors associated with maintenance activities at the organizational, intermediate, and depot/supplier levels can be accomplished. Repair time factors include localization, fault isolation, disassembly, interchange, reassembly, alignment, and checkout. The quantity of replaceable items, which may be a mixture of assemblies, subassemblies, and component parts, is limited only by disk storage space. MPP, using a personal computer, can be utilized to accomplish a maintainability prediction in accordance with MIL-HDBK-472 (Notice 1, Procedure V, Method B), *Maintainability Prediction,* Department of Defense, Washington, DC. *Reference:* Powertronic Systems, Inc., 13700 Chef Menteur Highway, P.O. Box 29109, New Orleans, LA 70189.

9. *Failure modes, effects, and criticality analysis (FME).* The FME can be used in the analysis and development of a FMECA for a system, and can be applied to a user-defined hierarchy including up to 25 levels of assemblies, subassemblies, components, and so on. The FME allows for the determination of failure modes, percent contributions, local effects and effects on the system, frequencies of occurrence, severity classifications, failure detection methods, and recommended compensating maintenance provisions. The data structure is presented in a "tree" form, and operations may be accomplished by highlighting individual tree elements. The FME can be implemented using a personal computer, and the input/output requirements need to be integrated with reliability and maintainability prediction programs (e.g., the results from the RPP and MPP models). This program includes all features necessary for developing the FMECA in accordance with MIL-STD-1629A (Notice 1), *Procedures for Performing a Failure Mode, Efects and Criticality Analysis*, Department of Defense, Washington, DC. *Reference:* Powertronic Systems, Inc., 13700 Chef Menteur Highway, P.O. Box 29109, New Orleans, LA 70189.

10. *Equipment designer's cost analysis system (EDCAS).* The EDCAS is a design tool that can be used in the accomplishment of a level-of-repair analysis (LORA). It includes a capability for the evaluation of repair versus discard-at-failure decisions, and it can handle up to 3500 unique items concurrently (i.e., 1500 line replaceable units, 2000 shop replaceable units). Repair-level analysis can be accomplished at two identure levels of the system, and the results can be used in determining optimum spare/repair part requirements and in the accomplishment of life-cycle cost analysis (LCCA). EDCAS is available through a simplified personal computer oriented University Edition (UE) and a full-scaled laboratory model edition and is compatible with the requirements of MIL-STD-1390B, *Level of Repair*, Department of Defense, Washington, DC. *Reference:* Systems Exchange, 170 17th St. Pacific Grove, CA 93950.

11. *Optimum repair level analysis (ORLA) model.* A level-of-repair analysis (LORA) model used to examine the economic feasibility of maintenance and support alternatives. Up to four alternatives can be evaluated, with life-cycle cost broken down into 13 distinct logistics areas. *Reference:* U.S. Army-MICOM, Code AMSMI-LC-TA-L, Redstone Arsenal, AL 35898.

12. *Level of repair (Navy).* This model is used in the performance of a level of repair (LOR) analysis during the design and development of new systems and equipment. Two objectives are to establish a least-cost maintenance policy, and to influence design in order to minimize logistic support costs. *Reference:* MIL-STD-1390B, *Level of Repair*, Department of Defense, Washington, DC.

13. *Network repair level analysis (NRLA).* The NRLA model is used in establishing equipment and component repair levels; and for making repair versus discard-at-failure decisions in both the design of new equipment and in provisioning. *Reference:*

U.S. Air Force, Air Force Acquisition Logistics Center, AFALC/LSS, Wright-Patterson AFB, OH 45433.

14. *Optimum supply and maintenance model (OSAMM).* A level of repair analysis (LORA) model used to determine the optimum economic maintenance policy for each item that fails, to identify items in terms of repair versus discard, to identify economic screening criteria, and to evaluate support equipment/repairmen options. Four levels of support and four indenture levels of equipment can be evaluated. *Reference:* U.S. ARMY-CECOM, Code AMSEL-PL-SA, Fort Monmouth, NJ 07703.

15. *Computed optimization of replenishment and initial spares based on demand and availability (CORIDA).* A model that computes initial and replenishment spare/repair part requirements, for multiple levels of maintenance, in terms of costs and distribution. It addresses organizational and budgeting constraints, and is utilized in support of provisioning activities. *Reference:* Thomson-CSF Systems Canada, 350 Sparks St., Suite 406, Ottawa, Ontario KIR 7S8, Canada.

16. *Opus model.* A versatile model used primarily for spare/repair parts and inventory optimization. It considers different operational scenarios and system utilization profiles in determining demand patterns for spares, and it aids in the evaluation of various design packaging schemes. Alternative support policies and structures may be evaluated on a cost-effectiveness basis. *Reference:* Systecon AB, Linnëgatan 5, S102 45 Stockholm, Sweden.

17. *VMetric.* This is a spares model that can be used to optimize system availability by determining the appropriate individual availabilities for system components and the stockage requirements for three indenture levels of equipment (e.g., line replaceable units, shop replaceable units, and subassemblies) at all echelons of maintenance. Outputs include optimum stock levels at each echelon of maintenance, EOQ quantities, and optimal reorder intervals. *Reference:* Systems Exchange, 170 17th St., Pacific Grove, CA 93950.

18. *Systems and logistics integration capability (SLIC).* This program is an integrated logistics support analysis data management system designed to respond to CALS objectives in producing a mini LSAR in accordance with MIL-STD-1388-2, *Requirements for a Logistic Support Analysis Record. Reference:* Integrated Micro Systems, Inc., 306 Pinecliff Dr., P.O. Box 1438, Seneca, SC 29679.

19. *Distributed integrated logistic support analysis (DILSA).* This program is a distributed database processor designed to produce a mini logistic support analysis record (LSAR). It incorporates reliability, maintainability, and logistics data, and creates LSAR reports in accordance with MIL-STD-1388-2, *Requirements for a Logistic*

Support Analysis Record. Reference: Logistic Engineering Associates, 2700 Navajo Road, Suite A, El Cajon, CA 92020.

20. *System design utility (SDU).* The SDU model can be applied during the early stages of system design in the accomplishment of requirements allocation (i.e., the top-down apportionment of system-level requirements to lower-level elements of the system), and in the later stages of development in the accomplishment of design trade-off studies. It incorporates the flexibility to allow for the evaluation of different system architectures, as well as providing database management capability for configuration control purposes. The output from the SDU model can be utilized in a variety of situations, to include providing an input for EDCAS. *Reference:* Systems Exchange, 170 17th St., Pacific Grove, CA 93950.

21. *Requirements driven development (RDD-100).* RDD is a modeling approach employed to evaluate various design alternatives and to assess system behavior. This approach utilizes functional block diagrams to describe sequential and concurrent activities, input-output requirements, data flow and physical item flow, and to assist in system decomposition. RDD is a system engineering method, supported by an executable graphic modeling language and computer-based tools developed to assist the engineer in designing a complex system through functional analysis and allocation. System platforms include Apollo, Apple Macintosh II, and Sun with 8 Mb RAM. *Reference:* Ascent Logic Corp., 180 Rose Orchard Way, Suite 200, San Jose, CA 95134.

22. *Repairable equipment population system (REPS).* The REPS model evaluates a homogenous population of repairable equipment deployed to meet a demand. As equipment units fail, they join a queue, are repaired in a multiple channel maintenance facility, and are returned to service. As equipment ages, the older units are removed from the system and replaced with new ones. The objective of REPS is to determine the number of equipment units to deploy, the number of parallel maintenance channels, and the replacement age of equipment so that the sum of all costs associated with the system will be minimized. *Reference:* Industrial and Systems Engineering Department, Virginia Tech, Blacksburg, VA 24061.

23. *Life-cycle cost calculator (LCCC).* The LCCC model aids in the accomplishment of a life-cycle cost analysis, using a cost breakdown structure (CBS) methodology. Objectives and activities are linked to resources, and constitute a logical subdivision of cost by functional activity area, major element of a system, or one or more discrete classes of common or like items. It provides a mechanism for the initial allocation of cost, cost categorization, and finally, cost summation. It has the capability of identifying cost contributions, in both real and discounted dollars, at any level in the CBS. *Reference:* Industrial and Systems Engineering Department, Virginia Tech, Blacksburg, VA 24061.

24. *Cost analysis strategy assessment (CASA).* The CASA model is utilized to develop life-cycle cost (LCC) estimates for a wide variety of systems and equipment. It incorporates various analysis tools into one functioning unit and allows the analyst to generate data files, perform life-cycle costing, sensitivity analysis, risk analysis, cost summaries, and the evaluation of alternatives. *Reference:* Defense Systems Management College, DSS Directorate (DRI-S), Fort Belvoir, VA 22060.

25. *Life-cycle cost model for defense materiel systems (Marine Corps).* This model provides the structure for the calculation of life-cycle cost (LCC) for a system or equipment being developed. *Reference:* MIL-HDBK-276-1 (MC), *Life-Cycle Cost Model for Defense Materiel Systems Data Collection Workbook*, Department of Defense, Washington, DC.

Utilization and Integration of Models

The models described thus far are primarily being utilized on a "stand-alone" basis. They were developed by independent suppliers for specific applications and, in many instances, these suppliers are in competition with others. As such, the models are independently operated and are not integrated into any higher-order information flow network.

At some point in the future, it will be necessary to develop an overall integrated approach. Referring to Figure 4.41, one can establish a structure and an information flow, tieing the models together in terms of input-output characteristics. From the example illustrated, the analyst may wish to accomplish a life-cycle cost analysis for the system (the output from block 5); perform a level of repair analysis independently (block 8); identify maintenance shop requirements based on the results of the level of repair analysis (blocks 8 and 9); and project the results from these lower levels of analysis into the system life-cycle cost analysis (block 5). It will be necessary to view the system as an entity; identify the high-cost areas of activity and optimize these for the purposes of improvement (i.e., reduction in LCC); and project the results back in the context of the whole! The interrelationships among the system components are numerous, and there are many interactions between the activities represented by the blocks in Figure 4.41 (also, refer to Figure 3.35).

As an ultimate objective, it is essential not only that the models "talk to each other," but that the collective group of models be appropriately integrated into the design workstation configuration concept as illustrated in Figure 5.11. If the logistics activity is to influence and be effective in system/product design, the appropriate communications must be established. An essential segment of this communications is through the transfer of information as noted.

With the objective above in mind, an initial attempt to properly integrate some of the models identified in the previous section has been made. Referring to Figure D.1, there have been some efforts to improve the communications between RPP, MPP, FME, and EDCAS. These efforts are continuing, with expansion, as this text edition goes to press. It is essential that the desired level of information flow be available in a timely and effective manner throughout the network.

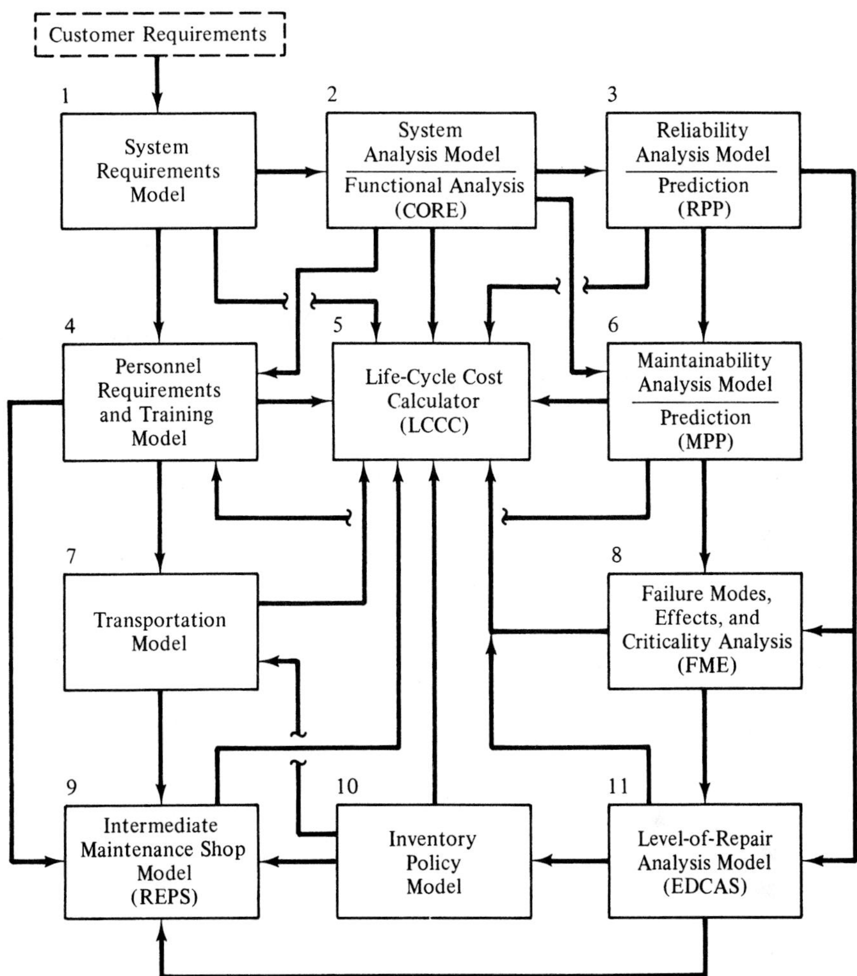

Figure D.1 Integration of logistic models.

APPENDIX E

LIFE-CYCLE COST ANALYSIS (LCCA)

As conveyed in the earlier chapters, logistics must be addressed from a *life-cycle* perspective. In the evaluation of alternatives, a life-cycle approach is assumed and, in the establishment of evaluation criteria, life-cycle considerations must be reflected. Thus, when considering economies in design, *life-cycle cost* becomes a major parameter.

Life-cycle costing, and the importance of acquiring full-cost *visibility* in design, is first introduced in Section 1.4 (Chapter 1). Answers to the questions of *what is it?* and *why is it important?* are addressed. In Section 2.11 (Chapter 2), the establishment of the appropriate economic factors as design criteria is discussed. A sample cost breakdown structure (Figure 2.26), a cost profile (Figure 2.27), a LCC summary (Figure 2.28), and an illustration showing LCC applications throughout the system life cycle (Figure 2.29) are included. In Section 4.2.1, a case study involving a life-cycle cost analysis is presented. Through an understanding of this case study (and the material presented earlier), the reader should be able to acquire some understanding relative to the accomplishment of a life-cycle cost analysis (LCCA). The purpose of this appendix is to provide supporting material relative to the analysis approach, the cost breakdown structure (CBS) and cost-estimating methods, the use of LCC analysis in design evaluation, and some LCC applications. For the purposes of discussion, the basic process illustrated in Figure E.1 serves as a point of reference. This process follows the steps for LCCA presented in Figure 4.11.[1]

[1] Life-cycle costing is covered further in Fabrycky, W. J., and Blanchard, B. S., *Life-Cycle Cost and Economic Analysis*, Prentice Hall, Inc., Upper Saddle River, N.J., 1991.

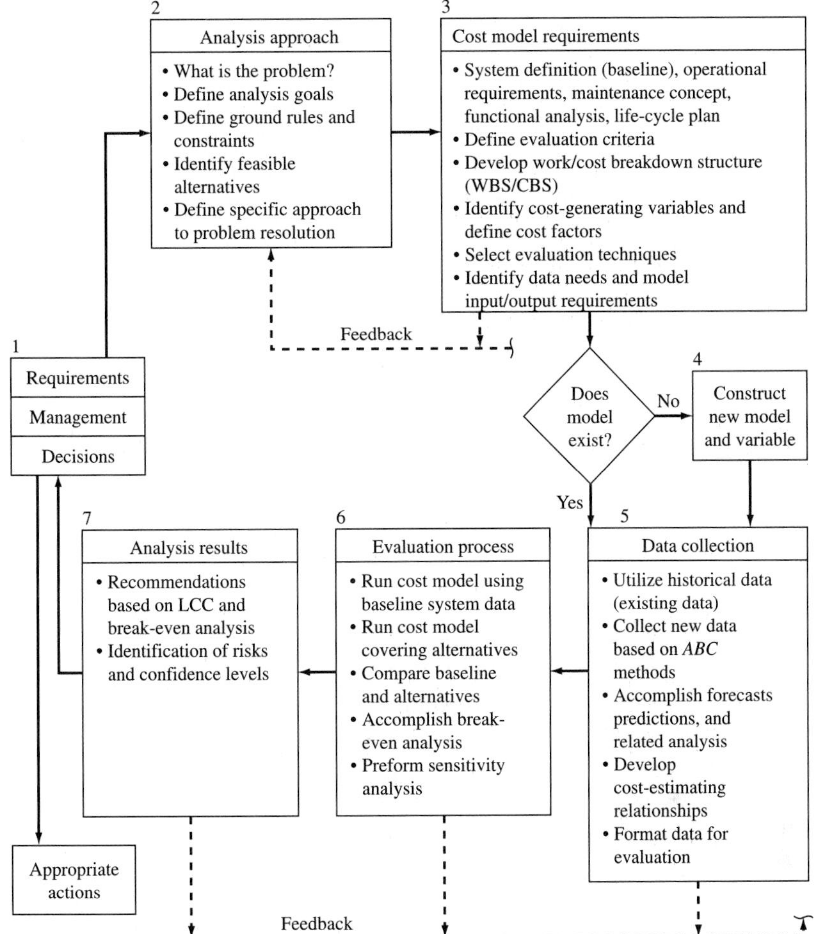

Figure E.1 Life-cycle cost analysis process.

Definition of System Requirements

In evaluating alternatives on the basis of life-cycle cost, the analyst needs to project each alternative in terms of life-cycle activities or events. Such activities (or events) generally evolve from a combination of stated requirements, concepts, plans, and so on, which have a significant impact on follow-on design, production, operation, and support requirements. In other words, the life-cycle cost analysis needs to be based on a definition of system operational requirements, a definition of the maintenance concept, and a program plan and profile illustrating major life-cycle activities and the projected operational horizon for the system (Figure E.1, blocks 2 and 3).

These requirements are covered in Chapters 3 and need to be addressed to the extent necessary in defining a specific problem. For a large complex system, the definition may be rather extensive involving a significant level of detail. Conversely, for small-scale systems, or components of a system, one still needs to address requirements in terms of the functions to be performed, quantities required and geographical distribution, reliability factors, maintenance and support concept, operational horizon period, and so on. However, the level of detail may be very limited.

In any event, regardless of the type of problem, the configuration(s) being evaluated must be projected in terms of *system-level requirements*. These requirements may change as the program evolves from phase to phase. However, an initial baseline must be established. From this point on, changes to this baseline may be evaluated systematically and in a controlled manner.

Cost Breakdown Structure (Cost Categories)

The functions described through the functional analysis can be broken down into subfunctions, categories of work, work packages, and ultimately the identification of physical elements. From a planning and management perspective, it is necessary to establish a top-down framework that will allow for the initial allocation and subsequent collecting, accumulating, organizing, and computing of costs. For a typical project, this may lead to the development of a work breakdown structure (WBS) prepared to show, in a hierarchical manner, all of the elements of work that are necessary to complete a given program. From Section 9.2, a summary work breakdown structure (SWBS) may be developed initially, followed by one or more individual contract work breakdown structures (CWBS) designed to address specific elements of work that are covered through some form of a contractual arrangement. It is the SWBS that provides a good basis for the development of a cost breakdown structure (CBS) used in life-cycle cost analyses, primarily because its intent is to cover *all* future activities and associated costs; that is, research and development, construction/production, distribution, operation and maintenance support, and retirement activities.

When accomplishing a life-cycle cost analysis, the analyst must develop a cost breakdown structure (i.e., cost tree) showing the numerous categories that are combined to provide the total cost. There is no set method for breaking down cost as long as the method used can be tailored to the specific application. However, the cost breakdown structure should exhibit the following basic characteristics:

1. *All* system cost elements must be considered.

2. Cost categories are generally identified with a function, a significant level of activity, or some major item of hardware/software. Cost categories must be well defined. The analyst, manager, customer, supplier, and so on, must all have the *same* understanding of what is included in a given category and what is not. Cost doubling (i.e., counting the same cost in two or more categories) and omissions must be eliminated. Lack of adequate definition causes inconsistencies in the evaluation process and could lead to a wrong decision.

3. The cost structure and categories should be coded in such a manner as to allow for the analysis of certain specific areas of interest (e.g., system operation, energy consumption, equipment design, spares, and maintenance personnel and support) while virtually ignoring other areas. In some instances, the analyst may wish to pursue a designated area in depth while covering other areas with gross top-level estimates. This will certainly occur from time to time as a system evolves through the different phases of its life cycle. The areas of concern (for decision-making purposes) will vary.

4. When related to a specific program, the cost structure should be compatible (through cross-indexing, coding, etc.) with the program work breakdown structure (WBS) and with the management accounting procedures used in collecting costs. Certain costs are derived from accounting records and should be a direct input into the life-cycle cost analysis.

5. For programs where subcontracting is prevalent, it is often desirable and necessary to separate supplier costs (i.e., initial bid price and follow-on program costs) from the other costs. The cost structure should allow for the identification of specific work packages that require close monitoring and control.

The CBS structure may vary from one program to the next. One example is presented in Figure 2.26 (Chapter 2), and another example is shown in Figure E.2.[2]

Referring to Figure 2.26, the various elements of the CBS must be broken down further to convey the method(s) by which the costs for each category are derived, along with key assumptions made in the cost-estimating process. An example of a breakout of several of the cost categories in Figure 2.26 is presented in Table E.1. The cost analyst needs to define all of the costs, the method(s) for determination, and a description of what is included in each category.

Cost Estimating

The next step is to estimate the costs, by category in the CBS, for each year in the system life cycle. Such estimates must consider the effects of inflation, learning curves when repetitive processes or activities occur, and any other factors that are likely to cause changes in cost, either upward or downward. Cost estimates may be derived from a combination of accounting records, cost projections, supplier proposals, and predictions in one form or another.

In Figures 1.10 and 2.25, it can be seen that the early stages in the system life cycle is the preferred time to commence with the estimation of costs, because it is at this point when the greatest impact on total system life-cycle cost can be realized. However, the availability of good historical cost data at this time is almost nonexistent in most organizations, particularly the type of data that pertain to the downstream activities of operations and support for singular systems in the past. Thus, one must depend heavily on the use of various cost-estimating methods in order to accomplish the end objectives.

[2] The cost breakdown structure (CBS) should be tailored to the system in question. If the system is very "software-intensive," then Category *Crs* (in Figure 2.26) should be broken down to show more detail. If the system is very "operator-intensive" (e.g., a ground radar tracking station requiring a large number of operating personnel), the Category C_{OO} should be expanded. On the other hand, if Category C_{OM} is too detailed for the purposes of a given analysis, then one can summarize the costs accordingly. The objective is to provide *visibility* relative to key functional activities.

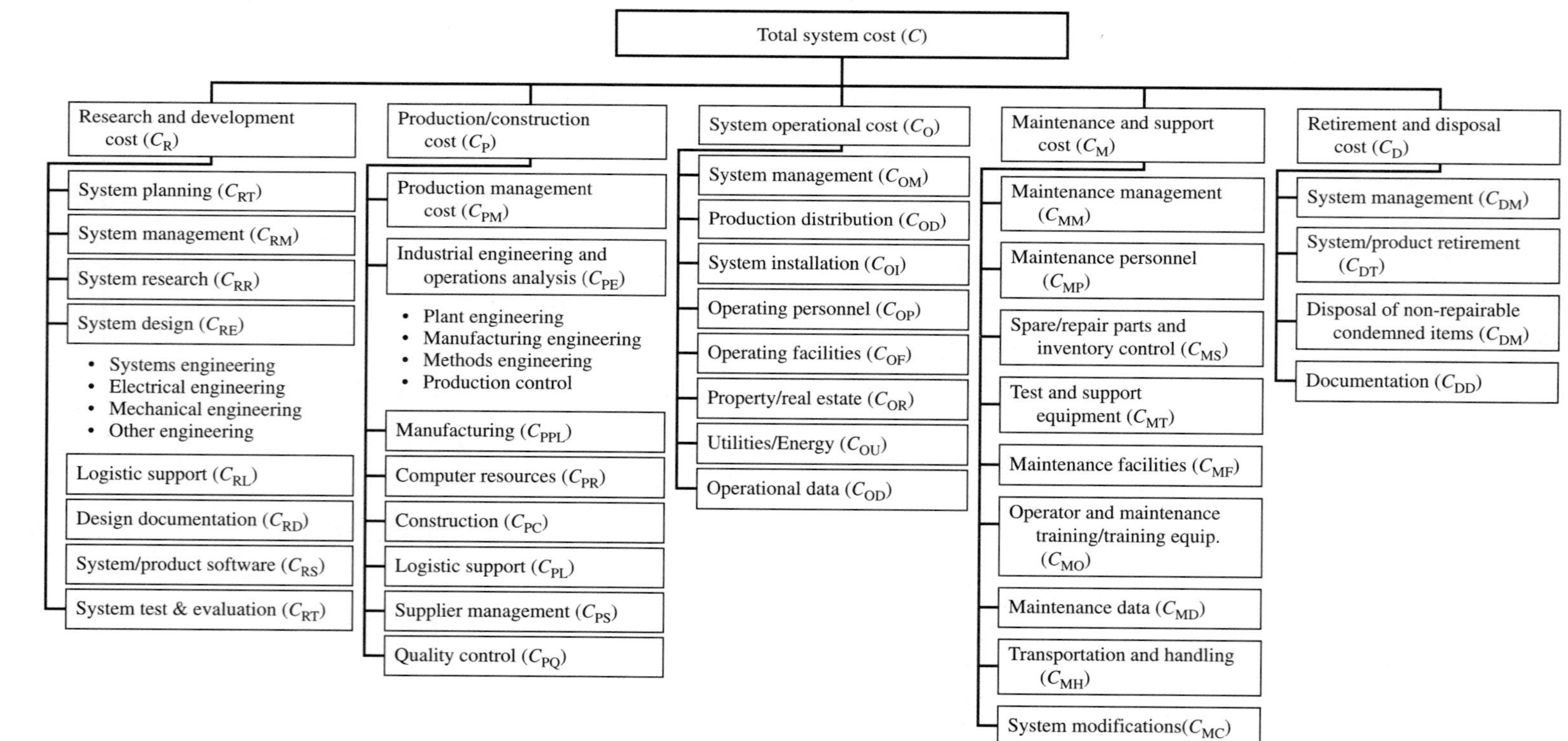

Figure E.2 Sample cost breakdown structure (reference: Figure 4.10).

TABLE E.1 Description of Cost Categories

Cost Category (Reference Figure 2.26)	Method of Determination (Quantitative Expression)	Cost Category Description and Justification
Maintenance cost (C_{OM})	$C_{OM} = [C_{OMM} + C_{OMX} + C_{OMS} + C_{OMT} + C_{OMP} + C_{OMF} + C_{OMD}]$ C_{OMM} = Maintenance personnel and support cost C_{OMX} = Cost of spare/repair parts C_{OMS} = Test and support equipment maintenance cost C_{OMT} = Transportation and handling cost C_{OMF} = Cost of maintenance facilities C_{OMD} = Cost of technical data	Includes all sustaining maintenance labor, spare/repair parts, test and support equipment, transportation and handling, replenishment training, support data, and facilities necessary to meet the maintenance needs of the prime equipment throughout its life-cycle. Such needs include both corrective and preventive maintenance requirements at all echelons—organizational, intermediate, depot, and factory.
Maintenance personnel and support cost (C_{OMM})	$C_{OMM} = [C_{OOU} + C_{OOS}]$ C_{OOU} = Cost of equipment corrective maintenance C_{OOS} = Cost of equipment preventive maintenance Total cost is the sum of the C_{OMM} values for each echelon of maintenance.	Includes corrective and preventive maintenance labor, associated material handling, and supporting documentation. When a system/equipment malfunction occurs or when a scheduled maintenance action is performed, personnel manhours are expended, the handling of spares and related material takes place, and maintenance action reports are completed. This category includes all directly related costs.
Corrective maintenance cost (C_{OOU})	$C_{OOU} = [(Q_{CA})(M_{MHC})(C_{OCP}) + (Q_{CA})(C_{MHC}) + (Q_{CA})(C_{DC})](N_{MS})$ Q_{CA} = Quantity of corrective maintenance actions (M_A). $Q_{CA} = (T_O)(\lambda)$ M_{MHC} = Corrective maintenance manhours/M_A C_{OCP} = Corrective maintenance labor cost ($/$M_{MHC}$) C_{MHC} = Cost of material handling/corrective M_A C_{DC} = Cost of documentation/corrective M_A N_{MS} = Number of maintenance sites Determine C_{OOU} for each appropriate echelon of maintenance.	This category includes the personnel activity costs associated with the accomplishment of corrective maintenance. Related spares, test and support equipment, transportation, training, and facility costs are covered in C_{OMX}, C_{OMS}, C_{OMT}, and C_{OMF}, respectively. Total cost includes the sum of individual costs for each maintenance action multiplied by the quantity of maintenance actions anticipated over the entire system life-cycle. A maintenance action includes any requirement for corrective maintenance resulting from catastrophic failures, dependent failures, operator/maintenance induced faults, manufacturing defects, etc. The cost per maintenance action considers the personnel labor expended for direct tasks (localization, fault isolation, remove and replace, repair, verification), associated administrative/logistic delay time, material handling, and maintenance documentation (failure reports, spares issue reports). The corrective maintenance labor cost, C_{OCP}, will of course vary with the personnel skill level required for task performance. Both direct labor and overhead costs are included.

(continued)

TABLE E.1 *(continued)* Description of Cost Categories

Cost Category (Reference Figure 2.26)	Method of Determination (Quantitative Expression)	Cost Category Description and Justification
Preventive maintenance cost (C_{OOS})	$C_{OOS} = [(Q_{PA})(M_{MHP})(C_{OPP}) + (Q_{PA})(C_{MHP}) + (Q_{PA})(C_{DP})](N_{MS})$ Q_{PA} = Quantity of preventive maintenance actions (M_A). Q_{PA} relates to fpt M_{MHP} = Preventive maintenance manhours/M_A C_{OPP} = Preventive maintenance labor cost ($/$M_{MHP}$) C_{MHP} = Cost of material handling/ preventive M_A	This category includes the personnel activity costs associated with the accomplishment of preventive or scheduled maintenance. Related spares/consumables, test and support equipment, transportation, training, and facility costs are covered in C_{OMX}, C_{OMS}, C_{OMT}, and C_{OMF}, respectively. Total cost includes the sum of individual costs for each preventive maintenance action multiplied by the quantity of maintenance actions anticipated over the system life-cycle. A maintenance action includes servicing, lubrica-?????
Spare/repair parts cost (C_{OI})	$C_{OI} = [C_{SO} + C_{SI} + C_{SD} + C_{SS} + C_{SC}]$ C_{SO} = Cost of organizational spare/repair parts C_{SI} = Cost of intermediate spare/repair parts C_{SD} = Cost of depot spare/repair parts C_{SS} = Cost of supplier spare/repair parts C_{SC} = Cost of consumables $C_{SO} = \sum_{N_{MS}} \left[(C_A)(Q_A) + \sum_{i=1} (C_{Mi})(Q_{Mi}) + \sum_{i=1} (C_{Hi})(Q_{Hi}) \right]$ C_A = Average cost of material purchase order ($/order) Q_A = Quantity of purchase orders C_M = Cost of spare item i Q_M = Quantity of i items required or demand C_H = Cost of maintaining spare item i in the inventory ($/$ value of the inventory) Q_H = Quantity of i items in the inventory N_{MS} = Number of maintenance sites	Initial spare/repair part costs are covered in C_{ILS}. This category includes all replenishment spare/repair parts and consumable materials (e.g., oil, lubricants, fuel, etc.) that are required to support maintenance activities associated with prime equipment, operational support and handling equipment (C_{OOE}), test and support equipment (C_{OMS}), and training equipment at each echelon (organizational, intermediate, depot, supplier). This category covers the cost of purchasing; the actual cost of the material itself; and the cost of holding or maintaining items in the inventory. Costs are assigned to the applicable level of maintenance. Specific quantitative requirements for spares (Q_M) are derived from the Supportability Analysis (SA) discussed in Chapter 4. These requirements are based on the criteria described in Chapter 2. The optimum quantity of purchase orders (Q_A) is based on the EOQ criteria described in Section 2.3. Support equipment spares are based on the same criteria used in determining spare part requirements for prime equipment.

(continued)

TABLE E.1 *(continued)* Description of Cost Categories

Cost Category (Reference Figure 2.26)	Method of Determination (Quantitative Expression)	Cost Category Description and Justification
Test and support equipment cost (C_{OMS})	$C_{OMS} = [C_{SEO} + C_{SEI} + C_{SED}]$ C_{SEO} = Cost of organizational test and support equipment C_{SEI} = Cost of intermediate test and support equipment C_{SED} = Cost of depot test and support equipment $C_{SEO} = [C_{OOU} + C_{OOS}]$ C_{OOU} = Cost of equipment corrective maintenance C_{OOS} = Cost of equipment preventive maintenance $C_{OOU} = [(Q_{CA})(M_{MHC})(C_{OCP}) + (Q_{CA})(C_{MHC}) + (Q_{CA})(C_{DC})](N_{MS})$ Q_{CA} = Quantity of corrective maintenance actions (M_A) or $Q_{CA} = (T_O)(\lambda)$ M_{MHC} = Corrective maintenance labor hours/M_A C_{OCP} = Corrective maintenance labor cost ($/$M_{MHC}$) C_{MHC} = Cost of material handling/corrective M_A C_{DC} = Cost of documentation/corrective M_A N_{MS} = Number of maintenance sites (involving organizational maintenance)	Initial acquisition cost for test and support equipment is covered in C_{ILX}. This category includes the annual recurring life-cycle maintenance cost for test and support equipment at each echelon. Support equipment operational costs are actually covered by the tasks performed in C_{OMM}. Maintenance constitutes both corrective and preventive maintenance, and the costs are derived on a similar basis with prime equipment (C_{OOU} and C_{OOS}). Spares and consumables are included in C_{OI}. In some instances, specific items of test and support equipment are utilized for more than one (1) system, and in such cases, associated costs are allocated proportionately to each system concerned.

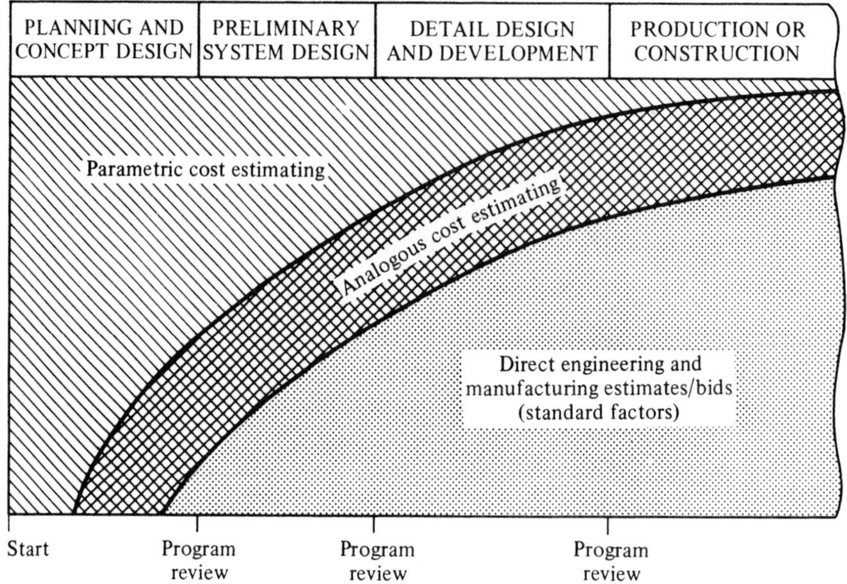

Figure E.3 Cost estimation by program phase.

From Figure E.3, as the system configuration becomes better defined in a developmental effort, the use of direct engineering and manufacturing standard factors based on past experience can be applied as is the case for any "cost-to-complete" projection on a typical project today (e.g., cost per labor hour). On the other hand, in the earlier stages of the life cycle when the system configuration has not been well-defined, the analyst must rely on the use of a combination of analogous or parametric methods developed from experience on similar systems in the past. The objective is to collect data on a "known entity," identify the major functions that have been accomplished and the costs associated with these functions, relate the costs in terms of some functional or physical parameter of the system, and then use this relationship in attempting to estimate the costs for a new system. As a goal, one should identify the applicable technical performance measures (TPMs) for the system in question and estimate the cost per a given level of performance (e.g., cost per unit of product output, cost per mile of range, cost per unit of weight, cost per volume of capacity used, cost per unit of acceleration, cost per functional output, etc.). Costs can be related to the appropriate blocks in the functional description of the system. Figures E.4 and E.5 provide some simple illustrations of considerations in cost estimating. However, care must be exercised to ensure that the historical information used in the development of cost-estimating relationships (CERs) is relevant to the system configuration being evaluated today. CERs based on the mission and performance characteristics of one system may not be appropriate for another system configuration, even if the configuration is similar in a physical sense. Thus, costs must be related from a *functional* perspective.

To be effective in total cost management (and in the accomplishment of cost-effectiveness analyses) requires full-cost visibility allowing for the traceability of all

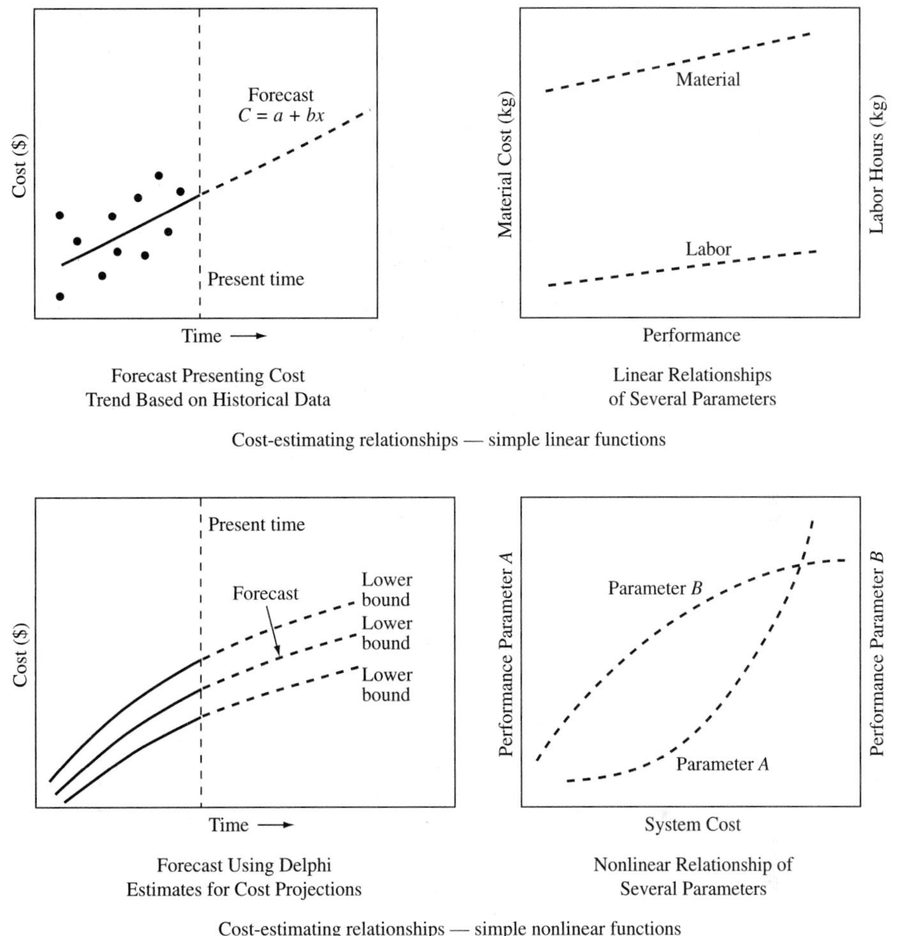

Figure E.4 Cost-estimating relationships (CERs).

costs back to the activities, processes, or products that generate these costs. In the traditional accounting structures employed in most organizations, a large percentage of the total cost cannot be traced back to the "causes." For example, "overhead" or "indirect" costs, which often constitute greater than 50% of the total, include a lot of management costs, supporting organization costs, and other costs that are difficult to trace and assign to specific objects. With these costs being allocated across the board, it is impossible to identify the actual "causes" and to pinpoint the *true* high-cost contributors. As a result, the concept of *activity-based costing* (ABC) has been introduced.[3]

Activity-based costing is a methodology directed toward the detailing and assignment of costs to the items that cause them to occur. The objective is to enable the

[3] Canada, J. R., Sullivan, W. G., and White, J. A., *Capital Investment Analysis for Engineering and Management*, 2nd Ed., Prentice Hall, Upper Saddle River, N.J., 1996; and Kidd, P. T., *Agile Manufacturing: Forging New Frontiers*, Addison-Wesley, Reading, Mass., 1994.

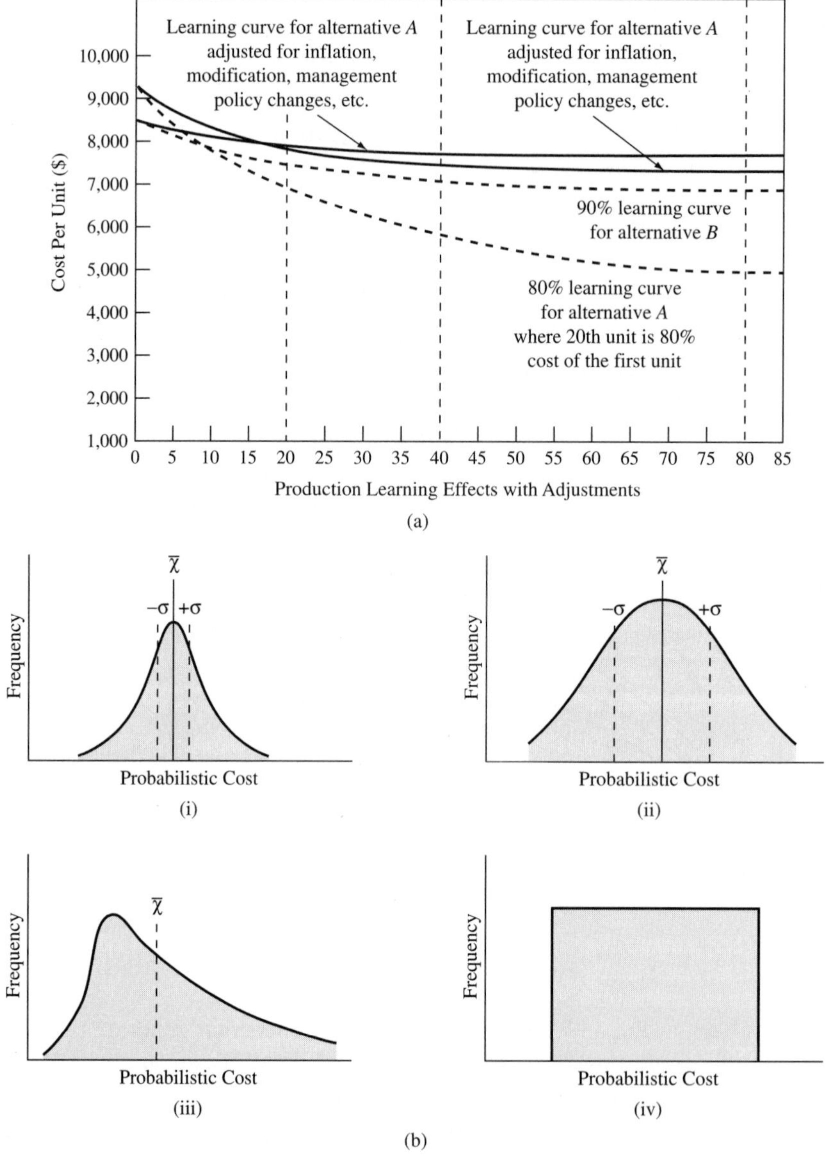

Figure E.5 (a) Learning curves and (b) the probabilistic aspects of costs.

"traceability" of *all* applicable costs to the process or product that generates these costs. The ABC approach allows for the initial allocation and later assessment of costs by function, and was developed to deal with the shortcomings of the traditional management accounting structure where large overhead factors are assigned to all elements of the enterprise across the board without concern for whether they directly apply or not. More specifically, the principles of ABC follow:

1. Cost are directly traceable to the applicable cost-generating process, product, or a related object. Cause-and effect relationships are established between a cost factor and a specific process or activity.
2. There is no distinction between direct and indirect (or overhead) costs. Although 80% to 90% of all costs are traceable, those nontraceable costs are not allocated across the board but are allocated directly to the organizational unit(s) involved in the project.
3. Costs can be easy allocated on *functional* basis—that is, the functions identified in Section 3.6. It is relatively easy to develop cost-estimating relationships in terms of the cost of activities per some activity measure (i.e., the cost per unit output).
4. The emphasis in ABC is on "resource consumption" (versus "spending"). Processes and products consume activities, and activities consume resources. With resource consumption being the objective, the ABC approach facilitates the evaluation of day-to-day decisions in terms of their impact on resource consumption downstream.
5. The ABC approach fosters the establishment of cause-and-effect relationships and, as such, enables the identification of the "high-cost contributors." Areas of risk can be identified with some specific activity and the decisions that are being made within.
6. The ABC approach tends to eliminate some of the cost doubling (or double counting) that occurs when attempting to differentiate on what should be included as a "direct" cost or as an "indirect" cost. By not having the necessary visibility, there is the potential of including the same costs in both categories.

Implementation of the ABC approach, or something of an equivalent nature, is essential if one is to do a good job of total cost management. Costs are tied to objects and viewed over the long term, and the use of such facilitates the life-cycle cost analysis process. An objective for the future is to convince the accounting organizations in various companies/agencies to supplement their current end-of-year financial reporting structure to include the objectives of ABC.

Evaluation of Alternatives

In the evaluation of alternatives, each alternative configuration is represented by a different cost profile (see Figures 2.27 and 4.13), because cost-generating activities will vary from one instance to the next, reliability and maintainability factors will be different, and the specific logistic support requirements will be unique for each situation. Figure E.6 reflects the cost profiles for three potential system configurations being considered for a single application.

The comparison of various possible courses of action requires that the costs associated with the alternatives in question be related on an equivalent basis (i.e., the point in time when the decision is to be made, which is generally considered as the *present time* or *now*). Thus, the costs for each year in the life cycle (for each profile being evaluated) are discounted to the present value.

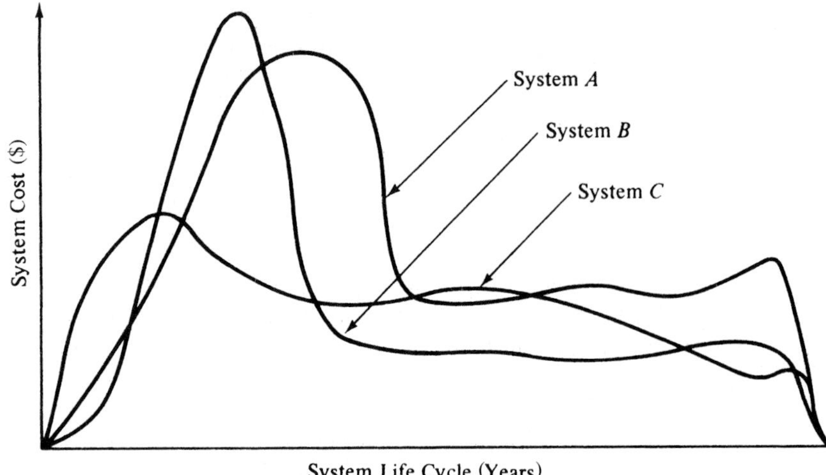

Figure E.6 Life-cycle cost profiles of alternatives.

The aspect of discounting refers to the application of a selected rate of interest to measure differences in importance or preference between dollars at the present time and anticipated dollars in the future. Discounting allows for the evaluation of the time-phased profiles of cost streams for various alternative configurations as if they occurred at one point (the only fair method of evaluation if one is to make a decision today) rather than spaced over the life of the system.

In every investment, one should recognize the fact that a dollar today is worth more than a dollar tomorrow because of the interest cost that is related to expenditures which occur over time; thus, a financial transaction (cash in-flow or cost) projected for tomorrow has a present value less than its undiscounted dollar value. Transactions that accrue in the future cannot be compared directly with investments made at the present because of this time value of money. Discounting is a technique for converting various cash flows/costs occurring over time to equivalent amounts at a common point in time to facilitate a valid comparison of alternative investments.

For the purposes of illustration, assume that costs are determined for a system with a life cycle of 5 years. Although costs actually occur continually or at discrete points throughout each year, they are usually treated either at the beginning or at the end of the applicable period. In this instance, the costs are determined at the end of each year throughout the life cycle. The object is to relate these costs in today's value, because this is the point in time when decisions related to equipment design and logistic support are made. Figure E.7 presents a simple illustration of the system life cycle, and the costs for each year must be conveyed in terms of the present value (decision point).

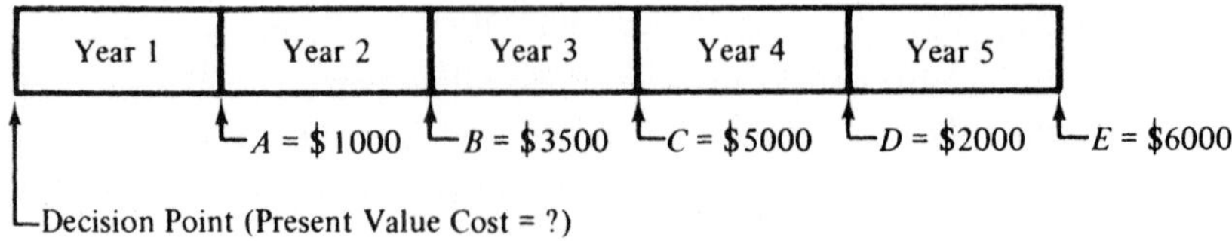

Figure E.7 System life-cycle costs.

For the first year, costs are estimated for each applicable item and are represented at point A. The question is: What is the value of that total cost at point A in terms of the decision point? This can be determined from the following single-payment present-value expression:[4]

$$P = F\left[\frac{1}{(1+i)^n}\right] \tag{E.1}$$

where

P = present value or present principal sum
F = future sum at some interest period hence
i = annual interest rate
n = interest period

Assuming an interest rate of 10%, the present-value cost at point A is

$$P = 100\left[\frac{1}{(1+0.1)}\right] = \$909.09$$

For point B, this value becomes

$$P = 3500\left[\frac{1}{(1+0.1)^2}\right] = \$2892.40$$

Next, consider the present value of the costs at point A and point B combined. This is calculated from the expression

$$P = F_A\left[\frac{1}{(1+i)}\right] + F_B\left[\frac{1}{(1+i)^2}\right] \tag{E.2}$$

Assuming the undiscounted estimates of \$1000 and \$3500 at points A and B, respectively, the present value is

$$P = 1000\left[\frac{1}{(1+0.1)}\right] + 3500\left[\frac{1}{(1+0.1)^2}\right] = \$3801.49$$

This continues until the costs for each year in the system life cycle are discounted to the present value and totaled. The total present-value cost for the system represented in Figure E.7 is \$12,649.74 (note that the undiscounted total value is \$17,500). When costs are different for each year in the life cycle, the present-value expression is a continuation of Equations (E.1) and (E.2), or

$$P = F_A\left[\frac{1}{(1+i)}\right] + F_B\left[\frac{1}{(1+i)^2}\right] + F_C\left[\frac{1}{(1+i)^3}\right] + \cdots + F_n\left[\frac{1}{(1+i)^n}\right] \tag{E.3}$$

Present-value cost calculations can be simplified by using standard interest tables and multiplying the future sum by the appropriate factor. Appendix F includes a set of interest tables with present-value factors.

[4]Thuesen G. J., and Fabrycky, W. J., *Engineering Economy*, 8th ed., Prentice Hall, Inc., Upper Saddle River, N.J., 1993. This reference presents an in-depth coverage of present value, future value, equal-payment series, interest tables, and related material.

As an example of the application of discounting in evaluating alternatives, suppose that a manufacturing firm is considering the possibility of introducing a new product into the market which is expected to meet sales projections for at least 10 years. The firm must invest in capital equipment in order to manufacture the product. Based on a survey of potential sources, there are two equipment alternatives considered to be feasible.

Table E.2 gives the projected cash flow for the two equipment alternatives (both revenues and costs are given). A 6-year life cycle and an interest rate of 10% are assumed. There is an initial investment of \$15,000 for equipment A and \$20,000 for equipment B. Benefits (in terms of anticipated net revenues) and costs are shown in the table together with the interest factors from Appendix F. The net present value (NPV) for equipment A is \$30,486 and the NPV for equipment B is \$7960. Thus, equipment A is preferred.

Before making a final choice, the analyst should perform a break-even analysis and establish the payback points for each project. Although other factors may affect the ultimate decision, the project exhibiting an early payback is usually desirable when considering risk and uncertainty. Figure E.8 illustrates cash flows and payback points, and equipment A retains its preferred status.

The discussion thus far has dealt with the conversion of anticipated future revenues and costs to the present value. This is necessary for the evaluation of proposals in terms of the *present* time. However, one may wish to relate money transactions to some future point in time, assuming this point to be at a time when a major investment decision will be made. In such instances, the single-payment present-value equation (E.1) can be transposed to

$$F = P[(1 + i)^n] \tag{E.4}$$

This equation can be used to determine the results of various investments. For instance, assume that \$6000 is invested today in some venture and the annual interest rate is 8%. Then the compound amount at the end of year 2 is

$$F = \$6000[(1 + 0.08)^2] = \$6998.40$$

With the application of present-value and future-value concepts, the analyst can relate to the life cycle and convert revenues and costs to any designated decision point. Figure E.9 illustrates this, where the anticipated decision point is at the end of year 2. Referring to the figure, the discounted value of the cost stream at 10% is

$$\begin{aligned} P_{(\text{decision point})} &= 8000 + 2000(1.100) + 1000(1.210) \\ &\quad + 12{,}000(0.9091) + 18{,}000(0.8265) \\ &\quad + 22{,}000(0.7513) = \$53{,}724.80 \end{aligned}$$

In summary, revenues and costs may be treated differently depending on the problem at hand. Money has time value and discounting is appropriate in the direct

TABLE E.2 Net-Present-Value Comparison

Equipment *A*

Year n	Cash Flow		Discount Factor (10%)	Present Value	
	Benefits	Costs		Benefits	Costs
0		$15,000	1.0000		$15,000.00
1		6,000	0.9091		5,454.60
2	$ 5,000	3,000	0.8264	$ 4,132.00	2,479.20
3	12,000		0.7513	9,015.60	
4	16,500		0.6830	11,269.50	
5	28,800		0.6209	16,019.22	
6	23,000		0.5645	12,983.50	
Total	$82,300	$24,000		$53,419.82	$22,933.80

Equipment *B*

Year n	Cash Flow		Discount Factor (10%)	Present Value	
	Benefits	Costs		Benefits	Costs
0		$20,000	1.0000		$20,000.00
1		12,000	0.9091		10,909.20
2	$ 4,000	6,000	0.8264	$ 3,305.60	4,958.40
3	13,000	5,000	0.7513	9,766.90	3,756.50
4	17,000	3,000	0.6830	11,611.00	2,049.00
5	22,000		0.6209	13,569.90	
6	20,000		0.5645	11,290.00	
Total	$76,000	$46,000		$49,633.30	$41,673.10

comparison of alternative cost profiles. For economic analyses, discounting is employed to relate all revenues and costs to a specific decision point, whether now or in the future.[5]

[5] If the analyst wishes to evaluate a preferred money profile on the basis of future *budget* requirements, then he or she may wish to convert the discounted values back to the inflated yearly values.

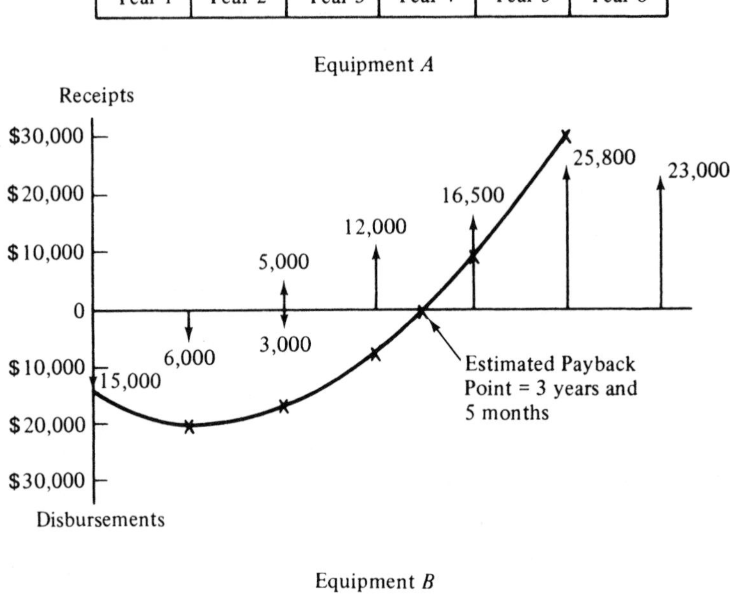

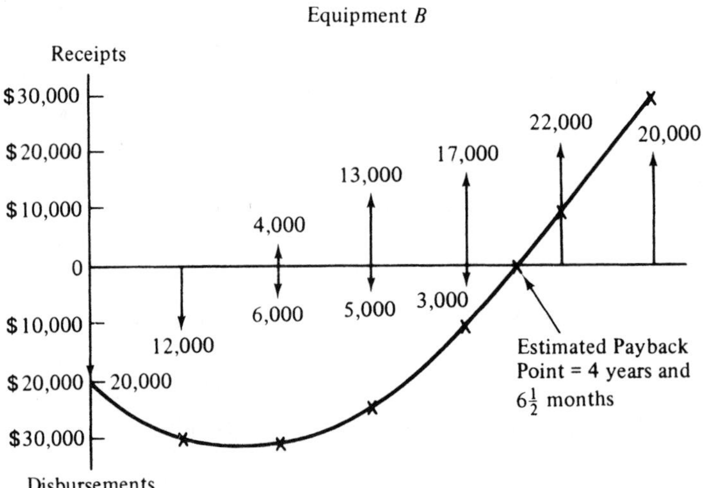

Figure E.8 Cash flow and payback comparisons.

Life-Cycle Cost Applications and Benefits

Inherent within the activities identified in Figure 2.29, there are many different and varied design and management decisions that are made and can have a significant impact on life-cycle cost. In particular, those early system-level decisions made during the conceptual and preliminary design phases will have a significant influence on the activities and associated costs of system operations, maintenance and support, and retirement and material disposal (refer to Figures 1.10 and 2.25). Thus, it is essential that designers, support personnel, managers, and others in the decision-making loop

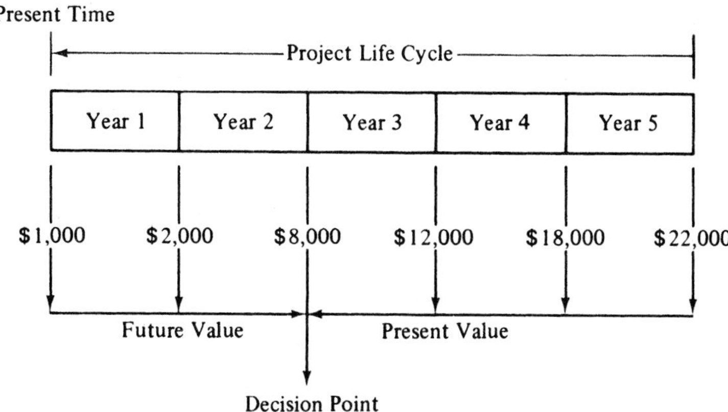

Figure E.9 Present value/future value cost projection.

consider the impact of their day-to-day actions on total life-cycle cost. In other words, the greatest benefits can be gained if the issues of economics and cost were addressed from the beginning in the design and development of new systems.

At the same time, many benefits can be gained through the application of life-cycle costing methods in the evaluation and subsequent improvement of systems that are already in the inventory and in operational use. The identification of high-cost contributors and the follow-on system modification for improvement (realizing a reduction in projected life-cycle cost), implemented on an iterative and continuing basis, can result in many benefits in this day and age where resources are limited, and there is a great deal of international competition. Success in this area is heavily dependent on the availability of good historical data and having the visibility needed for the implementation of a *continuous product/process improvement* capability. With regard to *visibility*, can you answer the following questions:

1. For your system, can you identify the high-cost contributors from a *functional* perspective? What is it costing to perform certain critical functions over the life cycle?
2. Can you identify the *causes* for these high-cost areas? What elements of the system, or segments of a process, are "driving" these costs?
3. What are the relationships between the high-cost areas and functions that are *critical* regarding accomplishment of the system mission?
4. What are the *high-risk* areas?

The life-cycle cost analysis process can be used as an aid in solving a wide variety of problems. Figure E.10 highlights a few areas where one can apply the steps identified in Figure E.1; however, the process reflected by the steps identified in the figure must be properly "tailored" to the problem at hand. In essence, one can accomplish a LCC analysis on a single sheet of paper in a few minutes, or can go to great depths in terms of data manipulations. The results of such an analysis must be to the proper depth, must be timely, and must be responsive to designer/manager involved in the decision-making process. The important issue here is to *think economics* and to *think life cycle*, the former being within the context of the latter.

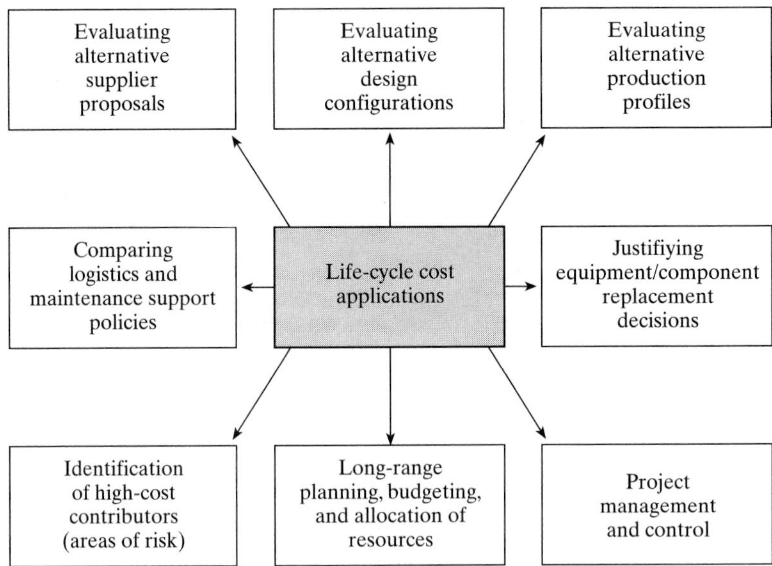

Figure E.10 Life-cycle cost applications.

Although the benefits are numerous, there are also some major impediments. Our current "thought processes," accounting practices, budgetary cycles, organizational objectives, and politically driven activities are more oriented to the "short term." Further, on many occasions the visibility from a LCC analysis is not desired because of the fear of exposure for one reason of another.

To be successful in this area requires that the proper *organizational environment* be established that will allow it to happen. There must be a committent to "life-cycle thinking" from the top down; the right type of data must be collected and available; and the analyst must have direct access to all applicable areas of activity. Given this, it is important to get "involved" by understanding the process, applying it to a known entity and evaluating the results, and establishing some cost-estimating relationships that can be applied to future LCC analyses.

APPENDIX F

SELECTED INTEREST TABLES

TABLE F.1 6% Interest Factors for Annual Compounding

	Single Payment		Equal Payment Series				Uniform gradient-series factor
n	Compound-amount factor	Present-worth factor	Compound-amount factor	Sinking-fund factor	Present-worth factor	Capital-recovery factor	
	To find *F* Given *P* *F/P, i, n*	To find *P* Given *F* *P/F, i, n*	To find *F* Given *A* *F/A, i, n*	To find *A* Given *F* *A/F, i, n*	To find *P* Given *A* *P/A, i, n*	To find *A* Given *P* *A/P, i, n*	To find *A* Given *G* *A/G, i, n*
1	1.060	0.9434	1.000	1.0000	0.9434	1.0600	0.0000
2	1.124	0.8900	2.060	0.4854	1.8334	0.5454	0.4854
3	1.191	0.8396	3.184	0.3141	2.6730	0.3741	0.9612
4	1.262	0.7921	4.375	0.2286	3.4651	0.2886	1.4272
5	1.338	0.7473	5.637	0.1774	4.2124	0.2374	1.8836
6	1.419	0.7050	6.975	0.1434	4.9173	0.2034	2.3304
7	1.504	0.6651	8.394	0.1191	5.5824	0.1791	2.7676
8	1.594	0.6274	9.897	0.1010	6.2098	0.1610	3.1952
9	1.689	0.5919	11.491	0.0870	6.8017	0.1470	3.6133
10	1.791	0.5584	13.181	0.0759	7.3601	0.1359	4.0220
11	1.898	0.5268	14.972	0.0668	7.8869	0.1268	4.4213
12	2.012	0.4970	16.870	0.0593	8.3839	0.1193	4.8113
13	2.133	0.4688	18.882	0.0530	8.8527	0.1130	5.1920
14	2.261	0.4423	21.015	0.0476	9.2950	0.1076	5.5635
15	2.397	0.4173	23.276	0.0430	9.7123	0.1030	5.9260
16	2.540	0.3937	25.673	0.0390	10.1059	0.0990	6.2794
17	2.693	0.3714	28.213	0.0355	10.4773	0.0955	6.6240
18	2.854	0.3504	30.906	0.0324	10.8276	0.0924	6.9597
19	3.026	0.3305	33.760	0.0296	11.1581	0.0896	7.2867
20	3.207	0.3118	36.786	0.0272	11.4699	0.0872	7.6052
21	3.400	0.2942	39.993	0.0250	11.7641	0.0850	7.9151
22	3.604	0.2775	43.392	0.0231	12.0416	0.0831	8.2166
23	3.820	0.2618	46.996	0.0213	12.3034	0.0813	8.5099
24	4.049	0.2470	50.816	0.0197	12.5504	0.0797	8.7951
25	4.292	0.2330	54.865	0.0182	12.7834	0.0782	9.0722
26	4.549	0.2198	59.156	0.0169	13.0032	0.0769	9.3415
27	4.822	0.2074	63.706	0.0157	13.2105	0.0757	9.6030
28	5.112	0.1956	68.528	0.0146	13.4062	0.0746	9.8568
29	5.418	0.1846	73.640	0.0136	13.5907	0.0736	10.1032
30	5.744	0.1741	79.058	0.0127	13.7648	0.0727	10.3422
31	6.088	0.1643	84.802	0.0118	13.9291	0.0718	10.5740
32	6.453	0.1550	90.890	0.0110	14.0841	0.0710	10.7988
33	6.841	0.1462	97.343	0.0103	14.2302	0.0703	11.0166
34	7.251	0.1379	104.184	0.0096	14.3682	0.0696	11.2276
35	7.686	0.1301	111.435	0.0090	14.4983	0.0690	11.4319
40	10.286	0.0972	154.762	0.0065	15.0463	0.0665	12.3590
45	13.765	0.0727	212.744	0.0047	15.4558	0.0647	13.1413
50	18.420	0.0543	290.336	0.0035	15.7619	0.0635	13.7964
55	24.650	0.0406	394.172	0.0025	15.9906	0.0625	14.3411
60	32.988	0.0303	533.128	0.0019	16.1614	0.0619	14.7910
65	44.145	0.0227	719.083	0.0014	16.2891	0.0614	15.1601
70	59.076	0.0169	967.932	0.0010	16.3846	0.0610	15.4614
75	79.057	0.0127	1300.949	0.0008	16.4559	0.0608	15.7058
80	105.796	0.0095	1746.600	0.0006	16.5091	0.0606	15.9033
85	141.579	0.0071	2342.982	0.0004	16.5490	0.0604	16.0620
90	189.465	0.0053	3141.075	0.0003	16.5787	0.0603	16.1891
95	253.546	0.0040	4209.104	0.0002	16.6009	0.0602	16.2905
100	339.302	0.0030	5638.368	0.0002	16.6176	0.0602	16.3711

Source: Tables F.1 to F.12 are from G. J. Thuesen and W. J. Fabrycky, *Engineering Economy*, 6th ed.,

TABLE F.2 7% Interest Factors for Annual Compounding

	Single Payment		Equal Payment Series				Uniform gradient-series factor
	Compound-amount factor	Present-worth factor	Compound-amount factor	Sinking-fund factor	Present-worth factor	Capital-recovery factor	
n	To find *F* Given *P* $F/P, i, n$	To find *P* Given *F* $P/F, i, n$	To find *F* Given *A* $F/A, i, n$	To find *A* Given *F* $A/F, i, n$	To find *P* Given *A* $P/A, i, n$	To find *A* Given *P* $A/P, i, n$	To find *A* Given *G* $A/G, i, n$
1	1.070	0.9346	1.000	1.0000	0.9346	1.0700	0.0000
2	1.145	0.8734	2.070	0.4831	1.8080	0.5531	0.4831
3	1.225	0.8163	3.215	0.3111	2.6243	0.3811	0.9549
4	1.311	0.7629	4.440	0.2252	3.3872	0.2952	1.4155
5	1.403	0.7130	5.751	0.1739	4.1002	0.2439	1.8650
6	1.501	0.6664	7.153	0.1398	4.7665	0.2098	2.3032
7	1.606	0.6228	8.654	0.1156	5.3893	0.1856	2.7304
8	1.718	0.5820	10.260	0.0975	5.9713	0.1675	3.1466
9	1.838	0.5439	11.978	0.0835	6.5152	0.1535	3.5517
10	1.967	0.5084	13.816	0.0724	7.0236	0.1424	3.9461
11	2.105	0.4751	15.784	0.0634	7.4987	0.1334	4.3296
12	2.252	0.4440	17.888	0.0559	7.9427	0.1259	4.7025
13	2.410	0.4150	20.141	0.0497	8.3577	0.1197	5.0649
14	2.579	0.3878	22.550	0.0444	8.7455	0.1144	5.4167
15	2.759	0.3625	25.129	0.0398	9.1079	0.1098	5.7583
16	2.952	0.3387	27.888	0.0359	9.4467	0.1059	6.0897
17	3.159	0.3166	30.840	0.0324	9.7632	0.1024	6.4110
18	3.380	0.2959	33.999	0.0294	10.0591	0.0994	6.7225
19	3.617	0.2765	37.379	0.0268	10.3356	0.0968	7.0242
20	3.870	0.2584	40.996	0.0244	10.5940	0.0944	7.3163
21	4.141	0.2415	44.865	0.0223	10.8355	0.0923	7.5990
22	4.430	0.2257	49.006	0.0204	11.0613	0.0904	7.8725
23	4.741	0.2110	53.436	0.0187	11.2722	0.0887	8.1369
24	5.072	0.1972	58.177	0.0172	11.4693	o.0872	8.3923
25	5.427	0.1843	63.249	0.0158	11.6536	0.0858	8.6391
26	5.807	0.1722	68.676	0.0146	11.8258	0.0846	8.8773
27	6.214	0.1609	74.484	0.0134	11.9867	0.0834	9.1072
28	6.649	0.1504	80.698	0.0124	12.1371	0.0824	9.3290
29	7.114	0.1406	87.347	0.0115	12.2777	0.0815	9.5427
30	7.612	0.1314	94.461	0.0106	12.4091	0.0806	9.7487
31	8.145	0.1228	102.073	0.0098	12.5318	0.0798	9.9471
32	8.715	0.1148	110.218	0.0091	12.6466	0.0791	10.1381
33	9.325	0.1072	118.933	0.0084	12.7538	0.0784	10.3219
34	9.978	0.1002	128.259	0.0078	12.8540	0.0778	10.4987
35	10.677	0.0937	138.237	0.0072	12.9477	0.0772	10.6687
40	14.974	0.0668	199.635	0.0050	13.3317	0.0750	11.4234
45	21.002	0.0476	285.749	0.0035	13.6055	0.0735	12.0360
50	29.457	0.0340	406.529	0.0025	13.8008	0.0725	12.5287
55	41.315	0.0242	575.929	0.0017	13.9399	0.0717	12.9215
60	57.946	0.0173	813.520	0.0012	14.0392	0.0712	13.2321
65	81.273	0.0123	1146.755	0.0009	14.1099	0.0709	13.4760
70	113.989	0.0088	1614.134	0.0006	14.1604	0.0706	13.6662
75	159.876	0.0063	2269.657	0.0005	14.1964	0.0705	13.8137
80	224.234	0.0045	3189.063	0.0003	14.2220	0.0703	13.9274
85	314.500	0.0032	4478.576	0.0002	14.2403	0.0702	14.0146
90	441.103	0.0023	6287.185	0.0002	14.2533	0.0702	14.0812
95	618.670	0.0016	8823.854	0.0001	14.2626	0.0701	14.1319
100	867.716	0.0012	12381.662	0.0001	14.2693	0.0701	14.1703

TABLE F.3 8% Interest Factors for Annual Compounding

	Single Payment		Equal Payment Series				Uniform gradient-series factor
n	Compound-amount factor	Present-worth factor	Compound-amount factor	Sinking-fund factor	Present-worth factor	Capital-recovery factor	
	To find F Given P $F/P, i, n$	To find P Given F $P/F, i, n$	To find F Given A $F/A, i, n$	To find A Given F $A/F, i, n$	To find P Given A $P/A, i, n$	To find A Given P $A/P, i, n$	To find A Given G $A/G, i, n$
1	1.080	0.9259	1.000	1.0000	0.9259	1.0800	0.0000
2	1.166	0.8573	2.080	0.4808	1.7833	0.5608	0.4808
3	1.260	0.7938	3.246	0.3080	2.5771	0.3880	0.9488
4	1.360	0.7350	4.506	0.2219	3.3121	0.3019	1.4040
5	1.469	0.6806	5.867	0.1705	3.9927	0.2505	1.8465
6	1.587	0.6302	7.336	0.1363	4.6229	0.2163	2.2764
7	1.714	0.5835	8.923	0.1121	5.2064	0.1921	2.6937
8	1.851	0.5403	10.637	0.0940	5.7466	0.1740	3.0985
9	1.999	0.5003	12.488	0.0801	6.2469	0.1601	3.4910
10	2.159	0.4632	14.487	0.0690	6.7101	0.1490	3.8713
11	2.332	0.4289	16.645	0.0601	7.1390	0.1401	4.2395
12	2.518	0.3971	18.977	0.0527	7.5361	0.1327	4.5958
13	2.720	0.3677	21.495	0.0465	7.9038	0.1265	4.9402
14	2.937	0.3405	24.215	0.0413	8.2442	0.1213	5.2731
15	3.172	0.3153	27.152	0.0368	8.5595	0.1168	5.5945
16	3.426	0.2919	30.324	0.0330	8.8514	0.1130	5.9046
17	3.700	0.2703	33.750	0.0296	9.1216	0.1096	6.2038
18	3.996	0.2503	37.450	0.0267	9.3719	0.1067	6.4920
19	4.316	0.2317	41.446	0.0241	9.6036	0.1041	6.7697
20	4.661	0.2146	45.762	0.0219	9.8182	0.1019	7.0370
21	5.034	0.1987	50.423	0.0198	10.0168	0.0998	7.2940
22	5.437	0.1840	55.457	0.0180	10.2008	0.0980	7.5412
23	5.871	0.1703	60.893	0.0164	10.3711	0.0964	7.7786
24	6.341	0.1577	66.765	0.0150	10.5288	0.0950	8.0066
25	6.848	0.1460	73.106	0.0137	10.6748	0.0937	8.2254
26	7.396	0.1352	79.954	0.0125	10.8100	0.0925	8.4352
27	7.988	0.1252	87.351	0.0115	10.9352	0.0915	8.6363
28	8.627	0.1159	95.339	0.0105	11.0511	0.0905	8.8289
29	9.317	0.1073	103.966	0.0096	11.1584	0.0896	9.0133
30	10.063	0.0994	113.283	0.0088	11.2578	0.0888	9.1897
31	10.868	0.0920	123.346	0.0081	11.3498	0.0881	9.3584
32	11.737	0.0852	134.214	0.0075	11.4350	0.0875	9.5197
33	12.676	0.0789	145.951	0.0069	11.5139	0.0869	9.6737
34	13.690	0.0731	158.627	0.0063	11.5869	0.0863	9.8208
35	14.785	0.0676	172.317	0.0058	11.6546	0.0858	9.9611
40	21.725	0.0460	259.057	0.0039	11.9246	0.0839	10.5699
45	31.920	0.0313	386.506	0.0026	12.1084	0.0826	11.0447
50	46.902	0.0213	573.770	0.0018	12.2335	0.0818	11.4107
55	68.914	0.0145	848.923	0.0012	12.3186	0.0812	11.6902
60	101.257	0.0099	1253.213	0.0008	12.3766	0.0808	11.9015
65	148.780	0.0067	1847.248	0.0006	12.4160	0.0806	12.0602
70	218.606	0.0046	2720.080	0.0004	12.4428	0.0804	12.1783
75	321.205	0.0031	4002.557	0.0003	12.4611	0.0803	12.2658
80	471.955	0.0021	5886.935	0.0002	12.4735	0.0802	12.3301
85	693.456	0.0015	8655.706	0.0001	12.4820	0.0801	12.3773
90	1018.915	0.0010	12723.939	0.0001	12.4877	0.0801	12.4116
95	1497.121	0.0007	18701.507	0.0001	12.4917	0.0801	12.4365
100	2199.761	0.0005	27484.516	0.0001	12.4943	0.0800	12.4545

TABLE F.4 9% Interest Factors for Annual Compounding

n	Single Payment		Equal Payment Series				Uniform gradient-series factor
	Compound-amount factor	Present-worth factor	Compound-amount factor	Sinking-fund factor	Present-worth factor	Capital-recovery factor	
	To find F Given P $F/P, i, n$	To find P Given F $P/F, i, n$	To find F Given A $F/A, i, n$	To find A Given F $A/F, i, n$	To find P Given A $P/A, i, n$	To find A Given P $A/P, i, n$	To find A Given G $A/G, i, n$
1	1.090	0.9174	1.000	1.0000	0.9174	1.0900	0.0000
2	1.188	0.8417	2.090	0.4785	1.7591	0.5685	0.4785
3	1.295	0.7722	3.278	0.3051	2.5313	0.3951	0.9426
4	1.412	0.7084	4.573	0.2187	3.2397	0.3087	1.3925
5	1.539	0.6499	5.985	0.1671	3.8897	0.2571	1.8282
6	1.677	0.5963	7.523	0.1329	4.4859	0.2229	2.2498
7	1.828	0.5470	9.200	0.1087	5.0330	0.1987	2.6574
8	1.993	0.5019	11.028	0.0907	5.5348	0.1807	3.0512
9	2.172	0.4604	13.021	0.0768	5.9953	0.1668	3.4312
10	2.367	0.4224	15.193	0.0658	6.4177	0.1558	3.7978
11	2.580	0.3875	17.560	0.0570	6.8052	0.1470	4.1510
12	2.813	0.3555	20.141	0.0497	7.1607	0.1397	4.4910
13	3.066	0.3262	22.953	0.0436	7.4869	0.1336	4.8182
14	3.342	0.2993	26.019	0.0384	7.7862	0.1284	5.1326
15	3.642	0.2745	29.361	0.0341	8.0607	0.1241	5.4346
16	3.970	0.2519	33.003	0.0303	8.3126	0.1203	5.7245
17	4.328	0.2311	36.974	0.0271	8.5436	0.1171	6.0024
18	4.717	0.2120	41.301	0.0242	8.7556	0.1142	6.2687
19	5.142	0.1945	46.018	0.0217	8.9501	0.1117	6.5236
20	5.604	0.1784	51.160	0.0196	9.1286	0.1096	6.7675
21	6.109	0.1637	56.765	0.0176	9.2923	0.1076	7.0006
22	6.659	0.1502	62.873	0.0159	9.4424	0.1059	7.2232
23	7.258	0.1378	69.532	0.0144	9.5802	0.1044	7.4358
24	7.911	0.1264	76.790	0.0130	9.7066	0.1030	7.6384
25	8.623	0.1160	84.701	0.0118	9.8226	0.1018	7.8316
26	9.399	0.1064	93.324	0.0107	9.9290	0.1007	8.0156
27	10.245	0.0976	102.723	0.0097	10.0266	0.0997	8.1906
28	11.167	0.0896	112.968	0.0089	10.1161	0.0989	8.3572
29	12.172	0.0822	124.135	0.0081	10.1983	0.0981	8.5154
30	13.268	0.0754	136.308	0.0073	10.2737	0.0973	8.6657
31	14.462	0.0692	149.575	0.0067	10.3428	0.0967	8.8083
32	15.763	0.0634	164.037	0.0061	10.4063	0.0961	8.9436
33	17.182	0.0582	179.800	0.0056	10.4645	0.0956	9.0718
34	18.728	0.0534	196.982	0.0051	10.5178	0.0951	9.1933
35	20.414	0.0490	215.711	0.0046	10.5668	0.0946	9.3083
40	31.409	0.0318	337.882	0.0030	10.7574	0.0930	9.7957
45	48.327	0.0207	525.859	0.0019	10.8812	0.0919	10.1603
50	74.358	0.0135	815.084	0.0012	10.9617	0.0912	10.4295
55	114.408	0.0088	1260.092	0.0008	11.0140	0.0908	10.6261
60	176.031	0.0057	1944.792	0.0005	11.0480	0.0905	10.7683
65	270.846	0.0037	2998.288	0.0003	11.0701	0.0903	10.8702
70	416.730	0.0024	4619.223	0.0002	11.0845	0.0902	10.9427
75	641.191	0.0016	7113.232	0.0002	11.0938	0.0902	10.9940
80	986.552	0.0010	10950.574	0.0001	11.0999	0.0901	11.0299
85	1517.932	0.0007	16854.800	0.0001	11.1038	0.0901	11.0551
90	2335.527	0.0004	25939.184	0.0001	11.1064	0.0900	11.0726
95	3593.497	0.0003	39916.635	0.0000	11.1080	0.0900	11.0847
100	5529.041	0.0002	61422.675	0.0000	11.1091	0.0900	11.0930

TABLE F.5 10% Interest Factors for Annual Compounding

	Single Payment		Equal Payment Series				Uniform gradient-series factor
n	Compound-amount factor	Present-worth factor	Compound-amount factor	Sinking-fund factor	Present-worth factor	Capital-recovery factor	
	To find F Given P $F/P, i, n$	To find P Given F $P/F, i, n$	To find F Given A $F/A, i, n$	To find A Given F $A/F, i, n$	To find P Given A $P/A, i, n$	To find A Given P $A/P, i, n$	To find A Given G $A/G, i, n$
1	1.100	0.9091	1.000	1.0000	0.9091	1.1000	0.0000
2	1.210	0.8265	2.100	0.4762	1.7355	0.5762	0.4762
3	1.331	0.7513	3.310	0.3021	2.4869	0.4021	0.9366
4	1.464	0.6830	4.641	0.2155	3.1699	0.3155	1.3812
5	1.611	0.6209	6.105	0.1638	3.7908	0.2638	1.8101
6	1.772	0.5645	7.716	0.1296	4.3553	0.2296	2.2236
7	1.949	0.5132	9.487	0.1054	4.8684	0.2054	2.6216
8	2.144	0.4665	11.436	0.0875	5.3349	0.1875	3.0045
9	2.358	0.4241	13.579	0.0737	5.7590	0.1737	3.3724
10	2.594	0.3856	15.937	0.0628	6.1446	0.1628	3.7255
11	2.853	0.3505	18.531	0.0540	6.4951	0.1540	4.0641
12	3.138	0.3186	21.384	0.0468	6.8137	0.1468	4.3884
13	3.452	0.2897	24.523	0.0408	7.1034	0.1408	4.6988
14	3.798	0.2633	27.975	0.0358	7.3667	0.1358	4.9955
15	4.177	0.2394	31.772	0.0315	7.6061	0.1315	5.2789
16	4.595	0.2176	35.950	0.0278	7.8237	0.1278	5.5493
17	5.054	0.1979	40.545	0.0247	8.0216	0.1247	5.8071
18	5.560	0.1799	45.599	0.0219	8.2014	0.1219	6.0526
19	6.116	0.1635	51.159	0.0196	8.3649	0.1196	6.2861
20	6.728	0.1487	57.275	0.0175	8.5136	0.1175	6.5081
21	7.400	0.1351	64.003	0.0156	8.6487	0.1156	6.7189
22	8.140	0.1229	71.403	0.0140	8.7716	0.1140	6.9189
23	8.954	0.1117	79.543	0.0126	8.8832	0.1126	7.1085
24	9.850	0.1015	88.497	0.0113	8.9848	0.1113	7.2881
25	10.835	0.0923	98.347	0.0102	9.0771	0.1102	7.4580
26	11.918	0.0839	109.182	0.0092	9.1610	0.1092	7.6187
27	13.110	0.0763	121.100	0.0083	9.2372	0.1083	7.7704
28	14.421	0.0694	134.210	0.0075	9.3066	0.1075	7.9137
29	15.863	0.0630	148.631	0.0067	9.3696	0.1067	8.0489
30	17.449	0.0573	164.494	0.0061	9.4269	0.1061	8.1762
31	19.194	0.0521	181.943	0.0055	9.4790	0.1055	8.2962
32	21.114	0.0474	201.138	0.0050	9.5264	0.1050	8.4091
33	23.225	0.0431	222.252	0.0045	9.5694	0.1045	8.5152
34	25.548	0.0392	245.477	0.0041	9.6086	0.1041	8.6149
35	28.102	0.0356	271.024	0.0037	9.6442	0.1037	8.7086
40	45.259	0.0221	442.593	0.0023	9.7791	0.1023	9.0962
45	72.890	0.0137	718.905	0.0014	9.8628	0.1014	9.3741
50	117.391	0.0085	1163.909	0.0009	9.9148	0.1009	9.5704
55	189.059	0.0053	1880.591	0.0005	9.9471	0.1005	9.7075
60	304.482	0.0033	3034.816	0.0003	9.9672	0.1003	9.8023
65	490.371	0.0020	4893.707	0.0002	9.9796	0.1002	9.8672
70	789.747	0.0013	7887.470	0.0001	9.9873	0.1001	9.9113
75	1271.895	0.0008	12708.954	0.0001	9.9921	0.1001	9.9410
80	2048.400	0.0005	20474.002	0.0001	9.9951	0.1001	9.9609
85	3298.969	0.0003	32979.690	0.0000	9.9970	0.1000	9.9742
90	5313.023	0.0002	53120.226	0.0000	9.9981	0.1000	9.9831
95	8556.676	0.0001	85556.760	0.0000	9.9988	0.1000	9.9889
100	13780.612	0.0001	137796.123	0.0000	9.9993	0.1000	9.9928

TABLE F.6 12% Interest Factors for Annual Compounding

n	Single Payment		Equal Payment Series				Uniform gradient-series factor
	Compound-amount factor	Present-worth factor	Compound-amount factor	Sinking-fund factor	Present-worth factor	Capital-recovery factor	
	To find *F* Given *P* *F/P, i, n*	To find *P* Given *F* *P/F, i, n*	To find *F* Given *A* *F/A, i, n*	To find *A* Given *F* *A/F, i, n*	To find *P* Given *A* *P/A, i, n*	To find *A* Given *P* *A/P, i, n*	To find *A* Given *G* *A/G, i, n*
1	1.120	0.8929	1.000	1.0000	0.8929	1.1200	0.0000
2	1.254	0.7972	2.120	0.4717	1.6901	0.5917	0.4717
3	1.405	0.7118	3.374	0.2964	2.4018	0.4164	0.9246
4	1.574	0.6355	4.779	0.2092	3.0374	0.3292	1.3589
5	1.762	0.5674	6.353	0.1574	3.6048	0.2774	1.7746
6	1.974	0.5066	8.115	0.1232	4.1114	0.2432	2.1721
7	2.211	0.4524	10.089	0.0991	4.5638	0.2191	2.5515
8	2.476	0.4039	12.300	0.0813	4.9676	0.2013	2.9132
9	2.773	0.3606	14.776	0.0677	5.3283	0.1877	3.2574
10	3.106	0.3220	17.549	0.0570	5.6502	0.1770	3.5847
11	3.479	0.2875	20.655	0.0484	5.9377	0.1684	3.8953
12	3.896	0.2567	24.133	0.0414	6.1944	0.1614	4.1897
13	4.364	0.2292	28.029	0.0357	6.4236	0.1557	4.4683
14	4.887	0.2046	32.393	0.0309	6.6282	0.1509	4.7317
15	5.474	0.1827	37.280	0.0268	6.8109	0.1468	4.9803
16	6.130	0.1631	42.753	0.0234	6.9740	0.1434	5.2147
17	6.866	0.1457	48.884	0.0205	7.1196	0.1405	5.4353
18	7.690	0.1300	55.750	0.0179	7.2497	0.1379	5.6427
19	8.613	0.1161	63.440	0.0158	7.3658	0.1358	5.8375
20	9.646	0.1037	72.052	0.0139	7.4695	0.1339	6.0202
21	10.804	0.0926	81.699	0.0123	7.5620	0.1323	6.1913
22	12.100	0.0827	92.503	0.0108	7.6447	0.1308	6.3514
23	13.552	0.0738	104.603	0.0096	7.7184	0.1296	6.5010
24	15.179	0.0659	118.155	0.0085	7.7843	0.1285	6.6407
25	17.000	0.0588	133.334	0.0075	7.8431	0.1275	6.7708
26	19.040	0.0525	150.334	0.0067	7.8957	0.1267	6.8921
27	21.325	0.0469	169.374	0.0059	7.9426	0.1259	7.0049
28	23.884	0.0419	190.699	0.0053	7.9844	0.1253	7.1098
29	26.750	0.0374	214.583	0.0047	8.0218	0.1247	7.2071
30	29.960	0.0334	241.333	0.0042	8.0552	0.1242	7.2974
31	33.555	0.0298	271.293	0.0037	8.0850	0.1237	7.3811
32	37.582	0.0266	304.848	0.0033	8.1116	0.1233	7.4586
33	42.092	0.0238	342.429	0.0029	8.1354	0.1229	7.5303
34	47.143	0.0212	384.521	0.0026	8.1566	0.1226	7.5965
35	52.800	0.0189	431.664	0.0023	8.1755	0.1223	7.6577
40	93.051	0.0108	767.091	0.0013	8.2438	0.1213	7.8988
45	163.988	0.0061	1358.230	0.0007	8.2825	0.1207	8.0572
50	289.002	0.0035	2400.018	0.0004	8.3045	0.1204	8.1597

TABLE F.7 15% Interest Factors for Annual Compounding

	Single Payment		Equal Payment Series				Uniform gradient-series factor
n	Compound-amount factor	Present-worth factor	Compound-amount factor	Sinking-fund factor	Present-worth factor	Capital-recovery factor	
	To find F Given P $F/P, i, n$	To find P Given F $P/F, i, n$	To find F Given A $F/A, i, n$	To find A Given F $A/F, i, n$	To find P Given A $P/A, i, n$	To find A Given P $A/P, i, n$	To find A Given G $A/G, i, n$
1	1.150	0.8696	1.000	1.0000	0.8696	1.1500	0.0000
2	1.323	0.7562	2.150	0.4651	1.6257	0.6151	0.4651
3	1.521	0.6575	3.473	0.2880	2.2832	0.4380	0.9071
4	1.749	0.5718	4.993	0.2003	2.8550	0.3503	1.3263
5	2.011	0.4972	6.742	0.1483	3.3522	0.2983	1.7228
6	2.313	0.4323	8.754	0.1142	3.7845	0.2642	2.0972
7	2.660	0.3759	11.067	0.0904	4.1604	0.2404	2.4499
8	3.059	0.3269	13.727	0.0729	4.4873	0.2229	2.7813
9	3.518	0.2843	16.786	0.0596	4.7716	0.2096	3.0922
10	4.046	0.2472	20.304	0.0493	5.0188	0.1993	3.3832
11	4.652	0.2150	24.349	0.0411	5.2337	0.1911	3.6550
12	5.350	0.1869	29.002	0.0345	5.4206	0.1845	3.9082
13	6.153	0.1625	34.352	0.0291	5.5832	0.1791	4.1438
14	7.076	0.1413	40.505	0.0247	5.7245	0.1747	4.3624
15	8.137	0.1229	47.580	0.0210	5.8474	0.1710	4.5650
16	9.358	0.1069	55.717	0.0180	5.9542	0.1680	4.7523
17	10.761	0.0929	65.075	0.0154	6.0472	0.1654	4.9251
18	12.375	0.0808	75.836	0.0132	6.1280	0.1632	5.0843
19	14.232	0.0703	88.212	0.0113	6.1982	0.1613	5.2307
20	16.367	0.0611	102.444	0.0098	6.2593	0.1598	5.3651
21	18.822	0.0531	118.810	0.0084	6.3125	0.1584	5.4883
22	21.645	0.0462	137.632	0.0073	6.3587	0.1573	5.6010
23	24.891	0.0402	159.276	0.0063	6.3988	0.1563	5.7040
24	28.625	0.0349	184.168	0.0054	6.4338	0.1554	5.7979
25	32.919	0.0304	212.793	0.0047	6.4642	0.1547	5.8834
26	37.857	0.0264	245.712	0.0041	6.4906	0.1541	5.9612
27	43.535	0.0230	283.569	0.0035	6.5135	0.1535	6.0319
28	50.066	0.0200	327.104	0.0031	6.5335	0.1531	6.0960
29	57.575	0.0174	377.170	0.0027	6.5509	0.1527	6.1541
30	66.212	0.0151	434.745	0.0023	6.5660	0.1523	6.2066
31	76.144	0.0131	500.957	0.0020	6.5791	0.1520	6.2541
32	87.565	0.0114	577.100	0.0017	6.5905	0.1517	6.2970
33	100.700	0.0099	664.666	0.0015	6.6005	0.1515	6.3357
34	115.805	0.0086	765.365	0.0013	6.6091	0.1513	6.3705
35	133.176	0.0075	881.170	0.0011	6.6166	0.1511	6.4019
40	267.864	0.0037	1779.090	0.0006	6.6418	0.1506	6.5168
45	538.769	0.0019	3585.128	0.0003	6.6543	0.1503	6.5830
50	1083.657	0.0009	7217.716	0.0002	6.6605	0.1501	6.6205

APPENDIX G

NORMAL DISTRIBUTION TABLES

TABLE G.1 Areas Under the Normal Curve

Proportion of total area under the curve that is under the portion of the curve from $-\infty$ to $\frac{X_i - \bar{X}'}{\sigma'}$. ($X_i$ represents any desired value of the variable X)

$Z = \frac{X_i - \bar{X}'}{\sigma'}$	*0.09*	*0.08*	*0.07*	*0.06*	*0.05*	*0.04*	*0.03*	*0.02*	*0.01*	*0.00*
−3.5	0.00017	0.00017	0.00018	0.00019	0.00019	0.00020	0.00021	0.00022	0.00022	0.00023
−3.4	0.00024	0.00025	0.00026	0.00027	0.00028	0.00029	0.00030	0.00031	0.00033	0.00034
−3.3	0.00035	0.00036	0.00038	0.00039	0.00040	0.00042	0.00043	0.00045	0.00047	0.00048
−3.2	0.00050	0.00052	0.00054	0.00056	0.00058	0.00060	0.00062	0.00064	0.00066	0.00069
−3.1	0.00071	0.00074	0.00076	0.00079	0.00082	0.00085	0.00087	0.00090	0.00094	0.00097
−3.0	0.00100	0.00104	0.00107	0.00111	0.00141	0.00118	0.00122	0.00126	0.00131	0.00135
−2 9	0.0014	0.0014	0.0015	0.0015	0.0016	0.0016	0.0017	0.0017	0.0018	0.0019
−2.8	0.0019	0.0020	0.0021	0.0021	0.0022	0.0023	0.0023	0.0024	0.0025	0.0026
−2.7	0.0026	0.0027	0.0028	0.0029	0.0030	0.0031	0.0032	0.0033	0.0034	0.0035
−2.6	0.0036	0.0037	0.0038	0.0039	0.0040	0.0041	0.0043	0.0044	0.0045	0.0047
−2.5	0.0048	0.0049	0.0051	0.0052	0.0054	0.0055	0.0057	0.0059	0.0060	0.0062
−2.4	0.0064	0.0066	0.0068	0.0069	0.0071	0.0073	0.0075	0.0078	0.0080	0.0082
−2.3	0.0084	0.0087	0.0089	0.0091	0.0094	0.0096	0.0099	0.0102	0.0104	0.0107
−2.2	0.0110	0.0113	0.0116	0.0119	0.0122	0.0125	0.0129	0.0132	0.0136	0.0139
−2.1	0.0143	0.0146	0.0150	0.0154	0.0158	0.0162	0.0166	0.0170	0.0174	0.0179
−2.0	0.0183	0.0188	0.0192	0.0197	0.0202	0.0207	0.0212	0.0217	0.0222	0.0228
−1.9	0.0233	0.0239	0.0244	0.0250	0.0256	0.0262	0.0268	0.0274	0.0281	0.0287
−1.8	0.0294	0.0301	0.0307	0.0314	0.0322	0.0329	0.0336	0.0344	0.0351	0.0359
−1.7	0.0367	0.0375	0.0384	0.0392	0.0401	0.0409	0.0418	0.0427	0.0436	0.0446
−1.6	0.0455	0.0465	0.0475	0.0485	0.0495	0.0505	0.0516	0.0526	0.0537	0.0548
−1.5	0.0559	0.0571	0.0582	0.0594	0.0606	0.0618	0.0630	0.0643	0.0652	0.0668
−1.4	0.0681	0.0694	0.0708	0.0721	0.0735	0.0749	0.0764	0.0778	0.0793	0.0808
−1.3	0.0823	0.0838	0.0853	0.0869	0.0885	0.0901	0.0918	0.0934	0.0951	0.0968
−1.2	0.0985	0.1003	0.1020	0.1038	0.1057	0.1075	0.1093	0.1112	0.1131	0.1151
−1.1	0.1170	0.1190	0.1210	0.1230	0.1251	0.1271	0.1292	0.1314	0.1335	0.1357
−1.0	0.1379	0.1401	0.1423	0.1446	0.1469	0.1492	0.1515	0.1539	0.1562	0.1587
−0.9	0.1611	0.1635	0.1660	0.1685	0.1711	0.1736	0.1762	0.1788	0.1814	0.1841
−0.8	0.1867	0.1894	0.1922	0.1949	0.1977	0.2005	0.2033	0.2061	0.2090	0.2119
−0.7	0.2148	0.2177	0.2207	0.2236	0.2266	0.2297	0.2327	0.2358	0.2389	0.2420
−0.6	0.2451	0.2483	0.2514	0.2546	0.2578	0.2611	0.2643	0.2676	0.2709	0.2743
−0.5	0.2776	0.2810	0.2843	0.2877	0.2912	0.2946	0.2981	0.3015	0.3050	0.3085
−0.4	0.3121	0.3156	0.3192	0.3228	0.3264	0.3300	0.3336	0.3372	0.3409	0.3446
−0.3	0.3483	0.3520	0.3557	0.3594	0.3632	0.3669	0.3707	0.3745	0.3783	0.3821
−0.2	0.3859	0.3897	0.3936	0.3974	0.4013	0.4052	0.4090	0.4129	0.4168	0.4207
−0.1	0.4247	0.4286	0.4325	0.4346	0.4404	0.4443	0.4483	0.4522	0.4562	0.4602
−0.0	0.4641	0.4681	0.4721	0.4761	0.4801	0.4840	0.4880	0.4920	0.4960	0.5000

Source: E. L. Grant and R. S. Leavenworth, *Statistical Quality Control*, 4th ed., McGraw-Hill Book Company, New York, N.Y., 1972. Reprinted by permission.

TABLE G.1 Areas Under the Normal Curve *(continued)*

$Z=\frac{X_i-\bar{X}'}{\sigma'}$	*0.00*	*0.01*	*0.02*	*0.03*	*0.04*	*0.05*	*0.06*	*0.07*	*0.08*	*0.09*
+0.0	0.5000	0.5040	0.5080	0.5120	0.5160	0.5199	0.5239	0.5279	0.5319	0.5359
+0.1	0.5398	0.5438	0.5478	0.5517	0.5557	0.5596	0.5636	0.5675	0.5714	0.5753
+0.2	0.5793	0.5832	0.5871	0.5910	0.5948	0.5987	0.6026	0.6064	0.6103	0.6141
+0.3	0.6179	0.6217	0.6255	0.6293	0.6331	0.6368	0.6406	0.6443	0.6480	0.6517
+0.4	0.6554	0.6591	0.6628	0.6664	0.6700	0.5736	0.6772	0.6808	0.6844	0.6879
+0.5	0.6915	0.6950	0.6985	0.7019	0.7054	0.7088	0.7123	0.7157	0.7190	0.7224
+0.6	0.7257	0.7291	0.7324	0.7357	0.7389	0.7422	0.7454	0.7486	0.7517	0.7549
+0.7	0.7580	0.7611	0.7642	0.7673	0.7704	0.7734	0.7764	0.7794	0.7823	0.7852
+0.8	0.7881	0.7910	0.7939	0.7967	0.7995	0.8023	0.8051	0.8079	0.8106	0.8133
+0.9	0.8159	0.8186	0.8212	0.8238	0.8264	0.8289	0.8315	0.8340	0.8365	0.8389
+1.0	0.8413	0.8438	0.8461	0.8485	0.8508	0.8531	0.8554	0.8577	0.8599	0.8621
+1.1	0.8643	0.8665	0.8686	0.8708	0.8729	0.8749	0.8770	0.8790	0.8810	0.8830
+1.2	0.8849	0.8869	0.8888	0.8907	0.8925	0.8944	0.8962	0.8980	0.8997	0.9015
+1.3	0.9032	0.9049	0.9066	0.9082	0.9099	0.9115	0.9131	0.9147	0.9162	0.9177
+1.4	0.9192	0.9207	0.9222	0.9236	0.9251	0.9265	0.9279	0.9292	0.9306	0.9319
+1.5	0.9332	0.9345	0.9357	0.9370	0.9382	0.9394	0.9406	0.9418	0.9429	0.9441
+1.6	0.9452	0.9463	0.9474	0.9484	0.9495	0.9505	0.9515	0.9525	0.9535	0.9545
+1.7	0.9554	0.9564	0.9573	0.9582	0.9591	0.9599	0.9608	0.9616	0.9625	0.9633
+1.8	0.9641	0.9649	0.9656	0.9664	0.9671	0.9678	0.9686	0.9693	0.9699	0.9706
+1.9	0.9713	0.9719	0.9726	0.9732	0.9738	0.9744	0.9750	0.9756	0.9761	0.9767
+2.0	0.9773	0.9778	0.9783	0.9788	0.9793	0.9798	0.9803	0.9808	0.9812	0.9817
+2.1	0.9821	0.9826	0.9830	0.9834	0.9838	0.9842	0.9846	0.9850	0.9854	0.9857
+2.2	0.9861	0.9864	0.9868	0.9871	0.9875	0.9878	0.9881	0.9884	0.9887	0.9890
+2.3	0.9893	0.9896	0.9898	0.9901	0.9904	0.9906	0.9909	0.9911	0.9913	0.9916
+2.4	0.9918	0.9920	0.9922	0.9925	0.9927	0.9929	0.9931	0.9932	0.9934	0.9936
+2.5	0.9938	0.9940	0.9941	0.9943	0.9945	0.9946	0.9948	0.9949	0.9951	0.9952
+2.6	0.9953	0.9955	0.9956	0.9957	0.9959	0.9960	0.9961	0.9962	0.9963	0.9964
+2.7	0.9965	0.9966	0.9967	0.9968	0.9969	0.9970	0.9971	0.9972	0.9973	0.9974
+2.8	0.9974	0.9975	0.9976	0.9977	0.9977	0.9978	0.9979	0.9979	0.9980	0.9981
+2.9	0.9981	0.9982	0.9983	0.9983	0.9984	0.9984	0.9985	0.9985	0.9986	0.9986
+3.0	0.99865	0.99869	0.99874	0.99878	0.99882	0.99886	0.99889	0.99893	0.99896	0.99900
+3.1	0.99903	0.99906	0.99910	0.99913	0.99915	0.99918	0.99921	0.99924	0.99926	0.99929
+3.2	0.99931	0.99934	0.99936	0.99938	0.99940	0.99942	0.99944	0.99946	0.99948	0.99950
+3.3	0.99952	0.99953	0.99955	0.99957	0.99958	0.99960	0.99961	0.99962	0.99964	0.99965
+3.4	0.99966	0.99967	0.99969	0.99970	0.99971	0.99972	0.99973	0.99974	0.99975	0.99976
+3.5	0.99977	0.99978	0.99978	0.99979	0.99980	0.99981	0.99981	0.99982	0.99983	0.99983

SELECTED BIBLIOGRAPHY

When addressing the subject of this text, one should become familiar not only with the available literature in the field of *logistics* itself but with some of the subject areas closely aligned to logistics. With this in mind, the author has included several key references pertaining to systems engineering and analysis, reliability and maintainability, human factors and safety, economic analysis and life-cycle costing, production and quality control, and management. These subject areas have been discussed extensively throughout this text, and it is hoped that the references listed will prove to be beneficial. This list is certainly not to be considered as being all inclusive.

H.1 Logistics, Logistics Engineering, and Integrated Logistic Support (ILS)

1. ADPA, *Commercial-Off-the-Shelf (COTS) Supportability Study*, Technical Report, American Defense Preparedness Association, Arlingtion, Va. 1994.
2. Allen, M. K., and Helferich, O. K., *Putting Expert Systems to Work in Logistics*, Council of Logistics Management, 2803 Butterfield Road, Oak Brook, IL 60521, 1990.
3. Ballou, R. H., *Business Logistics Management*, 3rd Ed., Prentice Hall, Inc., Upper Saddle River, N.J., 1992.
4. Banks, J., and Fabrycky, W. J., *Procurement and Inventory Systems Analysis*, Prentice Hall, Inc., Upper Saddle River, N.J., 1987.
5. Barnes, T. A., *Logistics Support Training: Design and Development*, McGraw-Hill Book Co., New York, N.Y., 1992.
6. Blanchard, B. S., *Logistics Engineering and Management*, 5th Ed., Prentice Hall, Inc., Upper Saddle River, N.J., 1998.
7. Bowersox, D., Closs, D., and Helferich, O., *Logistical Management*, Macmillan Publishing Co., New York, N.Y., 1986.
8. Council of Logistics Management, *Journal of Business Logistics*, 2803 Butterfield Road, Oak Brook, IL 60521.

9. Council of Logistics Management, *Logistics Software*, Annual Ed., Anderson Consulting, 1345 Avenue of the Americas, New York, N.Y., 1993.
10. Coyle, J. J., Bardi, E. J., and Langley, C. J., *The Management of Business Logistics*, 5th Ed., West Publishing Co., St. Paul, Minn., 1992.
11. Coyle, J. J., Bardi, E. J., and Cavinato, J. L., *Transportation*, 4th Ed., West Publishing Co., St. Paul, Minn., 1993.
12. Defense Systems Management College (DSMC), *Integrated Logistics Support Guide*, DSMC, Fort Belvoir, VA 22060.
13. Glaskowsky, N. A., Hudson, D. R., and Ivie, R. M., *Business Logistics*, 3rd Edition, The Dryden Press, Harcourt Brace Jovanovich Co., Orlando, Fa., 1992.
14. Green, L. L., *Logistics Engineering*, John Wiley & Sons, Inc., New York, N.Y., 1991.
15. Hutchinson, N. E., *An Integrated Approach to Logistics Management*, Prencice Hall, Inc., Upper Saddle River, N.J., 1987.
16. International Society of Logistics (SOLE), *Annals*, 8100 Professional Place, Suite 211, Hyattsville, MD 20785.
17. International Society of Logistics (SOLE), *Logistics Spectrum*, 8100 Professional Place, Suite 211, Hyattsville, MD 20785.
18. International Society of Logistics (SOLE), Proceedings, *Annual Symposium*, 8100 Professional Place, Suite 211, Hyattsville, MD 20785.
19. John, J. C., and Wood, D. F., *Contemporary Logistics*, 4th Ed., Macmillian Publishing Co., New York N.Y., 1990.
20. Jones, J. V., *Integrated Logistics Support Handbook*, Tab Books, Inc., Blue Ridge Summit, PA 17294, 1987.
21. Jones, J. V., *Logistic Support Analysis Handbook*, Tab Books, Inc., Blue Ridge Summit, PA 17294, 1989.
22. Langford, J. W., *Logistics Principles and Practices*, McGraw-Hill Book Co., New York, N.Y., 1995.
23. Magee, J. F., Copacino, W. C., and Rosenfield, D. B., *Modern Logistic Management*, John Wiley & Sons, Inc., New York, N.Y., 1985.
24. MIL-HDBK-59A, Military Handbook, *Computer-Aided Acquisition Logistic Support (CALS) Implementation Guide*, Department of Defense, Washington, D.C.
25. MIL-HDBK-226, Military Handbook, *Application of Reliability-Centered Maintenance to Naval Aircraft, Weapon Systems, and Support Equipment*, Department of Defense, Washington, D.C.
26. MIL-HDBK-502, *DOD Handbook: Acquisition Logistics,* Department of Defense, Washington, D.C., May 1997.
27. MIL-PRF-49506, Performance Specification, *Logistics Management Information,* Department of Defense, Washington, D.C., November 1996.
28. MIL-STD-1388-1A, Military Standard, *Logistic Support Analysis*, Department of Defense, Washington, D.C.
29. MIL-STD-1388-2B, Military Standard, *Department of Defense Requirements for a Logistic Support Analysis Record*, Department of Defense, Washington, D.C.
30. MIL-STD-1840A, Military Standard, *Automated Interchange of Technical Information*, Department of Defense, Washington, D.C.
31. Orsburn, D. K., *Introduction to Spares Management*, Academy Printing & Publishing Co., 16202 S. Orange Avenue, P.O. Box 560, Paramount, CA 90723, 1985.

32. Patton, Jr., J. D., *Logistics Technology and Management—The New Approach*, The Solomon Press, New York, N.Y., 1986.
33. Stock, J. R., and Lambert, D. M., *Strategic Logistics Management*, 3rd Ed., Richard D. Irwin, Inc., Homewood, Ill., 1992.
34. Tersine, R. J., *Principles of Inventory and Materials Management*, 4th Ed., Prentice Hall, Inc., Upper Saddle River, N.J., 1994.

H.2 Systems, Systems Engineering, and Systems Analysis

1. Ackoff, R. L., *Redesigning the Future*, John Wiley & Sons, Inc., New York, N.Y., 1974.
2. Beam, W. R., *Systems Engineering: Architecture and Design*, McGraw-Hill Book Co., New York, N.Y., 1990.
3. Belcher, R., and Aslaksen, E., *Systems Engineering*, Prentice Hall of Australia, Sydney, Australia, 1992.
4. Blanchard, B. S., *System Engineering Management*, 2nd Ed., John Wiley & Sons, Inc., New York, N.Y., 1998.
5. Blanchard, B. S., "The Systems Engineering Process: An Application for the Identification of Resource Requirements," *Systems Engineering*, Journal of the International Council on Systems Engineering (INCOSE), Seattle Washington, Volume I, Number I, July/September 1994.
6. Blanchard, B. S., and Fabrycky, W. J., *Systems Engineering and Analysis*, 3rd Ed., Prentice Hall, Inc., Upper Saddle River, N.J., 1998.
7. Blanchard, B. S., Fabrycky, W. J., and Verma, D. (Eds.), *Application of the Systems Engineering Process to Define Requirements for Computer-Based Design Tools*, Monograph, International Society of Logistic (SOLE), Hyattsville, Md., 1994.
8. Boulding, K., "General Systems Theory: The Skeleton of Science," *Management Science*, April 1956.
9. DSMC, *Systems Engineering Management Guide*, Defense Systems Management College, Fort Belvoir, V., latest edition.
10. EIA/IS-632, *Systems Engineering,* Electronic Industries Association (EIA), 2001 Pennsylvania Avenue, N.W., Washington, D.C., 1994.
11. Fabrycky, W. J., "Modeling and Indirect Experimentation in Design Evaluation," *Systems Engineering*, Journal of the International Council on Systems Engineering (INCOSE), Seattle, Washington, Volume I, Number I, July/September 1994.
12. Forrester, J. W., *Principles of Systems*, The MIT Press, Cambridge, Mass., 1968.
13. Grady, J. O., *Systems Requirements Analysis*, McGraw-Hill Book Co., New York, N.Y. 1993.
14. Grady, J. O., *System Engineering Planning and Enterprise Identify*, CRC Press, Boca Raton, Fla., 1995.
15. Hall, A. D., *A Methodology for Systems Engineering*, D. Van Nostrand Co., Ltd., Princeton, N.J., 1962.
16. IEEE P1220, *Standard for Application and Management of the Systems Engineering Process,* Institute of Electrical and Electronics Engineers (IEEE), 345 East 47th Street, New York, N.Y. 1994.
17. INCOSE, *Journal of the International Council on Systems Engineering*, United Airlines Bldg., Suite 804, 2033 Sixth Ave., Seattle, WA 98121.
18. INCOSE, *Annual Conference*, Proceedings, International Council on Systems Engineering, United Airlines Bldg., Suite 804, 2033 Sixth Ave., Seattle, WA 98121.

19. Lacy, J. A., *Systems Engineering Management: Achieving Total Quality*, McGraw-Hill Book Co., New York, N.Y., 1992.
20. Martin, J. N., *Systems Engineering Guidebook: A Process for Developing Systems and Products*, CRC Press, Boca Raton, Fla., 1997.
21. Oliver, D. W., *Engineering Complex Systems with Models and Objects*, McGraw-Hill Book Co., New York, N.Y., 1997.
22. Rechtin, E., *System Architecting: Creating and Building Complex Systems*, Prentice Hall, Inc., Upper Saddle River, N.J., 1991.
23. Rechtin, E., and Maier, M., *The Art of Systems Architecting*, CRC Press, Inc., Boca Raton, Fla., 1996.
24. Reilly, Norman B., *Successful Systems Engineering for Engineers and Managers*, Van Nostrand Reinhold, New York, N.Y., 1993.
25. Sage, A. P., *Decision Support Systems Engineering*, John Wiley & Sons, Inc., New York, N.Y., 1991.
26. Sage, A. P., *Systems Engineering*, John Wiley & Sons, Inc., New York, N.Y., 1992.
27. Sage, A. P., *Systems Management for Information Technology and Software Engineering*, John Wiley & Sons, New York, N.Y., 1995.
28. SECMM-95-01, "A Systems Engineering Capability Maturity Model (SE-CMM)," Version 1.1, Software Engineering Institute (SEI), Carnegie Melon University, Pittsburgh, Pa., 1995.
29. Shishko, R., *NASA Systems Engineering Handbook*, NASA, Washington, D.C., 1995.
30. SPC-95075-CMC, "Systems Engineering Maturity and Benchmarking," Version 01.00.06, Software Productivity Consortium, SPC Building, 2214 Rock Hill Road, Herndon, Va., 1996.
31. Thome', Bernhard (Ed.), *Systems Engineering: Principles and Practice of Computer-Based Systems Engineering*, John Wiley & Sons, New York, N.Y., 1993.
32. Truxal, J. G., *Introductory System Engineering*, McGraw-Hill Book Co., New York, N.Y., 1972.
33. Von Bertalanffy, L., "General Systems Theory: A New Approach to Unity of Science," *Human Biology*, December 1951.
34. Von Bertalanffy, L., *General Systems Theory*, George Braziller Press, New York, 1968.
35. Wiener, N., *Cybernetics*, John Wiley & Sons, Inc., New York, N.Y., 1948.
36. Wymore, A. W., *Model-Based Systems Engineering*, CRC Press, Inc., Boca Raton, Fla., 1993.

H.3 Concurrent Engineering

1. Kusiak, A., *Concurrent Engineering: Automation, Tools, and Techniques*, John Wiley & Sons, Inc., New York, N.Y., 1992.
2. Miller, L. C. G., *Concurrent Engineering Design: Integrating the Best Practices for Process Improvement*, Society of Manufacturing Engineers, Dearborn, MI, 48121, 1993.
3. Prasad, B., *Concurrent Engineering Fundamentals: Integrated Product and Process Organization*, Prentice Hall, Inc., Upper Saddle River, N.J., 1996.
4. Shina, S. G. (Ed.), *Successful Implementation of Concurrent Engineering Products and Processes*, Van Nostrand Reinhold, New York, N.Y., 1994.
5. Winner, R. I., Pennell, J. P., Bertrand, H. E., and Slusarczuk, M. M. G., *The Role of Concurrent Engineering in Weapons Systems Acquisition*, Report R-338, Institute for Defense Analysis, Arlington, Va., 1988.

H.4 Software and Computer-Aided Systems

1. Boehm, B. W., *Software Engineering Economics*, Prentice Hall, Inc., Upper Saddle River, N.J., 1981.
2. Eisner, H., *Computer-Aided Systems Engineering*, Prentice Hall, Inc., Upper Saddle River, N.J., 1988.
3. Humphrey, W. S., *A Discipline for Software Engineering*, Addison-Wesley Publishing Co., Reading, Mass., 1995.
4. Krouse, J. K., *What Every Engineer Should Know about Computer-Aided Design and Computer-Aided Manufacturing*, Marcel Dekker, Inc., New York, N.Y., 1982.
5. MIL-STD-498, *Software Development And Documentation*, Department of Defense, Washington, D.C., 1994.
6. Pressman, R. S., *Software Engineering: A Practitioner's Approach*, 3rd Ed., McGraw-Hill Book Co., New York, N.Y., 1992.
7. Sage, A. P., and Palmer, J. D., *Software Systems Engineering*, John Wiley & Sons, Inc., New York, 1990.
8. Vick, C. R., and Ramamcorthy, C. V., *Handbook of Software Engineering*, Van Nostrand Reinhold Co., New York, N.Y., 1984.
9. Zeid, I., *CAD/CAM Theory and Practice*, McGraw-Hill Book Co., New York, N.Y., 1991.

H.5 Reliability Engineering

1. *Annual Reliability and Maintainability Symposium*, Proceedings, Evans Associates, Durham, N.C.
2. Barlow, R. E., *Mathematical Theory of Reliability*, John Wiley & Sons, New York, N.Y., 1965.
3. EPRD-97, "Electronic Parts Reliability Data," Reliability Analysis Center, Rome, N.Y., 1997.
4. Instruction Manual, *Potential Failure Mode and Effects Analysis*, Chrysler, Ford, and General Motors Corp., Rev. 1993.
5. Intruction Manual, *Failure Mode and Effects Analysis*, Saturn Quality System, Saturn Corporation, 1990.
6. Ireson, W. G., and Coombs, C. F. (Eds), *Handbook of Reliability Engineering and Management*, McGraw-Hill Book Co., New York, N.Y., 1988.
7. Kececioglu, D., *Reliability Engineering Handbook*, Volumes I and II, Prentice Hall, Inc., Upper Saddle River, N.J., 1991.
8. Klion, J., *Practical Electronic Reliability Engineering*, Van Nostrand Reinhold, New York, N.Y., 1992.
9. Knezevic, J., *Reliability, Maintainability, and Supportability: A Probabilistic Approach*, McGraw-Hill Book Co., New York, N.Y., 1933.
10. Lloyd, D. K., and Lipow, M., *Reliability: Management, Methods, and Mathematics*, 2nd Ed., published by the authors, Defense and Space Systems Group, TRW Systems and Energy, Redondo Beach, Calif., 1984.
11. MIL-HDBK-189, Military Handbook, *Reliability Growth Management*, Department of Defense, Washington, D.C.
12. MIL-HDBK-217F, Military Handbook, *Reliability Predictions of Electronic Equipment*, Department of Defense, Washington, D.C.
13. MIL-STD-781D, Military Standard, *Reliability Testing for Engineering Development, Qualification, and Production*, Department of Defense, Washington, D.C.

14. MIL-STD-785B, Military Standard, *Reliability Program for Systems and Equipment Development and Production*, Department of Defense, Washington, D.C.
15. MIL-STD-1629A, Military Standard, *Procedures for Performing a Failure Mode, Effects and Critical Analysis*, Department of Defense, Washington, D.C.
16. MIL-STD-2155, Military Standard, *Failure Reporting, Analysis, and Corrective Action System (FRACAS)*, Department of Defense, Washington, D.C.
17. MIL-STD-2164, Military Standard, *Environmental Stress Screening Process for Electronic Equipment*, Department of Defense, Washington, D.C.
18. Modarres, M., *What Every Engineer Should Know about Reliability and Risk Analysis*, Marcel Dekker, Inc., New York, N.Y., 1992.
19. Musa, J. D., Lannino, A., Okumoto, K., *Software Reliability: Measurement, Prediction, Application*, McGraw-Hill Book Co., New York, N.Y., 1987.
20. NPRD-95, "Nonelectronic Parts Reliability Data," Reliability Analysis Center, Rome, N.Y., 1995.
21. O' Connor, P. D. T., *Practical Reliability Engineering*, 3rd Ed., John Wiley & Sons, Inc., New York, N.Y., 1991.
22. Pham, H., *Software Reliability and Testing*, IEEE Computer Society, Los Alamitos, CA, 1995.
23. RAC, *Failure Mode, Effects and Criticality Analysis (FMECA)*, Reliability Analysis Center, Rome, N.Y., 1992.
24. *RADC, Reliability Engineer's Toolkit*, Rome Air Development Center, RADC/RBE, Rome, N.Y., 1988.
25. Raheja, D. G., *Assurance Technologies: Principles and Practices*, McGraw-Hill, Book Co., New York, N.Y., 1991.
26. Rao, S. S., *Reliability-Based Design*, McGraw-Hill Book Co., New York, N.Y., 1992.

H.6 Maintainability Engineering and Maintenance

1. Anderson, R. T., and Neri, L., *Reliability-Centered Maintenance*, Elsevier Science Publishing, Ltd., London, England, 1990.
2. Blanchard, B. S., Verma, D., and Peterson, E., *Maintainability: A Key to Effective Serviceability and Maintenance Management*, John Wiley & Sons, New York, N.Y., 1995.
3. Bray, D. E., and McBride, D., (Eds.), *Nondestructive Testing Techniques*, John Wiley & Sons, Inc., New York, N.Y., 1992.
4. Faulkenberry, L. M. (Ed.), *Systems Troubleshooting Handbook*, John Wiley & Sons, Inc., New York, N.Y., 1986.
5. Gotoh, F., *Equipment Planning for TPM*, Productivity Press, Portland, Ore., 1991.
6. Higgins, L. R., *Maintenance Engineering Handbook*, 5th Ed., McGraw-Hill Book Co., New York, N.Y., 1994.
7. MIL-STD-470B, Military Standard, *Maintainability Program for Systems and Equipment*, Department of Defense, Washington, D.C.
8. MIL-STD-417A, Military Standard, *Maintainability Verification, Demonstration, Evaluation*, Department of Defense, Washington, D.C.
9. MIL-HDBK-472, Military Handbook, *Maintainability Prediction*, Department of Defense, Washington, D.C.
10. MIL-STD-1390D (Navy), Military Standard, *Level of Repair Analysis*, Department of Defense, Washington D.C.

11. Mobley, R. K., *An Introduction to Predictive Maintenance*, Van Nostrand Reinhold, New York, N.Y., 1990.
12. Moubray, J., *Reliability-Centered Maintenance*, Butterworth-Heinemann Ltd., Boston, Mass., 1992.
13. Nakajiima, S. (Ed.), *TPM Development Program: Implementing Total Productive Maintenance*, Productivity Press, Inc., Portland, Ore., 1989.
14. Nakajiima, S., *Total Productive Maintenance (TPM)*, Productivity Press, Inc., Portland, Ore., 1988.
15. Niebel, B. W., *Engineering Maintenance Management*, 2nd Ed., Marcel Dekker, Inc., New York, N.Y., 1994.
16. Nowlan, F. S., and Heap, H. F., *Reliability-Centered Maintenance*, United Airlines (MDA 903-75-C-0349), San Francisco, CA 94128, 1978.
17. Patton, J. D., *Maintainability and Maintenance Management*, 2nd Ed., Instrument Society of America, 67 Alexandria Drive, P.O. Box 12277, Research Triangle Park, NC 27709, 1994.
18. Smith, A. M., *Reliability-Centered Maintenance*, McGraw-Hill Co., Inc., New York, N.Y., 1993.
19. Suzuki, T. (Ed.), *TPM in Process Industries*, Productivity Press, P.O. Box 13390, Portland, Ore., 1994.
20. Wireman, T., *World Class Maintenance Management*, Industrial Press, New York, N.Y., 1990.

H.7 Human Factors and Safety Engineering

1. Hammer, W., *Occupational Safety Management and Engineering*, 4th Ed., Prentice Hall, Inc., Upper Saddle River, N.J., 1988.
2. Kroemer, K., Kroemer, H., and Kroemer-Elbert, K., *Ergonomics—How to Design for Ease and Efficiency*, Prentice Hall, Upper Saddle River, N.J., 1994.
3. Meister, D., *Behavioral Analysis and Measurement Methods*, John Wiley & Sons, Inc., New York, N.Y., 1985.
4. MIL-STD-882B, Military Standard, *System Safety Program Requirements*, Department of Defense, Washington, D.C.
5. MIL-STD-1472D, Military Standard, *Human Engineering Design Criteria for Military Systems, Equipment, and Facilities*, Department of Defense, Washington, D.C.
6. MIL-STD-1800, *Human Factors Engineering Performance Requirements for Systems*, Department of Defense, Washington, D.C.
7. Roland, H.E., and Moriarty, B., *System Safety Engineering and Management*, 2nd Ed., John Wiley & Sons, Inc., New York, N.Y., 1990.
8. Salvendy, G. (Ed), *Handbook of Human Factors*, John Wiley & Sons, Inc., New York, N.Y., 1987.
9. Sanders, M. S., and McCormick, E. J., *Human Factors in Engineering and Design*, 7th Ed., McGraw-Hill Book Co., New York, N.Y., 1992.
10. Van Cott, H. P., and Kinkade, R. G., (Eds.), *Human Engineering Guide to Equipment Design*, U.S. Government Printing Office, Washington, D.C., 1972.
11. Woodson, W. E., Tillman, B., and Tillman, P., *Human Factors Design*, 2nd Ed., McGraw-Hill Book Co., New York, N.Y., 1992.

H.8 Production, Quality, Quality Control, and Quality Assurance

1. Akao, Y. (Ed.), *Quality Function Deployment: Integrating Customer Requirements into Product Design*, Productivity Press, P.O. Box 13390, Portland, Ore., 1990.

2. Besterfield, D. H., *Quality Control*, 4th Ed., Prentice Hall, Inc., Upper Saddle River, N.J., 1993.
3. Besterfield, D. H., *Total Quality Management*, Prentice Hall, Inc., Upper Saddle River, N.J., 1994.
4. Cohen, L., *Quality Function Deployment: How to Make QFD Work for You*, Addison-Wesley Publishing Co., Reading, Mass., 1995.
5. Deming., W. E., *Out of the Crisis*, Massachusetts Institute of Technology Press, Cambridge, Mass., 1986.
6. DOD 5000.51G, "Total Quality Management: A Guide for Implementation," Department of Defense, Washington, D.C.
7. Duncan, A. J., *Quality Control and Industrial Statistics*, 5th Ed., Richard D. Irwin, Inc., Homewood, Il, 1986.
8. Feigenbaum, A. V., *Total Quality Control*, 3rd Ed., McGraw-Hill Book Co., New York, N.Y., 1991.
9. Garvin, D. A., *Managing Quality: The Strategic and Competitive Edge*, The Free Press, Macmillan Publishing Co., Inc., New York, N.Y., 1988.
10 Grant, E. L., and Leavenworth, R. S., *Statistical Quality Control*, 6th Ed., McGraw-Hill Book Co., New York, N.Y., 1988.
11. Ishikawa, K., *Introduction to Quality Control*, Chapman and Hall, London, England, 1991.
12. ISO 9000-9004, International Standards Series, "Quality Standards," International Standards Organization.
13. Juran, J. M., and Gryna, F. M., *Quality Planning and Analysis*, 3rd Ed., McGraw-Hill Book Co., New York, N.Y., 1993.
14. Juran, J. M., and Gryna, F. M. (Ed.), *Quality Control Handbook*, 4th Ed., McGraw-Hill Book Co., New York, N.Y., 1988.
15. Ross, P. J., *Taguchi Techniques for Quality Engineering*, 2nd Ed., McGraw-Hill Book Co., New York, N.Y., 1995.
16. Saylor, J. H., *TQM Field Manual*, McGraw-Hill Book Co., New York, N.Y., 1992.
17. Taguchi, G., Elsayed, E. A., and Hsiang, T. C., *Quality Engineering in Production Systems*, McGraw-Hill Book Co., New York, N.Y., 1989.
18. Thompkins, J. A., and White, J. A., *Facilities Planning*, John Wiley & Sons, New York, N.Y., 1981.

H.9 Operations Research

1. Buffa, E. S., and Sarin, R. K., *Modern Production and Operations Management*, 8th Ed., John Wiley & Sons, Inc., New York, N.Y., 1987.
2. Churchman, C. W., Ackoff, R. L., and Arnoff, E. L., *Introduction to Operations Research*, John Wiley & Sons, Inc., New York, N.Y., 1957.
3. Fabrycky, W. J., Ghare, P. M., and Torgerson, P. E., *Applied Operations Reserach and Management Science*, Prentice Hall, Inc., Upper Saddle River, N.J., 1984.
4. Hillier, F. S., and Lieberman, G. J., *Introduction to Operations Research*, 6th Ed., McGraw-Hill Book Co., New York, N.Y., 1995.
5. Taha, H. A., *Operations Research: An Introduction*, 5th Ed., Macmillan Publishing Co., Inc., New York, N.Y., 1992.

Engineering Economy, Life-Cycle Cost Analysis, and Cost Estimating

1. Canada, J. R., Sullivan, W. G., and White, J. A., *Capital Investment Analysis for Engineering and Management*, 2nd Ed., Prentice Hall, Inc. Upper Saddle River, N.J., 1996.
2. DOD-HDBK-766, Military Handbook, *Design to Cost*, Department of Defense, Washington, D.C.
3. Fabrycky, W. J., and Blanchard, B. S., *Life-Cycle Cost and Economic Analysis*, Prentice Hall, Inc., Upper Saddle River, N.J., 1991.
4. Fabrycky, W. J., Thuesen, G. J., and Verma, D., *Economic Decision Analysis*, 3rd Ed., Prentice Hall, Inc., Upper Saddle River, New Jersey, 1998.
5. Grant, E. L., Ireson, W. G., and Leavenworth R. S., *Principles of Engineering Economy*, 8th Ed., The Ronald Press Co., New York, N.Y., 1990.
6. MIL-HDBK-259, Military Handbook, *Life Cycle Cost in Navy Acquisitions*, Department of Defense, Washington, D.C.
7. MIL-STD-337, Military Standard, *Design to Cost*, Department of Defense, Washington, D.C.
8. Ostwald, P. F., *Engineering Cost Estimating*, 3rd Ed., Prentice Hall, Inc., Upper Saddle River, N.J., 1992.
9. Stewart, R. D., and Wyskida, R. M., *Cost Estimator's Reference Manual*, 2nd Ed., John Wiley & Sons, Inc., New York, N.Y., 1995.
10. Stewart, R. D., and Stewart, A. L., *Cost Estimating with Microcomputers*, McGraw-Hill Book Co., New York, N.Y., 1980.
11. Stewart, R. D., *Cost Estimating*, 2nd Ed., John Wiley & Sons, Inc., New York, 1991.
12. Thusesen, G. J., and Fabrycky, W. J., *Engineering Economy*, 8th Ed., Prentice Hall, Inc., Upper Saddle River, N.J., 1993.

H.11 Management and Supporting Areas

1. American Productivity and Quality Center, *The Benchmarking Management Guide*, Productivity Press, Portland, Ore., 1993.
2. Balm., G. J., *Benchmarking: A Practitioner's Guide for Becoming and Staying the Best of the Best*, QPMA Press, Schaumburg, Ill., 1992.
3. Camp, R. C., *Benchmarking—The Search for Industry Best Practices That Lead to Superior Performance*, ASQC Press, Milwaukee, Wisc., 1989.
4. Cleland, D. I., and King, W. R., *Project Management Handbook*, 2nd Ed., Van Nostrand Reinhold Co., Inc., New York, N.Y., 1989.
5. Defense Systems Management College (DSMC), *Manufacturing Management: Guide for Program Managers*, DSMC, Fort Belvoir, VA 22060.
6. Defense Systems Management College (DSMC), *Test and Evaluation Management Guide*, DSMC, Fort Belvoir, VA 22060.
7. DOD Regulation 5000.2, "Mandatory Procedures for Major Defense Acquisition Programs (MDAPs) and Major Automated Information System (MAIS) Acquisition Programs," Department of Defense, Washington, D.C., 1996.
8. EIA-IS-649, "Configuration Management," Electronic Industries Association, Arlington, Va., 1997.
9. Kerzner, H., *Project Management: A Systems Approach to Planning, Scheduling, and Controlling*, 5th Ed., Van Nostrand Reinhold, New York, N.Y., 1995.

10. Kidd, P. T., *Agile Manufacturing: Forging New Frontiers*, Addison-Wesley, Reading, Mass., 1994.
11. Koontz, H., O'Donnell, C., and Weihrich, H., *Essentials of Management*, 5th Ed., McGraw-Hill Book Co., New York, N.Y., 1990.
12. Magrab, E. B., *Integrated Product and Process Design and Development*, CRC Press, Boca Raton, Fla., 1997.
13. MIL-HDBK-61, "Configuration Management," Department of Defense, Washington, D.C. (latest revision).
14. MIL-STD-1521B, *Technical Reviews and Audits for Systems, Equipments, and Computer Software*, Department of Defense, Washington, D.C.
15. SD-2, "Buying Commercial and Nondevelopmental Items Handbook," Department of Defense, Washington, D.C., April 1996.
16. SD-5, "Market Research," Department of Defense, Washington, D.C., July 1997.
17. SD-15, "Performance Specification Guide," Department of Defense, Washington, D.C., June 1995.
18. Shtub, A., Bard, J. F., and Globerson, S., *Project Management: Engineering, Technology, and Implementation*, Prentice Hall, Inc., Upper Saddle River, N.J., 1994.
19. Stewart, R. D., and Stewart, A. L., *Proposal Preparation*, 2nd Ed., John Wiley & Sons, Inc., New York, N.Y., 1992.
20. Thamhain, H. J., *Engineering Management*, John Wiley & Sons, Inc., New York, N.Y., 1992.
21. Ullmann, J. E., Christman, D. A., Holtje, B. (Eds.), *Handbook of Engineering Management*, John Wiley & Sons, Inc., New York, New York, N.Y., 1986.
22. Wall, W. C., *Proposed Preparation Guide*, John Wiley & Sons, Inc., New York, N.Y., 1990.

INDEX

B

C

D

E

M

N

O

P

T